Universitext

Universitext

Universitext is a series of textbooks that presents material from a wide variety of mathematical disciplines at master's level and beyond. The books, often well class-tested by their author, may have an informal, personal even experimental approach to their subject matter. Some of the most successful and established books in the series have evolved through several editions, always following the evolution of teaching curricula, to very polished texts.

Thus as research topics trickle down into graduate-level teaching, first textbooks written for new, cutting-edge courses may make their way into *Universitext*.

For further volumes:
http://www.springer.com/series/223

Martin Arkowitz

Introduction to Homotopy Theory

Martin Arkowitz
Department of Mathematics
Dartmouth College
Hanover, NH 03755-3551
USA
Martin.A.Arkowitz@Dartmouth.edu

ISSN 0172-5939 e-ISSN 2191-6675
ISBN 978-1-4419-7328-3 e-ISBN 978-1-4419-7329-0
DOI 10.1007/978-1-4419-7329-0
Springer New York Dordrecht Heidelberg London

Library of Congress Control Number: 2011933473

Mathematics Subject Classification (2010): 55Q05, 55R05, 55S35, 55S45, 55U30

Printed on acid-free paper

Springer is part of Springer Science+Business Media (www.springer.com)

To Eleanor, my bashert,
and to
Dylan, Jake, and Gregory Arkowitz

Preface

This book deals with homotopy theory, which is one of the main branches of algebraic topology. The ideas and methods of homotopy theory have pervaded many parts of topology as well as many parts of mathematics. A general approach in these areas has been to reduce a geometric, analytic, or topological problem to a homotopy problem, and to then attempt to solve the homotopy problem, usually by algebraic methods. Thus, in addition to being interesting and important in its own right, homotopy theory has been successfully applied to geometry, analysis, and other parts of topology. There are several treatments of homotopy theory in general categories. However we confine ourselves to a study of classical homotopy theory, that is, homotopy theory of topological spaces and continuous functions. There are a number of books devoted to classical homotopy theory as well as extensive expositions of it in books on algebraic topology. This book differs from those in that the unifying theme by which the subject is developed is the Eckmann–Hilton duality theory.

The Eckmann–Hilton theory has been around for about fifty years but there appears to be no book-length exposition of it, apart from the early lecture notes of Hilton [40]. There are advantages, both expository and pedagogical, to presenting homotopy theory in this way. Dual concepts occur in pairs, such as H-space and co-H-space, fibration and cofibration, loop space and suspension, and so on, and so do many theorems. We often give complete details in describing one of these and only sketch its dual. This is done when the latter can essentially be derived by dualization. In this way we shorten the exposition by reducing the amount of repetitious material. This also allows the reader to test his or her understanding of the subject by supplying the missing details.

There is another advantage to studying Eckmann–Hilton duality theory. Frequently the dual of a result is known or trivial. But from time to time the dual result is neither of these and is in fact an interesting problem. This could give the reader material to work on.

A feature of this book is that it is designed primarily for students to learn the subject. The proofs in the text contain a great deal of detail. We also try to supplement the discussion of several of the concepts by explaining them intuitively. We provide many pictures and include a large number of exercises of varying degrees of difficulty at the end of each chapter. The exercises that have been used in the text are marked with a dagger (†) and the more difficult exercises are marked with an asterisk ($*$). It is generally regarded as important to do the exercises in order to learn the material. It has been said many times that mathematics is not a spectator sport.

This book has been written so that it can be used as a text for a university course in algebraic topology. We assume that the reader has been exposed to the basic ideas of the fundamental group, homology theory, and cohomology theory, material that is often covered in a first algebraic topology course. We state explicitly the results from these areas that we use and summarize the essential facts in the appendices. The text could also be used by mathematicians who wish to learn some homotopy theory. However, the book is not intended to introduce readers to current research in topology. There are many texts and survey articles that do this. Instead it is hoped that this book will provide a solid foundation for those who wish to work in topology or to learn more advanced homotopy theory.

We now summarize the text chapter by chapter. The first chapter contains a discussion of the notion of homotopy and its variations and related notions. We consider homotopy relative to a subset, homotopy of pairs, retracts, sections, homotopy equivalence, contractibility, and so on. Most of these should be familiar to the reader, but we present them for the sake of completeness. If X and Y are based spaces, we define the homotopy set $[X, Y]$ to be the set of homotopy classes of based maps $X \to Y$. Next CW complexes are introduced and some of their elementary properties established. These spaces play a major role in the rest of the book. Finally, there is a short section indicating some of the reasons for studying homotopy theory.

The next chapter deals with grouplike spaces and cogroups. The former is a group object in the category of based spaces and homotopy classes of maps. The latter is the categorical dual of a group object in this category. We consider loop spaces and suspensions, important examples of grouplike spaces and cogroups, respectively. This leads to a discussion of basic properties of the homotopy groups $[S^n, Y]$, where S^n is the n-sphere. We then define and construct spaces with a single nonvanishing homology group, called Moore spaces, and spaces with a single nonvanishing homotopy group, called Eilenberg–Mac Lane spaces. These give rise to homotopy groups with coefficients and to cohomology groups with coefficients. This gives a homotopical interpretation of cohomology groups. The chapter ends with a discussion of Eckmann–Hilton duality.

In Chapter 3 we discuss two dual classes of maps, fiber maps and cofiber maps. Fiber maps are defined by the covering homotopy property which is a well-known feature of covering spaces and fiber bundles. Cofiber maps appear often in topology because the inclusion map of a subcomplex of a CW complex into the complex is a cofiber map. A fiber map $E \to B$ determines a three term fiber sequence $F \to E \to B$, where $F \subseteq E$ is the fiber over the base point of B. A cofiber map $i : A \to X$ determines a cofiber sequence $A \to X \to Q$, where $Q = X/i(A)$ is the cofiber. We then study fiber bundles. We give examples of fiber bundles and these provide many examples of fiber sequences. We conclude the chapter by showing that any map can be factored as the composition of a homotopy equivalence and a fiber map or as the composition of a cofiber map and a homotopy equivalence.

The next chapter deals with exact sequences of homotopy sets. The main sequences are a long exact sequence associated to a fiber sequence and one associated to a cofiber sequence. By specializing these sequences we obtain the exact homotopy sequence of a fibration and the exact cohomology sequence of a cofibration. We next study the action of a grouplike space on a space and the coaction of a cogroup on a space. These give additional information on the exact sequences of homotopy sets. We then consider homotopy groups and define the relative homotopy groups of a pair of spaces. We discuss the exact homotopy sequence of a pair and the relative Hurewicz homomorphism. We conclude the chapter by introducing certain excision maps which are used in Chapter 6.

Chapter 5 is devoted to some applications of the exact sequences of the preceding chapter. We begin with two universal coefficient theorems. The first relates the cohomology groups with coefficients of a space to the integral cohomology groups of the space and the second relates the homotopy groups of a space with coefficients to the homotopy groups. Then we show how the operation of homotopy sets in Chapter 4 can be specialized to yield an operation of the fundamental group $\pi_1(Y)$ on the homotopy set $[X, Y]$. This operation is used to compare the based homotopy set $[X, Y]$ with the unbased homotopy classes of maps $X \to Y$. Finally we calculate some homotopy groups of several spaces including spheres, Moore spaces, and topological groups.

Chapter 6 contains the statement and proof of many of the important theorems of classical homotopy theory such as (1) the Serre theorem on the exact cohomology sequence of a fibration, (2) the Blakers–Massey theorem on the exact homotopy sequence of a cofibration, (3) the Hurewicz theorems which relates homology and homotopy groups, and (4) Whitehead's theorem regarding the induced homology homomorphism and the induced homotopy homomorphism. In the first part of the chapter we define homotopy pushouts and homotopy pullbacks and derive some of their properties. A major result that is used to prove both the Serre and Blakers–Massey theorems is that a certain homotopy-commutative square is a homotopy pushout square.

In Chapter 7 we discuss two basic and dual techniques for approximating a space by a sequence of simpler spaces. The obstruction theory developed in Chapter 9 is based on these approximations. The first technique, called the homotopy decomposition, assigns a sequence of spaces $X^{(n)}$ to a space X such that the ith homotopy group of $X^{(n)}$ is zero for $i > n$ and is isomorphic to the ith homotopy group of X for $i \leqslant n$. From the point of view of homotopy groups, the spaces $X^{(n)}$ approach X as n increases. The second technique, called the homology decomposition, is similar with homology groups in place of homotopy groups. We consider several properties and applications of these decompositions. In the last section of the chapter we generalize these decompositions from spaces to maps.

In Chapter 8 we derive some general results for the homotopy set $[X, Y]$. We give hypotheses in terms of cohomology and homotopy groups that imply that the set is countably infinite or finite. We consider some properties of the group $[X, Y]$ when X is a cogroup or Y is a grouplike space. We show that if Y is a grouplike space, then $[X, Y]$ is a nilpotent group whose nilpotency class is bounded above by the Lusternik–Schnirelmann category of X.

In the final chapter we consider two basic problems for mappings. In the first, called the extension problem, we seek to extend a map defined on a subspace to the whole space. In the second, called the lifting problem, we seek to lift a map into the base of a fibration to a map into the total space. These are two special cases of the extension-lifting problem. We develop an obstruction theory for this problem which gives a step-by-step procedure for obtaining the desired map. We present two approaches to the theory. For the first, we take a homotopy decomposition of the fiber map and assume that the desired map exists at the nth step. This determines an element in a cohomology group, whose vanishing is a necessary and sufficient condition for the map to exist at the $(n+1)$st step. In the final section we discuss a method for obtaining obstruction elements by taking homology decompositions. These elements are in homotopy groups with coefficients.

After Chapter 9 there are six appendices. These are of two types. One type consists of results whose proofs in the text would be a digression of the topics being treated. The proofs of these results appear in the appendix. The other type provides a summary and reference for those basic results about point-set topology, the fundamental group, homology theory, and category theory that are used in the text. Definitions are given, the results are stated, and in some cases the proof is either given or sketched.

In conclusion, I would like to acknowledge the many helpful suggestions of the following people: Robert Brown, Vladimir Chernov, Dae-Woong Lee, Gregory Lupton, John Oprea, Nicholas Scoville, Jeffrey Strom, and Dana Williams. I would like to express my appreciation to the following people at Springer: Katie Leach for editorial assistance, Rajiv Monsurate for advice on Tex, and Brian Treadway for drawing the figures. Finally, I am particularly indebted to Peter Hilton for having introduced me to this material and tutored me in it while I was a graduate student. To all these people, many thanks.

Contents

Chapter 1
Basic Homotopy

1.1 Introduction

In topology we study topological spaces and continuous functions from one topological space to another. In algebraic topology these objects are studied by assigning algebraic invariants to them. We assign groups, rings, vector spaces, or other algebraic objects to topological spaces and we assign homomorphisms of these objects to continuous functions. A basic equivalence relation called homotopy on the set of continuous functions from one topological space into another naturally arises in the study of these invariants. By investigating this relation we obtain interesting, deep, and sometimes surprising information about topological spaces and continuous functions and their algebraic representations. In addition, the relation of homotopy leads to new algebraic invariants for topological spaces and continuous functions.

We begin this chapter by recalling the notion of homotopy and its properties. We discuss derivative concepts such as homotopy retract, homotopy lift, homotopy equivalence and contractibility. We describe the homotopy category whose objects are spaces and whose morphisms are homotopy classes of continuous functions. In Section 1.5 we introduce CW complexes and derive many of their properies. These spaces are of fundamental importance in homotopy theory. They are broad enough to include most of the spaces that are of interest to topologists. But they are sufficiently restricted so that many important results which do not hold for arbitrary spaces such as the homotopy extension property and Whitehead's theorems hold for CW complexes. Furthermore, the definition of a CW complex is recursive, and so inductive arguments can often be used. From Chapter 2 on we assume that all spaces under consideration have the homotopy type of CW complexes. In Section 1.6 we briefly discuss some reasons for studying homotopy theory by indicating a few applications of homotopy theory to other areas of topology.

As stated in the Preface, we assume that the reader has some familiarity with the main ideas of homology theory and the fundamental group. These are summarized in Appendices B and C.

1.2 Spaces, Maps, Products, and Wedges

We consider topological spaces, often denoted by roman letters $X, Y, \ldots$, and functions $f : X \to Y$. Unless otherwise stated, spaces are path-connected and functions are continuous. We usually assume that a fixed point has been chosen in each topological space X which is called the *basepoint* of X and denoted by $*_X$ or just $*$. Such topological spaces are called *based spaces* or, more commonly, *spaces*. Subspaces of based spaces contain the basepoint. Topological spaces without a basepoint or spaces in which the basepoint is ignored are called *unbased spaces* or *free spaces*. We usually assume that any continuous function $f : X \to Y$ of spaces carries the basepoint of X to the basepoint of Y, $f(*_X) = *_Y$, and we refer to such functions as *based maps* or just *maps*. Continuous functions of unbased spaces or those of based spaces that are not required to preserve the basepoint are called *continuous functions*, *free maps*, or *unbased maps*.

The *product* of two spaces X and Y is just their cartesian product $X \times Y$ with the product topology and basepoint $(*, *)$. There are then *projection maps* $p_1 : X \times Y \to X$ and $p_2 : X \times Y \to Y$ defined by $p_1(x, y) = x$ and $p_2(x, y) = y$, for all $x \in X$ and $y \in Y$. Clearly maps $f : A \to X$ and $g : A \to Y$ determine a unique map $(f, g) : A \to X \times Y$ such that $p_1(f, g) = f$ and $p_2(f, g) = g$. Let $\mathrm{id} = \mathrm{id}_X : X \to X$ be the *identity map* defined by $\mathrm{id}(x) = x$, for $x \in X$. Then the *diagonal map* $\Delta = \Delta_X : X \to X \times X$ which is defined by $\Delta(x) = (x, x)$, can be expressed as $\Delta = (\mathrm{id}, \mathrm{id})$. Also, if $f : X \to X'$ and $g : Y \to Y'$ are maps, then there is a map $f \times g : X \times Y \to X' \times Y'$ defined by $(f \times g)(x, y) = (f(x), g(y))$ for all $x \in X$ and $y \in Y$ or by $f \times g = (fp_1, gp_2)$.

We next define the wedge of two spaces. We first introduce some notation that is used throughout the book. If Z is a space or an unbased space with an equivalence relation and $z \in Z$, then $\langle z \rangle$ denotes the *equivalence class* containing z, and so $\langle z \rangle$ is an element in the set of equivalence classes. Now we let X and Y be two spaces and define their wedge $X \vee Y$. We form the *disjoint union* of X and Y which we write as $X \sqcup Y$ and give it the *weak topology* with respect to $\{X, Y\}$ (Appendix A). Thus $U \subseteq X \sqcup Y$ is open if and only if $U \cap X$ is open in X and $U \cap Y$ is open in Y. But $X \sqcup Y$ does not have a natural basepoint. We therefore consider the subspace $S = \{*_X, *_Y\} \subseteq X \sqcup Y$ and set the *wedge* $X \vee Y$ equal to the quotient space $X \sqcup Y/S$ obtained by identifying S to a point (Appendix A). The basepoint of $X \vee Y$ is $\langle *_X \rangle = \langle *_Y \rangle$. There are maps $i_1 : X \to X \vee Y$ and $i_2 : Y \to X \vee Y$ called *injections* defined by $i_1(x) = \langle x \rangle$ and $i_2(y) = \langle y \rangle$, for all $x \in X$ and $y \in Y$. Maps $f : X \to B$ and $g : Y \to B$ determine a unique map $\{f, g\} : X \vee Y \to B$ such that

$\{f, g\}i_1 = f$ and $\{f, g\}i_2 = g$. The *folding map* $\nabla = \nabla_X : X \vee X \to X$ is defined by $\nabla = \{\mathrm{id}, \mathrm{id}\}$. Furthermore, maps $f : X' \to X$ and $g : Y' \to Y$ determine a map $f \vee g : X' \vee Y' \to X \vee Y$ defined by $f \vee g = \{i_1 f, i_2 g\}$.

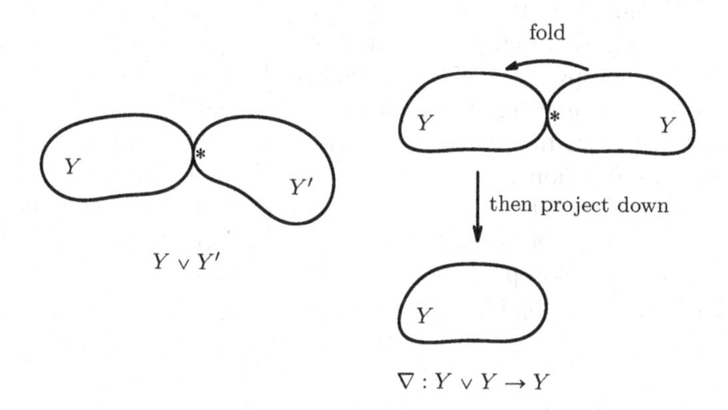

Figure 1.1

For spaces X and Y, there is a natural map $j : X \vee Y \to X \times Y$ defined by $j\langle x \rangle = (x, *)$ and $j\langle y \rangle = (*, y)$, for all $x \in X$ and $y \in Y$. Then j is a homeomorphism of $X \vee Y$ onto the subspace $X \times \{*\} \cup \{*\} \times Y$ of $X \times Y$. Because of this we usually regard $X \vee Y$ as $X \times \{*\} \cup \{*\} \times Y$ and j as an inclusion map.

NOTATION *For the remainder of the book, we consistently use* $p_1 : X \times Y \to X$ *and* $p_2 : X \times Y \to Y$ *for the projections and* $i_1 : X \to X \vee Y$ *and* $i_2 : Y \to X \vee Y$ *for the injections. In addition,* $q_1 = p_1 j : X \vee Y \to X$ *and* $q_2 = p_2 j : X \vee Y \to Y$ *denote the maps from the wedge and* $j_1 = j i_1 : X \to X \times Y$ *and* $j_2 = j i_2 : Y \to X \times Y$ *denote inclusions into the product.*

1.3 Homotopy I

Our main example of an unbased space is the *closed unit interval*

$$I = [0, 1] = \{t \mid t \text{ a real number, } 0 \leqslant t \leqslant 1\}.$$

This space plays a special role in homotopy theory. If X is a space or unbased space, then the unbased space $X \times I$ is called the *cylinder* over X. Occasionally, we would like the cylinder $X \times I$ to be a based space when X is a based space. To do this we choose $0 \in I$ or $1 \in I$ as basepoint. We always note when this is done.

Definition 1.3.1 Suppose maps $f, g : X \to Y$ are maps. We say that f is *homotopic* to g, written $f \simeq g$, if there is a continuous function $F : X \times I \to Y$

such that

$$F(x,0) = f(x), \quad F(x,1) = g(x), \quad \text{and} \quad F(*,t) = *,$$

for all $x \in X$ and $t \in I$. We call F a *homotopy* between f and g, and write $f \simeq g$ or $f \simeq_F g$. We describe the condition $F(*,t) = *$ by saying that $*$ is *fixed* by the homotopy F, that the homotopy F *preserves the basepoint*, or that $\{*\}$ is *stationary* during the homotopy. If X and Y are unbased spaces and $f, g : X \to Y$ are free maps, then f and g are *freely homotopic* if there is a continuous function $F : X \times I \to Y$ such that $F(x,0) = f(x)$ and $F(x,1) = g(x)$. We write this as $f \simeq_{\text{free}} g$. We sometimes refer to such a homotopy as a *free* or *unbased homotopy.* If Y is contained in a Euclidean space $\mathbb{R}^n$, then a homotopy F is called *linear* if for each $x \in X$ and $t \in I$, $F(x,t) = (1-t)f(x) + tg(x)$.

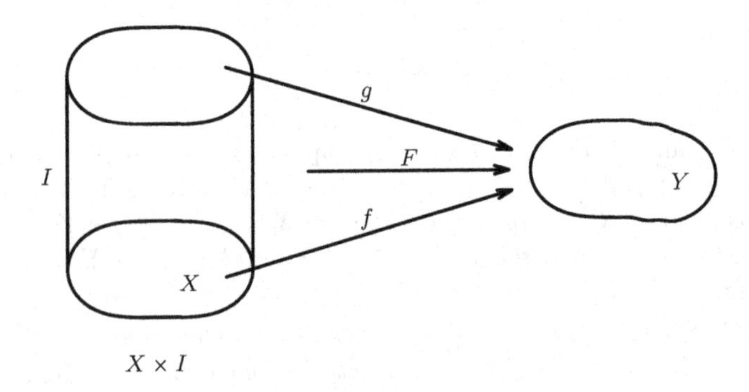

Figure 1.2

Another way to think of a homotopy is as a collection of maps $f_t : X \to Y$ indexed by $t \in I$ such that $f_0 = f$ and $f_1 = g$. We simply define $f_t(x) = F(x,t)$ and note that $f_t(x)$ is jointly continuous in x and t. It is often very useful to consider homotopies this way for stating definitions or proving results (e.g., Lemma 1.3.6), although it is necessary to verify that any homotopy constructed in this way is jointly continuous. We can carry this description of homotopy further by imagining that t represents time and that the homotopy provides a temporal deformation of the subspace $f(X)$ of Y into the subspace $g(X)$ of Y. That is, at time $t = 0$ we have $f_0(X) = f(X) \subseteq Y$ and at time $t = 1$ we have $f_1(X) = g(X) \subseteq Y$. At intermediate times t, $0 < t < 1$, $f(X)$ has been moved within Y to $f_t(X)$. See Figure 1.3. Strictly speaking, however, it is the map f that is being deformed into the map g rather than their images. The collection of all maps f_t for $t \in I$ is also called a *homotopy* between f and g.

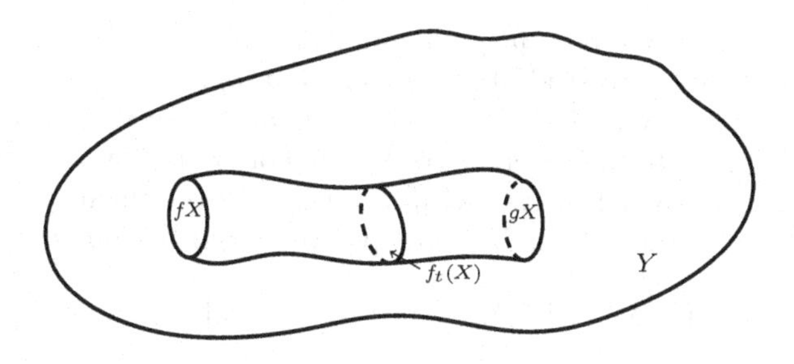

Figure 1.3

Proposition 1.3.2 *For all spaces X and Y, homotopy is an equivalence relation on the set of maps from X to Y. Furthermore, if $f \simeq g : X \to Y$ and $k : A \to X$ and $l : Y \to B$ are maps, then $fk \simeq gk : A \to Y$ and $lf \simeq lg : X \to B$. These results also hold for free maps and free homotopies.*

Proof. If $f : X \to Y$, then $f \simeq_F f$, where $F(x,t) = f(x)$, for all $x \in X$ and $t \in I$. If $f \simeq_F g$, then $g \simeq_G f$, where $G(x,t) = F(x, 1 - t)$. Thus the relation of homotopy is reflexive and symmetric. For transitivity, suppose that $f \simeq_F g$ and $g \simeq_G h$. Then define a homotopy H between f and h by

$$H(x,t) = \begin{cases} F(x, 2t) & \text{if } 0 \leqslant t \leqslant \frac{1}{2} \\ G(x, 2t-1) & \text{if } \frac{1}{2} \leqslant t \leqslant 1. \end{cases}$$

Because $F(x,1) = g(x) = G(x,0)$, H is a well-defined function $X \times I \to Y$. By the pasting lemma in Appendix A, H is continuous, and thus is the desired homotopy. Therefore homotopy is an equivalence relation. Finally, to prove that homotopy is compatible with composition we use the alternative view of a homotopy and let f_t be a homotopy between f and g. Then $f_t k$ is a homotopy between fk and gk and lf_t is a homotopy between lf and lg. □

When two homotopies F and G such that $F(x,1) = G(x,0)$ are put together to form a homotopy H as in the preceding proof, then H is said to be a *concatenation* of F and G and we write $H = F + G$.

Definition 1.3.3 For a map $f : X \to Y$, we let $[f]$ denote the equivalence class containing f, called the *homotopy class* of f. The collection of all homotopy classes of maps $X \to Y$ is denoted $[X, Y]$ and called the *homotopy set* (of homotopy classes from X to Y).

We make a few simple observations.

• We first note that we can compose homotopy classes. If $\alpha = [f] \in [Y, Z]$ and $\beta = [g] \in [X, Y]$, then by Proposition 1.3.2 we can define $\alpha\beta = [fg] \in [X, Z]$.

• It follows that there is a category (see Appendix F) whose objects are spaces and whose morphisms are *homotopy classes* of maps. This category is called the (based) *homotopy category* and is denoted $HoTop_*$. A major goal of homotopy theory is to study the category $HoTop_*$.

• For any spaces X and Y, there is the *constant map* $*_{X,Y} = * : X \to Y$ which takes all of X to the basepoint of Y. The homotopy class of $*$ is denoted $0 \in [X, Y]$, and so $[X, Y]$ is a set with a distinguished element 0.

• If $k : A \to X$ and $l : Y \to B$ are maps, we obtain the *induced functions*

$$k^* : [X, Y] \to [A, Y] \quad \text{and} \quad l_* : [X, Y] \to [X, B]$$

defined by $k^*[f] = [fk]$ and $l_*[f] = [lf]$ for $[f] \in [X, Y]$, such that $k^*(0) = 0$ and $l_*(0) = 0$. Furthermore, if $k \simeq k'$, then $k^* = k'^*$ and if $l \simeq l'$, then $l_* = l'_*$.

There is an equivalent way to define homotopy. For a space Y, let Y^I be the set of all continuous functions $I \to Y$. Then Y^I is a space with the compact–open topology (Appendix A). An element of Y^I is called a *path*. We regard Y^I as a based space with basepoint the constant path $*$ which takes all of I to the basepoint of Y.

Proposition 1.3.4 *Let $f, g : X \to Y$ be maps. Then $f \simeq g$ if and only if there is a map $G : X \to Y^I$ such that $G(x)(0) = f(x)$ and $G(x)(1) = g(x)$, for all $x \in X$.*

Proof. Suppose $F : X \times I \to Y$ and $G : X \to Y^I$ are functions such that for all $x \in X$ and $t \in I$,

$$F(x, t) = G(x)(t).$$

Then by Appendix A, F is continuous if and only if G is continuous. The proposition now follows. □

Although this characterization of homotopy is equivalent to Definition 1.3.1, the latter is more frequently used. The functions F and G in the above proof are said to be *adjoint* to each other (see Appendix F). In particular, we say that F is the adjoint of G and that G is the adjoint of F.

Remark 1.3.5 If $F : X \times I \to Y$ is a homotopy between maps f and g, we require in Definition 1.3.1 that $F(*, t) = *$, for all $t \in I$. Then F induces a map $F' : X \ltimes I \to Y$, where $X \ltimes I = X \times I/\{(*, t) \mid t \in I\}$. We call $X \ltimes I$ the *reduced cylinder of* X. Therefore we can regard a homotopy between f and g as a map $F' : X \ltimes I \to Y$ such that $F'\langle x, 0\rangle = f(x)$ and $F'\langle x, 1\rangle = g(x)$, where $\langle x, t\rangle$ denotes the equivalence class of (x, t). In the equivalence between the definition of a homotopy as a function $X \times I \to Y$ and as a function $X \to Y^I$ in Proposition 1.3.4, it would be more consistent to replace $X \times I$ with $X \ltimes I$ as the domain of the homotopy. This is because $X \ltimes I$ is a based space and $X \times I$ is not, whereas Y^I does have a natural basepoint. We would then have based maps $X \ltimes I \to Y$ in one–one correspondence with based maps $X \to Y^I$. However, it is customary to regard $X \times I$ as the domain of a homotopy, and we usually do so.

We next give a simple but useful result.

Lemma 1.3.6

1. *Let A, X, and Y be spaces. If $f_0 \simeq f_1 : A \to X$ and $g_0 \simeq g_1 : A \to Y$, then $(f_0, g_0) \simeq (f_1, g_1) : A \to X \times Y$.*
2. *Let B, X, and Y be spaces. If $f_0 \simeq f_1 : X \to B$ and $g_0 \simeq g_1 : Y \to B$, then $\{f_0, g_0\} \simeq \{f_1, g_1\} : X \vee Y \to B$.*

Proof. For (1), let f_t be the homotopy between f_0 and f_1 and g_t the homotopy between g_0 and g_1. Then $(f_t, g_t) : A \to X \times Y$, which is continuous as a function $A \times I \to X \times Y$, is a homotopy between (f_0, g_0) and (f_1, g_1). The proof of (2) is similar. □

Corollary 1.3.7

1. *There is a bijection $\theta : [A, X] \times [A, Y] \to [A, X \times Y]$ defined by $\theta([f], [g]) = [(f, g)]$.*
2. *There is a bijection $\rho : [X, B] \times [Y, B] \to [X \vee Y, B]$ defined by $\rho([f], [g]) = [\{f, g\}]$.*

Proof. The inverse of θ is defined by $\lambda[h] = ([p_1 h], [p_2 h])$, for $[h] \in [A, X \times Y]$. The inverse of ρ is defined by $\tau[k] = ([ki_1], [ki_2])$, for $[k] \in [X \vee Y, B]$. □

1.4 Homotopy II

We begin by defining some important notions both for maps and for homotopy classes.

Definition 1.4.1 Let X be a space and A a subspace (containing the base-point of X). A map $r : X \to A$ such that $r(a) = a$, for all $a \in A$, is called a *retraction* of X onto A and A is then called a *retract* of X. This can be restated as follows. If $i : A \to X$ is the inclusion map, then $r : X \to A$ is a map such that $ri = \mathrm{id}_A$. If, in addition, $ir \simeq \mathrm{id}_X$, we call r a *deformation retraction* and A a *deformation retract* of X. Similarly, if $f : X \to Y$ is a map, then a *section* of f is a map $s : Y \to X$ such that $fs = \mathrm{id}_Y$. A *homotopy retraction* of f, also called a *left homotopy inverse of* f, is a map $g : Y \to X$ such that $gf \simeq \mathrm{id}_X$. In this case we say that X is *dominated* by Y. A *homotopy section* of f, also called a *right homotopy inverse of* f, is a map $h : Y \to X$ such that $fh \simeq \mathrm{id}_Y$. Clearly g is a homotopy retraction of f if and only if f is a homotopy section of g. Also, we consider maps $f : X \to Y$ and $p : E \to Y$. A map $f' : X \to E$ such that $pf' = f$ is called a *lift* or *lifting of* f. If $pf' \simeq f$, then f' is called a *homotopy lift*. If $f : X \to Y$ is a map and $i : X \to C$ is an inclusion map, then a map $f'' : C \to Y$ such that $f''i = f$ is called an *extension* of f. If now f and i are arbitrary maps and $f''i \simeq f$, then we call f'' a *homotopy extension*. In the first case we say that f *factors* or *homotopy factors* through E and in the second case that f *factors* or *homotopy factors* through C. The analogous notions exist for unbased spaces and free maps.

We frequently use the following notation. E^n is the *unit ball* in Euclidean n-space $\mathbb{R}^n$ that is defined by $E^n = \{x \in \mathbb{R}^n \mid |x| \leqslant 1\}$. Then S^{n-1} is the *unit sphere* in $\mathbb{R}^n$, that is, the boundary of E^n, and is defined by $S^{n-1} = \{x \in \mathbb{R}^n \mid |x| = 1\}$, $n \geqslant 1$. We take $* = (1, 0, \ldots, 0)$ as the basepoint of S^{n-1} and of E^n. Furthermore, the *upper cap* (of S^n) is defined by $E^n_+ = \{(x_1, x_2, \ldots, x_{n+1}) \in S^n \mid x_{n+1} \geqslant 0\}$. Similarly, the *lower cap* (of S^n) is defined by replacing the condition $x_{n+1} \geqslant 0$ with the condition $x_{n+1} \leqslant 0$. It is easily shown that there are homeomorphisms $E^n_+ \cong E^n \cong E^n_-$.

Example 1.4.2 We give examples of retractions, deformation retractions, and homotopy retractions. Clearly these examples yield examples of sections and homotopy sections.

First suppose that $A \subseteq X$ is a retract, and so the retraction $r : X \to A$ is a continuous surjection. Thus if X is connected or compact, so is A. This observation yields examples of nonretracts by taking X connected or compact and the subspace A nonconnected or noncompact, respectively. For instance, it applies to the (nonpath-connected) 0-sphere $S^0 = \{\pm 1\} \subseteq E^1 = [-1, 1]$. This is the lowest-dimensional case of the well-known result that S^{n-1} is not a retract of E^n [61, Chap. 8, Prop. 2.3]. This latter result is usually proved with elementary homology theory and is equivalent to the Brouwer fixed point theorem [61, Chap. 8, Prop. 2.5].

Next we easily give examples of retracts that are not deformation retracts. For instance, if X is a space and $P = \{*\} \subseteq X$, then P is a retract of X. But P is not a deformation retract if X is not contractible (see Definition 1.4.3). Another class of examples is obtained by taking the wedge $X \vee Y$ of X and Y. We consider the subspace $X \subseteq X \vee Y$ and define a retraction $r = q_1 : X \vee Y \to X$ that is a projection onto X. If r is a deformation retraction, it would follow that $X \vee Y$ and X have isomorphic homology groups. But for a large class of spaces the homology of $X \vee Y$ is the direct sum of the homologies of X and Y. This implies that Y has trivial homology groups. Thus r is not a deformation retraction when Y has nontrivial homology.

However, many retracts are deformation retracts, and we next give a few examples of the more common ones.

1. Consider the inclusion map $i : E^n \to \mathbb{R}^n$ and define $r : \mathbb{R}^n \to E^n$ by

$$r(x) = \begin{cases} \dfrac{x}{|x|} & \text{if } |x| \geqslant 1 \\ x & \text{if } |x| \leqslant 1. \end{cases}$$

Then r is a retraction. This map fixes points in E^n and pushes points x outside of E^n along a straight line from the origin to x onto S^{n-1}. The homotopy $F : \mathbb{R}^n \times I \to \mathbb{R}^n$ given by

$$F(x, t) = \begin{cases} (1-t)x + tx/|x| & \text{if } |x| \geqslant 1 \\ x & \text{if } |x| \leqslant 1 \end{cases}$$

yields id $\simeq_F ir$, and so r is a deformation retraction. The homotopy F fixes E^n and moves points x outside of E^n linearly from x to $r(x)$ along the straight line determined by x and the origin. In this example, the basepoint of $\mathbb{R}^n$ and E^n could either be the origin or the point $(1, 0, \ldots, 0)$.

2. Similarly $S^{n-1} \subseteq E^n - \{0\}$ is a deformation retract. Let $i : S^{n-1} \to E^n - \{0\}$ be the inclusion. Define a retraction by $s : E^n - \{0\} \to S^{n-1}$ by

$$s(x) = \frac{x}{|x|}.$$

This map pushes points $x \in E^n - \{0\}$ to the boundary S^{n-1} along a straight line from x to the origin. Then the linear homotopy

$$G(x, t) = (1 - t)x + t\frac{x}{|x|}$$

yields id $\simeq_G is$. This homotopy moves linearly along the straight line mentioned above from x to $r(x)$. It follows from (1) that $r|\mathbb{R}^n - \{0\} : \mathbb{R}^n - \{0\} \to E^n - \{0\}$ is a deformation retraction. This, together with Exercise 1.5, shows that $S^{n-1} \subseteq \mathbb{R}^n - \{0\}$ is a deformation retract.

3. Consider $(S^{n-1} \times I) \cup (E^n \times \{0\}) \subseteq E^n \times I$. A deformation retraction r is obtained from the straight line projection from the point $(0, \ldots, 0, 2)$ that maps $E^n \times I$ onto $(S^{n-1} \times I) \cup (E^n \times \{0\})$ as indicated in Figure 1.4. Because this retraction is frequently used, we define it explicitly as follows.

$$r(x, s) = \begin{cases} (x/|x|, 2 - (2 - s)/|x|) & \text{if } |x| \geqslant 1 - \frac{s}{2} \\ (2x/(2 - s), 0) & \text{if } |x| \leqslant 1 - \frac{s}{2}, \end{cases}$$

where $x \in E^n$ and $s \in I$. As in the previous cases, the deformation homotopy is given by the straight line between (x, s) and $r(x, s)$. Since $E^n \times I$ is unbased, the deformation retraction is a free map. However, by taking $0 \in I$ to be the basepoint, we obtain a based deformation retraction. In addition, there is a homeomorphism $E^n \cong I^n$, and under this homeomorphism S^{n-1} corresponds to ∂I^n, where ∂I^n is the boundary of I^n. Thus the deformation retraction r yields a deformation retraction $I^n \times I \to (\partial I^n \times I) \cup (I^n \times \{0\})$.

Definition 1.4.3 A map $f : X \to Y$ is called *nullhomotopic* if $f \simeq * : X \to Y$, where $* : X \to Y$ is the constant map. A homotopy between any map f and $*$ is called a *nullhomotopy.* A space X is *contractible* if id_X is nullhomotopic and the nullhomotopy is called a *contracting homotopy.* More generally, if A is a subset of X with inclusion map $i : A \to X$, then A *is contractible in* X if i is nullhomotopic. An unbased space X is contractible if id_X is homotopic to a constant function, that is, any function which carries all of X to single point.

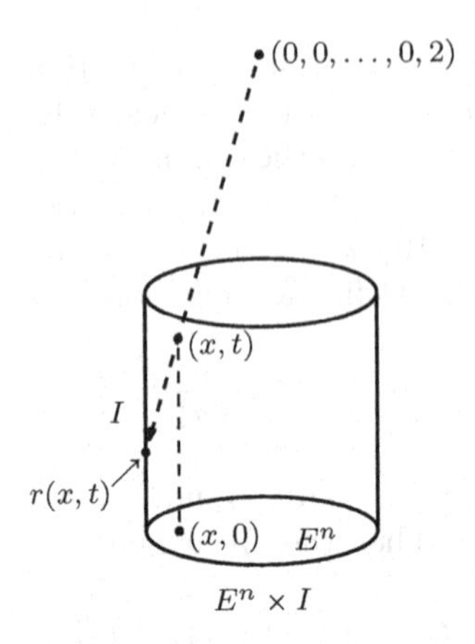

Figure 1.4

If either space X or Y is contractible, then any map $f : X \to Y$ is nullhomotopic. This is because f can be written as $\mathrm{id}_Y f$ or $f\,\mathrm{id}_X$. If $f_t : X \to X$ is a contracting homotopy, then f_t can be regarded as a deformation within X of the entire space $f_0(X) = X$ to the basepoint $f_1(X) = *$ of X (with $*$ fixed during the homotopy). One often says that a contractible space "can be shrunk to a point". The previous remarks also hold in the unbased case.

We use the following lemma in the next example.

Lemma 1.4.4 *If X is a contractible space, then for every neighborhood W of $*$, there is a neighborhood U of $*$ such that $U \subseteq W$ and U is contractible in W.*

Proof. Let $\mathrm{id} \simeq_F * : X \to X$. Then $F(*, t) = * \in W$, and so by continuity of F, there exist neighborhoods U_t of $*$ and V_t of t such that $F(U_t \times V_t) \subseteq W$. Clearly $\{V_t\}_{t \in I}$ covers I. By compactness of I, there is a finite subcover $V_{t_1}, \ldots, V_{t_n}$, and we set $U = \bigcap U_{t_i}$. It follows that $F|U \times I : U \times I \to W$ is the desired nullhomotopy of the inclusion map. □

We give some examples of contractible spaces.

Example 1.4.5

1. Let X be a subset of $\mathbb{R}^n$ with any point $*$ taken as basepoint in X. We say that X is $*$-*convex* if for every $x \in X$, the line segment from x to $*$ lies in X. If X is $*$-convex, then X is contractible. The contracting linear homotopy $F : X \times I \to X$ is given by $F(x, t) = (1 - t)x + t\,*$. The notion of $*$-convexity is weaker than that of convexity, therefore any convex subspace of $\mathbb{R}^n$ is contractible.
2. Let S^n be the unit n-sphere and choose $p = -*$ in S^n, the antipode of $*$. Then $S^n - \{p\}$ is contractible. The contracting homotopy $F : S^n - \{p\} \times I \to S^n - \{p\}$ is defined by

$$F(x,t) = \frac{(1-t)x + t*}{|(1-t)x + t*|},$$

where $x \in S^n - \{p\}$ and $t \in I$.

3. Other examples are the cones CX and the path spaces EY for any spaces X and Y that are defined in Definition 1.4.6 and proved to be contractible in Proposition 1.4.8.
4. An interesting example is the *comb space* C. This is the subspace of $\mathbb{R}^2$ consisting of the union of $[0,1] \times \{0\}$, $\{0\} \times [0,1]$ and all $\{1/n\} \times [0,1]$ for $n = 1, 2, \ldots$. We give C the induced topology as a subset of $\mathbb{R}^2$. We choose two possible basepoints for C, namely, $*_0 = (0,0)$ and $*_1 = (0,1)$.

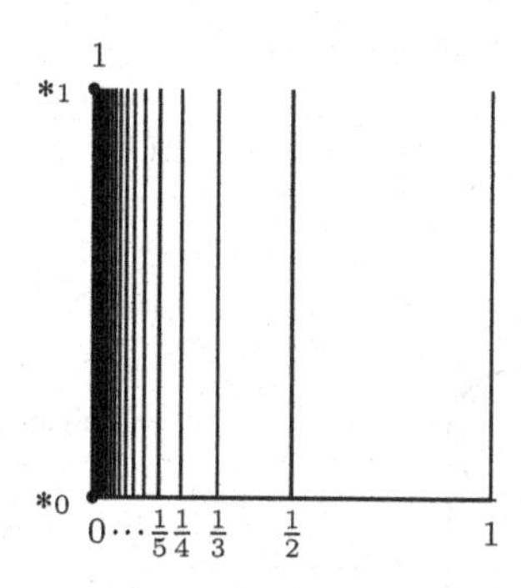

Figure 1.5

We show that C with basepoint $*_0$ is contractible, but C with basepoint $*_1$ is not. Let $c_i : C \to C$ be the constant map defined by $c_i(x) = *_i$, for $i = 0, 1$. We first show $\mathrm{id}_C \simeq c_0$ (where the homotopy fixes the basepoint $*_0$). The homotopy is given by concatenating two homotopies. The first homotopy projects C vertically downward onto $[0,1] \times \{0\}$ by a linear homotopy; the second projects $[0,1] \times \{0\}$ horizontally to the left onto $*_0$ by a linear homotopy. The resulting homotopy fixes $*_0$, but moves $*_1$. We next show $\mathrm{id}_C \not\simeq c_1$ (with basepoint $*_1$ fixed). We assume that $\mathrm{id}_C \simeq c_1$, that is, C with basepoint $*_1$ is contractible. Thus if W is a sufficiently small neighborhood of $*_1$ (e.g., $([0,\frac{1}{2}) \times (\frac{1}{2},1]) \cap C$), then by Lemma 1.4.4 there is a neighborhood $U \subseteq W$ of $*_1$ and a nullhomotopy $U \times I \to W$ of the inclusion map. We choose $x \in U - (\{0\} \times [0,1])$ and restrict the nullhomotopy to $\{x\} \times I$. This gives a path in W from x to $*_1$. This is not possible because the path would lie in several different path-connected components of W, and so $\mathrm{id}_C \not\simeq c_1$.

Definition 1.4.6 Let X be a space and $X \times I$ the cylinder over X. By identifying $X \times \{1\} \cup \{*\} \times I$ in $X \times I$ to a single point, we form an identification space which is denoted CX and called the *cone* on X (sometimes called the *reduced cone*). See Figure 1.6. There is a map $i : X \to CX$ defined by $i(x) = \langle x, 0\rangle$, where $\langle x, 0\rangle$ denotes the equivalence class of $(x, 0)$. Also, for any space Y, a subspace EY of Y^I (with the subspace topology) is defined by $EY = \{l \in Y^I \mid l(1) = *\}$ and called the *path space* of Y. The constant path defined by $*(t) = *_Y$, for $t \in I$, is the basepoint of EY. There is a map $p : EY \to Y$ defined by $p(l) = l(0)$. Elements of EY or of Y^I are called *paths* in Y. The *initial point* of a path l is $l(0)$ and the *terminal point* is $l(1)$. We also say that the path l begins or starts at $l(0)$ and ends at $l(1)$.

The cone and path space constructions are functorial: if $f : X \to Y$ is any map, then there are maps $Cf : CX \to CY$ and $Ef : EX \to EY$ defined by $Cf\langle x, t\rangle = \langle f(x), t\rangle$ and $Ef(l) = fl$ (the composition of f and l), where $\langle x, t\rangle \in CX$ and $l \in EY$. Also a map $f : CX \to Y$ and a map $g : X \to EY$ are said to be *adjoint* (to each other) if

$$f\langle x, t\rangle = g(x)(t),$$

for $x \in X$ and $t \in I$.

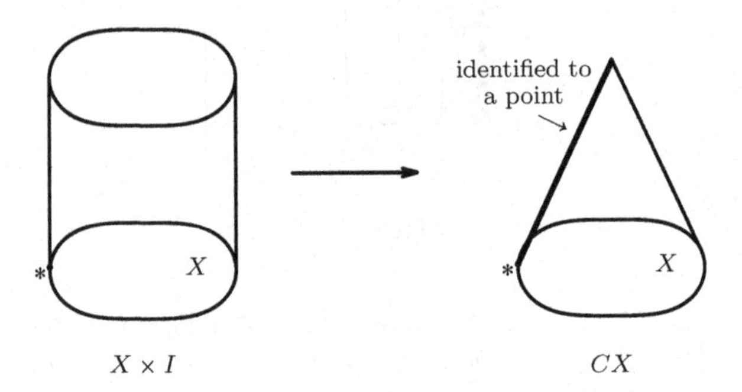

Figure 1.6

Remark 1.4.7 We note that certain pairs of paths in Y^I can be added together. If $l, l' \in Y^I$ are two paths such that $l(1) = l'(0)$, then $l + l'$ is the path in Y^I defined by

$$(l + l')(t) = \begin{cases} l(2t) & \text{if } 0 \leqslant t \leqslant \frac{1}{2} \\ l'(2t - 1) & \text{if } \frac{1}{2} \leqslant t \leqslant 1. \end{cases}$$

Some authors write this multiplicatively as $l \cdot l'$, $l * l'$ or ll'. Furthermore, $-l$ is the reverse path defined by $(-l)(t) = l(1 - t)$, for all $t \in I$. We return to these path operations in Chapter 2.

Proposition 1.4.8 *For every space X, the cone CX and path space EX are contractible.*

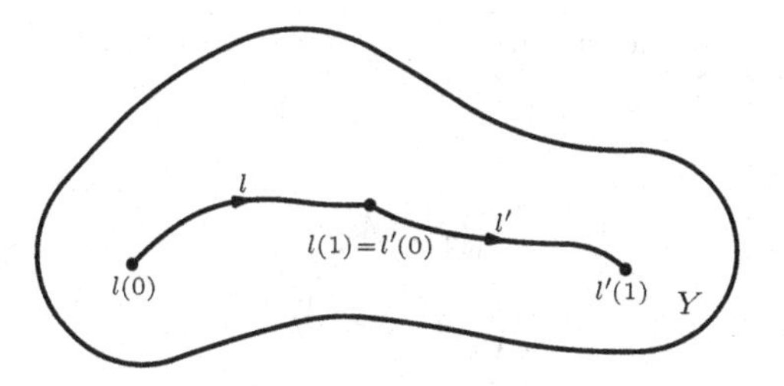

Figure 1.7

Proof. Define $F : CX \times I \to CX$ by

$$F(\langle x, t\rangle, s) = \langle x, s + (1 - s)t\rangle,$$

for $\langle x, t\rangle \in CX$ and $s \in I$. Then F is a well-defined continuous function that is the desired nullhomotopy. The nullhomotopy $G : EX \times I \to EX$ is similarly defined by $G(l, t)(s) = l(s + (1 - s)t)$ for $l \in EX$ and $s, t \in I$. □

If $0 \leqslant s \leqslant 1$, then $s + (1 - s)t$ is the line segment from t to 1. With $\langle x, 1\rangle$ the apex of the cone, the nullhomotopy F may be described as "pulling the cone up to its apex".

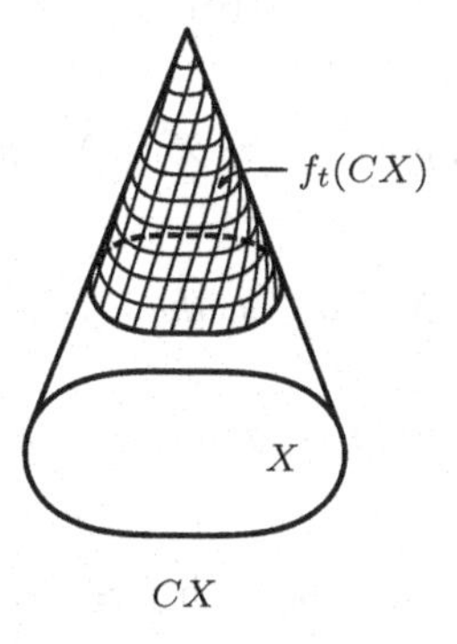

Figure 1.8

The nullhomotopy G may be visualized as follows. The elements of EX are paths in X ending at $*$. We can regard the images of these paths as strands of spaghetti sticking out of a mouth (identified with $*$) at time $t = 0$. The spaghetti is then sucked into the mouth. At time $t = 1$, all the spaghetti has disappeared and we are left with $*$. This is indicated in Figure 1.9. This description should be taken with a grain of salt because the strands of spaghetti are of different lengths and the points on each strand are not moving with the same speed. For example, at time $t = \frac{1}{2}$, each point on a strand has moved halfway to the mouth.

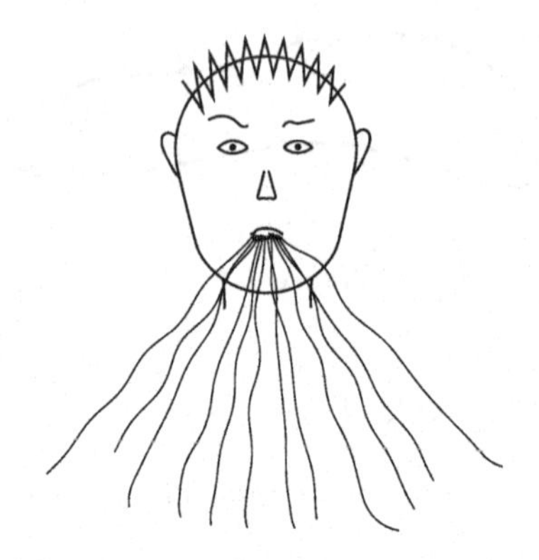

Figure 1.9

In the definition of the cone CX we identified $X \times \{1\} \cup \{*\} \times I$ to a single point. Let us denote this cone by C_1X. If instead we identify $X \times \{0\} \cup \{*\} \times I$ in $X \times I$ to a point, we get the cone C_0X which is homeomorphic to C_1X. We use either of these cones, but CX always denotes C_1X. Similarly we have $E_1X = EX$ and $E_0X = \{l \in X^I \mid l(0) = *\}$ which are homeomorphic path spaces.

Proposition 1.4.9 *For any map $f : X \to Y$, the following are equivalent:*

1. *f is nullhomotopic.*
2. *f can be extended to CX, that is, there is a map $\widehat{f} : CX \to Y$ such that $\widehat{f}\,i = f : X \to Y$.*
3. *f can be lifted to EY, that is, there is a map $\widetilde{f} : X \to EY$ such that $p\widetilde{f} = f : X \to Y$.*

Proof. Suppose f is nullhomotopic with $f \simeq_F *$. Then $F(X \times \{1\} \cup \{*\} \times I) = *$, and so F induces $\widehat{f} : CX \to Y$ such that $\widehat{f}\,i = f$. Thus (1) implies (2). Conversely, if $\widehat{f}\,i = f$, then, because CX is contractible, $f = \widehat{f}\,(\mathrm{id}_{CX})\,i \simeq \widehat{f}(*_{CX})i = *$. Thus (1) and (2) are equivalent. The proof that (3) is equivalent to (1) is similar, and hence omitted. □

Lemma 1.4.10 *For all $n \geqslant 1$, the cone on the unit $(n-1)$-sphere CS^{n-1} is homeomorphic to the unit n-ball E^n.*

Proof. Define a continuous function $K : S^{n-1} \times I \to E^n$ by

$$K(x,t) = t * +(1-t)x,$$

where $x \in S^{n-1}$ and $t \in I$. Then $K\big((S^{n-1} \times \{1\}) \cup (\{*\} \times I)\big) = *$ and so K induces a map

$$\widetilde{K} : CS^{n-1} = S^{n-1} \times I/(S^{n-1} \times \{1\} \cup \{*\} \times I) \to E^n,$$

which is a homeomorphism. □

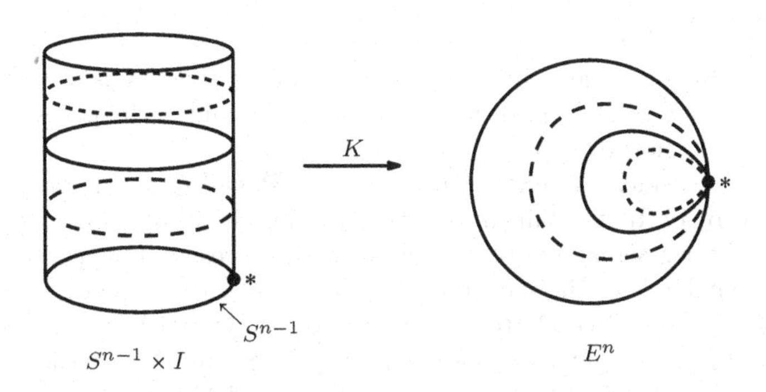

Figure 1.10

By putting Proposition 1.4.9 and Lemma 1.4.10 together we have that a map $f : S^{n-1} \to Y$ is nullhomotopic if and only if f is extendable to E^n. For later use we give an unbased version of this.

Corollary 1.4.11 *Let Y be an unbased space and $f : S^{n-1} \to Y$ a free map. Then f is homotopic to a constant function if and only if f is extendable to a free map $\widehat{f} : E^n \to Y$.*

Proof. We only show that if $c : S^{n-1} \to Y$ is a constant function and $c \simeq_{\text{free}} f$ with homotopy F, then there is an extension $\widehat{f}$. We represent points in E^n as tx for $x \in S^{n-1}$ and $0 \leqslant t \leqslant 1$ set $\widehat{f}(tx) = F(x, t)$. □

Next we consider homotopy equivalences.

Definition 1.4.12 A map $f : X \to Y$ is called a *homotopy equivalence* if there is a map $g : Y \to X$ that is a left and right homotopy inverse of f, that is, $fg \simeq \mathrm{id}_Y$ and $gf \simeq \mathrm{id}_X$. Two spaces X and Y have the *same homotopy type* if there is a homotopy equivalence from one to the other. We write this as $X \simeq Y$.

Clearly, any map that is a homeomorphism is a homotopy equivalence. Also, a map g that is a left and right homotopy inverse of f is unique up to homotopy. For if g' is another such map, then

$$g' \simeq g'(fg) = (g'f)g \simeq g.$$

We often denote the map g by f^{-1}. Then $f^{-1} : Y \to X$ is a homotopy equivalence whose homotopy class is uniquely determined. The composition of homotopy equivalences is a homotopy equivalence, thus same homotopy type is an equivalence relation among spaces. In fact, homotopy equivalences play the role of isomorphisms in homotopy theory, and their homotopy classes are categorical isomorphisms in the homotopy category $HoTop_*$. We regard two spaces of the same homotopy type as essentially the same.

Example 1.4.13

1. Let $P = \{*\}$ be a one-point space. Then clearly a space X is contractible if and only if $X \simeq P$. In particular, all cones CX and path spaces EY have the same homotopy type.
2. Next we consider the image of a curve in $\mathbb{R}^2$ called the *theta curve* and denoted by Θ. It is obtained from two disjoint circles by identifying points on the upper semicircle of one circle with corresponding points on the upper semicircle of the second circle. The theta curve is so-called because it looks like the Greek letter theta. We now show that $\Theta \simeq S^1 \vee S^1$. This follows from the fact that both Θ and $S^1 \vee S^1$ are deformation retracts of the same elliptical region in $\mathbb{R}^2$ with foci deleted (for details, see [73, p. 110]). Alternatively, the horizontal bar of Θ is homeomorphic to I and hence contractible. By Corollary 1.5.19 below, $\Theta \simeq \Theta/I \simeq S^1 \vee S^1$.
3. If $i : A \to X$ is an inclusion map and $r : X \to A$ a deformation retraction, then i and r are homotopy equivalences and $A \simeq X$. Therefore all the deformation retractions in Example 1.4.2 are homotopy equivalences.
4. For any space X, the set $[X, X]$ with the operation of composition of homotopy classes is a monoid with identity element $[\mathrm{id}_X]$. This means that $[X, X]$ has an associative binary operation with two-sided identity $[\mathrm{id}_X]$. The group of units or invertible elements of this monoid consists of all homotopy classes of homotopy equivalences of X to itself, and is called the *group of (homotopy classes of) self-homotopy equivalences* of X. This group has been studied and calculations have been made for Moore spaces, Eilenberg–Mac Lane spaces, projective spaces, and low-dimensional Lie groups (see [3]).

There are variations of the notion of homotopy that are useful, and we next discuss some of them.

The first case occurs when X and Y are spaces with subspaces $A \subseteq X$ and $B \subseteq Y$. Then (X, A) and (Y, B) are called *pairs* of spaces. A *map of pairs* $f : (X, A) \to (Y, B)$ is just a map $f : X \to Y$ such that $f(A) \subseteq B$. Two maps of pairs $f, g : (X, A) \to (Y, B)$ are homotopic, written $f \simeq g : (X, A) \to (Y, B)$ or $f \simeq g$ if there is a homotopy F as in Definition 1.3.1 with the additional restriction that $F(A \times I) \subseteq B$. In particular, a map of pairs $f : (X, A) \to (Y, B)$ is a homotopy equivalence (of pairs) if there is a map of pairs $g : (Y, B) \to (X, A)$ such that $fg \simeq \mathrm{id} : (Y, B) \to (Y, B)$ and $gf \simeq \mathrm{id} : (X, A) \to (X, A)$. This discussion of pairs can be extended to n subspaces in an obvious way. Next suppose $f, g : X \to Y$, $A \subseteq X$, and $f|A = g|A$. If $f \simeq_F g$ and $F(a,t) = f(a) = g(a)$ for all $a \in A$ and $t \in I$, then we say that f is homotopic to g *relative to* A and write $f \simeq g$ rel A. We say that A is *fixed* during the homotopy or that the homotopy is *stationary* on A. If $f, g : X \to Y$ are maps and $* \in X$ is the basepoint, then clearly there is no difference between $f \simeq g$ and $f \simeq g$ rel$\{*\}$. A special case of homotopy rel A occurs when $i : A \to X$ is an inclusion map, $r : X \to A$ is a retraction, and $ir \simeq \mathrm{id}$ rel A. Then r is called a *strong deformation retraction.* (Note that some authors refer to a strong deformation retraction as a deformation retraction.) These notions also exist for unbased spaces and free maps.

A simple extension of the argument in Proposition 1.3.2 yields the following result.

Proposition 1.4.14 *Homotopy of pairs and relative homotopy are equivalence relations. Homotopy of pairs is compatible with composition of the appropriate maps of pairs.*

Next we illustrate some of these concepts.

Example 1.4.15

1. The comb space C of Example 1.4.5 serves to illustrate several points. Recall that there are two possible basepoints in C, $*_0$ and $*_1$ with corresponding constant maps c_0 and c_1. We have seen that $\mathrm{id}_C \simeq c_0 : C \to C$ rel $\{*_0\}$. Since there is a path in C from $*_0$ to $*_1$, it follows that $c_0 \simeq_{\text{free}} c_1$ (Exercise 1.1). Therefore $\mathrm{id}_C \simeq_{\text{free}} c_1$, but by 1.4.5, $\mathrm{id}_C \not\simeq c_1$ rel $\{*_1\}$. This shows that based maps which are freely homotopic are not necessarily based homotopic. We show in Section 5.5 that they are based homotopic when the codomain is a simply connected CW complex. This example also shows that a contractible space cannot necessarily be contracted onto any preassigned basepoint.
2. More generally, it is easy to show that if $f, g : X \to Y$ are two maps which agree on $A \subseteq X$ and are homotopic, they are not necessarily homotopic rel A. For example, let Y be the region between two concentric circles in $\mathbb{R}^2$ with center at the origin and radii 1 and 3, respectively. Let $f, g : I \to Y$ be given by the two different semicircular paths in Y from $(-2, 0)$ to $(2, 0)$ with center the origin and radius 2. If $A = \{0, 1\}$, then $f|A = g|A$. But $f \not\simeq g$ rel A. This is clear intuitively, but also follows from Exercise 2.18. But if $c : I \to Y$ is the constant map defined by $c(t) = (-2, 0)$, then clearly $f \simeq c \simeq g$ rel $\{0\}$.
3. Finally, we give an example of two maps of pairs $f, g : (X, A) \to (Y, B)$ such that $f \simeq g : X \to Y$ and $f|A \simeq g|A : A \to B$, but f and g are not homotopic as maps of pairs. Let $f : (E^n, S^{n-1}) \to (S^n, \{*\})$ be the quotient map that identifies S^{n-1} to a point (see Section 4.5) and let g be the constant map. Then $f \simeq g : E^n \to S^n$ since E^n is contractible and $f|S^{n-1} \simeq g|S^{n-1} : S^{n-1} \to \{*\}$. But $f \not\simeq g : (E^n, S^{n-1}) \to (S^n, \{*\})$ because f and g induce different homomorphisms of relative homology

groups. That is, $f_* : H_n(E^n, S^{n-1}) \to H_n(S^n, \{*\})$ is an isomorphism [61] and $g_* = 0 : H_n(E^n, S^{n-1}) \to H_n(S^n, \{*\})$.

1.5 CW Complexes

In this section we introduce the notion of a CW complex and discuss its homotopical properties. This is continued in Sections 2.4 and 4.5. We show that a CW pair has the homotopy extension property. We discuss the cellular approximation theorem which asserts that every map of CW complexes is homotopic to one which preserves the cellular structure. For more details on CW complexes, see the following references: [33], [57], [20], [39, pp. 519–529] and [37, Chaps. 14, 16].

NOTE *In this section we are mainly concerned with unbased spaces and functions. Recall that such functions are called continuous functions, free maps, or unbased maps and their homotopies are called free homotopies or unbased homotopies.*

We begin by defining arbitrary unions, adjunction spaces, spaces obtained by attaching cells, and arbitrary wedges.

Definition 1.5.1

1. Let X_α be a collection of unbased topological spaces, for $\alpha \in A$. Then the disjoint union $\coprod_{\alpha \in A} X_\alpha$ is given the weak topology with respect to the collection of subsets $\{X_\alpha \,|\, \alpha \in A\}$ (Appendix A).
2. Let X and Y be unbased topological spaces, let $W \subseteq X$ and let $\phi : W \to Y$ be a free map. In the space $Y \sqcup X$ introduce the equivalence relation $w \sim \phi(w)$ for all $w \in W$. The resulting identification space is the *adjunction space* $Y \cup_\phi X$.
3. Suppose that X is an unbased space and $\phi_\alpha : S^{n-1}_\alpha \to X$ are free maps for $\alpha \in A$, where $S^{n-1}_\alpha = S^{n-1}$. Then the ϕ_α determine a free map $\phi : \coprod_{\alpha \in A} S^{n-1}_\alpha \to X$. Since the space $\coprod_{\alpha \in A} S^{n-1}_\alpha \subseteq \coprod_{\alpha \in A} E^n_\alpha$, where $E^n_\alpha = E^n$, we can form the adjunction space $X \cup_\phi \coprod_{\alpha \in A} E^n_\alpha$ which is called the space obtained from X by *attaching n-cells* by the functions ϕ_α.
4. Let X_α be a space with basepoint $*_\alpha$, for every $\alpha \in A$. Consider the unbased space $\coprod_{\alpha \in A} X_\alpha$ and let $S = \{*_\alpha \mid \alpha \in A\}$ be the subspace of all basepoints. Then the *wedge* $\bigvee_{\alpha \in A} X_\alpha$ is the quotient space $\coprod_{\alpha \in A} X_\alpha / S$. Note that for each $\alpha \in A$, there is an *injection* $i_\alpha : X_\alpha \to \bigvee_{\alpha \in A} X_\alpha$ defined by $i_\alpha(x_\alpha) = \langle x_\alpha \rangle$, for all $x_\alpha \in X_\alpha$. Futhermore, maps $f_\alpha : X_\alpha \to Y$ determine a unique map $f : \bigvee_{\alpha \in A} X_\alpha \to Y$ such that $f i_\alpha = f_\alpha$.

Remark 1.5.2 Suppose X is a based space and $\phi_\alpha : S^{n-1}_\alpha \to X$ are maps, for all $\alpha \in A$. Then the ϕ_α determine a map $\phi' : \bigvee_{\alpha \in A} S^{n-1}_\alpha \to X$ and $\bigvee_{\alpha \in A} S^{n-1}_\alpha \subseteq \bigvee_{\alpha \in A} E^n_\alpha$. Therefore the space obtained from X by attaching

n-cells, $X \cup_\phi \coprod_{\alpha \in A} E^n_\alpha$, equals $(X \vee \bigvee_{\alpha \in A} E^n_\alpha)/\sim$, where $(*, z) \sim (\phi'(z), *)$, for $z \in \bigvee_{\alpha \in A} S^{n-1}_\alpha$. We write this as $X \cup_{\phi'} \bigvee_{\alpha \in A} E^n_\alpha$.

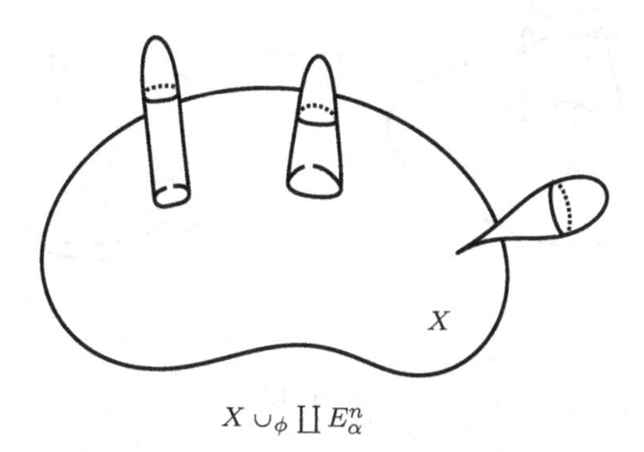

$X \cup_\phi \coprod E^n_\alpha$

Figure 1.11

The following lemma shows that the homotopy type of a space formed by attaching cells depends on the homotopy class of the attaching map. A more general result is given by Proposition 3.2.15.

Lemma 1.5.3 *If $\phi, \phi' : S^{n-1} \to X$ are free maps and $\phi \simeq_{\text{free}} \phi'$, then there is a free homotopy equivalence $\Phi : X \cup_\phi E^n \to X \cup_{\phi'} E^n$. The homotopy inverse Φ' has the property that $\Phi\Phi' \simeq \text{id}$ rel X and $\Phi'\Phi \simeq \text{id}$ rel X.*

Proof. If F is the free homotopy between ϕ and ϕ', then define Φ by $\Phi\langle x \rangle = \langle x \rangle$ and

$$\Phi\langle s, t \rangle = \begin{cases} \langle F(s, 2t) \rangle \text{ if } 0 \leqslant t \leqslant \frac{1}{2} \\ \langle s, 2t - 1 \rangle \text{ if } \frac{1}{2} \leqslant t \leqslant 1, \end{cases}$$

where $x \in X$, $s \in S^{n-1}$, $t \in I$, and $E^n = S^{n-1} \times I / S^{n-1} \times \{1\}$. See Figure 1.12. If F' is the opposite homotopy to F, that is, $F'(s,t) = F(s, 1-t)$, then F' determines a continuous function $\Phi' : X \cup_{\phi'} E^n \to X \cup_\phi E^n$ analogous to the definition of Φ. It is left as an exercise (Exercise 1.19) to show that Φ' is a homotopy inverse to Φ. □

Now we turn to the definition of a CW complex. We allow the empty set to be a CW complex and define a nonempty CW complex next.

Definition 1.5.4 A nonempty *CW complex* X is an unbased Hausdorff space together with a sequence of unbased subspaces called *skeleta*

$$X^0 \subseteq X^1 \subseteq \cdots \subseteq X^n \subseteq X^{n+1} \subseteq \cdots,$$

whose union is X. There are two conditions for X to be a CW complex.

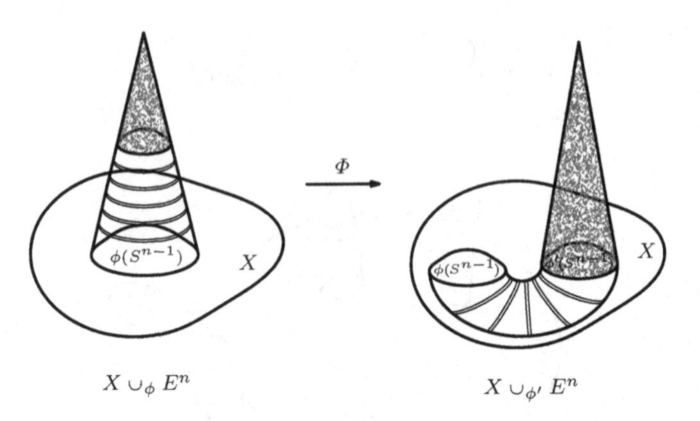

Figure 1.12

1. The skeleta are inductively defined as follows: X^0 is a nonempty discrete set of points. The elements of X^0 are called *0-cells* or *vertices*. We assume that X^{n-1} is defined for $n \geqslant 1$ and obtain X^n from X^{n-1} by attaching n-cells. That is, we assume that for some index set B, there exist free maps $\phi_\beta : S^{n-1}_\beta \to X^{n-1}$, where S^{n-1}_β denotes the $(n-1)$-sphere S^{n-1}, for each $\beta \in B$. The ϕ_β determine a free map $\phi : \coprod_{\beta \in B} S^{n-1}_\beta \to X^{n-1}$. Then X^n is defined as
$$X^n = X^{n-1} \cup_\phi \coprod_{\beta \in B} E^n_\beta,$$
where $E^n_\beta = E^n$.
The second condition will be stated after the following notation. The composition of the inclusion $E^n_\beta \to X^{n-1} \sqcup \coprod_{\beta \in B} E^n_\beta$ and the quotient map $X^{n-1} \sqcup \coprod_{\beta \in B} E^n_\beta \to X^n$ gives a continuous function of pairs $\Phi_\beta = \Phi^n_\beta : (E^n_\beta, S^{n-1}_\beta) \to (X^n, X^{n-1})$. The subset $\Phi_\beta(E^n_\beta) \subseteq X^n \subseteq X$ is called an *n-cell* or a *closed n-cell* and is denoted $\bar{e}^n_\beta$.
2. The space X has the weak topology with respect to the set $\{\bar{e}^n_\beta\}$ of all closed cells. Thus $U \subseteq X$ is open (respectively, closed) if and only if $U \cap \bar{e}^n_\beta$ is open (respectively, closed) in $\bar{e}^n_\beta$, for each closed cell $\bar{e}^n_\beta$.

We next introduce some standard terminology. The free maps $\phi_\beta : S^{n-1}_\beta \to X^{n-1}$ for $\beta \in B$ are called *attaching functions*, the free maps Φ_β are called *characteristic functions*, and $\Phi_\beta | S^{n-1}_\beta = \phi_\beta : S^{n-1}_\beta \to X^{n-1}$. The image $\Phi_\beta(E^n_\beta - S^{n-1}_\beta)$, denoted e^n_β, is called an *open n-cell* (although it need not be open in X) and $\Phi_\beta | E^n_\beta - S^{n-1}_\beta : E^n_\beta - S^{n-1}_\beta \to e^n_\beta$ is a homeomorphism. By Exercise 1.16, the closure of e^n_β is the n-cell $\Phi_\beta(E^n_\beta)$ (and thus the notation $\bar{e}^n_\beta$ is justified) and the topological boundary ∂e^n_β of e^n_β is $\Phi_\beta(S^{n-1}_\beta)$. Then $\Phi_\beta : (E^n_\beta, S^{n-1}_\beta) \to (\bar{e}^n_\beta, \partial e^n_\beta)$ is a continuous function of pairs. Moreover,

$X^n - X^{n-1}$ is a disjoint union of the e^n_β, for $\beta \in B$, and we write $X^n = X^{n-1} \cup \bigcup_\beta e^n_\beta$ or $X^n = X^{n-1} \cup \bigcup_\beta \bar{e}^n_\beta$. The CW complex X is called *finite* if it has finitely many cells and it is called *finite-dimensional* if, for some N, the N-skeleton $X^N = X$. The smallest such N is called the *dimension* of X. We observe that condition (2) in Definition 1.5.4 always holds if there are finitely many cells.

Remark 1.5.5 To make a nonempty CW complex into a based space, we choose a vertex as basepoint.

The following simple lemma is very useful.

Lemma 1.5.6 *Let X be a CW complex.*

1. *Every characteristic function $\Phi^n_\beta : E^n_\beta \to \bar{e}^n_\beta$ is an identification function.*
2. *X has the weak topology with respect to the set of skeleta $\{X^n\}$.*
3. *If C is a compact subset of X, then C is contained in a finite union of open cells.*

Proof. (1) This follows from the fact that X^n is a quotient space of $X^{n-1} \sqcup \coprod_\beta E^n_\beta$.

(2) Suppose $F \subseteq X$ and $F \cap X^n$ is closed in X^n, for every n. Then $F \cap \bar{e}^n_\beta$ is closed in $\bar{e}^n_\beta$, for every n and every β. Therefore F is closed. Conversely suppose $F \subseteq X$ and $F \cap \bar{e}^n_\beta$ is closed in $\bar{e}^n_\beta$, for every n and every β. Thus $(\Phi_\beta)^{-1}(F \cap \bar{e}^n_\beta)$ is closed in E^n_β, for every β. We argue by induction on n that $F \cap X^n$ is closed in X^n. Suppose $F \cap X^{n-1}$ is closed in X^{n-1}. If $q : X^{n-1} \sqcup \coprod_\beta E^n_\beta \to X^n$ is the quotient function, it follows that $q^{-1}(F \cap X^n)$ is closed in $X^{n-1} \sqcup \coprod_\beta E^n_\beta$. Therefore $F \cap X^n$ is closed in X^n.

(3) We suppose that C meets an infinite number of open cells of X and choose an infinite set $S = \{x_1, x_2, \ldots\} \subseteq C$ such that each element of S is in a different open cell. For any subset $A \subseteq S$, we show that A is closed in X. By (2), it suffices to prove by induction on n that $A \cap X^n$ is closed in X^n, for all $n \geqslant 0$. Assume that $A \cap X^{n-1}$ is closed in X^{n-1} and let $\bar{e}^n_\alpha$ be any n-cell with characteristic function $\Phi_\alpha : E^n_\alpha \to \bar{e}^n_\alpha$ and attaching function $\phi_\alpha : S^{n-1}_\alpha \to X^{n-1}$. Then $\phi^{-1}_\alpha(A)$ is closed in S^{n-1}_α. Because $\Phi^{-1}_\alpha(A)$ consists of $\phi^{-1}_\alpha(A)$ union with at most one additional point, $\Phi^{-1}_\alpha(A)$ is closed in E^n_α. From this it follows, as in the proof of (2), that $A \cap X^n$ is closed in X^n, and so A is closed in X. Thus S is a discrete space since every subset of S is closed. Also S is closed in X and contained in the compact space C. Therefore S is compact. But a compact discrete space is finite. Hence we conclude that C is contained in a finite union of open cells. □

If e^n is an open cell of a CW complex, then the closed cell $\bar{e}^n$ is compact. It then follows from the previous lemma that $\bar{e}^n$ is contained in a finite union of open cells. This is called the *closure-finite* condition. The letters C and W that appear in the term CW complex are to indicate that a CW complex is closure finite and has the weak topology of Definition 1.5.4(2).

We now have the following result.

Lemma 1.5.7 *Let X be a CW complex and $f : X \to Y$ a function. Then f is continuous $\Longleftrightarrow$ $f|\bar{e}^n_\beta : \bar{e}^n_\beta \to Y$ is continuous for each closed cell $\bar{e}^n_\beta$ $\Longleftrightarrow$ $f\Phi^n_\beta : E^n_\beta \to Y$ is continuous, for each characteristic map Φ^n_β. Moreover, f is continuous if and only if $f|X^n : X^n \to Y$ is continuous, for each n.*

Proof. The first assertion is an immediate consequence of the weak topology condition for CW complexes and the fact that Φ^n_β is a quotient function. The last sentence follows from Lemma 1.5.6(2). □

Remark 1.5.8 If X and Y are CW complexes, we next give CW structure to the product $X \times Y$ such that the closed $(m+n)$-cells of $X \times Y$ are of the form $\bar{e}^m \times \bar{e}^n$, where $\bar{e}^m$ is a closed m-cell of X and $\bar{e}^n$ is a closed n-cell of Y. If Φ and Ψ are characteristic functions for $\bar{e}^m$ and $\bar{e}^n$, then

$$E^{m+n} \cong E^m \times E^n \xrightarrow{\Phi \times \Psi} \bar{e}^m \times \bar{e}^n,$$

and this function carries $S^{m+n-1} \cong E^m \times S^{n-1} \cup S^{m-1} \times E^n$ to the topological boundary $\partial(e^m \times e^n) \cong e^m \times \partial e^n \cup \partial e^m \times e^n$. It follows that this is a characteristic function for $\bar{e}^m \times \bar{e}^n$. The skeleta are given by $(X \times Y)^k = \cup_{i+j=k} X^i \times Y^j$. If we give $X \times Y$ the weak topology described in Definition 1.5.4(2), then $X \times Y$ is a CW complex. We note that this topology may differ from the product topology on $X \times Y$. For an example, see [25]. If X or Y is locally compact or if X and Y both have countably many cells, then it has been proved [39, Thm. A.6] that the two topologies coincide. Whenever we discuss the product of CW complexes X and Y we assume that $X \times Y$ has the weak topology.

A special case of the product occurs when X is a CW complex and Y is the closed unit interval I. We give I the structure of a CW complex with two vertices 0 and 1 and one closed 1-cell I. Because I is compact, the weak topology and the product topology on $X \times I$ agree. Therefore $X \times I$ with the product topology is a CW complex. The following lemma is useful in determining the continuity of a homotopy.

Lemma 1.5.9 *Let X be a CW complex and $F : X \times I \to Y$ a function. Then F is continuous $\Longleftrightarrow$ $F|\bar{e}^n \times I : \bar{e}^n \times I \to Y$ is continuous for every closed cell $\bar{e}^n$ $\Longleftrightarrow$ $F(\Phi \times \mathrm{id}) : E^n \times I \to Y$ is continuous for every characteristic function Φ $\Longleftrightarrow$ $F|X^n \times I : X^n \times I \to Y$ is continuous for each n.*

Proof. The first equivalence is a consequence of Lemma 1.5.7. The second is based on the fact that Φ and hence $(\Phi \times \mathrm{id})$ is a quotient function (Lemma 1.5.6 and Appendix A). The third follows from Lemma 1.5.7. □

We next give some well-known examples of CW complexes.

Example 1.5.10

1. A one-dimensional CW complex is a graph in which we allow the ends of an edge to have the same vertex.

2. A compact connected surface can be made into a CW complex with one vertex, one 2-cell, and several 1-cells. This is described in detail in [61, Chap. 8].
3. Every simplicial complex is a CW complex. An n-simplex is an n-cell since every n-simplex is homeomorphic to E^n.
4. A compact differentiable manifold has the homotopy type of a CW complex. This appears to have first been proved by Radó [77] and is also a basic result of Morse theory [71].
5. The n-sphere S^n can be given simple CW structure consisting of one vertex e^0 and one open n-cell e^n, so that $S^n = e^0 \cup e^n$. The characteristic function $\Phi : E^n \to S^n$ sends the origin in E^n to the point $(-1, 0, \ldots, 0) \in S^n$ and stretches E^n over S^n so that the boundary S^{n-1} of E^n goes to the basepoint $* = (1, 0, \ldots, 0) \in S^n$. This function is explicitly defined just before Proposition 4.5.1 as the function h. Then S^n is a CW complex. As based space we must choose basepoint $e^0 = *$ because it is the only vertex.
6. Let $\mathbb{R}\mathrm{P}^n$ be *real projective n-space.* This can be defined by considering Euclidean space minus the origin, $\mathbb{R}^{n+1} - \{0\}$, and introducing the equivalence relation $x \simeq y$ if there exists $\lambda \in \mathbb{R} - \{0\}$ such that $y = \lambda x$, where $x, y \in \mathbb{R}^{n+1} - \{0\}$. Then $\mathbb{R}\mathrm{P}^n$ is the resulting identification space. Note that $l \in \mathbb{R}\mathrm{P}^n$ if and only if l is a one-dimensional vector subspace of $\mathbb{R}^{n+1}$ (i.e., a line through the origin) minus $\{0\}$. Now $S^n \subseteq \mathbb{R}^{n+1} - \{0\}$ and we introduce the following equivalence relation on S^n: $x \approxeq y$ if and only if $y = \pm x$, for $x, y \in S^n$. Then the inclusion $S^n \subseteq \mathbb{R}^{n+1} - \{0\}$ induces a continuous bijection θ from $S^n/\approxeq$ to $\mathbb{R}\mathrm{P}^n$. Since $S^n/\approxeq$ is compact and $\mathbb{R}\mathrm{P}^n$ is Hausdorff (Exercise 1.23), θ is a homeomorphism. Furthermore, by restricting θ to the upper cap $E^n_+ = \{(x_1, x_2, \ldots, x_{n+1}) \in S^n \mid x_{n+1} \geqslant 0\} \subseteq S^n$, we have that $\mathbb{R}\mathrm{P}^n \cong E^n_+/\sim$, where $x \sim y$ if and only if $y = \pm x$ and $x, y \in S^{n-1} \subseteq E^n_+$. By mapping $(x_1, \ldots, x_k) \in E^{k-1}_+$ to $(x_1, \ldots, x_k, 0) \in E^k_+$, we see that $E^{k-1}_+ \subseteq E^k_+$ and that $\mathbb{R}\mathrm{P}^{k-1} \cong E^{k-1}_+/\sim$ is homeomorphic to a subspace of $\mathbb{R}\mathrm{P}^k \cong E^k_+/\sim$. We put a CW structure on $X = \mathbb{R}\mathrm{P}^n$ by defining each skeleton X^k inductively so that (1) $X^k = \mathbb{R}\mathrm{P}^k$ and (2) $X^k - X^{k-1}$ consists of one open k-cell, for $0 \leqslant k \leqslant n$. Since $\mathbb{R}\mathrm{P}^0$ is a point $\{*\}$, set $X^0 = \{*\}$. Now assume that the CW structure has been defined on $\mathbb{R}\mathrm{P}^{k-1}$. Define a characteristic function $\Phi : (E^k, S^{k-1}) \to (\mathbb{R}\mathrm{P}^k, \mathbb{R}\mathrm{P}^{k-1})$ as the composition of the homeomorphism $E^k \cong E^k_+$ with the quotient function $E^k_+ \to E^k_+/\sim$ $\cong \mathbb{R}\mathrm{P}^k$. Then $\Phi(x) = \langle x, \sqrt{1 - |x|^2}\rangle$, for $x \in E^k$, where $\langle - \rangle$ denotes the equivalence class in E^k_+. Hence $\Phi|E^k - S^{k-1} : E^k - S^{k-1} \to \mathbb{R}\mathrm{P}^k - \mathbb{R}\mathrm{P}^{k-1}$ is a homeomorphism with inverse function $\Psi : \mathbb{R}\mathrm{P}^k - \mathbb{R}\mathrm{P}^{k-1} \to E^k - S^{k-1}$ defined by $\Psi\langle x_1, x_2, \ldots, x_{k+1}\rangle = (x_1, x_2, \ldots, x_k)$. Furthermore, (1) of Definition 1.5.4 holds for X^k, so this completes the induction. $\mathbb{R}\mathrm{P}^n$ has finitely many cells, therefore the weak topology condition (2) is satisfied. Thus $\mathbb{R}\mathrm{P}^n = X = X^n$ is a CW complex. In addition, we can define real projective infinite-dimensional space $\mathbb{R}\mathrm{P}^\infty = \bigcup \mathbb{R}\mathrm{P}^n$. With the weak topology, this is the CW complex whose nth skeleton is real projective n-space.

7. Similarly, if $\mathbb{C}$ denotes the complex numbers and $\mathbb{H}$ denotes the quaternions, we can define *complex projective n-space* $\mathbb{C}\mathrm{P}^n$ and *quaternionic projective n-space* $\mathbb{H}\mathrm{P}^n$ and give them CW structure. We just do this for $\mathbb{C}\mathrm{P}^n$. An equivalence relation on $\mathbb{C}^{n+1} - \{0\}$ is given by $x \simeq y$ if there is a nonzero complex number λ such that $y = \lambda x$, for $x, y \in \mathbb{C}^{n+1} - \{0\}$. We set $\mathbb{C}\mathrm{P}^n = (\mathbb{C}^{n+1} - \{0\})/\simeq$. Consider $S^{2n+1} = \{(z_1, z_2, \ldots, z_{n+1}) \mid z_i \in \mathbb{C}, \sum |z_i|^2 = 1\} \subseteq \mathbb{C}^{n+1} - \{0\}$. We define an equivalence relation on S^{2n+1} by $x \approxeq y$ if and only if there is a complex number λ with $|\lambda| = 1$ such that $y = \lambda x$, where $x, y \in S^{2n+1}$. Then $\mathbb{C}\mathrm{P}^n \cong S^{2n+1}/\approxeq$. We put CW structure on $X = \mathbb{C}\mathrm{P}^n$ by defining the skeleta X^k inductively such that (1) $X^{2k} = X^{2k+1} = \mathbb{C}\mathrm{P}^k$ and (2) $X^{2k} - X^{2k-1}$ consists of one open $2k$-cell, for $0 \leqslant k \leqslant n$. Since $\mathbb{C}\mathrm{P}^0$ is a single point $\{*\}$, set $X^0 = \{*\}$. Now assume that CW structure has been defined on $X^{2k-2} = X^{2k-1}$ satisfying (1) of Definition 1.5.4. Define a function $\Phi : E^{2k} \to \mathbb{C}\mathrm{P}^k$ by $\Phi(x) = \langle x, \sqrt{1 - |x|^2} \rangle$, for $x \in E^{2k}$. It can be verified that Φ maps S^{2k-1} onto $\mathbb{C}\mathrm{P}^{k-1}$ and that it is a homeomorphism from $E^{2k} - S^{2k-1}$ to $\mathbb{C}\mathrm{P}^k - \mathbb{C}\mathrm{P}^{k-1}$. Thus $\Phi : (E^{2k}, S^{2k-1}) \to (\mathbb{C}\mathrm{P}^k, \mathbb{C}\mathrm{P}^{k-1})$ is a characteristic function, and the skeleta $X^{2k} = X^{2k+1}$ are defined. For more details, see [87, p. 67]. This completes the induction and shows that $\mathbb{C}\mathrm{P}^n$ is a CW complex. Quaternionic projective n-space $\mathbb{H}\mathrm{P}^n$ is treated similarly. It has one cell in dimensions $0, 4, 8, \ldots, 4n$ and no other cells. The $4k$-skeleton is equal to $\mathbb{H}\mathrm{P}^k$. Complex and quaternionic projective spaces of infinite dimensions can also be defined as the union of complex and quaternionic finite-dimensional projective spaces, respectively, with the weak topology. We also note for consideration in Chapter 3 that for any $k \geqslant 1$, the attaching functions for complex and quaternionic projective spaces are functions $S^{2n+1} \to \mathbb{C}\mathrm{P}^n$ and $S^{4n+3} \to \mathbb{H}\mathrm{P}^n$.

The following lemma is used often.

Lemma 1.5.11 *Let X^n be the n-skeleton of a CW complex X with attaching functions $\phi_\beta : S^{n-1}_\beta \to X^{n-1}$, for all $\beta \in B$, and let $f : X^{n-1} \to Y$ be any free map. Then f can be extended to a free map $\widetilde{f} : X^n \to Y \iff f\phi_\beta \simeq_{\text{free}} c_\beta : S^{n-1}_\beta \to Y$, for all $\beta \in B$, where c_β is a constant function.*

Proof. If $\widetilde{f}$ is an extension of f and $i_\beta : S^{n-1}_\beta \to E^n_\beta$ and $j : X^{n-1} \to X^n$ are inclusions, then

$$f\phi_\beta = \widetilde{f}j\phi_\beta = \widetilde{f}\Phi_\beta i_\beta \simeq_{\text{free}} c_\beta,$$

for some constant function c_β, because the identity function of E^n_β is homotopic to the constant function. Conversely, suppose $f\phi_\beta \simeq_{\text{free}} c_\beta$ for all $\beta \in B$. Then there is a free map $F_\beta : E^n_\beta \to Y$ such that $F_\beta i_\beta = f\phi_\beta$ by Corollary 1.4.11. If $\phi : \coprod_{\beta \in B} S^{n-1}_\beta \to X^{n-1}$ is the free map determined by the ϕ_β, then the F_β determine a free map $F : \coprod_{\beta \in B} E^n_\beta \to Y$ such that $F(\coprod_{\beta \in B} i_\beta) = f\phi$. It now follows that F and f yield a free map $\widetilde{f} : X^n \to Y$ which is an extension of f. □

Definition 1.5.12 Let X be an unbased space and let A be a subspace. We say that the pair (X, A) has the *homotopy extension property* if for every space Y, every continuous function $f_0 : X \to Y$, and free homotopy $g_t : A \to Y$ such that $g_0 = f_0|A$, there exists a free homotopy $f_t : X \to Y$ of f_0 such that $f_t|A = g_t$.

This notion also exists in the based case. For this we require all spaces, functions, and homotopies to be based. In the terminology of Chapter 3, the homotopy extension property for the based pair (X, A) is just the assertion that the inclusion map $i : A \to X$ is a cofibration.

Proposition 1.5.13

1. *If $X \times \{0\} \cup A \times I$ is an unbased retract of $X \times I$ and A is a closed subspace of X, then the pair (X, A) has the homotopy extension property.*
2. *If the pair (X, A) has the homotopy extension property, then $X \times \{0\} \cup A \times I$ is an unbased retract of $X \times I$.*

Proof. Let $r : X \times I \to X \times \{0\} \cup A \times I$ be a free retraction and assume that $f_0 : X \to Y$ and $g_t : A \to Y$ are given such that $g_0 = f_0|A$. Since A is closed, the function f_0 and homotopy g_t determine a free map $H : X \times \{0\} \cup A \times I \to Y$ (Appendix A). Then $F = Hr : X \times I \to Y$ is the desired homotopy of f_0.

Next let $Y = X \times \{0\} \cup A \times I$ and define $f_0 : X \to Y$ by $f_0(x) = (x, 0)$ and $g_t : A \to Y$ by $g_t(a) = (a, t)$. By hypothesis, there exists a continuous function $F : X \times I \to Y$ such that $F(x, 0) = (x, 0)$ and $F(a, t) = (a, t)$, for all $x \in X$, $a \in A$ and $t \in I$. Therefore F is a free retraction of $X \times I$ onto $X \times \{0\} \cup A \times I$. □

We next define a subcomplex of a CW complex.

Definition 1.5.14 Let X be a CW complex and $A \subseteq X$. Then A is a *subcomplex* of X if A is a union of open cells of X such that if $e^n \subseteq A$ is an open cell, then $\bar{e}^n \subseteq A$. Clearly A is then a CW complex. We say that the pair (X, A) is a *CW pair*.

It is not hard to show that A is a closed subspace of X (Exercise 1.22) and that the CW topology of A coincides with the induced topology of A.

The notion of a CW pair can be generalized.

Definition 1.5.15 A pair of unbased spaces (X, A) with X Hausdorff and A a closed subspace of X is called a *relative CW complex* if there is a sequence of subspaces of X, called *skeleta* or *relative skeleta,*

$$(X, A)^0 \subseteq (X, A)^1 \subseteq \cdots \subseteq (X, A)^n \subseteq (X, A)^{n+1} \subseteq \cdots$$

whose union is X. These subspaces are inductively defined as follows.

1. $(X, A)^0$ is the union of A and a discrete set of points disjoint from A. We allow A or the discrete set to be empty. An element of the discrete set is called a *relative vertex* or *relative 0-cell*.

2. We assume that $(X,A)^{n-1}$ has been defined for $n \geqslant 1$ and that there are free maps $\phi_\beta : S_\beta^{n-1} \to (X,A)^{n-1}$ for $\beta \in B$, where $S_\beta^{n-1} = S^{n-1}$. The ϕ_β determine a free map $\phi : \coprod_{\beta \in B} S_\beta^{n-1} \to (X,A)^{n-1}$. Then $(X,A)^n$ is the space obtained from $(X,A)^{n-1}$ by attaching n-cells, that is,

$$(X,A)^n = (X,A)^{n-1} \cup_\phi \coprod_{\beta \in B} E_\beta^n.$$

As in Definition 1.5.4, we obtain characteristic functions $\Phi_\beta^n : E_\beta^n \to (X,A)^n$, *relative n-cells* $\bar{e}_\beta^n = \Phi_\beta^n(E_\beta^n)$ and *relative open n-cells* $e_\beta^n = \Phi_\beta^n(E_\beta^n - S_\beta^{n-1})$.

We also require that the topology of X is the weak topology with respect to the set consisting of A and all the relative n-cells $\bar{e}_\beta^n$. If $(X,A)^n = X$ for some n, then (X,A) is called *finite-dimensional.* The *dimension* of (X,A), written $\dim(X,A)$, is the smallest integer n such that $(X,A)^n = X$.

Remark 1.5.16

1. Some authors write $(X,A)^{-1} = A$ (although we do not). We note that for $i \leqslant n$, $(X,A)^i = A$ could occur.
2. If (X,A) is a relative CW complex, and we choose a basepoint in A, then (X,A) is a based pair of spaces and is called a based, relative CW complex.
3. If (X,A) is a CW pair, then (X,A) is a relative CW complex with $(X,A)^n = X^n \cup A$.
4. Let (X,A) be a relative CW complex. When A is the empty set, (X,A) is just the CW complex X. If (X,A) is a relative CW complex, then (X',A) is called a *subrelative CW complex* if (1) $X' \subseteq X$, (2) X' is a union of A and relative open cells of X, and (3) if e^k is a relative open cell of X contained in X', then $\bar{e}^k$ is contained in X'. Clearly (X',A) is a relative CW complex.
5. Many of the results for ordinary CW complexes easily carry over to relative CW complexes. For example, if (X,A) is a relative CW complex, then it follows that a function $f : X \to Y$ is continuous if and only if $f|(X,A)^n$ is continuous for all n. In addition, $F : X \times I \to Y$ is continuous if and only if $F|(X,A)^n \times I : (X,A)^n \times I \to Y$ is continuous for every n.

Proposition 1.5.17 *If (X,A) is a relative CW complex, then (X,A) has the homotopy extension property. If (X,A) is a based, relative CW complex, then (X,A) has the based homotopy extension property.*

Proof. For the first assertion, it suffices to prove that there is a free retraction $r : X \times I \to X \times \{0\} \cup A \times I$ by Proposition 1.5.13. We inductively define free retractions $r_n : X \times \{0\} \cup (X,A)^n \times I \to X \times \{0\} \cup A \times I$ such that $r_n | X \times \{0\} \cup (X,A)^{n-1} \times I = r_{n-1}$. For $n = 0$, define r_0 as the identity on $X \times \{0\} \cup A \times I$ and by $r_0(x,t) = (x,0)$, where x is a relative vertex and $t \in I$. Now assume that r_{n-1} has been defined and let $\bar{e}_\beta^n$ be a relative cell in $(X,A)^n$, $\beta \in B$, with characteristic function

$$\Phi_\beta : (E^n, S^{n-1}) \to (\bar{e}^n_\beta, \partial e^n_\beta) \subseteq (X, (X,A)^{n-1}).$$

Consider the function

$$E^n \times I \longrightarrow X \times \{0\} \cup A \times I$$

which is the composition of

$$E^n \times I \xrightarrow{s} E^n \times \{0\} \cup S^{n-1} \times I \xrightarrow{\Phi_\beta \times \mathrm{id}} X \times \{0\} \cup (X,A)^{n-1} \times I$$

with

$$X \times \{0\} \cup (X,A)^{n-1} \times I \xrightarrow{r_{n-1}} X \times \{0\} \cup A \times I,$$

where s is the free retraction of Example 1.4.2(3). This function induces $t^n_\beta : \bar{e}^n_\beta \times I \to X \times \{0\} \cup A \times I$. Thus the t^n_β, for $\beta \in B$, define a free retraction

$$X \times \{0\} \cup (X,A)^n \times I \xrightarrow{r_n} X \times \{0\} \cup A \times I$$

which extends r_{n-1}. It follows that the r_n can be extended to a continuous function $r : X \times I \to X \times \{0\} \cup A \times I$, and so r is a free retraction.

For the second assertion of the proposition note that the free retraction $r : X \times I \to X \times \{0\} \cup A \times I$ defined above satisfies $r(*, t) = (*, t)$, for $t \in I$. It now follows as in the proof of Proposition 1.5.13 that (X, A) has the based homotopy extension property. □

Proposition 1.5.18 *Let the pair (X, A) have the based homotopy extension property. If A is contractible, then the quotient map $q : X \to X/A$ is a homotopy equivalence.*

Proof. By hypothesis, there is a homotopy $g_t : A \to A$ such that $g_0 = \mathrm{id}_A$ and $g_1 = *$. If $i : A \to X$ is the inclusion, then $i\, g_t : A \to X$ is a homotopy and $i\, g_0 = \mathrm{id}_X\, i$

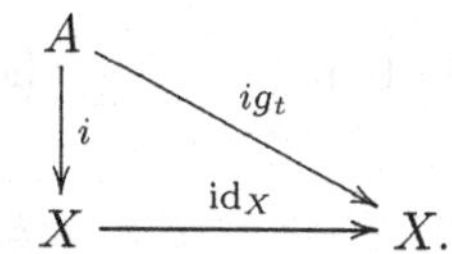

By the based homotopy extension property, there is a homotopy $f_t : X \to X$ with $f_0 = \mathrm{id}_X$ and $f_t\, i = i\, g_t$. Therefore $f_1\, i = i\, g_1 = *$, and so f_1 induces $f_1' : X/A \to X$ such that

$$f_1' q = f_1 \simeq f_0 = \mathrm{id}_X,$$

where $q : X \to X/A$ is the projection. Next we show $q f_1' \simeq \mathrm{id}_{X/A}$. Since $f_t\, i = i\, g_t$, the homotopy f_t induces a homotopy $h_t : X/A \to X/A$ such that $q f_t = h_t q$. Then $h_0 q = q f_0 = q$, so $h_0 = \mathrm{id}_{X/A}$. Also $h_1 q = q f_1 = q f_1' q$, and so $h_1 = q f_1'$. Therefore

$$qf_1' = h_1 \simeq h_0 = \mathrm{id}_{X/A}.$$

Thus f_1' is a homotopy inverse of q, and the proof is complete. □

From Proposition 1.5.17 we obtain the following corollary.

Corollary 1.5.19 *If (X, A) is a based, relative CW complex and A is contractible, then $q : X \to X/A$ is a homotopy equivalence.*

This corollary is useful in determining if two CW complexes have the same (based) homotopy type.

We next remark on the quotient of a relative complex.

Remark 1.5.20 If (X, A) is a relative CW complex, then the space X/A is a based CW complex. The cells of X/A are those of $X - A$ projected onto X/A and one additional vertex (the basepoint) to take the place of A. We apply this to an arbitrary wedge. For every $\beta \in B$, let X_β be a CW complex with basepoint $*_\beta$. Then by Definition 1.5.1 the wedge $\bigvee_{\beta \in B} X_\beta$ is the quotient space $(\coprod_{\beta \in B} X_\beta)/A$, where $A = \{*_\beta \mid \beta \in B\}$ is the subspace of all basepoints. Since $X = \coprod_{\beta \in B} X_\beta$ is a CW complex and A is a subcomplex, $\bigvee_{\beta \in B} X_\beta = X/A$ is a based CW complex.

Definition 1.5.21 If X and Y are CW complexes and $g : X \to Y$ is a free map, then g is a *cellular function* if g carries the n-skeleton of X into the n-skeleton of Y, that is, $g(X^n) \subseteq Y^n$, for all n. For relative CW complexes (X, A) and (Y, B), a continuous function $g : (X, A) \to (Y, B)$ of pairs is cellular if $g((X, A)^n) \subseteq (Y, B)^n$, for all n.

The following theorem is one version of the cellular approximation theorem.

Theorem 1.5.22 *Let (X, A) and (Y, B) be relative CW complexes and let $f : (X, A) \to (Y, B)$ be a continuous function. Then there is a cellular function $g : (X, A) \to (Y, B)$ such that $f \simeq_{\text{free}} g : (X, A) \to (Y, B)$ rel A. This result also holds for based complexes and maps.*

We give a sketch of the proof in Remark 4.5.8. The theorem has the following corollary.

Corollary 1.5.23 *If X and Y are CW complexes and $f : X \to Y$ is a free map, then f is homotopic to a cellular function. If A is a subcomplex of X and $f|A$ is cellular, then the homotopy can be chosen to be stationary on A. This result also holds for based complexes and maps.*

It is interesting to compare this corollary to the simplicial approximation theorem ([87, pp. 76–77] or [39, pp. 177–179]).

Theorem 1.5.22 has many consequences. We give one next.

Proposition 1.5.24 *Let X be a based CW complex and let (Y, B) be a based, relative CW complex. If $i : (Y, B)^n \to Y$ is the inclusion map, then the induced function $i_* : [X, (Y, B)^n] \to [X, Y]$ is injective if $\dim X < n$ and* *surjective if* $\dim X \leqslant n$. *In particular, if* X *and* Y *are based CW complexes, then the function* $i_* : [X, Y^n] \to [X, Y]$ *induced by the inclusion map is injective if* $\dim X < n$ *and surjective if* $\dim X \leqslant n$.

Proof. If $f : X \to Y$ is a map, then $f \simeq g : (X, \{*\}) \to (Y, B)$ by the cellular approximation theorem 1.5.22, where $g : (X, \{*\}) \to (Y, B)$ is a cellular map. If $\dim X \leqslant n$, then $g(X) \subseteq (Y, B)^n$, and so $g = ig'$ for some map $g' : X \to (Y, B)^n$. Therefore in $[X, Y]$, $[f] = [g] = i_*[g']$, and so i_* is surjective.

Suppose next that $f, g : X \to (Y, B)^n$ and $if \simeq_F ig : X \to Y$. Since $(X \times I, X \times \partial I \cup \{*\} \times I)$ is a relative CW complex, by the cellular approximation theorem 1.5.22, there exists a continuous function $G : X \times I \to Y$ such that $F \simeq G \operatorname{rel} (X \times \partial I \cup \{*\} \times I)$ and G is cellular. If $\dim X < n$, then $G(X \times I) \subseteq (Y, B)^n$. Therefore G determines a homotopy $G' : X \times I \to (Y, B)^n$ such that $f \simeq_{G'} g$, and so i_* is injective. □

1.6 Why Study Homotopy Theory?

Homotopy theory is based on the concept of homotopy of maps. The following is a list of some of the reasons for studying this notion.

• Homotopy is a geometrically intuitive relation. It is natural to consider two subspaces A and B of a larger space Y and ask if A can be continuously deformed into B within Y. As we have seen in Section 1.3 this occurs when $f \simeq g$, $A = f(X)$, and $B = g(X)$. A special case appears in the study of the fundamental group when $f, g : I \to Y$ are two paths with the same initial point and the same terminal point. Then one wants to know if the path f can be deformed into the path g with end points fixed, that is, if $f \simeq g \operatorname{rel}\{0, 1\}$.

• The set of homotopy classes $[X, Y]$ can be thought of as an approximation to the set of all maps of X into Y. The latter set is sometimes extremely large and unwieldy. This is often not the case for the homotopy set.

• We show that the notion of homotopy is related to the question of extending a continuous, unbased function. Let X be an unbased space with subspace $A \subseteq X$ and let $g : A \to Y$ be a function. If there exists a continuous, unbased function $f : X \to Y$ such that $f|A = g$, then f is called an extension of g and g is said to be *extendible* to X. This gives a diagram

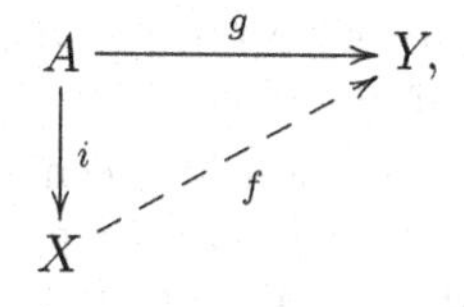

where i is the inclusion function and f is an extension (which may or may not exist). If we put mild restrictions on A and X, such as requiring that the pair (X, A) be a relative CW complex, then the following result holds. If $g \simeq g'$:

$A \to Y$ and g is extendible to X, then g' is extendible to X. This is because the relative CW complex (X, A) has the homotopy extension property (Definition 1.5.12) by Proposition 1.5.17. Thus the extendibility or nonextendibility of a function is a property of the homotopy class of the function. More generally, this result holds whenever i is a cofiber map (Chapter 3). There is another relationship between homotopy and extendibility. Two functions $f, g : X \to Y$ are homotopic if and only if the function $F' : X \times \partial I \to Y$ defined by $F'(x, 0) = f(x)$ and $F'(x, 1) = g(x)$, for all $x \in X$ is extendible to $X \times I$. A notion analogous to extendibility (in fact, dual to it) is that of lifting a function. Let $p : E \to B$ and $g : X \to B$ be functions. If there exists a function $f : X \to E$ such that $pf = g$, then f is called a lift or lifting of g and g is said to be *liftable* to E,

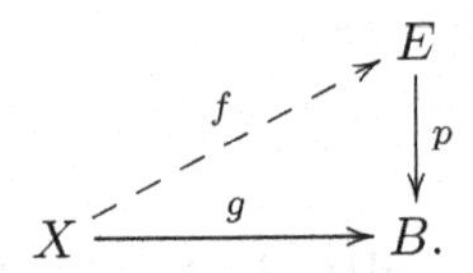

With certain restrictions on the function $p : E \to B$ (namely, that it is a fiber map; see Chapter 2), we have that if $g \simeq g' : X \to B$ and g is liftable, then so is g'. The previous discussion has been made in the unbased case and we note that it holds as well for based spaces, maps, and homotopies. In Chapter 9 we give a detailed and systematic treatment of extensions and liftings.

• Many of the invariants of algebraic topology for maps depend only on the homotopy class of the map. For example, homotopic maps induce the same homomorphism of homology groups (or cohomology groups). This is in fact one of the Eilenberg-Steenrod axioms for homology [32, pp. 10–12]. It is because of this that other invariants such as the degree of a map $S^n \to S^n$ (see Definition 2.4.14) or the Lefschetz number of a map $X \to X$ [39, p. 179] are invariants of the homotopy class of the map.

• We show in Section 2.3, that the homotopy set $[X, Y]$ has natural algebraic structure in certain cases. For example, if $X = S^n$, an n-sphere, and Y is any space, we obtain the homotopy groups $\pi_n(Y)$ (Definition 2.4.1). If X is any space and Y is an Eilenberg–Mac Lane space, we obtain the homotopical cohomology groups of X with coefficients in G (see 2.5.10). If X is a CW complex, then these groups are naturally isomorphic to the singular cohomology groups of X (2.5.11) with coefficients in G. Other spaces Y give rise to various topological K-theory groups of X. The algebraic structure on the set $[X, Y]$ makes it more interesting.

• Sometimes a continuous function between spaces that have additional structure is homotopic to a function which preserves that structure. The following two examples illustrate this. (1) If $f : X \to Y$ is a continuous function, where X and Y are CW complexes, then f is homotopic to a cellular function as was stated in the cellular approximation theorem 1.5.22. (2) If $f : M \to N$ is a continuous function of differentiable manifolds, then f

is homotopic to a differentiable function [14, Chap. II, Cor. 11.9]. Thus in these cases every homotopy class contains a "structure-preserving" function. Results of this sort enable one to apply the techniques of areas such as cellular topology or differential topology to the study of homotopy classes of maps.

• Homotopy and its analogues appear in other parts of algebraic topology and even other parts of mathematics. We give just a small sample of this.

1. Consider the set of n-dimensional real vector bundles over X and introduce the relation of vector bundle equivalence. If $\text{Vect}_n(X)$ denotes the set of equivalence classes, then there is a space $BO(n)$ (the classifying space of the orthogonal group $O(n)$) such that there is a natural bijection between $\text{Vect}_n(X)$ and $[X, BO(n)]$ (Remark 3.4.17 and [12, Thm. 23.12]). Thus a study of the set $\text{Vect}_n(X)$ is equivalent to a study of the homotopy set $[X, BO(n)]$. Similar results hold for real oriented vector bundles, complex vector bundles, and even more general bundles.
2. Let $\mathcal{N}_n$ denote the $\mathbb{Z}_2$-vector space of nonoriented cobordism classes of smooth, closed n-manifolds. Then there is the Thom space TO(q) with the property that $\mathcal{N}_n = \pi_{n+q}(\text{TO(q)}) = [S^{n+q}, \text{TO(q)}]$ for q sufficiently large [66, pp. 216–221]. This reduces the study of cobordism to a study of certain homotopy groups. Similar results hold for oriented cobordism, unitary cobordism, and so on. (see [87, Chap. 12]).

Exercises

Exercises marked with (∗) may be more difficult than the others. Exercises marked with (†) are used in the text.

1.1. (†) Let X and Y be unbased spaces and assume that Y is path-connected. Let $y_0, y_1 \in Y$ and define constant functions $c_0, c_1 : X \to Y$ by $c_0(x) = y_0$ and $c_1(x) = y_1$, for all $x \in X$. Prove that $c_0 \simeq_{\text{free}} c_1$.

1.2. State in what sense homotopy rel A is compatible with composition. Prove your assertion.

1.3. Prove that if $f, g : X \to S^n$ are two maps such that $f(x) \neq -g(x)$ for all $x \in X$, then $f \simeq g$.

1.4. Prove without using the cellular approximation theorem that S^{n-1} is contractible in S^n.

1.5. (†) Let $B \subseteq A \subseteq X$. If $r : A \to B$ and $s : X \to A$ are deformation retractions, then prove that $rs : X \to B$ is a deformation retraction. If r and s are strong deformation retractions, then so is rs.

1.6. Show that $S^{n-1} \subseteq S^n - \{N, S\}$ is a strong deformation retract, where N is the North Pole $(0, 0, \ldots, 1)$ and S is the South Pole $(0, 0, \ldots, -1)$.

1.7. Let V be a vector subspace of the vector space $\mathbb{R}^n$. Show that V is a strong deformation retract of the topological space $\mathbb{R}^n$.

1.8. Prove that the central circle $\{0\} \times S^1$ of the *solid torus* $E^2 \times S^1$ is a strong deformation retract. Generalize this result.

1.9. Write the homotopy for the deformation retraction in Example 1.4.2(3).

1.10. (*) Prove that if A is a retract of a Hausdorff space X, then A is closed.

1.11. (†) Prove that if $f : X \to Y$ has a right homotopy inverse and a left homotopy inverse, then f is a homotopy equivalence.

1.12. Suppose $f \simeq g : X \to Y$. Prove: (1) $Cf \simeq Cg : CX \to CY$; (2) $Ef \simeq Eg : EX \to EY$.

1.13. 1. Let X be a space and let l, l', and l'' be paths in X such that $l(1) = l'(0)$ and $l'(1) = l''(0)$. Prove that $l+(l'+l'') \simeq (l+l')+l'' : I \to X \operatorname{rel} \{0,1\}$.
2. If l is a path in X, c_0 is the constant path at $l(0)$, and c_1 is the constant path at $l(1)$, then prove that $c_0 + l \simeq l : I \to X \operatorname{rel} \{0,1\}$ and $l + c_1 \simeq l : I \to X \operatorname{rel} \{0,1\}$.
3. Prove that $l + \bar{l} \simeq c_0 : I \to X \operatorname{rel} \{0,1\}$ and $\bar{l} + l \simeq c_1 : I \to X \operatorname{rel} \{0,1\}$, where $\bar{l}(t) = l(1-t)$.

1.14. Let X be a triangle contained in $\mathbb{R}^2$ (with the induced topology) having its base [0,1] of the x-axis. The interior of the triangle is an open 2-cell and the 0-cells consist of the vertices of the triangle and all points on the base with x-coordinate $1/n$ for $n = 2, 3, \ldots$.. The 1-cells consist of the edges between adjacent 0-cells. Is X a CW complex?

1.15. Let C be the comb space (Example 1.4.5) with topology induced by $\mathbb{R}^2$. Show that C satisfies all the requirements for being a CW complex except the weak topology condition.

1.16. (†) Let X be a CW complex, let $\Phi : E^n \to X^n$ be a characteristic function and let $e^n = \Phi(E^n - S^{n-1})$. Prove that the closure of $\Phi(E^n - S^{n-1})$ equals $\Phi(E^n)$ and the boundary of $\Phi(E^n - S^{n-1})$ equals $\Phi(S^{n-1})$.

1.17. (*) Let $\mathcal{E}(X)$ be the group of homotopy classes of homotopy equivalences of X with itself and let G be any finite group. Show that there exists a finite CW complex X such that $G \subseteq \mathcal{E}(X)$ as a subgroup.

1.18. Let (X, A) have the based homotopy extension property and assume that the inclusion $i : A \to X$ is a homotopy retract. Then prove that A is a retract of X.

1.19. (*) (†) Show that Φ' is a homotopy inverse to Φ in Lemma 1.5.3.

1.20. Let X be an n-dimensional based CW complex consisting of an n-cell e^n attached to the $(n-1)$-skeleton with characteristic map $\Phi : (E^n, S^{n-1}) \to (X, X^{n-1})$ and let $x_0 = \Phi(0)$. Show that $X^{n-1} \subseteq X - \{x_0\}$ is a strong deformation retract. Generalize this to an arbitrary based n-dimensional CW complex.

1.21. Show condition (2) of Definition 1.5.4 holds if X is a space that satisfies condition (1) of Definition 1.5.4 and $X^n = X$ for some n.

1.22. Let K be a CW complex, not necessarily path-connected.

1. Prove that if L is a subcomplex of K, then L is a closed subset of K.
2. Prove that the path components of K are subcomplexes of K.
3. Show that K is path-connected if and only if K is connected.

1.23. $(*)$ (†) Prove that $\mathbb{R}\mathrm{P}^n$ is Hausdorff.

1.24. $(*)$ Prove that if X is a CW complex and C is a compact subset, then C is contained in a finite subcomplex. Consequently, $C \subseteq X^n$, for some n.

1.25. Let S^2 be the 2-sphere and let $A \subseteq S^2$ consist of n distinct points. Prove that S^2/A has the homotopy type of $S^2 \vee B$, where B is the wedge of $n-1$ circles S^1.

1.26. $(*)$ Show that $S^1 \times S^1 / * \times S^1 \simeq S^2 \vee S^1$.

1.27. $(*)$ Let X be a based CW complex that is the union $K \cup L$ of two subcomplexes K and L with the basepoint in $K \cap L$. Prove that if K, L, and $K \cap L$ are contractible, then X is contractible.

1.28. Consider $S^n \subseteq \mathbb{R}^{n+1}$ and $E^n \subseteq \mathbb{R}^n \subseteq \mathbb{R}^{n+1}$ and let $X = S^n \cup E^n$. Prove that $X \simeq S^n \vee S^n$. Note that if $n = 1$, then X is the θ-curve (Example 1.4.13).

1.29. Prove that every map $f : S^i \to S^n$ is homotopic to the constant map if $i < n$.

1.30. Prove or disprove by example the following assertion. If (X, A) is a relative CW complex and A is a CW complex, then X is a CW complex.

1.31. (†) Prove that if (K, L) is a relative CW complex and $C \subseteq K$ is a compact space, then $C \subseteq (K, L)^k$, for some k.

Chapter 2
H-Spaces and Co-H-Spaces

2.1 Introduction

NOTATION AND STANDING ASSUMPTIONS

• *From this chapter on, most of the spaces that we consider will be based and path-connected and have the based homotopy type of based CW complexes. Some notable exceptions to path-connectedness are the 0-sphere* S^0 *and the 0-skeleton of a CW complex. Unless otherwise stated, all functions under consideration will be continuous and based and all homotopies will preserve the base point. These restrictions are sometimes asserted explicitly for emphasis. We discuss unbased spaces, functions and homotopies from time to time. However, whenever doing so, we explicitly make note of the fact.*

• *We take all homology and cohomology to be reduced, so that a space has trivial zero-dimensional homology and cohomology.*

• *We adopt the following notation throughout: "*$\simeq$*" for homotopy of maps or same homotopy type of spaces, "*$\cong$*" for homeomorphism of spaces or isomorphism of groups and "*$\sim$*" for the relation of equivalence. Furthermore, if* X *is a set with an equivalence relation and* $x \in X$*, then* $\langle x \rangle$ *denotes the equivalence class containing* x*.*

There are reasons for the restrictions on spaces listed above. First of all, nearly all of the spaces that are of interest to us are of this type. Second, these assumptions avoid having to add additional hypotheses to several theorems since CW complexes satisfy many of these hypotheses. But because of these assumptions, we must ensure that the constructions that we perform on spaces of the homotopy type of CW complexes yield spaces of the homotopy type of CW complexes. This is so, but the proofs in some instances are long and difficult. Presenting this material would take us far afield, and so we describe some proofs and give references for the others.

In this chapter we discuss the important notions of H-space and grouplike space and of co-H-space and cogroup. A grouplike space is the homotopy analogue of a group. It is a group object in the homotopy category. An H-space is defined in the same way but without the assumption of associativity. Cogroups and co-H-spaces are the categorical duals of these in the homotopy category. We show that the set of homotopy classes of maps of any space into a grouplike space has an induced group structure as does the set of homotopy classes of maps of a cogroup into any space. We then consider the set of homotopy classes of maps from a cogroup into a grouplike space and show that the two group structures agree and are abelian. Loop spaces ΩY are examples of grouplike spaces and suspensions ΣX are examples of cogroups. We prove that there is a fundamental isomorphism $[\Sigma X, Y] \cong [X, \Omega Y]$. Since an n-sphere is a suspension, the set of homotopy classes of maps $[S^n, Y]$ is a group. These are the homotopy groups of Y, denoted $\pi_n(Y)$, and discussed in Section 2.4 and later in Section 4.5. Of particular interest in this section is a theorem which we call Whitehead's First Theorem which asserts that a map of CW complexes is a homotopy equivalence if and only if it induces an isomorphism of all homotopy groups. A natural generalization of spheres is Moore spaces which are spaces with a single nonvanishing homology group. Dually, Eilenberg–Mac Lane spaces are spaces with a single nonvanishing homotopy group. The existence and uniqueness up to homotopy type of these spaces are discussed. Homotopy groups with coefficients are then defined by using Moore spaces and (homotopical) cohomology groups with coefficients by using Eilenberg–Mac Lane spaces. The chapter ends with a discussion of Eckmann–Hilton duality.

2.2 H-Spaces and Co-H-Spaces

Before discussing H-spaces and co-H-spaces, we introduce some terminology that appears in the rest of the book. We assume that the reader is familiar with the concept of a commutative diagram of groups and homomorphisms and of spaces and maps. In commutative diagrams there is the initial point (a group or space), a terminal point (a group or space), and two compositions of homomorphisms or maps from the initial point to the terminal point. In the case of abelian groups, if one of the compositions is the negative of the other, then we say that the diagram *anticommutes* or is an *anticommutative diagram*. In the case of spaces, if the two compositions are homotopic, then we say that the diagram *homotopy-commutes*, *commutes up to homotopy*, or is a *homotopy-commutative diagram*.

Now we turn to the notions of a grouplike space and an H-space. Let Y be a space and recall that $j_1 : Y \to Y \times Y$ and $j_2 : Y \to Y \times Y$ are defined by $j_1(y) = (y, *)$ and $j_2(y) = (*, y)$ for all $y \in Y$.

Definition 2.2.1 A *grouplike space* consists of a space Y and two maps $m : Y \times Y \to Y$ and $i : Y \to Y$ such that

1. $mj_1 \simeq \mathrm{id} \simeq mj_2 : Y \to Y$,

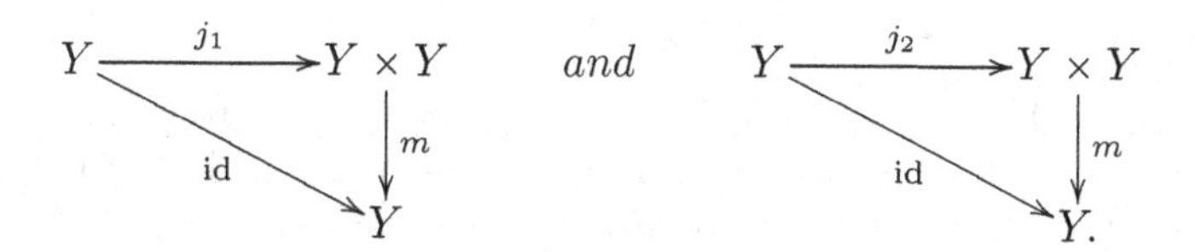

2. $m(m \times \mathrm{id}) \simeq m(\mathrm{id} \times m) : Y \times Y \times Y \to Y$,

$$\begin{array}{ccc} Y \times Y \times Y & \xrightarrow{m \times \mathrm{id}} & Y \times Y \\ \Big\downarrow{\scriptstyle \mathrm{id} \times m} & & \Big\downarrow{\scriptstyle m} \\ Y \times Y & \xrightarrow{m} & Y. \end{array}$$

3. $m(\mathrm{id}, i) \simeq * \simeq m(i, \mathrm{id}) : Y \to Y$,

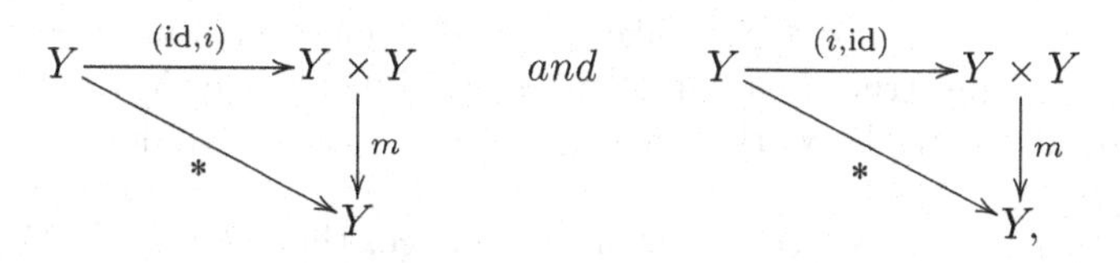

where $(\mathrm{id}, i), (i, \mathrm{id}) : Y \to Y \times Y$ are defined by $(\mathrm{id}, i)(y) = (y, i(y))$ and $(i, \mathrm{id})(y) = (i(y), y)$, for $y \in Y$.

A grouplike space is sometimes referred to as an *H-group*. The map m is called a *multiplication* and i is called a *homotopy inverse*. If only (1) holds, then Y (or more properly, the pair (Y, m)) is called an *H-space*. A space that is an H-space and a CW complex is called an *H-complex* and a grouplike space that is a CW complex is called a *grouplike complex*. We sometimes do not explicitly mention the multiplication or homotopy inverse and refer to a space Y as an H-space or grouplike space. Condition (2) is called *homotopy-associativity*. A homotopy-associative H-space is one in which (1) and (2) hold. In terms of the addition of maps defined below, condition (3) asserts that $\mathrm{id} + i \simeq * \simeq i + \mathrm{id}$. Therefore $[i]$ is the homotopy inverse of $[\mathrm{id}]$ in the group $[Y, Y]$. From this we obtain the inverse of any $\alpha = [f] \in [X, Y]$ defined as $i_*(\alpha) = [if]$. We show in Proposition 8.4.4 that a homotopy-associative H-complex always has a homotopy inverse, and so is a grouplike complex. The H-space (Y, m) is *homotopy-commutative* if $mt \simeq m : Y \times Y \to Y$ where $t : Y \times Y \to Y \times Y$ is defined by $t(y, y') = (y', y)$, for $y, y' \in Y$.

Definition 2.2.2 Let (Y, m) and (Y', m') be H-spaces and $h : Y \to Y'$ a map. We call h an *H-map* if the following diagram is homotopy-commutative,

$$\begin{array}{ccc} Y \times Y & \xrightarrow{h \times h} & Y' \times Y' \\ \Big\downarrow{\scriptstyle m} & & \Big\downarrow{\scriptstyle m'} \\ Y & \xrightarrow{h} & Y'. \end{array}$$

This is written $h : (Y, m) \to (Y', m')$.

The space Y is a *topological group* if (Y, m, i) is a grouplike space such that equality holds instead of homotopy in all parts of Definition 2.2.1. In this case, it is customary to write $m(y, y')$ as yy' and $i(y)$ as y^{-1}. A grouplike space is thus the analogue of a group in homotopy theory. Similarly an H-map is the analogue of a homomorphism of groups. We give a class of examples in Section 2.3 of spaces that are grouplike, but not topological groups. For now we note that the spheres S^1, S^3, and S^7 are all H-spaces. The first two are in fact topological groups. Multiplication of complex numbers induces a multiplication on S^1 which makes it into a topological group and quaternionic multiplication does the same for S^3. The sphere S^7 inherits its multiplication from the multiplication of octonions or Cayley numbers [49, pp. 448–449]. But the latter is not associative, and so S^7 is an H-space that is not a topological group. It has been proved [51] that this multiplication on S^7 is not homotopy-associative, so S^7 is not a grouplike space. The question of whether any other spheres have the structure of an H-space is a difficult one. A negative answer has been given by the work of several people with the major result due to Adams [1].

If (Y, m) is an H-space and X is any space, then the set $[X, Y]$ can be given an additively written binary operation which is defined as follows. Let $f, g : X \to Y$ and define $f + g = m(f \times g)\Delta = m(f, g)$

$$X \xrightarrow{\Delta} X \times X \xrightarrow{f \times g} Y \times Y \xrightarrow{m} Y,$$

where Δ is the diagonal map. Then if $\alpha = [f]$ and $\beta = [g] \in [X, Y]$, we set $\alpha + \beta = [f + g]$. This is a well-defined binary operation on the set $[X, Y]$. We make some simple remarks about this operation.

- By (1), $f + * = m(f \times *)\Delta = mj_1 f \simeq f$. Therefore $\alpha + 0 = \alpha$, and similarly $0 + \alpha = \alpha$, where 0 is the homotopy class of the constant map. Thus for an H-space (Y, m), the element $0 \in [X, Y]$ is a two-sided identity for the binary operation.
- If (3) holds, then, as mentioned earlier, $i_*(\alpha)$ is the inverse of α in $[X, Y]$.
- Clearly $m = p_1 + p_2 : X \times X \to X$, where $p_1, p_2 : X \times X \to X$ are the two projections, since $m(p_1 \times p_2)\Delta = m$.
- We obtain the *category of H-spaces* denoted $\mathcal{H}$ consisting of H-spaces and homotopy classes of H-maps and the *category of grouplike spaces* denoted $\mathcal{HG}$ consisting of grouplike spaces and homotopy classes of H-maps (see Appendix F).

We recall some categorical language and notation (Appendix F). Let $HoTop_*$ be the based homotopy category (consisting of spaces and homotopy classes of maps), let Gr be the category of groups, and let $Sets_*$ be the category of based sets. Furthermore, let $\mathcal{B}_*$ be the category of based sets with a binary operation for which the basepoint is a two-sided identity and the morphisms are based functions preserving the binary operation (called homomorphisms). In addition, $\mathcal{AB}_*$ is the full subcategory of $\mathcal{B}_*$ consisting of based sets for which the binary operation is commutative. Then there are forgetful functors $\mathcal{B}_* \to Sets_*$ and $Gr \to \mathcal{B}_*$ (see Appendix F).

Now let Y be a fixed space and define a contravariant functor $\mathcal{F}_Y : HoTop_* \to Sets_*$ by $\mathcal{F}_Y(X) = [X, Y]$ and $\mathcal{F}_Y(f) = f^* : [X', Y] \to [X, Y]$, where $f : X \to X'$. To say that $\mathcal{F}_Y : HoTop_* \to Sets_*$ factors through $\mathcal{B}_*$ means that for every space X, the set $[X, Y]$ is a based set having a binary operation with the homotopy class of the constant map a two-sided identity and that $f^* : [X', Y] \to [X, Y]$ is a homomorphism for every map $f : X \to X'$. Similarly $\mathcal{F}_Y : HoTop_* \to Sets_*$ factors through Gr means that $[X, Y]$ is a group for every X with unit the homotopy class of the constant map and that $f^* : [X', Y] \to [X, Y]$ is a homomorphism.

Proposition 2.2.3

1. *Y is an H-space if and only if $\mathcal{F}_Y : HoTop_* \to Sets_*$ factors through $\mathcal{B}_*$.*
2. *Y is a homotopy-commutative H-space if and only if $\mathcal{F}_Y : HoTop_* \to Sets_*$ factors through $\mathcal{AB}_*$.*
3. *Y is a grouplike space if and only if $\mathcal{F}_Y : HoTop_* \to Sets_*$ factors through Gr.*

Proof. (1) Let (Y, m) be an H-space. We have already noted that $[X, Y]$ is a set with binary operation for which 0 is a two-sided identity. If $f : X \to X'$ is a map and $[a], [b] \in [X', Y]$, then

$$(a + b)f = m(a \times b)\Delta_{X'} f = m(af \times bf)\Delta_X = af + bf,$$

and so $f^*([a] + [b]) = f^*[a] + f^*[b]$. Thus $\mathcal{F}_Y : HoTop_* \to Sets_*$ factors through $\mathcal{B}_*$. Conversely, suppose $[X, Y]$ is an object of $\mathcal{B}_*$ for every X with the property that $f^* : [X', Y] \to [X, Y]$ is a homomorphism for every map $f : X \to X'$. Let the binary operation be denoted by $+$ and let $[*]$ be the two-sided identity, where $*$ is the constant map. Now define $m : Y \times Y \to Y$ by $[m] = [p_1] + [p_2] \in [Y \times Y, Y]$, where p_1 and p_2 are the two projections of $Y \times Y$ onto Y. Then, if $j_1, j_2 : Y \to Y \times Y$ are the two inclusions,

$$j_1^*[m] = [p_1 j_1] + [p_2 j_1] = [\mathrm{id}_Y] + [*] = [\mathrm{id}_Y],$$

and so $m j_1 \simeq \mathrm{id}_Y$. Similarly, $m j_2 \simeq \mathrm{id}_Y$. Therefore (Y, m) is an H-space.

(2) If (Y, m) is homotopy-commutative and $[a], [b] \in [X, Y]$, then

$$a + b = m(a \times b)\Delta \simeq mt(a \times b)\Delta = m(b \times a)\Delta = b + a,$$

where $t : Y \times Y \to Y \times Y$ interchanges coordinates, and so $[X, Y]$ is commutative. Conversely, suppose $[X, Y]$ is commutative for all X. Let the multiplication m on Y be as defined in (1). Then

$$[mt] = t^*[m] = t^*([p_1]+[p_2]) = [p_1 t]+[p_2 t] = [p_2]+[p_1] = [p_1]+[p_2] = [m],$$

and so m is homotopy-commutative.

(3) We omit the proof which is like (1) and (2) but we record the following for later use. If $\mathcal{F}_Y$ factors through Gr, then the multiplication m and the homotopy inverse i are defined by

$$[m] = [p_1] + [p_2] \quad \text{and} \quad [i] = -[\mathrm{id}_Y]. \qquad \square$$

We next introduce a definition and corollary of Proposition 2.2.3.

Definition 2.2.4 A *contravariant binary operation induced by* Y is a binary operation on $[X, Y]$ for every space X such that $0 \in [X, Y]$ is a two-sided identity and for every $f : X \to X'$, the function $f^* : [X', Y] \to [X, Y]$ is a homomorphism. A *contravariant group operation* induced by Y is similarly defined.

Then we have the following immediate consequence of Proposition 2.2.3.

Corollary 2.2.5 *1. There is a one–one correspondence between the set of homotopy classes of multiplications of Y and the set of contravariant binary operations induced by Y.*

2. There is a one–one correspondence between the set of homotopy classes of grouplike multiplications of Y and the set of contravariant group operations induced by Y.

The following result is frequently used.

Proposition 2.2.6 *If (Y, m) and (Y', m') are H-spaces and $h : (Y, m) \to (Y', m')$ an H-map, then $h_* : [X, Y] \to [X, Y']$ is a homomorphism of based sets with a binary operation. In particular, if Y and Y' are grouplike spaces, then $h_* : [X, Y] \to [X, Y']$ is a group homomorphism.*

Proof. Let $[a], [b] \in [X, Y]$; then

$$h(a + b) = hm(a \times b)\Delta \simeq m'(h \times h)(a \times b)\Delta = ha + hb.$$

Therefore h_* is a homomorphism. $\square$

To obtain the notion which is dual to that of a grouplike space, we reverse the arrows and replace the product with the wedge in Definition 2.2.1. As noted in Section 1.2, we regard $X \vee X \subseteq X \times X$ so that every element of $X \vee X$ is of the form $(x, *)$ or $(*, x')$, for $x, x' \in X$. Recall that $q_1 = p_1 | X \vee X : X \vee X \to X$ and $q_2 = p_2 | X \vee X : X \vee X \to X$, where $p_1, p_2 : X \times X \to X$ are the projections.

Definition 2.2.7 A *cogroup* consists of a space X and two maps $c : X \to X \vee X$ and $j : X \to X$ such that

1. $q_1 c \simeq \mathrm{id} \simeq q_2 c : X \to X$.

2. $(c \vee \mathrm{id})c \simeq (\mathrm{id} \vee c)c : X \to X \vee X \vee X$

$$\begin{array}{ccc} X & \xrightarrow{c} & X \vee X \\ \downarrow{\scriptstyle c} & & \downarrow{\scriptstyle c\vee \mathrm{id}} \\ X \vee X & \xrightarrow{\mathrm{id}\vee c} & X \vee X \vee X. \end{array}$$

3. $\{\mathrm{id}, j\}c \simeq * \simeq \{j, \mathrm{id}\}c : X \to X$, where $\{\mathrm{id}, j\} : X \vee X \to X$ is defined by $\{\mathrm{id}, j\}(x, *) = x$ and $\{\mathrm{id}, j\}(*, x) = j(x)$, for all $x \in X$, and $\{j, \mathrm{id}\}$ is similarly defined.

A cogroup is also called a co-H-group, an H-cogroup, or a cogrouplike space. The map c is the *comultiplication* and j the *homotopy inverse.* If only (1) holds, then (X, c) or X is called a *co-H-space.* A co-H-space which is a CW complex is called a *co-H-complex.* Condition (2) is called *homotopy-associativity* (sometimes homotopy-coassociativity). We show in Proposition 8.4.4 that every simply connected, homotopy-associative co-H-complex has a homotopy inverse. The co-H-space X is called *homotopy-commutative* if $sc \simeq c : X \to X \vee X$, where $s : X \vee X \to X \vee X$ is defined by $s(x, *) = (*, x)$ and $s(*, x) = (x, *)$. We give examples of cogroups in Section 2.3 and show that all spheres and wedges of spheres of dimension $\geqslant 1$ are cogroups. There are spaces that are co-H-spaces but not cogroups (see [9]). In addition, a co-H-space in the topological category (defined by equality of maps instead of homotopy of maps) is a one point space (see Exercise 2.4).

Definition 2.2.8 Let (X, c) and (X', c') be co-H-spaces and $g : X \to X'$ a map. We call g a *co-H-map* if there is a homotopy-commutative diagram

$$\begin{array}{ccc} X & \xrightarrow{g} & X' \\ \downarrow{\scriptstyle c} & & \downarrow{\scriptstyle c'} \\ X \vee X & \xrightarrow{g\vee g} & X' \vee X'. \end{array}$$

This is written $g : (X, c) \to (X', c')$.

The set $[X, Y]$ has a binary operation when X is a co-H-space and Y is any space: let $f, g : X \to Y$ and let $\nabla : Y \vee Y \to Y$ be the folding map defined by $\nabla(y, *) = y$ and $\nabla(*, y) = y$, for $y \in Y$. We define $f + g = \nabla(f \vee g)c = \{f, g\}c$,

$$X \xrightarrow{c} X \vee X \xrightarrow{f\vee g} Y \vee Y \xrightarrow{\nabla} Y.$$

Then for $\alpha = [f]$ and $\beta = [g] \in [X, Y]$, we set $\alpha + \beta = [f + g]$.

As before $\alpha + 0 = \alpha = 0 + \alpha$ and $c = i_1 + i_2 : X \to X \vee X$, where $i_1, i_2 : X \to X \vee X$ are the two injections. In addition, if (3) holds, $j^*(\alpha) + \alpha = 0 = \alpha + j^*(\alpha)$, and so $j^*(\alpha)$ is the inverse of $\alpha \in [X, Y]$. We obtain the *category of co-H-spaces* $\mathcal{CH}$ whose objects are co-H-spaces and whose morphisms are homotopy classes of co-H-maps and a full *(sub)category of cogroups* $\mathcal{CG}$. Now let X be a fixed space and define a covariant functor $\mathcal{K}_X : HoTop_* \to Sets_*$ by $\mathcal{K}_X(Y) = [X, Y]$ and $\mathcal{K}_X(g) = g_* : [X, Y] \to [X, Y']$, where $g : Y \to Y'$. Then $\mathcal{K}_X : HoTop_* \to Sets_*$ factors through $\mathcal{B}_*$ means that for every space Y, the set $[X, Y]$ is a based set with a binary operation with the homotopy class of the constant map a two-sided identity and that $g_* : [X, Y] \to [X, Y']$ is a homomorphism for every map $g : Y \to Y'$. Similarly $\mathcal{K}_X : HoTop_* \to Sets_*$ factors through Gr means that $[X, Y]$ is a group for every Y with unit the homotopy class of the constant map and that $g_* : [X, Y] \to [X, Y']$ is a homomorphism.

Proposition 2.2.9

1. *X is a co-H-space if and only if $\mathcal{K}_X : HoTop_* \to Sets_*$ factors through $\mathcal{B}_*$.*
2. *X is a homotopy-commutative co-H-space if and only if $\mathcal{K}_X : HoTop_* \to Sets_*$ factors through $\mathcal{AB}_*$.*
3. *X is a cogroup if and only if $\mathcal{K}_X : HoTop_* \to Sets_*$ factors through Gr.*
4. *If (X, c) and (X', c') are co-H-spaces and $h : (X', c') \to (X, c)$ is a co-H-map, then $h^* : [X, Y] \to [X', Y]$ is a homomorphism of based sets with a binary operation. In particular, if X and X' are cogroups, then $h^* : [X, Y] \to [X, Y']$ is a group homomorphism.*
5. *If X is a co-H-space and $f, g : X \to Y$, then $(f + g)_* = f_* + g_* : H_n(X; G) \to H_n(Y; G)$ and $(f + g)^* = f^* + g^* : H^n(Y; G) \to H^n(X; G)$, for all $n \geqslant 0$ and abelian groups G.*

Proof. The proofs of (1) – (3) are analogous to the proof of Proposition 2.2.3, therefore we omit them. We do note, however, that in (3), if $\mathcal{K}_X$ factors through Gr, then the comultiplication c and homotopy inverse j are defined as follows,

$$c = i_1 + i_2 \quad \text{and} \quad j = -\mathrm{id}_X,$$

where i_1 and i_2 are the two injections of $X \to X \vee X$. The proof of (4) is parallel to the proof of Proposition 2.2.6, and also omitted. We only prove (5) for homology. Let $\mu_X : H_n(X \vee X) \to H_n(X) \oplus H_n(X)$ be the isomorphism given by $\mu_X(z) = (q_{1*}(z), q_{2*}(z))$ for $z \in H_n(X \vee X)$. Consider the commutative diagram

$$\begin{array}{ccccccc}
H_n(X) & \xrightarrow{c_*} & H_n(X \vee X) & \xrightarrow{(f \vee g)_*} & H_n(Y \vee Y) & \xrightarrow{\nabla_*} & H_n(Y) \\
& \searrow^{\Delta} & \downarrow{\scriptstyle \mu_X} & & \downarrow{\scriptstyle \mu_Y} & \nearrow_{\delta} & \\
& & H_n(X) \oplus H_n(X) & \xrightarrow{f_* \oplus g_*} & H_n(Y) \oplus H_n(Y), & &
\end{array}$$

where Δ is the diagonal and $\delta(u, u') = u + u'$, for $u, u' \in H_n(Y)$. Then

$$(f + g)_* = \nabla_*(f \vee g)_* c_* = \delta(f_* \oplus g_*)\Delta = f_* + g_*,$$

and the result follows. □

Analogous to Definition 2.2.4, we have the following for co-H-spaces.

Definition 2.2.10 A *covariant binary operation induced by* X is a binary operation on $[X,Y]$ for every space Y such that $0 \in [X,Y]$ is a two-sided identity and for every map $g : Y \to Y'$, the function $g_* : [X,Y] \to [X,Y']$ is a homomorphism. A covariant group operation induced by X is similarly defined.

We then have the following immediate consequence of Proposition 2.2.9.

Proposition 2.2.11 *1. There is a one–one correspondence between the set of homotopy classes of comultiplications of* X *and the set of covariant binary operations induced by* X*.*

2. There is a one–one correspondence between the set of homotopy classes of cogroup comultiplications of X *and the set of covariant group operations induced by* X*.*

An interesting situation arises when (X,c) is a co-H-space and (Y,m) is an H-space. Then the comultiplication c and the multiplication m each induce a binary operation in $[X,Y]$.

Proposition 2.2.12 *If* (X,c) *is a co-H-space and* (Y,m) *is an H-space, then the binary operation* $+_c$ *in* $[X,Y]$ *obtained from* c *equals the binary operation* $+_m$ *in* $[X,Y]$ *obtained from* m*. In addition, this binary operation is abelian.*

Proof. For every $\alpha = [f]$, $\beta = [g]$, $\gamma = [h]$, $\delta = [k] \in [X,Y]$, we prove

$$(\alpha +_m \beta) +_c (\gamma +_m \delta) = (\alpha +_c \gamma) +_m (\beta +_c \delta). \tag{2.1}$$

With $\Delta = \Delta_X$ and $\nabla = \nabla_Y$, the left-hand side of Equation 2.1 is represented by

$$\nabla(m(f \times g)\Delta \vee m(h \times k)\Delta)c = m\nabla_{Y\times Y}\big((f \times g) \vee (h \times k)\big)(\Delta \vee \Delta)c$$

and the right-hand side of Equation 2.1 is represented by

$$m\big((\nabla(f \vee h)c) \times (\nabla(g \vee k)c)\big)\Delta = m(\nabla \times \nabla)\big((f \vee h) \times (g \vee k)\big)\Delta_{X\vee X}\, c.$$

But it is easily checked that

$$\nabla_{Y\times Y}\big((f \times g) \vee (h \times k)\big)(\Delta \vee \Delta) = (\nabla \times \nabla)\big((f \vee h) \times (g \vee k)\big)\Delta_{X\vee X},$$

and so Equation 2.1 is established. Now take $\beta = 0 = \gamma$ in Equation 2.1, getting

$$\alpha +_c \delta = \alpha +_m \delta.$$

This shows that the two binary operations agree. Next set $\alpha = 0 = \delta$ in Equation 2.1, getting

$$\beta +_c \gamma = \gamma +_m \beta.$$

This shows that the operation is abelian. □

2.3 Loop Spaces and Suspensions

In this section we study loop spaces which are a class of grouplike spaces and suspensions which are a class of cogroups.

Definition 2.3.1 For a space B, the *loop space* ΩB is the subspace of B^I consisting of all paths l in B such that $l(0) = * = l(1)$. The loop space ΩB has the subspace topology of the space of paths B^I with the compact–open topology (see Appendix A). The elements of ΩB are called *loops* in B. If $g : B \to B'$ is a map, then $\Omega g : \Omega B \to \Omega B'$ is defined by $\Omega g(l) = g\,l$ (the composition of g and l).

Clearly if $g \simeq g' : B \to B'$, then $\Omega g \simeq \Omega g' : \Omega B \to \Omega B'$. We next define a map $m : \Omega B \times \Omega B \to \Omega B$ by

$$m(l, l')(t) = \begin{cases} l(2t) & \text{if } 0 \leqslant t \leqslant \frac{1}{2} \\ l'(2t-1) & \text{if } \frac{1}{2} \leqslant t \leqslant 1, \end{cases}$$

for $l, l' \in \Omega B$ and $t \in I$. We also define $i : \Omega B \to \Omega B$ by $i(l)(t) = l(1-t)$, for $l \in \Omega B$ and $t \in I$.

The loop $m(l, l')$ consists of the loop l followed by the loop l'. That is, $m(l, l')$ is obtained by traversing the loop l at double speed followed by the loop l' also at double speed. The loop $i(l)$ is the loop l traversed in the opposite direction. We note that $m(l, l')$ is just the sum of paths $l + l'$ and $i(l)$ is $-l$, both of which were defined in Remark 1.4.7. We will see that the map m provides ΩB with grouplike structure.

If B has the homotopy type of a CW complex, then so does ΩB by a theorem of Milnor [70]. It also follows from Milnor's result that many of the path spaces such as B^I or EB also have the homotopy type of a CW complex whenever B does.

Proposition 2.3.2 *If B is a space, then ΩB is a grouplike space with multiplication m and homotopy inverse i. For any map $f : B \to B'$, the map $\Omega f : \Omega B \to \Omega B'$ is an H-map.*

Proof. We must first verify the three conditions in Definition 2.2.1.

(1) We show $\mathrm{id} \simeq m j_1 : \Omega B \to \Omega B$ by defining a homotopy $F : \Omega B \times I \to \Omega B$. For $l \in \Omega B$ and $s, t \in I$, we set

$$F(l, s)(t) = \begin{cases} l\big(2t/(2-s)\big) & \text{if } 0 \leqslant t \leqslant \frac{2-s}{2} \\ * & \text{if } \frac{2-s}{2} \leqslant t \leqslant 1. \end{cases}$$

The other homotopy for (1) is similar.

(2) We show $m(m \times \mathrm{id}) \simeq m(\mathrm{id} \times m) : \Omega B \times \Omega B \times \Omega B \to \Omega B$ by defining a homotopy $G : \Omega B \times \Omega B \times \Omega B \times I \to \Omega B$. For $l, l', l'' \in \Omega B$ and $s, t \in I$, we set

$$G(l, l', l'', s)(t) = \begin{cases} l(4t/(1+s)) & \text{if } 0 \leqslant t \leqslant \frac{s+1}{4} \\ l'(4t - 1 - s) & \text{if } \frac{s+1}{4} \leqslant t \leqslant \frac{s+2}{4} \\ l''((4t - s - 2)/(2 - s)) & \text{if } \frac{s+2}{4} \leqslant t \leqslant 1. \end{cases}$$

(3) We show $* \simeq m(\mathrm{id}, i) : \Omega B \to \Omega B$ by defining a homotopy $H : \Omega B \times I \to \Omega B$. For $l \in \Omega B$ and $s \in I$, we set

$$H(l, s)(t) = \begin{cases} l(2st) & \text{if } 0 \leqslant t \leqslant \frac{1}{2} \\ l(2s(1-t)) & \text{if } \frac{1}{2} \leqslant t \leqslant 1. \end{cases}$$

The other homotopy for (3) is similar.

Finally, $m'(\Omega f \times \Omega f) = (\Omega f)m : \Omega B \times \Omega B \to \Omega B'$, where m' is the multiplication of $\Omega B'$. Therefore Ωf is an H-map. □

In the proof of the previous proposition formal definitions of the required homotopies were given. However, it is helpful in understanding these homotopies to visualize them and say what they actually do.

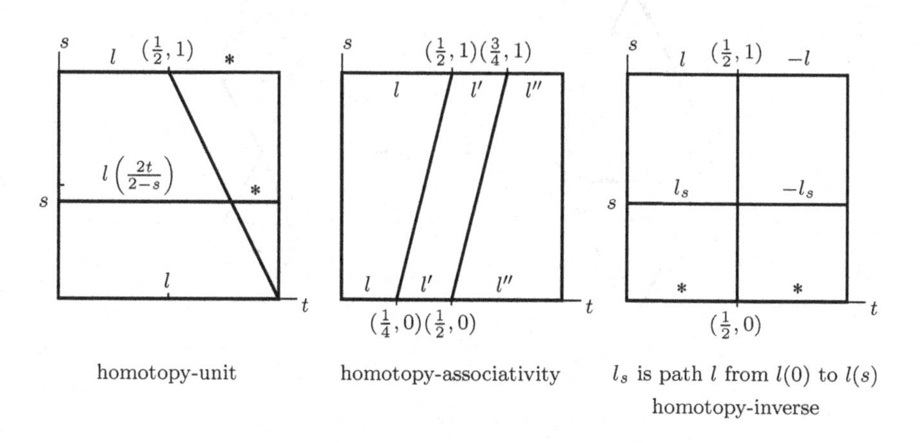

Figure 2.1

For example, in (3) we see that at time s the homotopy H applied to the path l is first the path l going from $l(0)$ to $l(s)$ and then is the path l in the opposite direction going from $l(s)$ to $l(0)$. Clearly this is the constant path $*$ when $s = 0$ and the path $m(l, il)$ when $s = 1$. A similar analysis can be made for the homotopies in (1) and (2).

Let $HoTop_*$ denote the homotopy category and let $\mathcal{HG}$ denote the category of grouplike spaces. Then $\Omega : HoTop_* \to \mathcal{HG}$ defined by $\Omega(X) = \Omega X$ and

$\Omega[f] = [\Omega f]$ is a well-defined functor. Clearly $(\Omega B, m)$ is a grouplike space that is not in general a topological group. From Propositions 2.2.3 and 2.3.2 it follows that for any space B, ΩB induces natural group structure on $[X, \Omega B]$. In addition, if $f : B \to B'$ is a map, then $(\Omega f)_* : [X, \Omega B] \to [X, \Omega B']$ is a homomorphism.

Next we turn to suspensions.

Definition 2.3.3 For any space A, define the *suspension* ΣA (sometimes called the *reduced suspension*) to be the identification space

$$(A \times I)/(A \times \{0\} \cup \{*\} \times I \cup A \times \{1\}).$$

There is a map $c : \Sigma A \to \Sigma A \vee \Sigma A$ defined by

$$c\langle a, t\rangle = \begin{cases} (\langle a, 2t\rangle, *) & \text{if } 0 \leqslant t \leqslant \frac{1}{2} \\ (*, \langle a, 2t-1\rangle) & \text{if } \frac{1}{2} \leqslant t \leqslant 1, \end{cases}$$

where $a \in A$, $t \in I$, and $*$ denotes the basepoint of ΣA. We also define $j : \Sigma A \to \Sigma A$ by $j\langle a, t\rangle = \langle a, 1-t\rangle$. If $f : A \to A'$, then $\Sigma f : \Sigma A \to \Sigma A'$ is given by $\Sigma f\langle a, t\rangle = \langle f(a), t\rangle$.

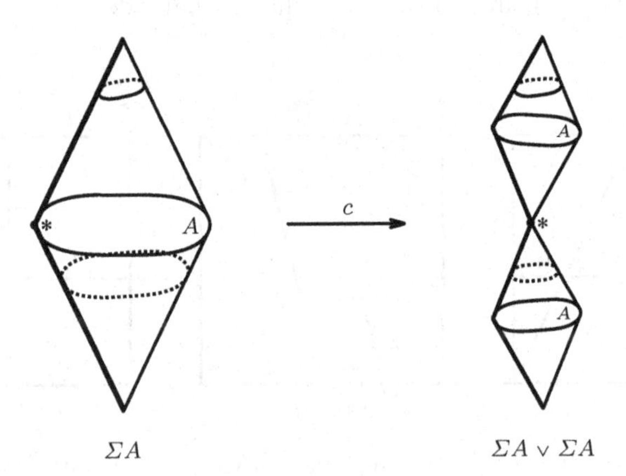

Figure 2.2

Clearly if $f \simeq f' : A \to A'$, then $\Sigma f \simeq \Sigma f' : \Sigma A \to \Sigma A'$.

There is another way to view the suspension. Let C_0X and C_1X be the two cones on X (see Section 1.4). Then $i_0 : X \to C_1X$ is defined by $i_0(x) = \langle x, 0\rangle$ and $i_1 : X \to C_0X$ is defined by $i_1(x) = \langle x, 1\rangle$. Then the suspension ΣX is homeomorphic to the identification space $C_0X \vee C_1X/{\sim}$, where $i_1(x) \sim i_0(x)$, for every $x \in X$.

Proposition 2.3.4 *For any space A, the space ΣA is a cogroup with comultiplication c and homotopy inverse j. For any $f : A \to A'$, the map $\Sigma f : \Sigma A \to \Sigma A'$ is a co-H-map.*

The proof of this is completely analogous to that of Proposition 2.3.2 and is left as an exercise. However, after we give the proof of Proposition 2.3.5 we show how a proof can be derived from Proposition 2.3.2.

If $\mathcal{CG}$ denotes the category of cogroups, it follows from Proposition 2.3.4, that $\Sigma : HoTop_* \to \mathcal{CG}$ is a functor defined by $\Sigma(A) = \Sigma A$ and $\Sigma(f) = \Sigma f$. By Proposition 2.2.9, for every space A, the set $[\Sigma A, Y]$ has group structure for every space Y such that a map $g : Y \to Y'$ induces a homomorphism $g_* : [\Sigma A, Y] \to [\Sigma A, Y']$. Moreover, a map $h : A' \to A$ induces a homomorphism $(\Sigma h)^* : [\Sigma A, Y] \to [\Sigma A', Y]$.

We have seen that if A and B are any two spaces, both $[\Sigma A, B]$ and $[A, \Omega B]$ are groups. If $f : \Sigma A \to B$ is a map, we define $\kappa(f) : A \to \Omega B$ by

$$\kappa(f)(a)(t) = f\langle a, t\rangle,$$

for $a \in A$ and $t \in I$.

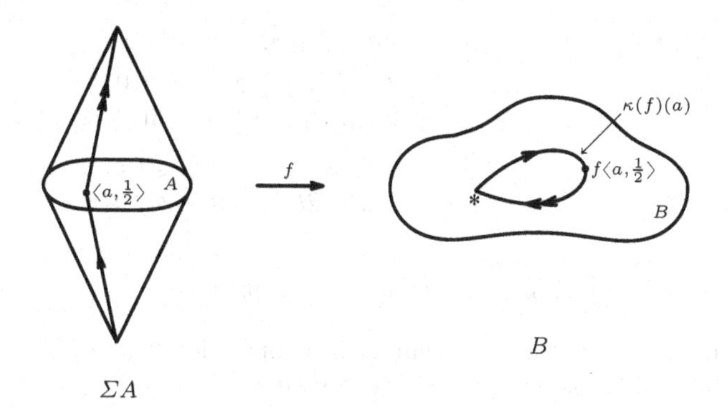

Figure 2.3

Clearly $\kappa(f)$ is well-defined and continuous (Appendix A). Furthermore, if f_t is a homotopy between $f : \Sigma A \to B$ and $f' : \Sigma A \to B$, then $\kappa(f_t)$ is a homotopy between $\kappa(f) : A \to \Omega B$ and $\kappa(f') : A \to \Omega B$. Thus κ induces $\kappa_* : [\Sigma A, B] \to [A, \Omega B]$. Similarly, if $g : A \to \Omega B$, we define $\overline{\kappa}(g) : \Sigma A \to B$ by $\overline{\kappa}(g)\langle a, t\rangle = g(a)(t)$, for $a \in A$ and $t \in I$. Then $\overline{\kappa}$ induces $\overline{\kappa}_* : [A, \Omega B] \to [\Sigma A, B]$. Now

$$\kappa(\overline{\kappa}(g))(a)(t) = (\overline{\kappa}(g))\langle a, t\rangle = g(a)(t),$$

and so $\kappa\overline{\kappa} = \mathrm{id}$. In a like manner, $\overline{\kappa}\kappa = \mathrm{id}$. Thus $\kappa_* : [\Sigma A, B] \to [A, \Omega B]$ is a bijection with inverse $\overline{\kappa}_* : [A, \Omega B] \to [\Sigma A, B]$. In addition, if $h : A' \to A$ and $k : B \to B'$ are maps, then

$$\kappa(f)\,h = \kappa(f\Sigma h) \quad \text{and} \quad (\Omega k)\,\kappa(f) = \kappa(kf),$$

for every $f : \Sigma A \to B$. Thus

$$h^* \kappa_* = \kappa_*\,(\Sigma h)^* \quad \text{and} \quad (\Omega k)_*\,\kappa_* = \kappa_*\,k_*.$$

Proposition 2.3.5 *For any spaces A and B, the bijection $\kappa_* : [\Sigma A, B] \to [A, \Omega B]$ is an isomorphism of groups.*

Proof. Let $f, g : \Sigma A \to B$ and consider $\kappa(f + g) = \kappa(\nabla(f \vee g)c) : A \to \Omega B$. Then for $a \in A$ and $t \in I$,

$$\begin{aligned}(\kappa(\nabla(f \vee g)c)(a))(t) &= \nabla(f \vee g)c\langle a, t\rangle \\ &= \begin{cases} \nabla(f \vee g)(\langle a, 2t\rangle, *) & \text{if } 0 \leqslant t \leqslant \frac{1}{2} \\ \nabla(f \vee g)(*, \langle a, 2t-1\rangle) & \text{if } \frac{1}{2} \leqslant t \leqslant 1 \end{cases} \\ &= \begin{cases} f\langle a, 2t\rangle & \text{if } 0 \leqslant t \leqslant \frac{1}{2} \\ g\langle a, 2t-1\rangle & \text{if } \frac{1}{2} \leqslant t \leqslant 1. \end{cases}\end{aligned}$$

On the other hand, $\kappa(f) + \kappa(g) = m(\kappa(f) \times \kappa(g))\Delta : A \to \Omega B$. Then

$$\begin{aligned}(m(\kappa(f) \times \kappa(g))\Delta(a))(t) &= m(\kappa(f)(a), \kappa(g)(a))(t) \\ &= \begin{cases} (\kappa(f)(a))(2t) & \text{if } 0 \leqslant t \leqslant \frac{1}{2} \\ (\kappa(g)(a))(2t-1) & \text{if } \frac{1}{2} \leqslant t \leqslant 1 \end{cases} \\ &= \begin{cases} f\langle a, 2t\rangle & \text{if } 0 \leqslant t \leqslant \frac{1}{2} \\ g\langle a, 2t-1\rangle & \text{if } \frac{1}{2} \leqslant t \leqslant 1. \end{cases}\end{aligned}$$

Thus $\kappa(f + g) = \kappa(f) + \kappa(g)$, and the result follows. □

Definition 2.3.6 The isomorphism κ_* in Proposition 2.3.5 or its inverse $\overline{\kappa}_*$ is called the *adjoint isomorphism*. We say that f and $\kappa(f)$ and also α and $\kappa_*(\alpha)$ are *adjoint* to each other.

Using the fact that κ_* is a bijection and that $(\Omega B, m, i)$ is grouplike for all B, we now show that $(\Sigma A, c, j)$ is a cogroup for all A, where c and j are the maps defined in Definition 2.3.3. We have that $[A, \Omega B]$ is a group, with binary operation denoted $+$, and so $\kappa_* : [\Sigma A, B] \to [A, \Omega B]$ induces group structure with two-sided identity $[*]$ on $[\Sigma A, B]$, for all B. We denote this binary operation in $[\Sigma A, B]$ by $+'$. Because any map $k : B \to B'$ induces a homomorphism $(\Omega k)_* : [A, \Omega B] \to [A, \Omega B']$, it follows from $(\Omega k)_*\,\kappa_* = \kappa_*\,k_*$ that $k_* : [\Sigma A, B] \to [\Sigma A, B']$ is a homomorphism. Therefore by Proposition 2.2.9(3), there exists a comultiplication $\widetilde{c}$ and a homotopy inverse $\widetilde{j}$ such that $(\Sigma A, \widetilde{c}, \widetilde{j})$ is a cogroup. We show that $c \simeq \widetilde{c}$ and $j \simeq \widetilde{j}$. By Definition 2.3.3, $\kappa(c) = \kappa(i_1) + \kappa(i_2)$, where $i_1, i_2 : \Sigma A \to \Sigma A \vee \Sigma A$ are the two injections. But $\widetilde{c} = i_1 +' i_2$ (see the proof of Proposition 2.2.9), and so $\kappa(\widetilde{c}) = \kappa(i_1) + \kappa(i_2)$. Thus $\kappa(c) = \kappa(\widetilde{c})$, and so $c \simeq \widetilde{c}$. Finally $j = -\mathrm{id}$ by definition and so $\kappa(j) = -\kappa(\mathrm{id})$. But $\kappa(\tilde{j}) = -\kappa(\mathrm{id})$ (proof of Proposition 2.2.9). Therefore $j \simeq \tilde{j}$, and so $(\Sigma A, c, j)$ is a cogroup.

The suspension and loop space constructions can be iterated.

Definition 2.3.7 For spaces A and B and define $\Sigma^0 A = A$ and $\Omega^0 B = B$ and for integers $n \geqslant 1$,

$$\Sigma^n A = \Sigma(\Sigma^{n-1} A) \quad \text{and} \quad \Omega^n B = \Omega(\Omega^{n-1} B).$$

We next consider homotopy commutativity of iterated suspensions and loop spaces.

Proposition 2.3.8 *For spaces A and B, $\Sigma^n A$ is a homotopy-commutative cogroup and $\Omega^n B$ is a homotopy-commutative grouplike space, if $n \geqslant 2$.*

Proof. We just show that $\Sigma^n A$ is homotopy commutative. For any space Y, we have the following isomorphism of groups, $[\Sigma^n A, Y] \cong [\Sigma^{n-1} A, \Omega Y]$, for $n \geqslant 2$, by Proposition 2.3.5. The latter group is abelian by Proposition 2.2.12. By Proposition 2.2.9(2), $\Sigma^n A$ is homotopy-commutative. $\square$

Recall that the upper cap E^n_+ of the unit n-sphere S^n is defined by $E^n_+ = \{(x_1, x_2, \ldots, x_{n+1}) \in S^n \mid x_{n+1} \geqslant 0\}$. The lower cap E^n_- of S^n is similarly defined by $x_{n+1} \leqslant 0$. Then $S^n = E^n_+ \cup E^n_-$ and $S^{n-1} = E^n_+ \cap E^n_-$.

Proposition 2.3.9 *For all $n \geqslant 1$, S^n is homeomorphic to ΣS^{n-1}.*

Proof. There are homeomorphisms $h_+ : E^n \to E^n_+$ and $h_- : E^n \to E^n_-$ defined by

$$h_+(x) = \left(x, \sqrt{1 - |x|^2}\right) \quad \text{and}$$
$$h_-(x) = \left(x, -\sqrt{1 - |x|^2}\right),$$

for $x \in E^n$. Recall that $C_0 X = (X \times I)/(X \times \{0\} \cup \{*\} \times I)$ and $C_1 X = (X \times I)/(X \times \{1\} \cup \{*\} \times I)$. By Lemma 1.4.10, there is a homeomorphism $\widetilde{K} : C_1(S^{n-1}) \to E^n$. Similarly by defining $L : S^{n-1} \times I \to E^n$ by $L(x,t) = (1-t)* + tx$, we obtain a homeomorphism $\widetilde{L} : C_0(S^{n-1}) \to E^n$ as in Lemma 1.4.10. We compose $\widetilde{K}$ with h_+ to obtain a homeomorphism $\tau : C_1(S^{n-1}) \to E^n_+$ and we compose $\widetilde{L}$ with h_- to obtain a homeomorphism $\lambda : C_0(S^{n-1}) \to E^n_-$. Each of τ and λ restricted to S^{n-1} is the identity map of S^{n-1}. We regard ΣS^{n-1} as $C_1(S^{n-1}) \cup_{S^{n-1}} C_0(S^{n-1})$, the disjoint union of $C_1(S^{n-1})$ and $C_0(S^{n-1})$ with $S^{n-1} \subseteq C_1(S^{n-1})$ identified with $S^{n-1} \subseteq C_0(S^{n-1})$. Then the maps τ and λ yield a homeomorphism (see Figure 2.4)

$$\Sigma S^{n-1} = C_1(S^{n-1}) \cup_{S^{n-1}} C_0(S^{n-1}) \cong E^n_+ \cup E^n_- = S^n. \qquad \square$$

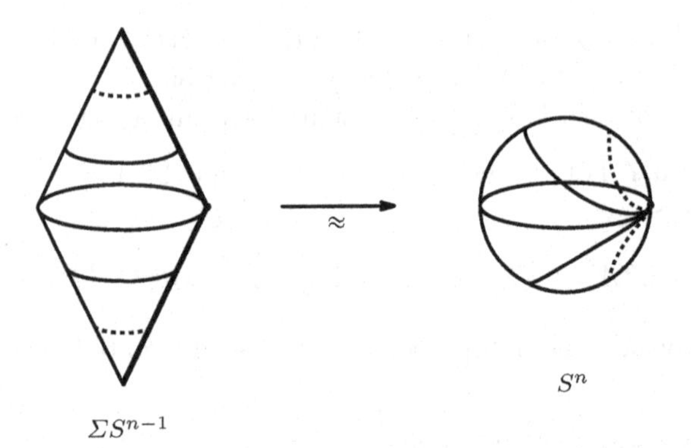

Figure 2.4

2.4 Homotopy Groups I

By Propositions 2.2.9, 2.3.4, and 2.3.9, the set $[S^n, Y]$ is a group for all spaces Y and all $n \geqslant 1$. These are the homotopy groups of Y.

Definition 2.4.1 For every space Y and $n \geqslant 0$, the set $[S^n, Y]$ is called the *nth (ordinary) homotopy group* of Y and is denoted $\pi_n(Y)$. For $n = 1$, it is called the *fundamental group* of Y.

We assume that the reader has had some exposure to the basic properties of fundamental groups. For review, we have presented the topics on the fundamental group that we use in Appendix B. If $n \geqslant 1$, then $\pi_n(Y)$ is a group for all Y and a map $f : Y \to Y'$ induces a homomorphism $f_* : \pi_n(Y) \to \pi_n(Y')$. In general, $\pi_0(Y)$ is a set with a distinguished element and $f_* : \pi_0(Y) \to \pi_0(Y')$ is a function that preserves the distinguished element. For another characterization of $\pi_0(Y)$, see Exercise 2.24.

We next give a few elementary properties of homotopy groups. We give more information on homotopy groups in Section 4.5 and compute some of these groups in Section 5.6.

- For $n \geqslant 2$, the groups $\pi_n(Y)$ are abelian. This follows from Proposition 2.3.8.
- The fundamental group $\pi_1(Y)$ is abelian if Y is an H-space by Proposition 2.2.12. In general, $\pi_1(Y)$ is not abelian (Appendix B). If Y is a grouplike space, then $\pi_0(Y)$ is a group (Exercise 2.24).
- If $n \geqslant 1$, then $\pi_n(Y) \cong \pi_{n-1}(\Omega Y)$ as groups by Proposition 2.3.5. In particular, ΩY is path-connected if and only if $\pi_1(Y) = 0$ by Exercise 2.24.
- If $f \simeq g : Y \to Y'$, then $f_* = g_* : \pi_n(Y) \to \pi_n(Y')$, for all $n \geqslant 0$.
- If $f : X \to Y$ is a homotopy equivalence, then $f_* : \pi_n(Y) \to \pi_n(Y')$ is an isomorphism, for all $n \geqslant 0$. For if $g : Y \to X$ is a homotopy inverse of

f, then $fg \simeq \mathrm{id}$. Therefore $f_*g_* = (fg)_* = \mathrm{id}_* = \mathrm{id}$. Similarly $gf \simeq \mathrm{id}$ implies that $g_*f_* = \mathrm{id}$. Therefore f_* is an isomorphism.

- Let $i : X \to Y$ be an inclusion and let $r : Y \to X$ be a retraction. Then $i_* : \pi_n(X) \to \pi_n(Y)$ is a monomorphism and $r_* : \pi_n(Y) \to \pi_n(X)$ is an epimorphism, for all n, since $r_*i_* = \mathrm{id}$. In fact, $\pi_n(Y) = i_*\pi_n(X) \oplus \mathrm{Ker}\, r_*$. This clearly holds if r is a homotopy retraction. It also holds if r is an arbitrary map and i is a section or homotopy section of r.
- If Y is contractible, then $\pi_n(Y) = 0$, for all $n \geqslant 0$. This follows because $\mathrm{id} \simeq * : Y \to Y$, and so $\mathrm{id} = (\mathrm{id})_* = *_* = 0 : \pi_n(Y) \to \pi_n(Y)$, for all $n \geqslant 0$.
- For spaces Y and Y', we have $\pi_n(Y \times Y') \cong \pi_n(Y) \oplus \pi_n(Y')$, for all $n \geqslant 0$. For, by Corollary 1.3.7, the function $\theta : \pi_n(Y) \oplus \pi_n(Y') \to \pi_n(Y \times Y')$ defined by $\theta([f],[g]) = [(f,g)]$, for $[f] \in \pi_n(Y)$ and $[g] \in \pi_n(Y')$, is a bijection with inverse function λ given by $\lambda[h] = (p_{1*}[h], p_{2*}[h])$. Thus λ is an isomorphism, and so $\pi_n(Y \times Y') \cong \pi_n(Y) \oplus \pi_n(Y')$. Furthermore, we define $\mu : \pi_n(Y) \oplus \pi_n(Y') \to \pi_n(Y \times Y')$ by $\mu(\alpha, \beta) = j_{1*}(\alpha) + j_{2*}(\beta)$, where $j_1 : Y \to Y \times Y'$ and $j_2 : Y' \to Y \times Y'$ are the two inclusions. Then $\lambda\mu = \mathrm{id}$, so μ is an isomorphism and equals θ. These results clearly extend to the product of finitely many spaces.
- If Y and Y' are spaces of the homotopy type of CW complexes, then the fundamental group of the wedge $Y \vee Y'$ is the free product $\pi_1(Y) * \pi_1(Y')$ of $\pi_1(Y)$ and $\pi_1(Y')$ (Appendix B).
- If Y is a nonpath-connected space and X is the path-connected component of Y containing the basepoint, then the inclusion $i : X \to Y$ induces an isomorphism $i_* : \pi_n(X) \to \pi_n(Y)$, for all $n \geqslant 1$. This is since for any map $f : S^n \to Y$, we have that $f(S^n) \subseteq X$ because $f(S^n)$ is a path-connected space containing $*$. Similarly, for any homotopy $F : S^n \times I \to Y$, we have that $F(S^n \times I) \subseteq X$.

The result that the fundamental group of an H-space is abelian is easy to prove. The result that the fundamental group of a co-H-space is free, which we prove next, is more difficult. It requires some facts about free groups and free products of groups (Appendix B).

Let G be a group that is not necessarily abelian. For notational convenience, we write $\overline{g}$ for the inverse g^{-1} of $g \in G$. We denote the free product of G with itself by $G * G$. If $g \in G$, then g regarded as an element of the first factor of $G * G$ is written g' and as an element of the second factor of $G * G$ is written g''. Thus an element $\xi \in G * G$ can be written

$$\xi = \prod_{i=1}^{p} g_i{}' \overline{\gamma}_i{}'', \quad \text{where } g_i, \gamma_i \in G.$$

Then there are projection homomorphisms $p_1, p_2 : G * G \to G$ given by $p_1(\xi) = \prod g_i$ and $p_2(\xi) = \prod \overline{\gamma}_i$. We introduce the following notation:

$$E_G = \{\xi \in G * G \mid p_1(\xi) = p_2(\xi)\}.$$

Thus $\xi = \prod_{i=1}^{p} g_i'\overline{\gamma}_i'' \in E_G$ if and only if $\gamma_p \cdots \gamma_1 g_1 \cdots g_p = 1$. Then $\pi : E_G \to G$ is defined by $\pi = p_1|E_G = p_2|E_G$, and so $\pi(\xi) = \prod g_i = \prod \overline{\gamma}_i$. Finally, let $\xi_u = u'u'' \in E_G$, where $u \in G$ and let $\Xi_G = \{\xi_u \mid u \neq 1\}$.

The following result, which appears in [7, Prop. 3.1], is based on ideas attributed to M. Kneser.

Lemma 2.4.2 *The group E_G is free with basis Ξ_G.*

Proof. It is clear that the set Ξ_G is an independent set. In order to write any expression $\xi = \prod g_i'\overline{\gamma}_i''$ that satisfies $\gamma_p \cdots \gamma_1 g_1 \cdots g_p = 1$ as a product of the ξ_u and their inverses, we use the following simple algorithm. For $1 \leqslant i \leqslant 2p$, define δ_i by the formulas

$$\delta_{2k} = \gamma_k \cdots \gamma_1 g_1 \cdots g_k \quad \text{and} \quad \delta_{2k+1} = \delta_{2k} g_{k+1}.$$

Thus $\delta_1 = g_1$, $\delta_2 = \gamma_1 g_1$, $\delta_3 = \gamma_1 g_1 g_2$, and so on, and $\delta_{2p} = 1$. Now one verifies that ξ is the alternating product

$$\xi = \prod_{i=1}^{2p} \xi_{\delta_i}^{(i)},$$

where $(i) = (-1)^{i+1}$ (Exercise 2.20). □

Proposition 2.4.3 *If X is a co-H-complex, then $\pi_1(X)$ is a free group.*

Proof. If $G = \pi_1(X)$, then as noted earlier, $\pi_1(X \vee X)$ is isomorphic to $G * G$, the free product of G with itself. Let $c : X \to X \vee X$ be a comultiplication and let $q_1, q_2 : X \vee X \to X$ be the projections. Because $q_i c \simeq \text{id}$, we have that c induces a homomorphism $s = c_* : G \to G * G$ such that $p_1 s = p_2 s = \text{id} : G \to G$. Thus s determines a homomorphism $\sigma : G \to E_G$ such that $\pi\sigma = \text{id}$. Therefore σ maps G isomorphically onto $\sigma(G) \subseteq E_G$. By Lemma 2.4.2, E_G is free. Since a subgroup of a free group is free [39, p. 85], $\sigma(G)$ is free. Hence $G = \pi_1(X)$ is free. □

Next we present additional results on homotopy groups. We begin with a definition.

Definition 2.4.4 A path-connected space Y is said to be *n-connected*, if $\pi_i(Y) = 0$, for all $i \leqslant n$. A 1-connected space Y is also called *simply connected.* A map $f : X \to Y$ is called an *n-equivalence* (also called an *n-connected map*), if $f_* : \pi_i(X) \to \pi_i(Y)$ is an isomorphism for all $i < n$ and an epimorphism for $i = n$. A map $f : X \to Y$ is a *weak (homotopy) equivalence* or an *∞-equivalence* if $f_* : \pi_n(X) \to \pi_n(Y)$ is an isomorphism for all n.

Lemma 2.4.5 *Let (X, A) be a based, relative CW complex with* $\dim(X, A) \leqslant n$*, let B and Y be spaces (not necessarily of the homotopy type of CW complexes), and let $e : B \to Y$ be an n-equivalence, $n \leqslant \infty$. Let $j : A \to X$ be the* *inclusion and assume that there are maps $f : X \to Y$ and $g : A \to B$ and a diagram*

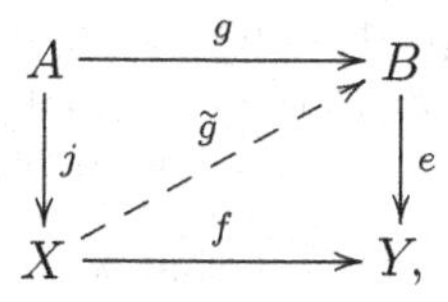

such that $eg \simeq_L fj$, for some homotopy $L : A \times I \to Y$. Then there exists a map $\widetilde{g} : X \to B$ such that $\widetilde{g}j = g$ and a homotopy $F : X \times I \to Y$ such that $e\widetilde{g} \simeq_F f$, where $F|A \times I = L$.

This lemma, which is the major step in proving Whitehead's theorem 2.4.7, follows from the HELP lemma 4.5.7 which is proved in Section 4.5 after we have discussed the relative homotopy groups.

From Lemma 2.4.5 we can easily prove the following proposition.

Proposition 2.4.6 *Let X be a based CW complex, let B and Y be spaces (not necessarily of the homotopy type of CW complexes), and let $e : B \to Y$ be an n-equivalence, $n < \infty$. Then $e_* : [X, B] \to [X, Y]$ is an injection if* $\dim X < n$ *and a surjection if* $\dim X \leqslant n$. *If $n = \infty$, then $e_* : [X, B] \to [X, Y]$ is a bijection for any based CW complex X.*

Proof. We first show that e_* is onto if $\dim X \leqslant n$. Let $[f] \in [X, Y]$, set $A = \{*\}$, and define $g : A \to B$ to be the constant map. We then apply Lemma 2.4.5 to f and g and obtain a map $\widetilde{g} \in [X, B]$ such that $e_*[\widetilde{g}] = [f]$. Thus e_* is onto.

Now assume that $\dim X < n$ and $eg_0 \simeq_F eg_1$ for $g_0, g_1 : X \to B$. Let $X' = X \times I$ and so $\dim X' \leqslant n$. We set $A' = X \times \partial I \cup \{*\} \times I$ and define $G : A' \to B$ by

$$G(x, i) = g_i(x) \quad \text{and} \quad G(*, t) = *,$$

for $x \in X$, $t \in I$, and $i = 0, 1$. Since $\dim (X', A') \leqslant n$, we can apply Lemma 2.4.5 to F and G. We get a homotopy $H : X \times I \to B$ such that $H|A' = G$. Then $g_0 \simeq_H g_1$, and so e_* is one–one. □

There are two important theorems due to J. H. C. Whitehead which we shall arbitrarily call Whitehead's first theorem and Whitehead's second theorem. We now prove Whitehead's first theorem [92].

Theorem 2.4.7 *If $f : X \to Y$ is a map of CW complexes, then f is a weak equivalence if and only if f is a homotopy equivalence.*

Proof. We only prove that if f is a weak equivalence, it is a homotopy equivalence, since the other implication has been proved. Consider the function $f_* : [Y, X] \to [Y, Y]$. By Proposition 2.4.6, f_* is a bijection. Therefore there is a map $g : Y \to X$ such that $fg \simeq \mathrm{id}_Y$. But $fgf \simeq f$ and so $f_*[gf] = f_*[\mathrm{id}_X]$, where $f_* : [X, X] \to [X, Y]$. This latter f_* is a bijection, and so $gf \simeq \mathrm{id}_X$. Thus f is a homotopy equivalence. □

Remark 2.4.8 Whitehead's first theorem is useful to show that a map is a homotopy equivalence. For this we would prove that the map induces isomorphisms of all homotopy groups. Because our spaces have the homotopy type of CW complexes, it would follow that the map is a homotopy equivalence. We frequently use this remark without comment.

We observe that it is not sufficient that $\pi_n(X) \cong \pi_n(Y)$, for all n, for X and Y to have the same homotopy type. By Whitehead's first theorem, there should be a map $f : X \to Y$ that induces an isomorphism of all homotopy groups. An example of the nonsufficiency is given in 5.6.2.

Theorem 2.4.9 *Let X and Y be path-connected spaces (not necessarily of the homotopy type of CW complexes), let $f : X \to Y$ be a map, and let $n \geqslant 0$. Then there is a space K such that (K, X) is a relative CW complex having relative cells of dimensions $\geqslant n+1$ with the following property. There exists a map $\bar{f} : K \to Y$ such that $\bar{f}|X = f$ and $\bar{f}_* : \pi_i(K) \to \pi_i(Y)$ is an isomorphism for $i > n$ and a monomorphism for $i = n$.*

Proof. In the proof we write h_{*i} for the induced homotopy homomorphism $h_* : \pi_i(W) \to \pi_i(Z)$, for any map $h : W \to Z$. The idea of the proof is to attach $(n+1)$-cells to X to kill $\mathrm{Ker} f_{*n}$ and then attach additional $(n+1)$-cells to map onto $\pi_{n+1}(Y)$. This process is then repeated. We begin by choosing generators $[g_\alpha]_{\alpha \in A}$ of $\mathrm{Ker} f_{*n}$, where $g_\alpha : S^n_\alpha \to X$ and $S^n_\alpha = S^n$. Then the g_α determine $g : \bigvee_{\alpha \in A} S^n_\alpha \to X$ and we attach $(n+1)$-cells to X by g to form the adjunction space $X' = X \cup_g \bigvee E^{n+1}_\alpha$. Since $fg \simeq *$, the map fg can be extended to $\bigvee E^{n+1}_\alpha$ by Lemma 1.4.10 and Proposition 1.4.9. This extension and f determine a map $f' : X' \to Y$ such that $f'|X = f$. Then $f'_{*n+1} : \pi_{n+1}(X') \to \pi_{n+1}(Y)$, and we choose elements $[h_\beta] \in \pi_{n+1}(Y)$ for $\beta \in B$ that are a set of generators. Then the h_β determine $h : \bigvee_{\beta \in B} S^{n+1}_\beta \to Y$ and we form $X^{n+1} = X' \vee \bigvee_{\beta \in B} S^{n+1}_\beta$ and define $f^{n+1} : X^{n+1} \to Y$ by $f^{n+1} = \{f', h\}$. Note that (X^{n+1}, X) is a relative CW complex.

Let $k : X \to X'$ and $l : X' \to X^{n+1}$ be inclusion maps and let $j = lk : X \to X^{n+1}$. Then there is a commutative diagram

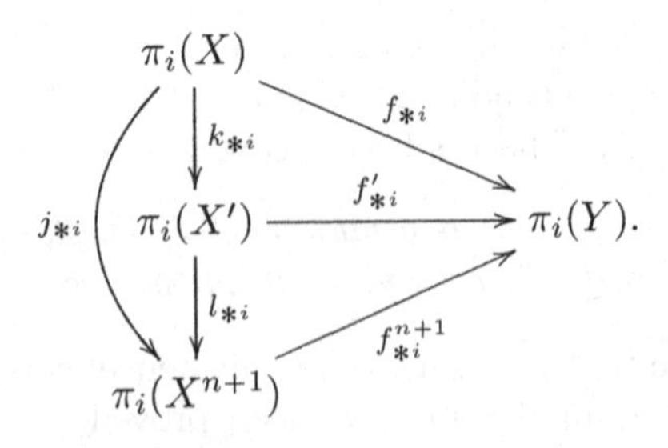

We claim that f^{n+1}_{*n} is a monomorphism and f^{n+1}_{*n+1} is an epimorphism. If $\gamma \in \mathrm{Ker} f^{n+1}_{*n}$, then $\gamma = j_{*n}(\delta)$ for some $\delta \in \mathrm{Ker} f_{*n}$, since j_{*n} is an epimorphism by Proposition 1.5.24. Therefore

$$\delta = \sum_{\alpha \in A'} n_\alpha [g_\alpha],$$

where $A' \subseteq A$ is a finite subset and $n_\alpha \in \mathbb{Z}$. But

$$k_{*n}(\delta) = \sum_{\alpha \in A'} n_\alpha [k g_\alpha] = 0,$$

because $kg_\alpha \simeq *$. Therefore $\gamma = j_{*n}(\delta) = l_{*n}(k_{*n}(\delta)) = 0$. Hence f_{*n}^{n+1} is a monomorphism.

Next we show that f_{*n+1}^{n+1} is an epimorphism. Given $\epsilon \in \pi_{n+1}(Y)$, we have $\epsilon = \sum_{\beta \in B'} m_\beta [h_\beta]$, where $B' \subseteq B$ is a finite subset and $m_\beta \in \mathbb{Z}$. Since $X^{n+1} = X' \vee \bigvee_{\beta \in B} S_\beta^{n+1}$, we let $i_\beta : S_\beta^{n+1} \to X^{n+1}$ be the inclusion maps and set $\xi = \sum_{\beta \in B'} m_\beta [i_\beta]$ in $\pi_{n+1}(X^{n+1})$. Then

$$f_{*n+1}^{n+1}(\xi) = \sum_{\beta \in B'} m_\beta [h_\beta] = \epsilon,$$

and so f_{*n+1}^{n+1} is an epimorphism. This proves the claim.

We then apply this construction to f^{n+1} and obtain an extension $f^{n+2} : X^{n+2} \to Y$ such that f_{*n+1}^{n+2} is a monomorphism and f_{*n+2}^{n+2} is an epimorphism. Because f_{*n+1}^{n+1} is an epimorphism, it follows that f_{*n+1}^{n+2} is an isomorphism and f_{*n+2}^{n+2} is an epimorphism.

We continue this process and obtain maps $f^k : X^k \to Y$, for all $k > n$. We set $X^n = X$ and form the space $K = \cup_{k \geqslant n} X^k$ with the weak topology determined by the X^k. Then the f^k determine a map $\bar{f} : K \to Y$ which is an extension of f. Therefore $\bar{f}_{*i}$ is an isomorphism for $i > n$ and a monomorphism for $i = n$ by Proposition 1.5.24. This completes the proof. □

The following corollary is frequently used.

Corollary 2.4.10

1. *Let X and Y be path-connected spaces and let $f : X \to Y$ be an n-equivalence, $n \geqslant 0$. Then there exists a space K obtained from X by attaching cells of dimensions $\geqslant n+1$ and there exists a map $\bar{f} : K \to Y$ such that $\bar{f}|X = f$ and $\bar{f}$ is a weak equivalence.*
2. *Let Y be a k-connected space (not necessarily of the homotopy type of a CW complex) with $k \geqslant 0$. Then there exists a CW complex K and a weak equivalence $f : K \to Y$ such that $K^k = \{*\}$. In particular, if Y is any path-connected space, there exists a CW complex K with $K^0 = \{*\}$ and a weak equivalence $f : K \to Y$.*
3. *If Y is a k-connected space of the homotopy type of a CW complex, $k \geqslant 0$, then there exists a CW complex K of the homotopy type of Y such that $K^k = \{*\}$. In particuliar, $H_i(Y) = 0$ for $i \leqslant k$.*

Proof. (1) Let K be the space constructed in Theorem 2.4.9. By Proposition 1.5.24, the inclusion map $i : X \to K$ is an n-equivalence. This and the fact that f is an n-equivalence implies that $\bar{f}$ is an n-equivalence. By Theorem 2.4.9, $\bar{f}$ is a weak equivalence.

(2) By hypothesis, the map $\{*\} \to Y$ is a k-equivalence. We then apply Part (1) to obtain the desired result.

(3) We apply Whitehead's first theorem 2.4.7 to (2). □

Definition 2.4.11 For any space Y (not necessarily of the homotopy type of a CW complex), a CW complex K together with a weak equivalence $K \to Y$ is called a *CW approximation* to Y.

The existence of a CW approximation for *any* space, gives some indication of the importance of CW complexes in homotopy theory. It has been shown in [69] how to construct a CW approximation functorially. We do not prove this. However, the following remark is a consequence.

Remark 2.4.12 If $a : K \to X$ and $b : L \to Y$ are two CW approximations and $f : X \to Y$ is a map, then there exists a map $h : K \to L$, unique up to homotopy, such that $fa \simeq bh$. It follows that the homotopy type of a CW approximation of a space is uniquely determined by the homotopy type of the space.

Proposition 2.4.6 gives conditions for an induced map of homotopy sets to be a bijection. The following similar result is very useful.

Proposition 2.4.13 *Let (X, A) be a relative CW complex such that all relative cells have dimension $\geqslant n+2$, let $i : A \to X$ be the inclusion map, and let Y be a space. Then $i^* : [X, Y] \to [A, Y]$ is an injection if $\pi_j(Y) = 0$ for $j > n+1$ and is a surjection if $\pi_j(Y) = 0$ for $j > n$.*

Proof. We first show that i^* is onto if $\pi_j(Y) = 0$ for $j > n$. Let $f : A \to Y$ be a map and consider the relative $(n+2)$-skeleton

$$(X, A)^{n+2} = A \cup \bigcup_{\gamma \in C} e_\gamma^{n+2},$$

for $\gamma \in C$. Let $\phi_\gamma : S_\gamma^{n+1} \to (X, A)^{n+1} = A$ be an attaching function. By Exercise 2.25, $f\phi_\gamma \simeq_{\text{free}} h_\gamma : S_\gamma^{n+1} \to Y$, for some based map h_γ. By hypothesis, $h_\gamma \simeq *$ and so $f\phi_\gamma$ is freely homotopic to a constant function. By Corollary 1.4.11, $f\phi_\gamma$ extends to a free map $\widetilde{f}_\gamma : E_\gamma^{n+2} \to Y$. These functions together with f determine a map $f^{n+2} : (X, A)^{n+2} \to Y$ that extends f. Next we write

$$(X, A)^{n+3} = (X, A)^{n+2} \cup \bigcup_{\delta \in D} e_\delta^{n+3}$$

with attaching maps $\psi_\delta : S_\delta^{n+2} \to (X, A)^{n+2}$, where $\delta \in D$. Then as before $f^{n+2}\psi_\delta$ is freely homotopic to a constant function, and so f^{n+2} extends to a map $f^{n+3} : (X, A)^{n+3} \to Y$. We continue in this way and obtain a map $g : X \to Y$ such that $gi = f$. Thus i^* is onto.

Next we show that i^* is one–one if $j > n+1$. Suppose $f, g : X \to Y$ and $fi \simeq_F gi$. Then f, g and F determine a map $F' : X \times \partial I \cup A \times I \to Y$. We then apply the previous argument to the relative CW complex $(X \times I, X \times \partial I \cup A \times I)$ and the map F' to obtain an extension $G : X \times I \to Y$ of F'. Thus $f \simeq_G g$, and so i^* is one–one. □

We next discuss a relation between the homotopy groups and the homology groups of a space. We begin by defining the *Hurewicz homomorphism* $h_n : \pi_n(Y) \to H_n(Y)$, for any space Y and integer $n \geqslant 1$. Let $\alpha = [f] \in \pi_n(Y)$. Then $f : S^n \to Y$ induces a homomorphism $f_* : H_n(S^n) \to H_n(Y)$. We fix a generator $\gamma_n \in H_n(S^n) \cong \mathbb{Z}$ for all $n \geqslant 1$ and set $h_n(\alpha) = f_*(\gamma_n) \in H_n(Y)$. Clearly h_n is well-defined. By Proposition 2.2.9, $(f+g)_*(\gamma_n) = f_*(\gamma_n) + g_*(\gamma_n)$, and thus h_n is a homomorphism. Also, it is easily seen that if $k : Y \to Y'$ is a map, then the following diagram is commutative

$$\begin{array}{ccc} \pi_n(Y) & \xrightarrow{k_*} & \pi_n(Y') \\ \downarrow{h_n} & & \downarrow{h'_n} \\ H_n(Y) & \xrightarrow{k_*} & H_n(Y'), \end{array}$$

where h_n and h'_n are Hurewicz homomorphisms.

We next wish to prove that the Hurewicz homomorphism is an isomorphism in a special case. For this, we first introduce the notion of the degree of a map.

Definition 2.4.14 Let $f : S^n \to S^n$ be a map, $n \geqslant 1$, and let $f_* : H_n(S^n) \to H_n(S^n)$ be the induced homology homomorphism. We define an integer, the *degree of* f, denoted $\deg f$, by $f_*(\gamma_n) = (\deg f)\gamma_n$, where $\gamma_n \in H_n(S^n) \cong \mathbb{Z}$ is a generator. The definition is clearly independent of the choice of generator.

Lemma 2.4.15 *Let* $f, g : S^n \to S^n$.

1. $f \simeq g \Rightarrow \deg f = \deg g$.
2. $\deg(fg) = (\deg f)(\deg g)$.
3. $\deg(f+g) = \deg f + \deg g$.

Proof. Only (3) requires proof and this follows from Proposition 2.2.9. □

Thus the degree yields a homomorphism $\deg : \pi_n(S^n) \to \mathbb{Z}$.

Proposition 2.4.16 *For* $n \geqslant 1$, *the homomorphism* $\deg : \pi_n(S^n) \to \mathbb{Z}$ *is an isomorphism and so* [id] *a generator of* $\pi_n(S^n) \cong \mathbb{Z}$.

Proof. Since $\deg(\text{id}) = 1$, it follows that deg is onto. We show that deg is one–one in Appendix D. □

This is an important result that plays a crucial role in what follows.

We introduce some notation before returning to the Hurewicz homomorphism. If G_α is an abelian group for $\alpha \in A$, then $\bigoplus_\alpha G_\alpha$ denotes the direct sum of the G_α. If $f_\alpha : G_\alpha \to H$ is a homomorphism of abelian groups for every α, we denote by $\{f_\alpha\} : \bigoplus_\alpha G_\alpha \to H$ the homomorphism determined by the f_α. Similarly, if the G_α are groups (not necessarily abelian), we let $*_\alpha G_\alpha$ denote the free product of the G_α (Appendix B). Homomorphisms $f_\alpha : G_\alpha \to H$ of groups determine a homomorphism $\{f_\alpha\} : *_\alpha G_\alpha \to H$. Now consider the wedge of n-spheres $\bigvee_\alpha S^n_\alpha$ for $\alpha \in A$, where A is any index set and let $i_\alpha : S^n_\alpha \to \bigvee_\alpha S^n_\alpha$ be the inclusion. Then it is known [39, p. 126] that $\{i_{\alpha *}\} : \bigoplus_\alpha H_n(S^n_\alpha) \to H_n(\bigvee_\alpha S^n_\alpha)$ is an isomorphism.

Lemma 2.4.17 *1. For $n \geqslant 2$, $\{i_{\alpha *}\} : \bigoplus_{\alpha \in A} \pi_n(S^n_\alpha) \to \pi_n(\bigvee_{\alpha \in A} S^n_\alpha)$ is an isomorphism.*

2. $\{i_{\alpha *}\} : *_{\alpha \in A}\ \pi_1(S^1_\alpha) \to \pi_1(\bigvee_{\alpha \in A} S^1_\alpha)$ is an isomorphism.

Proof. We assume that each sphere S^n_α is a CW complex with two cells (Example 1.5.10(5)).
(1) The result is clear if A consists of one element. Now let $A = \{\alpha_1, \ldots, \alpha_k\}$ be a finite set with $k \geqslant 2$. Let $W = \bigvee_{i=1}^k S^n_{\alpha_i}$ and $P = \prod_{i=1}^k S^n_{\alpha_i}$ and let $j : W \to P$ be the inclusion. Then P is a CW complex and W is a subcomplex such that the $n+1$-skeleton $P^{n+1} = W$. By Proposition 1.5.24, $j_* : \pi_n(W) \to \pi_n(P)$ is an isomorphism. But if $j_{\alpha_i} : S^n_{\alpha_i} \to \prod_{i=1}^k S^n_{\alpha_i}$ is the inclusion, then $\{j_{\alpha_i *}\} : \bigoplus_{i=1}^k \pi_n(S^n_{\alpha_i}) \to \pi_n(P)$ is an isomorphism by the discussion at the beginning of this section. From this (1) follows when A is finite. Now let A be infinite and let $f : S^n \to \bigvee_\alpha S^n_\alpha$ be a map. Since $f(S^n)$ is compact, there is a finite set $\{\alpha_1, \ldots, \alpha_k\}$ such that $f(S^n) \subseteq \bigvee_{i=1}^k S^n_{\alpha_i}$ by Lemma 1.5.6. Therefore $[f] \in \pi_n(\bigvee_\alpha S^n_\alpha)$ is in the image of $\pi_n(\bigvee_{i=1}^k S^n_{\alpha_i}) \to \pi_n(\bigvee_\alpha S^n_\alpha)$, for some set $\{\alpha_1, \ldots, \alpha_k\}$. Consequently $\{i_{\alpha *}\} : \bigoplus_\alpha \pi_n(S^n_\alpha) \to \pi_n(\bigvee_\alpha S^n_\alpha)$ is onto. To show that $\{i_{\alpha *}\}$ is one–one, we observe that any homotopy $F : S^n \times I \to \bigvee_\alpha S^n_\alpha$ has compact image and so factors through a homotopy $F' : S^n \times I \to \bigvee_{i=1}^k S^n_{\alpha_i}$ for some finite set $\{\alpha_1, \ldots, \alpha_k\}$. This completes the proof of (1).
(2) This is proved in Appendix B as Proposition B.3. □

The following proposition contains a special case of the Hurewicz theorem for a wedge of spheres of the same dimension. The full Hurewicz theorem is proved in Section 6.4 as Theorem 6.4.8.

Proposition 2.4.18

1. *If $n \geqslant 1$, then $\pi_i(S^n) = 0$ for $i < n$ and $h_n : \pi_n(S^n) \to H_n(S^n) \cong \mathbb{Z}$ is an isomorphism.*
2. *Let $S^n_\alpha = S^n$ for all α in some set A.*
   a. *If $n \geqslant 2$, then the Hurewicz homomorphism $h_n : \pi_n(\bigvee_{\alpha \in A} S^n_\alpha) \to H_n(\bigvee_{\alpha \in A} S^n_\alpha)$ is an isomorphism.*

b. The Hurewicz homomorphism $h_1 : \pi_1(\bigvee_{\alpha \in A} S^1_\alpha) \to H_1(\bigvee_{\alpha \in A} S^1_\alpha)$ *is an epimorphism.*

Proof. (1) Let $i < n$ and give each of S^i and S^n the CW decomposition with two cells. If $f : S^i \to S^n$ is any map, then by Corollary 1.5.23, f is homotopic to a cellular map $f' : S^i \to S^n$. Thus $f'(S^i) = *$ and so $f \simeq *$. Therefore $\pi_i(S^n) = 0$. To determine h_n, consider the commutative diagram

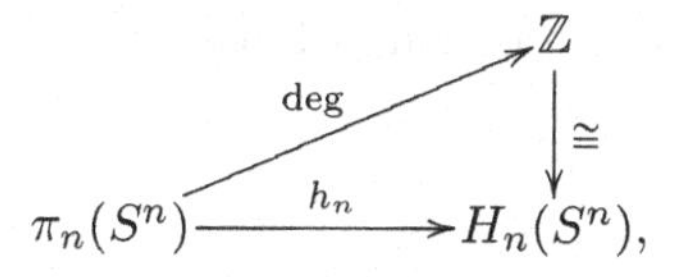

where the vertical arrow is the isomorphism that assigns to the integer k the element $k\gamma_n \in H_n(S^n)$, for γ_n a generator of $H_n(S^n)$. The result follows from Proposition 2.4.16.

(2) For Part (a) consider the commutative diagram

$$\begin{array}{ccc} \bigoplus_\alpha \pi_n(S^n_\alpha) & \xrightarrow{\{i_{\alpha *}\}} & \pi_n(\bigvee_\alpha S^n_\alpha) \\ \downarrow{\scriptstyle \bigoplus_\alpha h_\alpha} & & \downarrow{\scriptstyle h_n} \\ \bigoplus_\alpha H_n(S^n_\alpha) & \xrightarrow{\{i_{\alpha *}\}} & H_n(\bigvee_\alpha S^n_\alpha), \end{array}$$

where $h_\alpha : \pi_n(S^n_\alpha) \to H_n(S^n_\alpha)$ and h_n are Hurewicz homomorphisms. Because the horizontal homomorphisms are isomorphisms and the h_α are isomorphisms, h_n is an isomorphism. Part (b) is a special case of Proposition B.5. □

The last result of this section is part of Whitehead's second theorem 6.4.15.

Proposition 2.4.19 *Let X and Y be path-connected CW complexes, let $f : X \to Y$ be a map, and let $n \geqslant 1$ be an integer. If f is an n-equivalence, then $f_* : H_i(X) \to H_i(Y)$ is an isomorphism for all $i < n$ and an epimorphism for $i = n$.*

Proof. By Corollary 2.4.10(1) and Exercise 2.26, there is a CW complex K containing X such that K is obtained from X by adjoining cells of dimensions $\geqslant n+1$ and there is a weak homotopy equivalence $\bar{f} : K \to Y$ such that the following diagram commutes

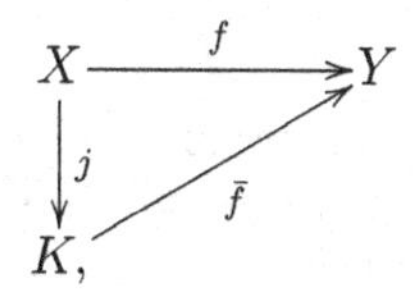

where j is the inclusion map. By Whitehead's first theorem 2.4.7, $\bar{f}$ is a homotopy equivalence, and so $f_* : H_i(X) \to H_i(Y)$ is an isomorphism for all $i < n$ and an epimorphism for $i = n$ if and only if the same holds for $j_* : H_i(X) \to H_i(K)$. All the cells of K of dimension $\leqslant n$ lie in X, thus the relative homology group $H_i(K, X) = 0$, for all $i \leqslant n$. From the exact homology sequence of the pair (K, X), it follows that $j_* : H_i(X) \to H_i(K)$ is an isomorphism for all $i < n$ and an epimorphism for $i = n$. □

The complete second theorem of Whitehead (Theorem 6.4.15) has another part in which the roles of homology and homotopy groups are interchanged. This theorem is proved in Section 6.4 as a consequence of the relative Hurewicz theorem.

2.5 Moore Spaces and Eilenberg–Mac Lane Spaces

The following two lemmas are useful in our discussion of Moore spaces and Eilenberg–Mac Lane spaces.

Lemma 2.5.1 *Let $n \geqslant 1$ and let X be a based CW complex with $(n-1)$-skeleton $X^{n-1} = \{*\}$ and $\dim X \leqslant n+1$, that is, $X = X^n \cup \bigcup_{\beta \in B} e_\beta^{n+1}$, for $X^n = \bigvee_{\alpha \in A} S_\alpha^n$, where S_α^n are n-spheres and e_β^{n+1} are open $(n+1)$-cells. Let Y be a space and let $\phi : \pi_n(X) \to \pi_n(Y)$ be a homomorphism. Then there exists a map $f : X \to Y$ such that $f_* = \phi : \pi_n(X) \to \pi_n(Y)$.*

Proof. Let $k : X^n \to X$ and $i_\alpha : S_\alpha^n \to X^n$ be the inclusions maps. Then there are homomorphisms

$$\pi_n(X^n) \xrightarrow{k_*} \pi_n(X) \xrightarrow{\phi} \pi_n(Y)$$

and we define $f_\alpha : S_\alpha^n \to Y$ by $\phi k_*[i_\alpha] = [f_\alpha]$. The f_α determine a map $f^n : X^n \to Y$ such that $f^n i_\alpha = f_\alpha$. Therefore

$$f_*^n[i_\alpha] = [f_\alpha] = \phi k_*[i_\alpha].$$

By Lemma 2.4.17, the $[i_\alpha]$ are generators of $\pi_n(X^n)$, and so $f_*^n = \phi k_*$. Let $h_\beta : S_\beta^n \to X^n$ be an attaching function for e_β^{n+1}. By Exercise 2.25 and Lemma 1.5.3 we may assume that h_β is a (based) map. Then $kh_\beta \simeq *$ since kh_β factors through the contractible space $E_\beta^{n+1} = E^{n+1}$. Hence $f_*^n[h_\beta] = \phi k_*[h_\beta] = 0$. Thus $f^n h_\beta \simeq *$ for every $\beta \in B$, and consequently f^n extends to a map $f : X \to Y$. But $\phi k_* = f_*^n = f_* k_*$ and $k_* : \pi_n(X^n) \to \pi_n(X)$ is onto by Proposition 1.5.24. Therefore $f_* = \phi$. □

Lemma 2.5.2 *For every abelian group G and $n \geqslant 1$, there exists a based CW complex L with the following properties: the $(n-1)$-skeleton $L^{n-1} = \{*\}$, $\dim L \leqslant n+1$, and for all $i \geqslant 0$,*

$$H_i(L) = \begin{cases} G \text{ if } i = n \\ 0 \text{ if } i \neq n. \end{cases}$$

Proof. We take a presentation $G = F/R$, where F is free-abelian and $R \subseteq F$. We choose bases $\{x_\alpha\}_{\alpha \in A}$ and $\{r_\beta\}_{\beta \in B}$ of F and R, respectively. Let S^n_α be the n-sphere indexed by $\alpha \in A$ and define L^n to be the wedge $\bigvee_{\alpha \in A} S^n_\alpha$. The Hurewicz homomorphism $h_n : \pi_n(L^n) \to H_n(L^n)$ is an isomorphism for $n \geqslant 2$ and an epimorphism for $n = 1$ by Proposition 2.4.18. If $\beta \in B$, then $r_\beta \in R \subseteq F \cong H_n(L^n)$. We choose a map $k_\beta : S^n_\beta \to L^n$ such that $h_n[k_\beta] = r_\beta$. Using k_β, we attach $(n+1)$-cells e^{n+1}_β to L^n and form $L = L^n \cup \bigcup_{\beta \in B} e^{n+1}_\beta$. To complete the proof we use the CW homology of L (Appendix C). The ith chain group $C_i(L)$ is the free-abelian group generated by the i-cells, and so

$$C_i(L) \cong \begin{cases} F \text{ if } i = n \\ R \text{ if } i = n+1 \\ 0 \text{ if } i \neq 0, n, n+1. \end{cases}$$

Furthermore, it is not difficult to show that the boundary homomorphism $C_{n+1}(L) \to C_n(L)$ can be identified with the inclusion $R \subseteq F$ (see [64, Prop. 8.2.12]). Thus L has the desired homology. □

We turn to Moore spaces.

Definition 2.5.3 Let G be an abelian group and n an integer $\geqslant 2$. A based CW complex X is called a *Moore space* of type (G, n) if X is 1-connected and

$$H_i(X) = \begin{cases} G \text{ if } i = n \\ 0 \text{ if } i \neq n. \end{cases}$$

In Lemma 2.5.2, a Moore space L of type (G, n) has been constructed. We denote this Moore space (or any space homeomorphic to it) by $M(G, n)$.

We note a few properties of $M(G, n)$.

Lemma 2.5.4

1. *The Hurewicz homomorphism $h_n : \pi_n(M(G,n)) \to H_n(M(G,n))$ is an isomorphism, and so $\pi_n(M(G,n)) = G$.*
2. *If $\phi : G \to H$ is a homomorphism of abelian groups, then there exists a map $f : M(G,n) \to M(H,n)$ such that $f_* = \phi : H_n(M(G,n)) \to H_n(M(H,n))$.*

Proof. (1) From the construction of $M(G,n)$ in Lemma 2.5.2 as $(\bigvee_{\alpha \in A} S^n_\alpha) \cup \bigcup_{\beta \in B} e^{n+1}_\beta$, we have a commutative diagram

$$\begin{array}{ccccccccc} 0 & \longrightarrow & \pi_n(\bigvee_{\beta \in B} S^n_\beta) & \xrightarrow{k_*} & \pi_n(\bigvee_{\alpha \in A} S^n_\alpha) & \xrightarrow{i_*} & \pi_n(M(G,n)) & \longrightarrow & 0 \\ & & \downarrow h'_n & & \downarrow h''_n & & \downarrow h_n & & \\ 0 & \longrightarrow & H_n(\bigvee_{\beta \in B} S^n_\beta) & \xrightarrow{k_*} & H_n(\bigvee_{\alpha \in A} S^n_\alpha) & \xrightarrow{i_*} & H_n(M(G,n)) & \longrightarrow & 0, \end{array}$$

where h_n, h'_n, and h''_n are Hurewicz homomorphisms, $k : \bigvee_{\beta\in B} S^n_\beta \to \bigvee_{\alpha\in A} S^n_\alpha$ is determined by the k_β, and i is the inclusion. The bottom row is exact, h'_n and h''_n are isomorphisms (2.4.18), $ik \simeq *$, and the upper i_* is onto by Proposition 1.5.24. Thus the top row is exact and so h_n is an isomorphism.

(2) Let $\phi : G \to H$ be a homomorphism and let $L = M(G,n)$ and $M = M(H,n)$. Consider the Hurewicz homomorphisms $h' : \pi_n(L) \to H_n(L) = G$ and $h : \pi_n(M) \to H_n(M) = H$ which are isomorphisms by (1). By Lemma 2.5.1, there exists a map $f : L \to M$ such that $f_* = h^{-1}\phi h' : \pi_n(L) \to \pi_n(M)$. Hence the induced homology homomorphism $f_* = \phi : H_n(L) \to H_n(M)$. □

Remark 2.5.5 We note that several choices have been made in the construction of $M(G,n)$ such as the presentation of G and the choice of generators of R and F. We show in Proposition 6.4.16 that the homotopy type of a Moore space of type (G,n) depends only on G and n. However, in spite of the notation, $M(G,n)$ is not functorial in G. For now, when we write $M(G,n)$ we assume that it has been constructed relative to a choice of presentation and of generators.

The spaces $M(G,n)$ can easily be described in special cases. We have $M(\mathbb{Z},n) = S^n$ and if F is a free-abelian group with a basis whose cardinality is the same as some set A, then $M(F,n) = \bigvee_{\alpha\in A} S^n_\alpha$. Furthermore, $M(\mathbb{Z}_m,n)$ can be taken to be the space $S^n \cup_{\mathrm{m}} e^{n+1}$ obtained by attaching an $(n+1)$-cell to S^n by a map $\mathrm{m} : S^n \to S^n$ of degree m. For $n = 1$, we define the Moore space $M(\mathbb{Z},1) = S^1$. We do not consider $M(G,1)$ for other groups G (see [89]). For $n \geqslant 3$, $M(G,n) \cong \Sigma M(G,n-1)$ (see Exercise 3.1). In addition, if L is the space of Lemma 2.5.2 with $n = 1$, then $M(G,2) \cong \Sigma L$. Thus all Moore spaces $M(G,n)$ are suspensions, in fact, $M(G,n)$ is a double suspension if $n \geqslant 3$ or if $n = 2$ and G is free-abelian. Hence $[M(G,n),X]$ is a group for all X. It is abelian if $n \geqslant 3$ or if $n = 2$ and G is free-abelian. At the end of this section we discuss the reason for making the definition of Moore space in terms of homology groups instead of cohomology groups. We compare Moore spaces with spaces with a single nonvanishing cohomology group.

The suspension structure of Moore spaces $M(G,n)$ enables us to define homotopy groups with coefficients.

Definition 2.5.6 Let G be an abelian group and let $n \geqslant 1$ (assuming $G = \mathbb{Z}$ if $n = 1$). Then for every space X, define the *nth homotopy group of X with coefficients in G* by

$$\pi_n(X;G) = [M(G,n),X].$$

Other notation (which we do not use) for this is $\pi_n(G;X)$ (see [40, p. 10]).

Note that $\pi_n(X;\mathbb{Z}) = \pi_n(X)$, the nth (ordinary) homotopy group of X. Furthermore, a map $h : X \to X'$ induces a homomorphism $h_* : \pi_n(X;G) \to$

$\pi_n(X'; G)$ defined as the induced homomorphism $h_* : [M(G,n), X] \to [M(G,n), X']$.

Homotopy groups with coefficients are discussed in the sequel. In particular, we present a universal coefficient theorem in Section 5.2 that expresses homotopy groups with coefficients in terms of ordinary homotopy groups.

We next turn to Eilenberg–Mac Lane spaces.

Definition 2.5.7 Let G be an abelian group and let n be an integer $\geqslant 1$. An *Eilenberg–Mac Lane space* of type (G, n) is a space X of the homotopy type of a based CW complex such that for every $i \geqslant 1$,

$$\pi_i(X) = \begin{cases} G \text{ if } i = n \\ 0 \text{ if } i \neq n. \end{cases}$$

An Eilenberg–Mac Lane space of type (G, n) is denoted $K(G, n)$.

The following lemma is used to show that Eilenberg–Mac Lane spaces exist.

Lemma 2.5.8 *If X is a space, then for every $m \geqslant 1$, there exist spaces $W^{(m)}$ and inclusion maps $j_m : X \to W^{(m)}$ such that*

1. $\pi_i(W^{(m)}) = 0$ *for* $i > m$.
2. $j_{m*} : \pi_i(X) \to \pi_i(W^{(m)})$ *is an isomorphism for* $i \leqslant m$.
3. $W^{(m)}$ *is obtained from X by attaching cells of dimension* $\geqslant m + 2$.

Proof. Parts (1) and (3) follow immediately from Theorem 2.4.9 by taking $Y = \{*\}$ and $n = m + 1$. Part (2) is a consequence of Proposition 1.5.24. □

Next we show that Eilenberg–Mac Lane spaces exist and obtain some of their properties.

Proposition 2.5.9 *Let G be an abelian group and n an integer $\geqslant 1$.*

1. *There exists an Eilenberg–Mac Lane space $K(G, n)$.*
2. *If $\phi : G \to H$ is a homomorphism, then there exists a map $h : K(G, n) \to K(H, n)$ such that $h_* = \phi : \pi_n(K(G, n)) \to \pi_n(K(H, n))$.*
3. *Any two Eilenberg–Mac Lane spaces of type (G, n) have the same homotopy type.*

Proof. (1) We first construct an Eilenberg–Mac Lane space of type (G, n) when $n \geqslant 2$. We apply Lemma 2.5.8 with X equal to the Moore space $M(G, n)$. Then $W^{(n)}$ is an Eilenberg–Mac Lane space of type (G, n) by Lemma 2.5.4. For $n = 1$, we set $K(G, 1) = \Omega K(G, 2)$.

(2) Let $\phi : G \to H$ be a homomorphism and let $K = K(G, n)$ and $L = K(H, n)$ be any Eilenberg–Mac Lane spaces. Let X be the Eilenberg–Mac Lane space of type (G, n) constructed in (1). We construct a map $f : X \to L$

such that $f_* = \phi : \pi_n(X) \to \pi_n(L)$. If $j : X^{n+1} \to X$ is the inclusion, then $j_* : \pi_n(X^{n+1}) \to \pi_n(X)$ is an isomorphism by Proposition 1.5.24. The $n+1$-skeleton $X^{n+1} = M(G,n)$ and L satisfy the hypotheses of Lemma 2.5.1, and so there exists $f^{n+1} : X^{n+1} \to L$ such that $f_*^{n+1} = \phi j_* : \pi_n(X^{n+1}) \to \pi_n(L)$. Now apply Proposition 2.4.13 to the relative CW complex (X, X^{n+1}) and the map f^{n+1}. We conclude that there is a map $f : X \to L$ such that $fj \simeq f^{n+1}$, and so $f_* = \phi$. Similarly the identity map $\mathrm{id} : G \to G$ yields a map $g : X \to K$ such that $g_* = \mathrm{id}$. Then g is a homotopy equivalence by Whitehead's first theorem. Thus if $h = fg^{-1} : K(G,n) \to K(H,n)$, we have $h_* = \phi$.

(3) If K and L are both Eilenberg–Mac Lane spaces of type (G,n), then the identity homomorphism $\mathrm{id}_G : G \to G$ induces a map $h : K \to L$ by (2). Then h is a homotopy equivalence. □

In general Eilenberg–Mac Lane spaces are infinite-dimensional complexes and are not easy to describe. There are a few that are familiar spaces and we mention these now: $K(\mathbb{Z},1) = S^1$, $K(\mathbb{Z}_2,1) = \mathbb{R}\mathrm{P}^\infty$, infinite-dimensional real projective space, and $K(\mathbb{Z},2) = \mathbb{C}\mathrm{P}^\infty$, infinite-dimensional complex projective space (see Exercise 5.19).

It can be shown that for any group G (not necessarily abelian), Eilenberg–Mac Lane spaces of type $(G,1)$ exist and are unique up to homotopy. However, we have no need to consider the spaces $K(G,1)$ when G is non-abelian (except in Exercise 2.31).

Now let $K(G,n+1)$ be an Eilenberg–Mac Lane space with $n \geqslant 1$. We apply the loop space functor to this space and have by Proposition 2.3.5 that

$$\pi_i(\Omega K(G,n+1)) \cong \pi_{i+1}(K(G,n+1)) = \begin{cases} G \text{ if } i = n \\ 0 \text{ if } i \neq n. \end{cases}$$

Hence, as previously noted in the case $n = 1$, $\Omega K(G,n+1)$ is an Eilenberg–Mac Lane space $K(G,n)$. In fact, $\Omega^k K(G,n+k)$ is an Eilenberg–Mac Lane space of type (G,n). Therefore, for any space X, the set $[X, K(G,n)]$ has abelian group structure.

We next indicate how Eilenberg–Mac Lane spaces give rise to cohomology groups.

Definition 2.5.10 For any space X, abelian group G and integer $n \geqslant 1$, we define the *nth homotopical cohomology group* of X with coefficients in G as

$$H^n(X;G) = [X, K(G,n)].$$

If $h : X' \to X$ is a map, then $h^* : H^n(X;G) \to H^n(X';G)$ is just the induced homomorphism $h^* : [X, K(G,n)] \to [X', K(G,n)]$.

Remark 2.5.11 It can be shown that if X is a CW complex, then the homotopical cohomology groups $H^n(X;G)$ are isomorphic to the singular cohomology groups $H^n_{\mathrm{sing}}(X;G)$. There are several proofs of this: one uses the

Brown representation theorem [39, p. 448] and another uses obstruction theory [91, p. 250]. In addition, in [46] Huber gives an isomorphism between the homotopical cohomology groups and the Cech cohomology groups. We give a simple proof of the isomorphism with singular cohomology in Section 5.3.

We note that we can define a function $\rho : H^n(X;G) \to H^n_{\text{sing}}(X;G)$ as follows. If $[f] \in H^n(X;G)$, then $f^*_{\text{sing}} : H^n_{\text{sing}}(K(G,n);G) \to H^n_{\text{sing}}(X;G)$ is the induced homomorphism of singular cohomology. Note that $H_n(K(G,n)) = G$ since $K(G,n)$ can be taken to be $M(G,n)$ with cells of dimension $\geqslant n+2$ attached. Thus there is an element $b^n \in H^n_{\text{sing}}(K(G,n);G)$, called the nth basic class, which is defined by $\mu(b^n) = \text{id}$, where $\mu : H^n_{\text{sing}}(K(G,n);G) \to \text{Hom}(H_n(K(G,n)),G) \cong \text{Hom}(G,G)$ is the epimorphism in the universal coefficient theorem for cohomology (Appendix C). Then set $\rho[f] = f^*_{\text{sing}}(b^n)$. It is shown that ρ is an isomorphism in Theorem 5.3.2.

We need some properties of homotopical cohomology groups in the following chapters. These properties are known to hold for CW or singular cohomology groups. But since we do not prove the equivalence of the latter cohomology groups with the homotopical cohomology groups until Theorem 5.3.2, we next establish these properties.

Lemma 2.5.12 *Let X and Y be path-connected, based CW complexes, let $f : X \to Y$ be a map and let $n \geqslant 1$ be an integer. If f is an n-equivalence, then, for every group G, $f^* : H^i(Y;G) \to H^i(X;G)$ is an isomorphism for $i < n$ and a monomorphism for $i = n$.*

Proof. (1) By Corollary 2.4.10(1), there is a CW complex K obtained from X by adjoining cells of dimensions $\geqslant n+1$ and a homotopy equivalence $\bar{f} : K \to Y$ such that $\bar{f}j = f$, where $j : X \to K$ is the inclusion map. Therefore $f^* : H^i(Y;G) \to H^i(X;G)$ is an isomorphism for $i < n$ and a monomorphism for $i = n$ if and only if the same holds for $j^* : H^i(K;G) \to H^i(X;G)$. But the latter follows at once from Proposition 2.4.13. □

Next we let X be a space, G an abelian group, and $n \geqslant 1$ an integer. We define a homomorphism $\eta_\pi : H^n(X;G) \to \text{Hom}(\pi_n(X),G)$ by $\eta_\pi[f] = f_* : \pi_n(X) \to \pi_n(K(G,n)) = G$.

Lemma 2.5.13 *If X is an $(n-1)$-connected based CW complex, $n \geqslant 1$, then $\eta_\pi : H^n(X;G) \to \text{Hom}(\pi_n(X),G)$ is an isomorphism. (For $n = 1$ we assume that $\pi_1(X)$ is abelian.)*

Proof. By Corollary 2.4.10(3), we may assume $X^{n-1} = \{*\}$. We first prove that $\eta_\pi : H^n(X;G) \to \text{Hom}(\pi_n(X),G)$ is an isomorphism when $X = X^{n+1}$. By Lemma 2.5.1, η_π is onto. Now let $f : X \to K$, where $K = K(G,n)$, and assume that $\eta_\pi[f] = 0$. Since X^n is a wedge of n-spheres, $f|X^n \simeq * : X^n \to K$. But (X,X^n) has the homotopy extension property, and so there is a map $f' : X \to K$ such that $f \simeq f'$ and $f'|X^n = *$. Therefore f' induces a map $\widetilde{f'} : X/X^n \to K$ such that $\widetilde{f'}q = f'$, where $q : X \to X/X^n$ is the projection. Since X/X^n is a wedge of $(n+1)$-spheres, $\widetilde{f'} \simeq *$. Hence $f' \simeq *$, and so $f \simeq *$. Therefore η_π is an isomorphism when $\dim X \leqslant n+1$. For an arbitrary $(n-1)$-connected CW complex X, the inclusion map $X^{n+1} \to X$ is an $(n+1)$-equivalence by the cellular approximation theorem. Thus $\pi_n(X) \cong \pi_n(X^{n+1})$ and, by Lemma 2.5.12, $H^n(X;G) \cong H^n(X^{n+1};G)$. Therefore $\eta_\pi : H^n(X;G) \to \mathrm{Hom}(\pi_n(X), G)$ is an isomorphism. □

We next prove the Hopf classification theorem [91, p. 244].

Theorem 2.5.14 *If X is a CW complex of dimension $\leqslant n$, then there is a bijection between $[X, S^n]$ and $H^n(X)$.*

Proof. Let $K(\mathbb{Z}, n)$ be the Eilenberg–Mac Lane space constructed in the proof of Proposition 2.5.9 with $G = \mathbb{Z}$ and let $i : S^n \to K(\mathbb{Z}, n)$ be the inclusion. Then i induces $i_* : [X, S^n] \to [X, K(\mathbb{Z}, n)] = H^n(X)$. The $n+1$-skeleton of $K(\mathbb{Z}, n)$ is S^n, therefore Proposition 1.5.24 shows that i_* is a bijection. □

NOTE We denote the homotopical cohomology groups by $H^n(X;G)$ and refer to them as cohomology groups.

In the Eckmann–Hilton duality theory (discussed in the next section), cohomology groups with coefficients are dual to homotopy groups with coefficients. The former are defined as homotopy classes of maps with codomain an Eilenberg–Mac Lane space and the latter as homotopy classes of maps with domain a Moore space. Eilenberg–Mac Lane spaces are spaces with a single nonvanishing homotopy group, thus it would appear that the dual notion should be a co-Moore space, that is, a space with a single nonvanishing cohomology group. However, we have instead taken the dual to be a Moore space, that is, a space with a single nonvanishing homology group. The reason for this is that co-Moore spaces of type (G, n) do not exist for every group G (see [39, pp. 318–319]). Therefore to ensure the existence of homotopy groups with coefficients for any abelian group G, they have been defined in terms of Moore spaces.

We carry this discussion of co-Moore spaces a bit further. For a finitely generated abelian group G, write $G = F \oplus T$, where F is a free-abelian group and T is a finite abelian group. If $C(G, n)$ denotes a co-Moore space of type (G, n), then a simple calculation of cohomology shows that $M(F, n) \vee M(T, n-1)$ is a $C(G, n)$. In particular, if $G = \mathbb{Z}_m$, then $M(\mathbb{Z}_m, n-1)$ is a $C(\mathbb{Z}_m, n)$. As noted above, we could have defined the homotopy groups of X with coefficients in $\mathbb{Z}_m$ using co-Moore spaces by

$$\widetilde{\pi}_n(X; \mathbb{Z}_m) = [C(\mathbb{Z}_m, n), X].$$

Then

$$\widetilde{\pi}_n(X; \mathbb{Z}_m) = [M(\mathbb{Z}_m, n-1), X] = \pi_{n-1}(X; \mathbb{Z}_m).$$

However, we use Definition 2.5.6 for homotopy groups with coefficients.

2.6 Eckmann–Hilton Duality I

Our exposition is based on the duality theory of Eckmann and Hilton that we have referred to several times without explanation. We now discuss this topic. The first appearance of the duality in the literature was in the papers [27, 28, 29] of Eckmann and Hilton and the book [40] by Hilton. This duality principle differs from certain other duality principles which are formal and automatic. For example, in projective geometry there is a duality which asserts that every definition remains meaningful and every theorem true if we interchange the words *point* and *line* (and consequently other pairs of words such as *colinear* and *concurrent*, *side* and *vertex*, and so on)[22, Chap. 3]. Parts of the Eckmann–Hilton duality are formal and automatic (usually those parts that can be described in categorical terms). However, much of it is intuitive, informal, and heuristic.

We begin with the aspect of the Eckmann–Hilton duality which depends on duality in a category and we refer to some fundamental facts about categories and functors from Appendix F. With any category $\mathcal{C}$, we associate a *dual category* $\mathcal{C}^{op}$ (also called the *opposite category*). The objects of $\mathcal{C}^{op}$ are precisely those of $\mathcal{C}$, but the set of morphisms from an object X in $\mathcal{C}^{op}$ to an object Y in $\mathcal{C}^{op}$, denoted $\mathcal{C}^{op}(X, Y)$, is defined to be $\mathcal{C}(Y, X)$. If composition of morphisms in $\mathcal{C}^{op}$ is denoted by $*$ and composition in $\mathcal{C}$ by juxtaposition, then $f * g = gf$.

Now suppose that Σ is a statement or concept that is meaningful in a category $\mathcal{C}$. Then we can apply it to the category $\mathcal{C}^{op}$ *and interpret it as a statement or concept in* $\mathcal{C}$. This latter statement or concept in $\mathcal{C}$ that is denoted Σ^* is the *dual* of Σ. If $\Sigma = \Sigma^*$, then we say that Σ is *self-dual*.

For example, recall from Appendix F the notion of categorical product. If X and Y are objects in a category $\mathcal{C}$, their categorical product is an object P in $\mathcal{C}$ together with morphisms $p_1 : P \to X$ and $p_2 : P \to Y$ such that the following holds. If $f : A \to X$ and $g : A \to Y$ are any morphisms, then there exists a unique morphism $\theta : A \to P$ such that $p_1\theta = f$ and $p_2\theta = g$. If Σ denotes this concept in category $\mathcal{C}$, then the dual concept Σ^* in $\mathcal{C}$ is the following. Suppose X and Y are objects in $\mathcal{C}$ and there is an object C in $\mathcal{C}$ together with morphisms $i_1 : X \to C$ and $i_2 : Y \to C$ such that if $f : X \to B$ and $g : Y \to B$ are any morphisms, then there exists a unique morphism $\theta : C \to B$ with $\theta i_1 = f$ and $\theta i_2 = g$. Then C is the coproduct in $\mathcal{C}$ of X and Y, and so the coproduct is dual to the product. We can investigate various categories to see if the product or the coproduct exists and, if so, if it is given by a well-known construction. This can of course be done for any Σ and Σ^*.

The main categories that we consider are the topological category Top_* and the homotopy category $HoTop_*$ (Appendix F). In Top_* the product of X and Y is just their cartesian product $X \times Y$ with p_1 and p_2 the two projections. The coproduct is the wedge $X \vee Y$ with i_1 and i_2 the two injections. For the product and coproduct in the homotopy category $HoTop_*$ we take the cartesian product $X \times Y$ with homotopy classes of p_1 and p_2 for the former and the wedge $X \vee Y$ with homotopy classes of i_1 and i_2 for the latter (see Lemma 1.3.6). Thus the product and the wedge as just defined are dual in $HoTop_*$. Other examples of dual concepts in $HoTop_*$ are homotopy retracts and homotopy sections, and H-spaces and co-H-spaces.

There is another aspect of Eckmann–Hilton duality that is more obscure and which often takes the form of a duality between functors or constructions. We illustrate this with an example. Suppose $f : X \to Y$ is a map. Then $f \simeq *$ if and only if there is a contractible space T such that f factors through T, that is, there are maps $i : X \to T$ and $\widetilde{f} : T \to Y$ such that the following diagram commutes

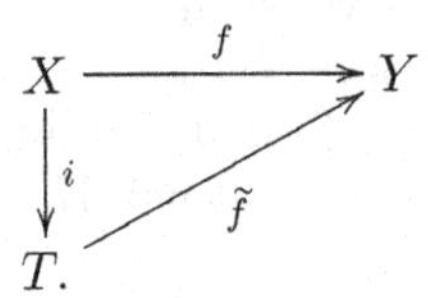

This happens if there is a functor C that assigns a contractible space $C(X)$ to every space X and a map $i_X : X \to C(X)$ with the above property. The cone on X does this, as we know by Proposition 1.4.9. If we dualize this by reversing the direction of the maps, then we seek a functor E such that $E(Y)$ is contractible for every space Y and a map $p_Y : E(Y) \to Y$ with the following property: $f \simeq * : X \to Y$ if and only if there is a map $\widehat{f} : X \to E(Y)$ such that the following diagram commutes

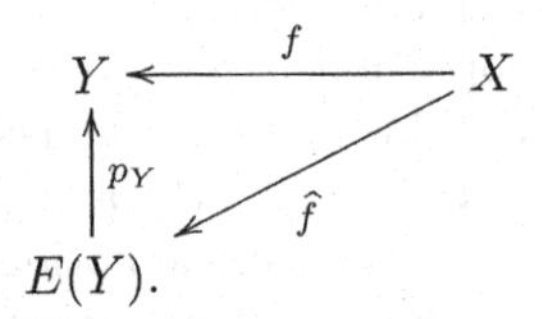

We have seen that the path space EY has this property (Proposition 1.4.9). One might raise the question of why the dual of a contractible space is a contractible space. This could be argued as follows. The dual of an identity morphism in a category is an identity morphism because the defining property of id_X is $\mathrm{id}_X f = f$ and $g\,\mathrm{id}_X = g$, for all morphisms f and g. The constant morphisms have a similar defining property, and so the dual of a constant map is a constant map. Thus identity morphisms and constant morphisms are self-dual. But in Top_* a contractible space is one in which the identity map is homotopic to the constant map. Thus it is reasonable to regard the dual of a contractible space to be a contractible space.

We digress briefly to comment further on this example in order to indicate the origin of the Eckmann–Hilton duality. The preceding discussion can be transfered to the category of (left) R-modules in the following way. A homomorphism $\phi : A \to B$ of R-modules is called i-nullhomotopic if it can be extended to some injective R-module Q that contains A. This could be regarded as the analogue of extending a map of spaces to the cone of the domain. We then say that two R-homomorphisms from A to B are i-homotopic if their difference is i-nullhomotopic. Alternatively, $\phi : A \to B$ is p-nullhomotopic if it can be factored through some projective R-module P that has B as a quotient. Then P could be regarded as the analogue in this category of the path space of the codomain. Two homomorphisms would then be p-homotopic if their difference is p-nullhomotopic. It can be shown that the notions of i-homotopy and p-homotopy in the category of R-modules do not agree. Furthermore, by taking Q/A we obtain an analogue of the suspension and by taking the kernel of $P \to B$ we obtain an analogue of the loop space. It was the realization that injective modules and their quotients play the role of cones and suspensions and that projective modules and their kernels play the role of path spaces and loop spaces in the category of left R-modules that was the beginning of the Eckmann–Hilton duality [40, Chap. 13].

We return to discussing the cone and path space functors in the category Top_*. We observe that they are adjoint functors. This implies that for a map $f : X \to Y$, there is a one–one correspondence between maps $F : CX \to Y$ such that $Fi_X = f$ and maps $\widetilde{F} : X \to EY$ such that $p_Y \widetilde{F} = f$. The correspondence is just the adjoint one given by

$$\widetilde{F}(x)(t) = F\langle x, t\rangle,$$

for $x \in X$ and $t \in I$ (Proposition 1.3.4). Thus the two notions of nullhomotopy in the category of R-modules become the single notion of ordinary nullhomotopy for the category of spaces and maps. We say that the cone functor C and the path space functor E are dual functors (in addition to being adjoint). The suspension ΣX is a quotient of the map $i_X : X \to CX$ and the loop space ΩY is a "kernel" of the map $p_Y : EY \to Y$, therefore we view the functors Σ and Ω as dual to each other. We have already noted that these two functors are adjoint in the homotopy category $HoTop_*$, namely,

$$[\Sigma X, Y] \cong [X, \Omega Y]$$

by Proposition 2.3.5. Therefore we regard the functors Σ and Ω as dual and adjoint in $HoTop_*$. In a similar way the reduced cylinder $X \ltimes I$ and the path space X^I can be regarded as dual and adjoint functors of X.

Furthermore, the homotopy groups appear to have properties dual to those of the cohomology groups. For example, the homotopy groups are covariant functors and the cohomology groups are contravariant functors. Moreover, there is a formula that expresses the homotopy groups of a product as a product of homotopy groups and an analogous formula for the cohomology groups of a wedge. In addition, the homotopy groups of the loop space of X are isomorphic to those of X (with a shift in degree) and a similar statement holds for the cohomology of a suspension. But there are also important differences. The fundamental group of a space is not necessarily abelian, but all cohomology groups are abelian. The homotopy groups for most common spaces such as CW complexes are not necessarily zero from some degree on (see Section 5.6), whereas the cohomology groups are zero above the dimension of the space (although this difference may not indicate a failure of duality). In spite of this, we do view ordinary homotopy groups and integral cohomology as being informally dual to each other. Although the duality becomes more tenuous when we assert that specific spaces are dual to each other, we regard Eilenberg–Mac Lane spaces and Moore spaces as duals in a weak sense. Therefore we think of homotopy groups with coefficients as dual to cohomology groups with coefficients.

We return to discussing duality in Section 6.5 after we have presented more material. There we also discuss some of the interesting, unusual, and anomalous features of duality.

Exercises

Exercises marked with (*) may be more difficult than the others. Exercises marked with (†) are used in the text.

2.1. (†) Let (Y, m) be an H-space and assume that $(Y \times Y, Y \vee Y)$ has the homotopy extension property. Prove that there a multiplication m' on Y that is homotopic to m and such that $m'(y, *) = y$ and $m'(*, y) = y$, for all $y \in Y$.

2.2. (†) Let $f : X \to Y$ be a map which has a left homotopy inverse. Prove that if Y is an H-space, then X is an H-space. With this multiplication on X, is f an H-map? What condition will ensure that if Y is homotopy-associative, then X is homotopy-associative?

2.3. Let Y be a grouplike space with multiplication m and homotopy inverse i. Define a *commutator map* $\phi : Y \times Y \to Y$ by $\phi = (p_1 + p_2) + (ip_1 + ip_2)$. Prove

1. (Y, m) is homotopy-commutative if and only if $\phi \simeq *$.
2. If $\alpha = [f]$, $\beta = [g] \in [X, Y]$, then the group commutator $[\alpha, \beta] = [\phi(f, g)]$.
3. If k is a positive integer, set $k\phi = \phi + \cdots + \phi$ (k terms) $: Y \times Y \to Y$. Show that $m + k\phi$ is a multiplication on Y.
4. Dualize (1)–(3) to cogroups.

2.4. Prove that if "$\simeq$" is replaced by equality in the definition that (X, c) is a co-H-space, then $X = \{*\}$.

2.5. Is S^0 a (nonpath-connected) co-H-space?

2.6. Prove that a space X admits a comultiplication if and only if the diagonal map $\Delta : X \to X \times X$ can be factored up to homotopy through $X \vee X$. Prove that a space X admits a multiplication if and only if the folding map $\nabla : X \vee X \to X$ can be extended up to homotopy to $X \times X$.

2.7. Let X be an $(n-1)$-connected space with $n \geqslant 2$.

1. Prove that if $\dim X \leqslant 2n-1$, then there is a comultiplication on X. If $\dim X \leqslant 2n-2$, prove that any two comultiplications on X are homotopic.
2. Prove that if $\pi_i(X) = 0$ for $i \geqslant 2n-1$, then there is a multiplication on X. Prove that any two multiplications on X are homotopic if $\pi_i(X) = 0$ for $i \geqslant 2n$

In Exercises 2.8–2.14 also consider the dual of the given problem.

2.8. Let (Y, m) be an H-space, let $f, g : X \to Y$ be maps and let $f+g : X \to Y$ be their sum. Prove for any space A, that $(f+g)_* = f_* + g_* : [A, X] \to [A, Y]$.

2.9. Let (X, c) and (X', c') be co-H-spaces and let $g : X' \to X$ be a map. Prove the following generalization of Proposition 2.2.9: g is a co-H-map if and only if for every space Y and every $\alpha, \beta \in [X, Y]$, we have $(\alpha+\beta)[g] = \alpha[g] + \beta[g]$.

2.10. ($*$) Let $j_{\Sigma X} : \Sigma X \to \Sigma X$ be the homotopy inverse map defined by $j_{\Sigma X}\langle x, t\rangle = \langle x, 1-t\rangle$, for $x \in X$ and $t \in I$. Consider the double suspension $\Sigma^2 X$ and the map $\tau : \Sigma^2 X \to \Sigma^2 X$ defined by $\tau\langle x, s, t\rangle = \langle x, t, s\rangle$. Prove that $j_{\Sigma^2 X} \simeq \tau \simeq \Sigma j_{\Sigma X}$.

2.11. (†) Define a map $\theta : \Sigma(X_1 \vee X_2) \to (\Sigma X_1) \vee (\Sigma X_2)$ by $\theta\langle(x_1, *), t\rangle = (\langle x_1, t\rangle, *)$ and $\theta\langle(*, x_2), t\rangle = (*, \langle x_2, t\rangle)$, for $x_1 \in X_1$, $x_2 \in X_2$ and $t \in I$. If $i_j : X_j \to X_1 \vee X_2$ and $\iota_j : \Sigma X_j \to (\Sigma X_1) \vee (\Sigma X_2)$ are inclusions and $q_j : X_1 \vee X_2 \to X_j$ and $\chi_j : (\Sigma X_1) \vee (\Sigma X_2) \to \Sigma X_j$ are projections, $j = 1, 2$, then prove that (1) $\theta\, \Sigma i_j = \iota_j$ and $\chi_j\, \theta = \Sigma q_j$, (2) θ is a homeomorphism with inverse $\{\Sigma i_1, \Sigma i_2\}$ and (3) $\theta \simeq \iota_1 \Sigma q_1 + \iota_2 \Sigma q_2$.

2.12. (†) Let (X, c_X) be a co-H-space and $\theta : \Sigma(X \vee X) \to \Sigma X \vee \Sigma X$ the homeomorphism of Exercise 2.11. Show that $\theta\, \Sigma c_X$ is a comultiplication that is homotopic to $c_{\Sigma X}$, the suspension comultiplication on ΣX.

2.13. Show that if X and Y are co-H-spaces, then $X \vee Y$ is a co-H-space. If X and Y are both homotopy-associative or both homotopy-commutative, does the same hold for $X \vee Y$? Let $\theta : \Sigma(X \vee Y) \to \Sigma X \vee \Sigma Y$ be the homeomorphism of Exercise 2.11. Show that θ is a co-H-map.

2.14. Prove that if X is a co-H-space, then ΣX is homotopy-commutative.

2.15. ($*$) Find another homotopy in the proof of Proposition 2.3.2(3).

2.16. For any space A, prove that $(A \times I)/(A \times \partial I)$ is homeomorphic to ΣB, for some space B. What is B?

2.17. Consider the cofiber sequence $X \to CX \to \Sigma X$, the fiber sequence $\Omega Y \to EY \to Y$ and a map $f : \Sigma X \to Y$. Find a map $\theta : CX \to EY$ such that the following diagram commutes

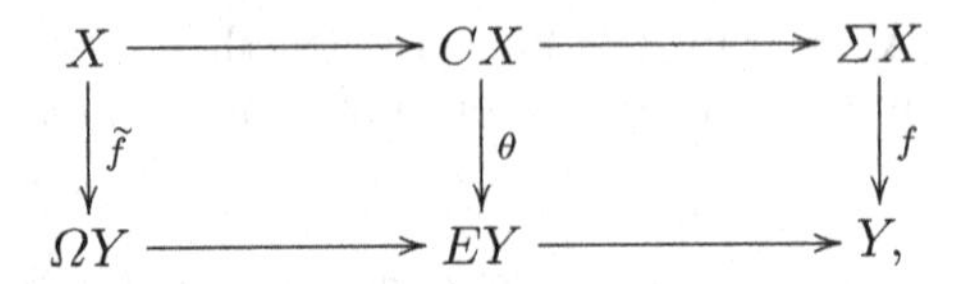

where $\widetilde{f}$ is the adjoint of f.

2.18. (*) (†) Prove: $\pi_1(X) = 0$ if and only if for paths $f, g : I \to X$ such that $f(0) = g(0) = *$ and $f(1) = g(1)$, we have $f \simeq g \operatorname{rel} \partial I$.

2.19. If G is a group we define a *comultiplication* on G to be a homomorphism $s : G \to G * G$ such that $p_1 s = \mathrm{id} = p_2 s : G \to G$.

1. Prove that G admits a comultiplication if and only if $\pi : E_G \to G$ has a right inverse.
2. Prove that G admits a comultiplication if and only if G is a free group.

Note that Proposition 2.4.3 asserts that the functor π_1 carries a space with a comultiplication to a group with a comultiplication.

2.20. (†) In the proof of Lemma 2.4.2 verify that

$$\xi = \prod_{i=1}^{2p} \xi_{\delta_i}^{(i)}.$$

2.21. If X is an H-space and a co-H-space, prove that $\pi_1(X)$ is 0 or $\mathbb{Z}$. Give examples of these spaces.

2.22. Let X and Y be co-H-complexes that are not simply connected. Prove that $X \times Y$ is not a co-H-space.

2.23. (†) Prove that a path-connected CW complex X is contractible if and only if $\pi_q(X) = 0$ for all $q \geqslant 1$.

2.24. (†) Let X be a space that is not necessarily path-connected. Show that there is a bijection μ from $\pi_0(X)$ to the set of path-components of X. Show that if (X, m) is a grouplike space, then m induces group structure on $\pi_0(X)$ and on the set of path-components of X such that μ is an isomorphism.

2.25. (†) Let X and Y be spaces with X a CW complex and Y path-connected. If $f : X \to Y$ is a free map, prove that there is a (based) map $g : X \to Y$ such that $f \simeq_{\text{free}} g$.

2.26. (†) In Corollary 2.4.10(1) show that if X is a CW complex, then K can be taken to be a CW complex containing X as a subcomplex.

2.27. Let $i : A \to X$ be a cofiber map, where A and X are not necessarily of the homotopy type of CW complexes. Prove that there exists a relative CW complex (K, A) and a weak equivalence $f : K \to X$ such that $f|A = i$.

2.28. Let X and Y be spaces (not necessarily of the homotopy type of CW complexes). We define $X \simeq Y$ if there exist spaces $X_1, X_2, \ldots, X_n$ such that $X_1 = X$, $X_n = Y$ and for $i = 1, 2, \ldots, n-1$, there exists a weak equivalence $X_i \to X_{i+1}$ or a weak equivalence $X_{i+1} \to X_i$. Prove that $X \simeq Y \iff X$ and Y have CW approximations of the same homotopy type. For this problem you can assume the result stated in Remark 2.4.12.

2.29. How many homotopy classes of homotopy retractions are there of the inclusion $i_1 : S^n \to S^n \vee S^n$?

2.30. (*) Let A be a set and let $F(A)$ be the free group generated by A. For every $\alpha \in A$, let $p_\alpha : F(A) \to F\{\alpha\} \cong \mathbb{Z}$ denote the projection. Prove that

$$[F(A), F(A)] = \bigcap_{\alpha \in A} \operatorname{Ker} p_\alpha,$$

where $[F(A), F(A)]$ is the commutator subgroup of $F(A)$. Note that $F(A) \cong \ast_{\alpha \in A} F\{\alpha\}$.

2.31. Let G be any group and let X be an Eilenberg–Mac Lane space of type $(G, 1)$. Prove that X is an H-space if and only if G is abelian. (You may assume existence and basic properties of $K(G, 1)$'s.)

2.32. (*) (†) Prove that $H_{n+1}(K(G, n)) = 0$, for $n > 1$.

2.33. Given a sequence of abelian groups $G_1, G_2, \ldots$, show that there exists a path-connected CW complex X such that $H_i(X) = G_i$ for all i. Show a similar result for homotopy groups.

2.34. (†) If G is an abelian group and $n \geqslant 1$, prove that the Hurewicz homomorphism $h_n : \pi_n(K(G, n)) \to H_n(K(G, n))$ is an isomorphism.

2.35. (*) (†) If $f, g : X \to K(G, n)$ are maps and X is $(n-1)$-connected, $n \geqslant 1$, then prove that $(f + g)_* = f_* + g_* : H_i(X) \to H_i(K(G, n))$ provided $i < 2n - 1$.

2.36. In analogy to $\rho : H^n(X; G) \to H^n_{\text{sing}}(X; G)$ defined after Remark 2.5.11, define a homomorphism $\rho' : \pi_n(X; G) \to H_n(X; G)$ and show that ρ' is the Hurewicz homomorphism when $G = \mathbb{Z}$.

2.37. (Cf. Lemma 2.5.13) If X is a space, then $[X, S^1]$ is an abelian group since S^1 is a commutative topological group. Show that the group $[X, S^1]$ contains no non-zero elements of finite order. (It may be helpful to use the covering space $\mathbb{R} \to S^1$.)

2.38. Prove the following generalization of the Hopf classification theorem. If X is a CW complex of dimension $\leqslant n$, then there is a bijection between $[X, M(G,n)]$ and $H^n(X;G)$. Formulate and prove the dual result.

2.39. Consider the natural map $j : X \vee Y \to X \times Y$.

1. Use the characterization of $X \vee Y$ as a categorical coproduct in $HoTop_*$ and the characterization of $X \times Y$ as a categorical product in $HoTop_*$ to define $[j]$.
2. Show that $[j]$ is self-dual.

Chapter 3
Cofibrations and Fibrations

3.1 Introduction

The notions of cofibration and fibration are central to homotopy theory. We show that the defining property of a cofiber inclusion map $i : A \to X$ is equivalent to the homotopy extension property of the pair (X, A). Thus the inclusion map of a subcomplex into a CW complex is a cofiber map, and so this concept is widespread in topology. We study cofiber maps in Section 3.2 and introduce pushout squares and mapping cones. In Section 3.3 we treat fiber maps as well as pullback squares and homotopy fibers and obtain some of their basic properties. We also consider fiber bundles which are defined by maps $p : E \to B$ for which there is a locally trivializing cover of B. We prove that fiber bundles are fibrations and thus obtain examples of fiber maps by exhibiting fiber bundles. In this way we obtain many diverse fibrations in which spheres, topological groups, Stiefel manifolds and Grassmannians appear. This is done in Section 3.4, and these results will be used in Chapter 5 to calculate homotopy groups. In the last section we introduce the mapping cylinder and its dual and use them to show that any map can be factored into the composition of a cofiber map followed by a homotopy equivalence or into the composition of a homotopy equivalence followed by a fiber map. These are important techniques and highlight the major role of cofiber and fiber maps in homotopy theory.

3.2 Cofibrations

In this chapter we consider both the based and unbased cases. We begin with the definition of a cofiber map.

Definition 3.2.1 A map $f : A \to X$ is called a *cofiber map* if for every space Z, maps $g_0 : A \to Z$ and $h_0 : X \to Z$ and homotopy $g_t : A \to Z$ of g_0 such that $h_0 f = g_0$, there exists a homotopy $h_t : X \to Z$ of h_0 such that $h_t f = g_t$,

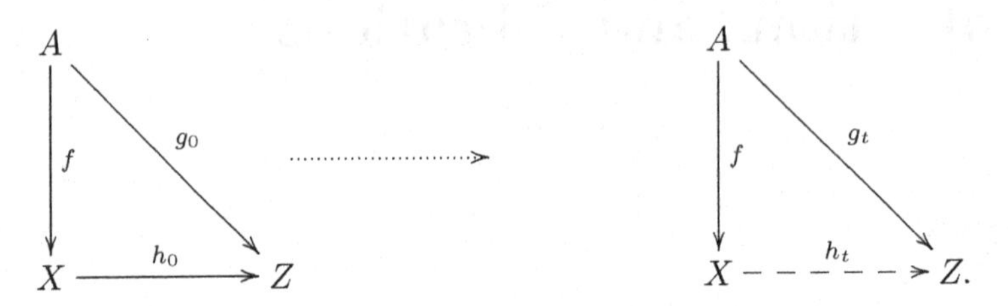

If f is a cofiber map, then $Q = X/f(A)$ is called the *cofiber* of f and the sequence

$$A \xrightarrow{f} X \xrightarrow{q} Q$$

is called a *cofiber sequence*, where q is the projection onto the quotient space. We refer to a cofiber map or to a cofiber sequence as a *cofibration*.

Remark 3.2.2 This definition also exists in the unbased case by taking unbased spaces and free maps and homotopies.

We give some examples of cofibrations next. Recall that the cone $CA = A \times I/(A \times \{1\} \cup \{*\} \times I)$ and $i : A \to CA$ is given by $i(a) = \langle a, 0 \rangle$.

Proposition 3.2.3 *For any space A, the map $i : A \to CA$ is a cofibration.*

Proof. Given $g_t : A \to Z$ and $h_0 : CA \to Z$ such that $h_0 i = g_0$, we seek a homotopy $h_t : CA \to Z$ of h_0 such that $h_t i = g_t$. Let $\partial I = \{0, 1\}$ and set $S = (I \times \{0\}) \cup (\partial I \times I) \subseteq I \times I$. We define $H'' : A \times S \to Z$ by

$$H''(a, s, 0) = h_0\langle a, s \rangle, \quad H''(a, 0, t) = g_t(a), \quad \text{and} \quad H''(a, 1, t) = *,$$

where $a \in A$ and $t, s \in I$. If $r : I \times I \to S$ is the retraction of Example 1.4.2(3), then an extension $H' : A \times I \times I \to Z$ of H'' is defined by $H'(a, s, t) = H''(a, r(s, t))$. By Appendix A, the map H' induces a map $H : CA \times I \to Z$, and $H(\langle a, s \rangle, t)$ is the desired homotopy $h_t\langle a, s \rangle$. □

It is clear from Definition 1.5.12 that a pair (X, A) has the based homotopy extension property if and only if the inclusion map $i : A \to X$ is a cofiber map. From Proposition 1.5.17, we then have the following result.

Proposition 3.2.4 *If (K, L) is a relative CW complex, then the inclusion map $L \to K$ is a cofibration. In particular, if L is a subcomplex of a CW complex K, then the inclusion $L \to K$ is a cofibration.*

This yields many cofibrations. The next lemma gives a few more.

Lemma 3.2.5

1. *For any spaces A and B, the injections $i_1 : A \to A \vee B$ and $i_2 : B \to A \vee B$ are cofiber maps.*

2. *If $f : A \to X$ and $f' : A' \to X'$ are cofiber maps, then $f \vee f' : A \vee A' \to X \vee X'$ is a cofiber map.*

Thus we have examples of cofiber sequences:

- $A \xrightarrow{i} CA \xrightarrow{q} \Sigma A$, for any space A.
- $L \longrightarrow K \xrightarrow{q} K/L$, for any relative CW complex (K, L).
- $A \xrightarrow{i_1} A \vee B \xrightarrow{q_2} B$, for any spaces A and B.
- $A \vee A' \xrightarrow{f \vee f'} X \vee X' \xrightarrow{q \vee q'} Q \vee Q'$, where the following cofiber sequences are given $A \xrightarrow{f} X \xrightarrow{q} Q$ and $A' \xrightarrow{f'} X' \xrightarrow{q'} Q'$.

We note that in all of these examples, the cofiber map is an embedding, that is, a homeomorphism onto its image. This is true for all cofibrations.

Proposition 3.2.6 *If $f : X \to Y$ is a cofiber map, then f is an embedding.*

Proof. If $i : X \to CX$ is the inclusion, then $* \simeq i$ since CX is contractible. Therefore we have the diagram

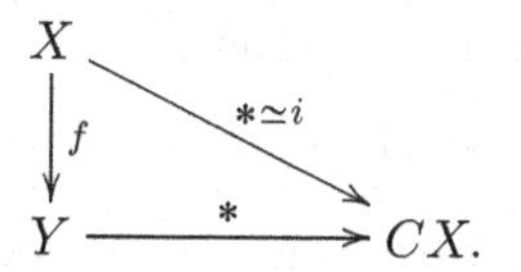

Since f is a cofibration, there is a map $h : Y \to CX$ such that $hf = i$. We consider the maps $f' : X \to f(X)$, $h' : f(X) \to i(X)$ and $i' : X \to i(X)$ induced by f, h and i, respectively. Since $i' = h'f'$ is a homeomorphism, it follows that f' is a homeomorphism. □

This shows that the cofibration property and the based homotopy extension property are essentially the same. More precisely, if $f : X \to Y$ is a cofiber map, then the inclusion of $f(X)$, the image of f, in Y has the based homotopy extension property and $X \cong f(X)$. We then obtain the following consequence of Proposition 1.5.18.

Proposition 3.2.7 *If $A \xrightarrow{f} X \xrightarrow{q} Q$ is a cofiber sequence and A is contractible, then q is a homotopy equivalence.*

Next we consider pushouts.

Definition 3.2.8 Given spaces and maps

$$Y \xleftarrow{g} A \xrightarrow{f} X \qquad (*)$$

a *pushout* of $(*)$ consists of a space P and maps $u : X \to P$ and $v : Y \to P$ such that $uf = vg$. In addition, we require the following. If Z is any space and if $r : X \to Z$ and $s : Y \to Z$ are maps such that $rf = sg$, then there is a unique map $t : P \to Z$ such that $tu = r$ and $tv = s$,

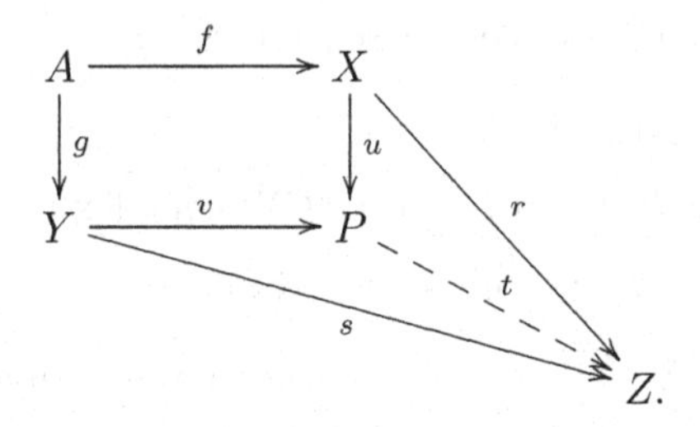

We call either P or the triple (P, u, v), the pushout of $(*)$. The square diagram above is then called a *pushout square.*

We show that any two pushouts of $(*)$ are homeomorphic. Suppose that (P, u, v) and (P', u', v') are both pushouts of $(*)$. Since P is a pushout, there is a map $t : P \to P'$ such that $tu = u'$ and $tv = v'$. Since P' is a pushout, there is a map $t' : P' \to P$ such that $t'u' = u$ and $t'v' = v$. Therefore $t'tu = u$ and $t'tv = v$. By the uniqueness of pushout maps, $t't = \mathrm{id}_P$. Similarly $t't = \mathrm{id}_{P'}$, and so t is a homeomorphism with inverse t'.

We now show the existence of pushouts.

Proposition 3.2.9 *Given* $(*)$, *there exists a pushout* (P, u, v).

Proof. Consider $X \vee Y$, regarded as a subspace of $X \times Y$, and introduce the equivalence relation on $X \vee Y$ defined by $(f(a), *) \sim (*, g(a))$, for every $a \in A$. Set $P = X \vee Y/\sim$ and let $q : X \vee Y \to P$ be the quotient map. Define u and v by $u = qi_1$ and $v = qi_2$, where $i_1 : X \to X \vee Y$ and $i_2 : Y \to X \vee Y$ are the two injections. Clearly $uf = vg$. Now we show that (P, u, v) is a pushout of $(*)$. If $r : X \to Z$ and $s : Y \to Z$ are maps with $rf = sg$, then there is a map $\{r, s\} : X \vee Y \to Z$ and

$$\{r, s\}(f(a), *) = rf(a) = sg(a) = \{r, s\}(*, g(a)).$$

Thus $\{r, s\}$ induces $t : P \to Z$ such that $tu = r$ and $tv = s$. To prove uniqueness of t, let $m : P \to Z$ be a map such that $mu = r$ and $mv = s$. Then

$$mqi_1 = mu = r = tu = tqi_1,$$

and similarly, $mqi_2 = tqi_2$. Therefore, $mq = tq$, and so $m = t$. □

Proposition 3.2.10 *Let*

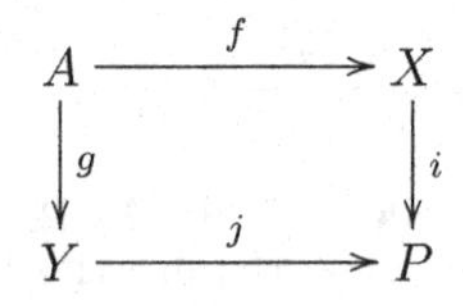

be a pushout square. If g is a cofiber map, then so is i. In this case, j induces a homeomorphism $j' : Y/g(A) \to P/i(X)$ of cofibers.

Proof. Consider the diagram

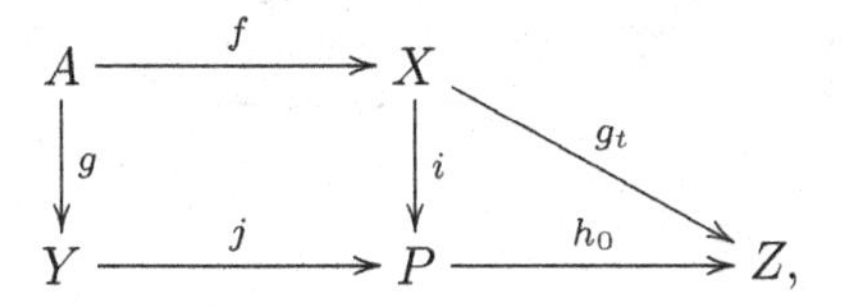

where g_t is a homotopy with $g_0 = h_0 i$. Then $g_0 f = h_0 j g$ and, because g is a cofibration, there is a homotopy $k_t : Y \to Z$ such that $k_t g = g_t f$ and $k_0 = h_0 j$. Then g_t and k_t induce a map $l_t : P \to Z$ such that $l_t i = g_t$ and $l_t j = k_t$. Then l_t is a homotopy since the map $L : P \times I \to Z$ obtained from l_t is continuous by Exercise 3.4. Furthermore, $l_0 i = g_0 = h_0 i$ and $l_0 j = k_0 = h_0 j$. By the uniqueness property of the pushout, $l_0 = h_0$, and so i is a cofiber map. For the second assertion of the proposition, note that j induces $j' : Y/g(A) \to P/i(X)$. We regard the pushout P as defined in the proof of Proposition 3.2.9. Then P/iX is obtained from $X \vee Y$ from the following relations: $(f(a), *) \sim (*, g(a))$, for every $a \in A$ and $(x, *) \sim *$, for every $x \in X$ (Appendix A). Thus $Y/g(A) \cong P/i(X)$, and the homeomorphism is j' defined by $j'\langle y \rangle = \langle *, y \rangle$ for $y \in Y$. □

We next consider a special case of a pushout.

Definition 3.2.11 Let $f : X \to Y$ be any map and consider the pushout square

$$\begin{array}{ccc} X & \xrightarrow{f} & Y \\ \downarrow{\scriptstyle i} & & \downarrow{\scriptstyle k} \\ CX & \xrightarrow{j} & P. \end{array}$$

We call P the *space obtained by attaching a cone on X by f* or, more briefly, the *mapping cone of f* or the *homotopy cofiber of f*. We write $P = C_f = Y \cup_f CX$. Because $X \longrightarrow CX \longrightarrow \Sigma X$ is a cofiber sequence,

$$Y \xrightarrow{k} C_f \xrightarrow{q} \Sigma X$$

is a cofiber sequence by Proposition 3.2.10 called the *principal cofibration induced by f* (see Figure 3.1).

The mapping cone is an important construction and we next present some of its properties. The first result is a generalization of Lemma 1.4.10 and Proposition 1.4.9.

Proposition 3.2.12 *Let $f : X \to Y$ be a map and C_f the mapping cone of f. Then a map $g : Y \to Z$ can be extended to C_f if and only if $gf \simeq * : X \to Z$.*

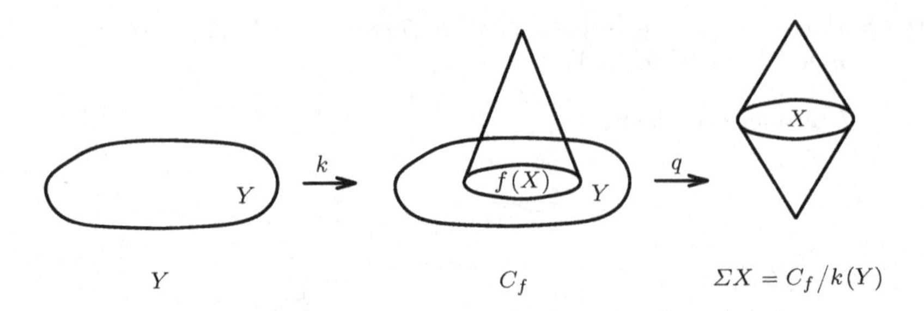

Figure 3.1

Proof. Let $k : Y \to C_f$ be the inclusion and let there exist $h : C_f \to Z$ such that $hk = g$,

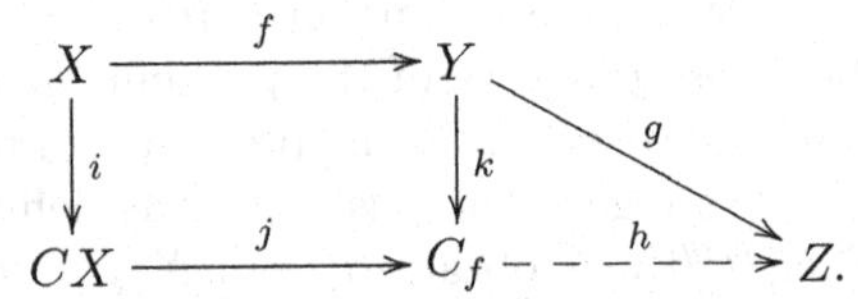

Then $gf = hkf = hji$. But $gf \simeq *$ because hji factors through the contractible space CX.

Now suppose $gf \simeq * : X \to Z$. By Proposition 1.4.9, gf can be extended to CX. Thus there is a map $F : CX \to Z$ such that $Fi = gf$. By Definition 3.2.8, F and g determine a map $h : C_f \to Z$ such that $hk = g$. □

Next we show how to obtain induced maps of mapping cones.

Proposition 3.2.13 *Given a homotopy-commutative diagram*

$$\begin{array}{ccc} X & \xrightarrow{\alpha} & X' \\ \downarrow{\scriptstyle f} & & \downarrow{\scriptstyle f'} \\ Y & \xrightarrow{\beta} & Y' \end{array}$$

with homotopy $F : X \times I \to Y'$ such that $\beta f \simeq_F f'\alpha$. Then there exists a map $\Phi_F : C_f \to C_{f'}$ such that in the following diagram

$$\begin{array}{ccccc} Y & \xrightarrow{k} & C_f & \xrightarrow{q} & \Sigma X \\ \downarrow{\scriptstyle \beta} & & \downarrow{\scriptstyle \Phi_F} & & \downarrow{\scriptstyle \Sigma\alpha} \\ Y' & \xrightarrow{k'} & C_{f'} & \xrightarrow{q'} & \Sigma X' \end{array}$$

the left square is commutative and the right square is homotopy-commutative, where k and k' are inclusions and q and q' are projections.

Proof. We define $\Phi_F\langle y\rangle = \langle \beta(y)\rangle$ and

$$\Phi_F\langle x,t\rangle = \begin{cases} \langle F(x,2t)\rangle & \text{if } 0 \leqslant t \leqslant \frac{1}{2} \\ \langle \alpha(x), 2t-1\rangle & \text{if } \frac{1}{2} \leqslant t \leqslant 1, \end{cases}$$

for all $y \in Y$, $x \in X$, and $t \in I$. □

For any space W, let $i_W : W \to CW$ be the inclusion. We next show the compatibility of the mapping cone construction with suspensions.

Proposition 3.2.14 *If $f : A \to X$ is a map and $\Sigma f : \Sigma A \to \Sigma X$ is its suspension, then there is a homeomorphism $\theta : C_{\Sigma f} \to \Sigma C_f$ such that the following diagram commutes*

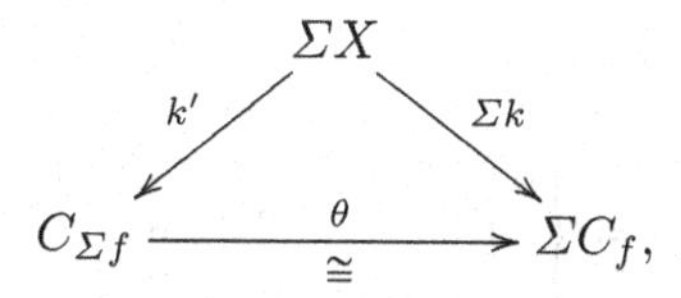

where k and k' are inclusions.

Proof. Consider the pushout square

$$\begin{array}{ccc} A & \xrightarrow{f} & X \\ \big\downarrow{\scriptstyle i_A} & & \big\downarrow{\scriptstyle u} \\ CA & \xrightarrow{v} & C_f. \end{array}$$

By Exercise 3.1,

$$\begin{array}{ccc} \Sigma A & \xrightarrow{\Sigma f} & \Sigma X \\ \big\downarrow{\scriptstyle \Sigma i_A} & & \big\downarrow{\scriptstyle \Sigma u} \\ \Sigma CA & \xrightarrow{\Sigma v} & \Sigma C_f \end{array}$$

is a pushout square. There is a homeomorphism $\rho : C\Sigma A \to \Sigma CA$ defined by $\rho\langle\langle a,t\rangle,u\rangle = \langle\langle a,u\rangle,t\rangle$, where $a \in A$ and $t,u \in I$. We then replace ΣCA by $C\Sigma A$ and Σi_A by $i_{\Sigma A}$ in the previous pushout square and obtain the pushout square

$$\begin{array}{ccc} \Sigma A & \xrightarrow{\Sigma f} & \Sigma X \\ \big\downarrow{\scriptstyle i_{\Sigma A}} & & \big\downarrow{\scriptstyle \Sigma u} \\ C\Sigma A & \xrightarrow{(\Sigma v)\rho} & \Sigma C_f. \end{array}$$

But $C_{\Sigma f}$ is the pushout of $C\Sigma A \xleftarrow{i_{\Sigma A}} \Sigma A \xrightarrow{\Sigma f} \Sigma X$. Because any two pushouts of the same maps are homeomorphic, $C_{\Sigma f} \cong \Sigma C_f$, with homeomorphism $\theta : C_{\Sigma f} \to \Sigma C_f$ such that $\theta\langle x,t\rangle = \langle x,t\rangle$ and $\theta\langle\langle a,t\rangle,u\rangle = \langle\langle a,u\rangle,t\rangle$, for $x \in X$, $a \in A$ and $t,u \in I$. The proposition now follows. □

We conclude this section by showing that the homotopy type of the mapping cone depends only on the homotopy class of the map.

Proposition 3.2.15 *If $f_0 \simeq f_1 : X \to Y$, then $C_{f_0} \simeq C_{f_1}$.*

Proof. We use $\langle - \rangle_0$ to denote elements of C_{f_0} and $\langle - \rangle_1$ to denote elements of C_{f_1}. Let F be the homotopy between f_0 and f_1 and let $\bar{F}$ defined by $\bar{F}(x,t) = F(x,1-t)$ be a homotopy between f_1 and f_0. We apply Proposition 3.2.13 with $\alpha = \mathrm{id}_X$ and $\beta = \mathrm{id}_Y$ and define $\rho : C_{f_0} \to C_{f_1}$ by $\rho = \Phi_F$. A similar definition using $\bar{F}$ gives $\sigma = \Phi_{\bar{F}} : C_{f_1} \to C_{f_0}$. We show that ρ is a homotopy equivalence with homotopy inverse σ. The map $\sigma\rho : C_{f_0} \to C_{f_0}$ is given by $\sigma\rho\langle y\rangle_0 = \langle y\rangle_0$ and

$$\sigma\rho\langle x,s\rangle_0 = \begin{cases} \langle F(x,2s)\rangle_0 & \text{if } 0 \leqslant s \leqslant \frac{1}{2} \\ \langle \bar{F}(x,4s-2)\rangle_0 & \text{if } \frac{1}{2} \leqslant s \leqslant \frac{3}{4}, \\ \langle x,4s-3\rangle_0 & \text{if } \frac{3}{4} \leqslant s \leqslant 1. \end{cases}$$

Then a homotopy $G : C_{f_0} \times I \to C_{f_0}$ between $\sigma\rho$ and id is defined by $G(\langle y\rangle_0, t) = \langle y\rangle_0$ and

$$G(\langle x,s\rangle_0, t) = \begin{cases} \langle F(x,2s)\rangle_0 & \text{if } 0 \leqslant s \leqslant \frac{1-t}{2} \\ \langle \bar{F}(x,4s+3t-2)\rangle_0 & \text{if } \frac{1-t}{2} \leqslant s \leqslant \frac{3-3t}{4}, \\ \langle x,(4s+3t-3)/(1+3t)\rangle_0 & \text{if } \frac{3-3t}{4} \leqslant s \leqslant 1. \end{cases}$$

A homotopy between $\rho\sigma$ and id is defined analogously. This completes the proof. □

Remark 3.2.16

1. We attempt to motivate the definition of the previous homotopy G. Fix $x \in X$ and define paths $l, \bar{l}$ and i in C_{f_0} by $l(s) = \langle f_s(x)\rangle$, $\bar{l}(s) = \langle \bar{f}_s(x)\rangle$, and $i(s) = \langle x,s\rangle$, where $f_s(x) = F(x,s)$ and $\bar{f}_s(x) = \bar{F}(x,s)$. Then the homotopy $G(\langle x,-\rangle,t)$ for fixed x and t is a path in C_{f_0} which consists of first traversing l from $l(0)$ to $l(1-t)$ $(0 \leqslant s \leqslant (1-t)/2)$ and then traversing l backwards from $l(1-t)$ to $l(0)$ $((1-t)/2 \leqslant s \leqslant (3-3t)/4)$ and then traversing i from $i(0)$ to $i(1)$. At $t = 0$, this is $\sigma\rho\langle x,s\rangle$. As t increases from 0 to 1, the first two paths get shorter until at $t = 1$ they are just the basepoint, and we have $i(s) = \langle x,s\rangle$.
2. If we regard Y as a subspace of both mapping cones C_{f_0} and C_{f_1}, then C_{f_0} and C_{f_1} have the same homotopy type rel Y by the proof of Proposition 3.2.15 (see Figure 3.2).

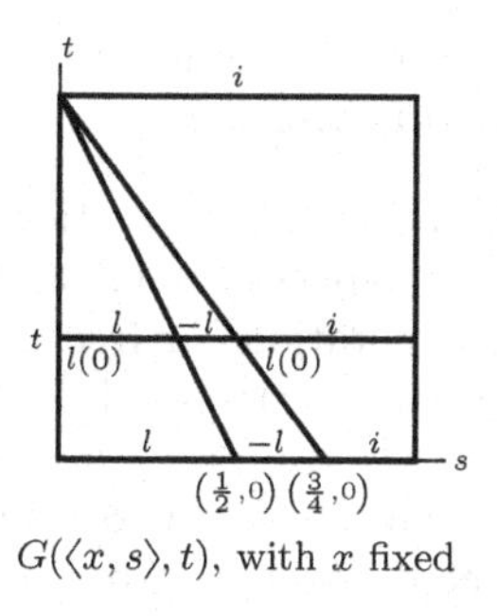

$G(\langle x, s\rangle, t)$, with x fixed

Figure 3.2

3.3 Fibrations

Most of our discussion of fibrations is just a straightforward dualization of the previous section on cofibrations. For this material, we state these results and occasionally indicate a proof, but our treatment is sketchy. However, we give details for the material on fibrations that are not duals of results in Section 3.2. We begin with some definitions.

Definition 3.3.1 A map $p: E \to B$ has the *homotopy lifting property* or the *covering homotopy property* with respect to a space W if for maps $g_0 : W \to B$ and $h_0 : W \to E$ and homotopy $g_t : W \to B$ of g_0 such that $ph_0 = g_0$, there exists a homotopy $h_t : W \to E$ of h_0 such that $ph_t = g_t$. This property is sometimes referred to as the HLP or the CHP

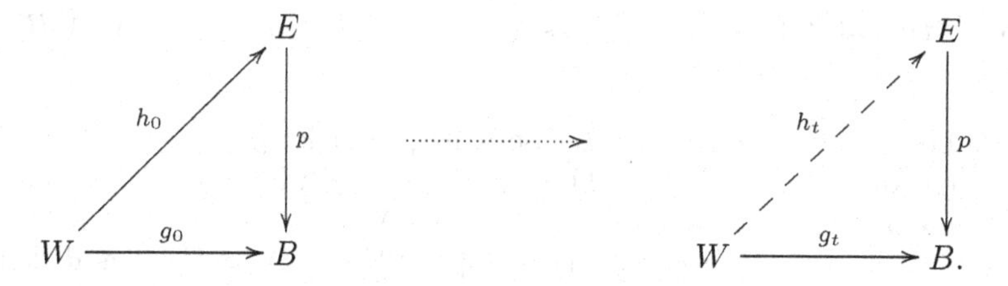

The homotopy h_t is called a *covering homotopy* of g_t. If $p : E \to B$ has the homotopy lifting property for all spaces W, then p is called a *fiber map*. This is sometimes referred to as a *Hurewicz fiber map*. If $p : E \to B$ has the homotopy lifting property for all CW complexes W, then p is called a *weak fiber map* and is sometimes referred to as a *Serre fiber map*. If $p : E \to B$ is a fiber map or weak fiber map, then $F = p^{-1}(*) \subseteq E$ is called the *fiber*, E is called the *total space* and B is called the *base space*. The sequence

$$F \xrightarrow{i} E \xrightarrow{p} B$$

is called a *fiber sequence*, where i is the inclusion. In addition, we refer to a fiber map or fiber sequence as a *fibration.*

We also consider unbased fibrations in this section. The definition is the same as Definition 3.3.1 except that spaces, maps, and homotopies are unbased. We refer to these as unbased fibrations or free fiber maps. Similarly we can consider unbased weak fibrations or free weak fiber maps. If $p : E \to B$ is an unbased fibration or unbased weak fibration and $b \in B$, then $p^{-1}(b)$ is called the *fiber over b*. It can be shown that any two fibers have the same homotopy type when B is path-connected. We now give some examples of fibrations.

- If B and F are any spaces, then $F \xrightarrow{j_2} B \times F \xrightarrow{p_1} B$ is a fiber sequence, where j_2 is the inclusion and p_1 is the projection. Similarly,
$$B \xrightarrow{j_1} B \times F \xrightarrow{p_2} F$$
is a fiber sequence.These are called *trivial fibrations*. Less trivial examples are given below.
- If $F \longrightarrow E \xrightarrow{p} B$ is fiber sequence and $A \subseteq B$, then the following sequence is a fiber sequence
$$F \longrightarrow p^{-1}(A) \longrightarrow A.$$

Next we give an important source of examples.

Let $A, B \subseteq X$ and consider the path space X^I. Define $E(X; A, B) \subseteq X^I$ by $E(X; A, B) = \{l \in X^I \mid l(0) \in A \text{ and } l(1) \in B\}$ and define $p : E(X; A, B) \to A \times B$ by $p(l) = (l(0), l(1))$.

Proposition 3.3.2 *For any spaces* $A, B \subseteq X$, *the map* $p : E(X; A, B) \to A \times B$ *is a fiber map.*

Proof. Let $f_0 : W \to E(X; A, B)$ be a map and let $g_s : W \to A \times B$ a homotopy such that $pf_0 = g_0$. Then $g_s = (g'_s, g''_s)$, where $g'_s : W \to A$ and $g''_s : W \to B$. Define the adjoint $h : W \times I \to X$ of f_0 by $h(w, t) = f_0(w)(t)$. for $w \in W$ and $t \in I$. Set $S = (I \times \{0\}) \cup (\partial I \times I) \subseteq I \times I$ and define $H' : W \times S \to X$ by

$$H'(w, t, 0) = h(w, t), \quad H'(w, 0, s) = g'_s(w), \quad \text{and} \quad H'(w, 1, s) = g''_s(w).$$

Since there is a retraction $r : I \times I \to S$ by Example 1.4.2(3), we define $H : W \times I \times I \to X$ by

$$H(w, t, s) = H'(w, r(t, s)).$$

We then obtain a homotopy $f_s : W \to E(X; A, B)$ of f_0 given by $f_s(w)(t) = H(w, t, s)$ which is a covering homotopy of g_s □

Remark 3.3.3 If we take $B = \{*\}$ and $A = X$ in Proposition 3.3.2, we obtain the fiber sequence

$$\Omega X \longrightarrow E_1 X \xrightarrow{p_0} X,$$

where $E_1 X = EX$ is the space of paths in X that end at the basepoint and $p_0(l) = l(0)$. If we take $A = \{*\}$ and $B = X$ in Proposition 3.3.2, we obtain the fiber sequence

$$\Omega X \longrightarrow E_0 X \xrightarrow{p_1} X,$$

where $E_0 X$ is the space of paths in X that begin at the basepoint and $p_1(l) = l(1)$.

Definition 3.3.4 Let $p : E \to B$ be a map and (K, L) a relative CW complex with inclusion map $i : L \to K$. Then p has the *covering homotopy extension property* with respect to (K, L) if for any maps $g_0 : K \to B$, $h_0 : L \to E$ and $f_0 : K \to E$ such that $f_0 i = h_0$ and $pf_0 = g_0$ and homotopies $g_t : K \to B$ of g_0 and $h_t : L \to E$ of h_0 such that $ph_t = g_t i$, there exists a homotopy $f_t : K \to E$ of f_0 such that $f_t i = h_t$ and $pf_t = g_t$. This property is sometimes called the CHEP

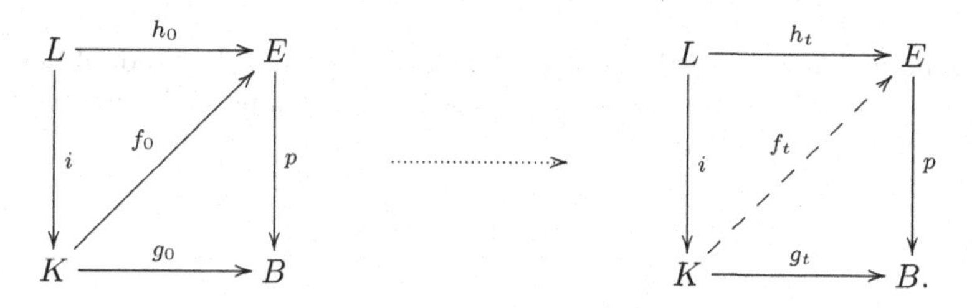

Since i is a cofiber map, there is a homotopy f_t' of f_0 such that $f_t' i = h_t$. If p is a fiber map, then there is homotopy f_t'' of f_0 such that $pf_t'' = g_t$. The CHEP says that we can take $f_t' = f_t''$ provided the given homotopies g_t and h_t are compatible. Note that if i is a cofiber map and p is a fiber map, the CHEP is a self-dual property. The CHEP is stronger than the CHP and can be formulated in either the based or unbased case.

We next characterize unbased weak fibrations by several equivalent conditions.

Proposition 3.3.5 *Let $p : E \to B$ be a free map. Then the following are equivalent:*

1. *The map p is an unbased weak fibration.*
2. *The map p has the unbased covering homotopy property with respect to n-balls E^n, for all $n \geqslant 0$.*
3. *The function p has the unbased covering homotopy extension property for all relative CW complexes (K, L).*

4. Let (K, L) be a relative CW complex with L an unbased strong deformation retract of K. If there are free maps $h : L \to E$ and $g : K \to B$ such that $ph = g|L$, then there is a free map $f : K \to E$ such that $f|L = h$ and $pf = g$

Proof. (1) $\Longrightarrow$ (2) is obvious.

(2) $\Longrightarrow$ (3) First note that there is a homeomorphism $(E^n \times I, E^n \times \{0\}) \cong (E^n \times I, E^n \times \{0\} \cup S^{n-1} \times I)$ of pairs. Figure 3.3 shows this homeomorphism for $n = 1$ (see also [14, p. 451]).

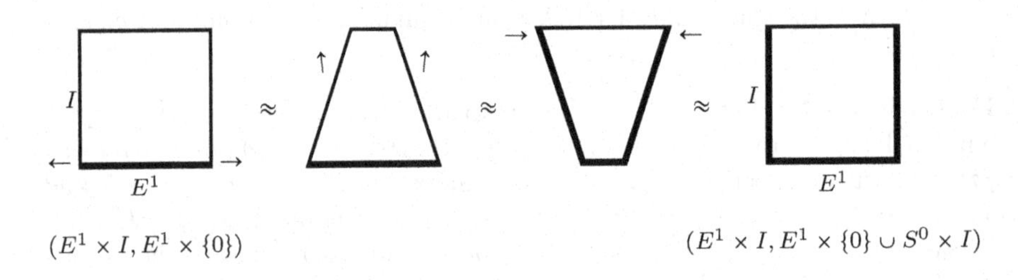

Figure 3.3

The general definition of the homeomorphism is given in the proof of Lemma 11.6 in [37, p. 81]. Now let (K, L) be a relative CW complex. We assume that there is a commutative square

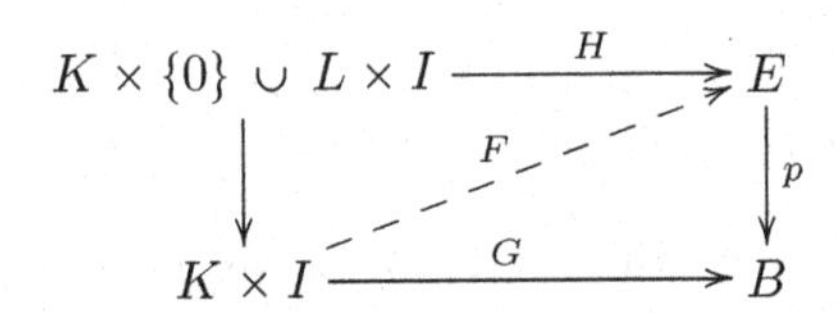

and we want to define F so that the triangles are commutative. The proof is by induction over the spaces $K \times \{0\} \cup (K, L)^n \times I$, where $(K, L)^n$ is the relative nth skeleton. We regard E^0 as a one point space and consider a relative vertex $v \in K - L$. Then by the covering homotopy property with respect to the space $\{v\}$, there is a map $F : \{v\} \times I \to E$ that extends $H|\{v\} \times \{0\}$ and such that $pF = G|\{v\} \times I$. We do this for all relative vertices $v \in K - L$, and this gives the result for $n = 0$. Now assume F defined on $K \times \{0\} \cup (K, L)^{n-1} \times I$ and let σ be a relative n-cell in $(K, L)^n$. Using the characteristic function $\Phi : E^n \to \sigma$, we obtain functions $G' = G(\Phi \times \mathrm{id}) : E^n \times I \to B$ and $F' = F(\Phi \times \mathrm{id}) : (E^n \times \{0\}) \cup (S^{n-1} \times I) \to E$ such that $pF' = G'|(E^n \times \{0\}) \cup (S^{n-1} \times I)$. Since $(E^n \times I, E^n \times \{0\}) \cong (E^n \times I, \big(E^n \times \{0\}) \cup (S^{n-1} \times I)\big)$, this yields a commutative diagram

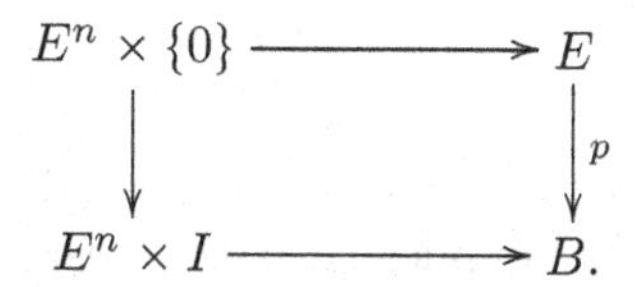

Then the covering homotopy property with respect to E^n and the existence of the homeomorphism $(E^n \times I, E^n \times \{0\}) \cong (E^n \times I, (E^n \times \{0\}) \cup (S^{n-1} \times I))$ imply that there exists $F'' : E^n \times I \to E$ such that $pF'' = G'$ and $F''|(E^n \times \{0\}) \cup (S^{n-1} \times I) = F'$. This induces $F_\sigma : \sigma \times I \to E$ which is an extension of $F|(\sigma \times \{0\}) \cup (\partial\sigma \times I)$ such that $pF_\sigma = G|\sigma \times I$. By doing this for all relative n-cells σ in $(K, L)^n$, we extend F to $K \times \{0\} \cup (K, L)^n \times I$. This completes the induction and yields the desired free homotopy $F : K \times I \to E$.

(3) $\Longrightarrow$ (4) Let $i : L \to K$ be the inclusion and let $r : K \to L$ be the free retraction. Denote by $g_t : K \to K$ the free homotopy of ir to id rel L. We assume that there is a commutative diagram

$$\begin{array}{ccc} L & \xrightarrow{h} & E \\ {\scriptstyle i}\downarrow & & \downarrow{\scriptstyle p} \\ K & \xrightarrow{g} & B. \end{array}$$

Now consider the free homotopy $h_t : L \to E$ defined by $h_t(x) = h(x)$, for $x \in L$ and set $f_0 = hr$. Then $ph_t = gg_ti$. By (3), there is a free homotopy $f_t : K \to E$ such that $f_ti = h_t$ and $pf_t = gg_t$. If we set $f = f_1$, then f satisfies the conclusion of (4).

(4) $\Longrightarrow$ (1) Given a CW complex W and a diagram

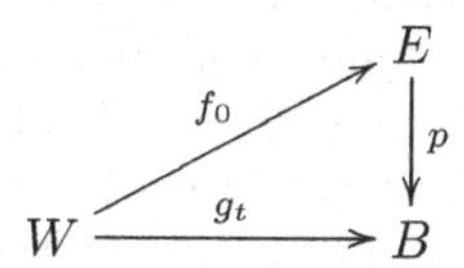

with $pf_0 = g_0$. We obtain

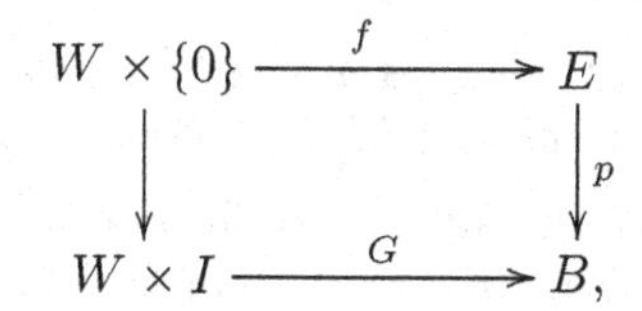

where $f(w, 0) = f_0(w)$ and $G(w, t) = g_t(w)$, for $w \in W$ and $t \in I$. The space $W \times \{0\}$ is a strong deformation retract of $W \times I$. Therefore, by (4), there is a function $F : W \times I \to E$ such that $pF = G$ and $F|W \times \{0\} = f$. Then F is the desired free homotopy. □

We then obtain the following corollary.

Corollary 3.3.6 *If $p : E \to B$ is a weak fibration, then p satisfies the (based)* CHEP.

Proof. Because every weak fibration is an unbased weak fibration (Exercise 3.9) and the unbased CHEP implies the CHEP, the result follows. □

Another consequence of Proposition 3.3.5 is the following corollary.

Corollary 3.3.7 *Let E and B be spaces and let $p : E \to B$ be a (based) map. If p is an unbased weak fibration, then p is a weak fibration.*

Proof. Assume that W is a based CW complex and that there are maps $h_0 : W \to E$ and $g_0 : W \to B$ and a based homotopy g_t of g_0, such that $ph_0 = g_0$. Then $(W, \{*\})$ is a relative CW complex and we apply the CHEP to the homotopies $h_t : \{*\} \to E$ and $g_t : W \to B$. Hence there is a homotopy $f_t : W \to E$ that covers g_t such that $f_t(*) = *$. Therefore f_t is a (based) homotopy. □

We next give a way to recognize when certain maps are weak fibrations.

Definition 3.3.8 Let $p : E \to B$ be a free map of unbased spaces. Then a *fiber bundle* consists of p and an unbased space F such that there exists an open cover $\{U_\alpha\}_{\alpha \in A}$ of B with the following property. For every $\alpha \in A$, there is a homeomorphism $\theta_\alpha : p^{-1}(U_\alpha) \to U_\alpha \times F$ such that the diagram

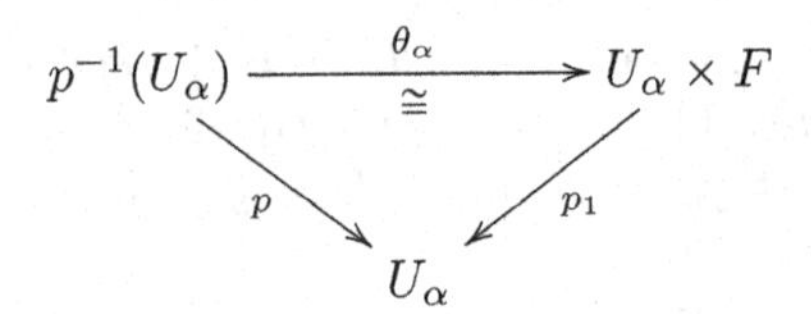

commutes, where p_1 is a projection onto the first factor. Then F is called the *fiber*, B the *base space*, E the *total space*, and p the *bundle map* of the fiber bundle. The homeomorphism θ_α is called a *local trivialization* and the cover $\{U_\alpha\}_{\alpha \in A}$ is called a *locally trivializing cover.* A fiber bundle is sometimes called a *locally trivial fibration.* If $p : E \to B$ is a bundle map with fiber F, we say that $F \longrightarrow E \xrightarrow{p} B$ is a fiber bundle.

As one might expect, a fiber bundle has the covering homotopy property.

Proposition 3.3.9 *If $p : E \to B$ is a based map that is a bundle map with fiber F, then p is a weak fibration with fiber homeomorphic to F.*

Proof. By Corollary 3.3.7, it suffices to show that p is an unbased weak fibration. By Proposition 3.3.5 it suffices to prove that p satisfies the covering homotopy property with respect to all n-balls E^n. Given the diagram

$$\begin{array}{ccc} E^n \times \{0\} & \xrightarrow{h} & E \\ \downarrow & \overset{H}{\nearrow} & \downarrow p \\ E^n \times I & \xrightarrow{G} & B \end{array}$$

with commutative square, we seek a homotopy $H : E^n \times I \to E$ such that the triangles in the above diagram are commutative.

We adopt the following terminology. Let $L \subseteq E^n$ and let S be any space with $L \times \{0\} \subseteq S \subseteq E^n \times I$. If $F : S \to E$ is a free map such that $pF = G|S$ and $F|(L \times \{0\}) = h|(L \times \{0\})$, then we call F a *partial lift* of G.

Let $\{U_\alpha\}_{\alpha \in A}$ be a locally trivializing cover of B. If $\theta : E^n \to I^n$ is a homeomorphism, then $\{(\theta \times \mathrm{id})G^{-1}(U_\alpha)\}$ is an open cover of the compact metric space $I^n \times I$. By the Lebesgue covering lemma (Appendix A), there is a positive number λ such that any subset of $I^n \times I$ of diameter $< \lambda$ is contained in some $(\theta \times \mathrm{id})G^{-1}(U_\alpha)$. We then subdivide each factor I of $I^n \times I$ into k subintervals

$$\left[0, \frac{1}{k}\right], \left[\frac{1}{k}, \frac{2}{k}\right], \ldots, \left[\frac{k-1}{k}, 1\right].$$

By choosing k sufficiently large, we have that $I^n \times I$ is a finite CW complex with each cell contained in some $(\theta \times \mathrm{id})G^{-1}(U_\alpha)$. Thus there is a CW decomposition K of E^n and a subdivision of I into k subintervals so that for every cell σ in K and for each $i = 0, 1, \ldots, k-1$,

$$G\Big(\sigma \times \Big[\frac{i}{k}, \frac{i+1}{k}\Big]\Big) \subseteq U_\alpha,$$

for some $\alpha \in A$. Fix i and assume that there exists a partial lift $H_i : K \times [0, i/k] \to E$ of G. We show there is a partial lift $H_{i+1} : K \times [0, (i+1)/k] \to E$ of G by induction on the skeleta K^j of K. The case $j = 0$ is straightforward and hence omitted. Now assume that for some $l > 0$ there is a partial lift $H_{i+1}^{l-1} : K^{l-1} \times [0, (i+1)/k] \to E$ of G such that $H_{i+1}^{l-1}|(K^{l-1} \times [0, i/k]) = H_i|(K^{l-1} \times [0, i/k])$. Let σ be an l-cell of K with boundary $\partial\sigma$ and suppose that $G(\sigma \times [i/k, (i+1)/k]) \subseteq U_\alpha$, for some α. Then H_{i+1}^{l-1} and H_i determine a map

$$H' : \sigma \times \Big\{\frac{i}{k}\Big\} \cup \partial\sigma \times \Big[\frac{i}{k}, \frac{i+1}{k}\Big] \longrightarrow p^{-1}(U_\alpha)$$

such that the square commutes in the following diagram

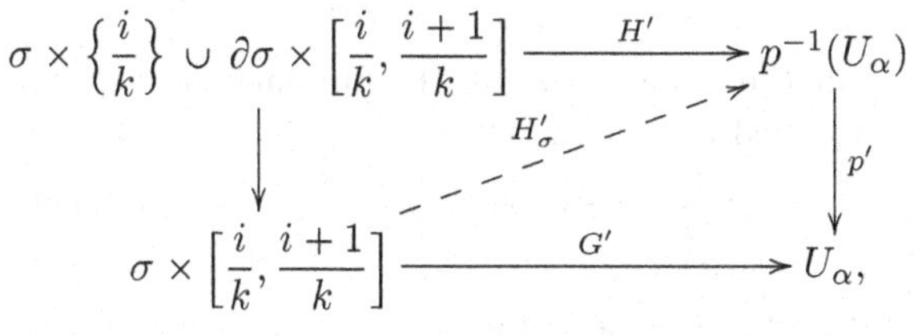

$$\begin{array}{ccc}
\sigma \times \Big\{\frac{i}{k}\Big\} \cup \partial\sigma \times \Big[\frac{i}{k}, \frac{i+1}{k}\Big] & \xrightarrow{H'} & p^{-1}(U_\alpha) \\
\downarrow & \overset{H'_\sigma}{\nearrow} & \downarrow p' \\
\sigma \times \Big[\frac{i}{k}, \frac{i+1}{k}\Big] & \xrightarrow{G'} & U_\alpha,
\end{array}$$

where G' and p' are the restrictions of G and p, respectively. Then p' is an unbased weak fibration by the locally trivializing property and $\sigma \times \{i/k\} \cup \partial\sigma \times [i/k, (i+1)/k]$ is a strong deformation retract of $\sigma \times [i/k, (i+1)/k]$. By Proposition 3.3.5, there exists

$$H'_\sigma : \sigma \times \left[\frac{i}{k}, \frac{i+1}{k}\right] \longrightarrow p^{-1}(U_\alpha)$$

such that the triangles in the above diagram commute. We do this for each l-cell in K and obtain a partial lift $H^l_{i+1} : K^l \times [0,(i+1)/k] \to E$ of G. Because $K = K^n = E^n$, this process yields a lift $H_{i+1} = H^n_{i+1} : K \times [0,(i+1)/k] \to E$ of G. Induction on i then gives the lift H of G. □

We note that it is proved in [83, p. 96] that if $p : E \to B$ is a bundle map with fiber F and paracompact base B, then p is a (Hurewicz) fibration with fiber homeomorphic to F. Moreover, an example is given in [16] of a weak fibration that is not a fibration.

By Proposition 3.3.9 we obtain examples of unbased weak fibrations by giving examples of bundle maps. We do this in the next section. The rest of this section is a straightforward dualization of the latter part of Section 3.2 and we give the definitions and state the propositions, usually without proof. We consider only based spaces and maps. We begin with pullbacks.

Definition 3.3.10 Given spaces and maps

$$Y \xrightarrow{g} A \xleftarrow{f} X, \qquad (**)$$

the *pullback* of $(**)$ consists of a space Q and maps $u : Q \to X$ and $v : Q \to Y$ such that $fu = gv$. In addition, we require the following. If $r : W \to X$ and $s : W \to Y$ are maps such that $fr = gs$, then there is a unique map $t : W \to Q$ such that $ut = r$ and $vt = s$,

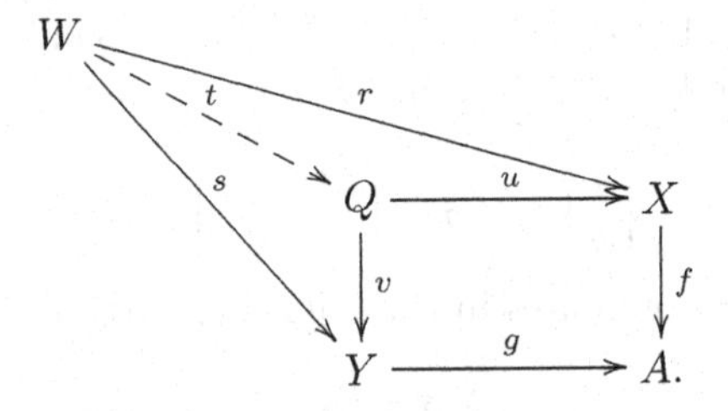

We call either Q or the triple (Q, u, v), the pullback of $(**)$. The rectangular diagram above is called a *pullback square.*

As in the dual case, it is easily seen that any two pullbacks of $(**)$ are homeomorphic.

Proposition 3.3.11 *Given* $(**)$, *there exists a pullback* (Q, u, v).

Proof. Let $Q \subseteq X \times Y$ be the set of pairs (x, y) with $x \in X$ and $y \in Y$ such that $f(x) = g(y)$. Then set $u(x, y) = x$ and $v(x, y) = y$. The verification that (Q, u, v) is a pullback is analogous to the proof of Proposition 3.2.9. □

The following proposition and its proof is dual to the corresponding one in Section 3.2, and hence omitted.

Proposition 3.3.12 *Let*

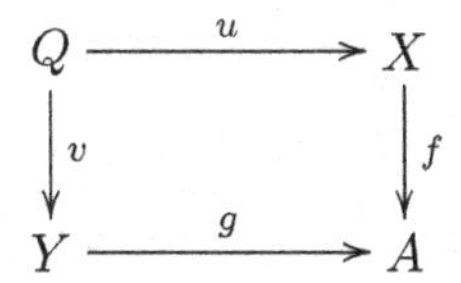

be a pullback square. If f is a fiber map, then so is v. In this case, u induces a homeomorphism of the fiber of v onto the fiber of f.

We next consider the dual of the mapping cone.

Definition 3.3.13 Let $f : X \to Y$ be any map and consider the pullback square

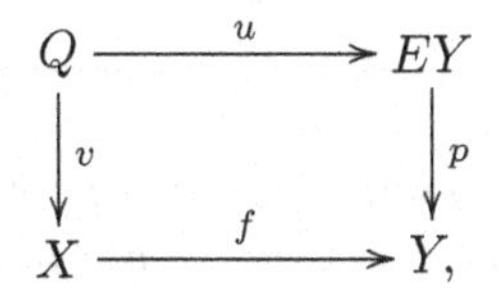

where $p : EY \to Y$ is the path space fibration. Then Q is called the *total space induced by f from the fibration p* or, more commonly, the *homotopy fiber of f* (see Remark 3.5.9). We write $Q = I_f$ and note that by Proposition 3.3.12,

$$\Omega Y \longrightarrow I_f \xrightarrow{v} X$$

is a fiber sequence called the *principal fibration induced by f*.

The next result is a basic one on lifting maps, and we give the proof.

Proposition 3.3.14 *Let $f : X \to Y$ be a map, let $v : I_f \to X$ be the principal fiber map, and let $g : W \to X$ be a map. Then there exists a map $g' : W \to I_f$ such that $vg' = g$ if and only if $fg \simeq *$.*

Proof. Suppose g' exists

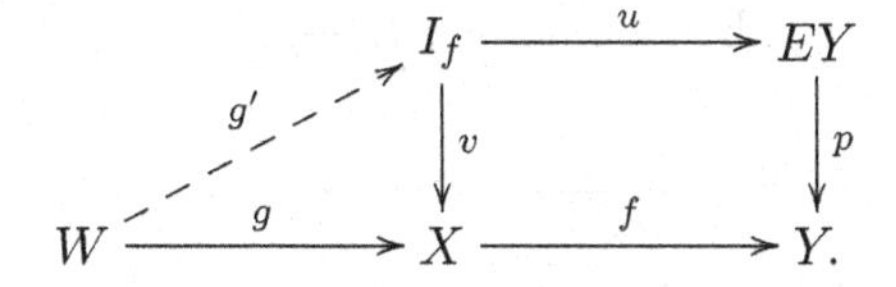

Then $fg = pug'$. This map factors through the contractible space EY, and therefore is nullhomotopic. Now assume that $fg \simeq *$. By Proposition 1.4.9, fg factors through EY and so there is a map $F : W \to EY$ such that $pF = fg$. Then g and F determine a map $g' : W \to I_f$ such that $vg' = g$. □

Next we state without proof the duals of several propositions of Section 3.2. We begin with induced maps of homotopy fibers.

Proposition 3.3.15 *Given a homotopy-commutative diagram*

$$\begin{array}{ccc} X & \xrightarrow{\alpha} & X' \\ \Big\downarrow{\scriptstyle f} & & \Big\downarrow{\scriptstyle f'} \\ Y & \xrightarrow{\beta} & Y' \end{array}$$

such that $f'\alpha \simeq_F \beta f$. Then there exists a map $\Psi_F : I_f \to I_{f'}$ such that in the following diagram

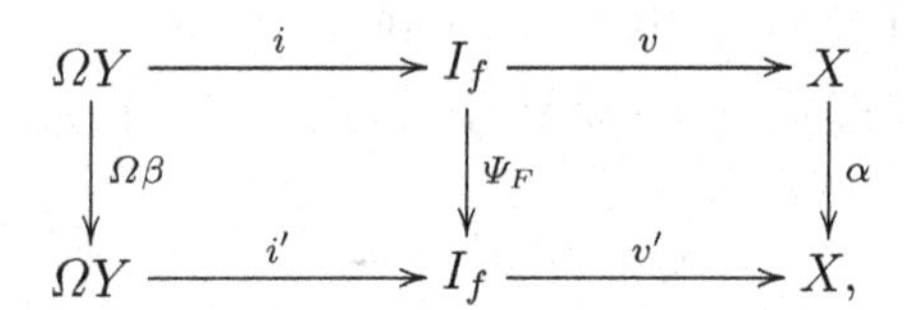

the left square is homotopy-commutative and the right square is commutative, where v and v' are fiber maps and i and i' are inclusions.

Let $v_X : I_f \to X$ denote the principal fiber map.

Proposition 3.3.16 *If $f : X \to Y$ and $\Omega f : \Omega X \to \Omega Y$, then there is a homeomorphism $\theta : \Omega I_f \to I_{\Omega f}$ such that the following diagram commutes*

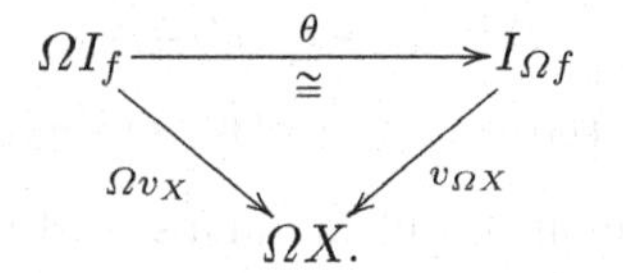

Proposition 3.3.17 *If $f_0 \simeq f_1 : X \to Y$, then $I_{f_0} \simeq I_{f_1}$.*

We conclude this section by discussing a variant of the notion of homotopy that applies to fibrations.

Definition 3.3.18 If $p : E \to B$ and $p' : E' \to B$ are two fiber maps over the same base space B, then a map $f : E \to E'$ is called *fiber-preserving* if $p'f = p$

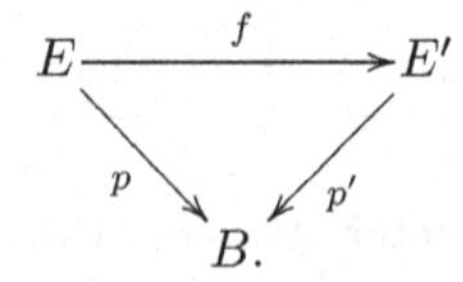

It follows that if F is the fiber of p and F' is the fiber of p', then $f|F : F \to F'$. Two fiber-preserving maps $f, g : E \to E'$ are *fiber-homotopic* if there is a homotopy $f_t : E \to E'$ such that $f_0 = f$, $f_1 = g$ and $p'f_t = p$, for every $t \in I$. A fiber-preseving map $f : E \to E'$ is a *fiber homotopy equivalence* if there is a fiber-preserving map $g : E' \to E$ such that fg is fiber-homotopic to $\mathrm{id}_{E'}$ and gf is fiber-homotopic to id_E.

For many results about fibrations in which it is asserted that two total spaces are homotopy equivalent, it can be shown that the homotopy equivalence is a fiber homotopy equivalence. For now we show that fiber homotopy and homotopy relative to a subspace are essentially dual notions.

Remark 3.3.19 By dualizing the concept of fiber homotopy we obtain the following provisional definitions. Given two cofibrations $i : A \to X$ and $i' : A \to X'$, a map $f : X \to X'$ is called cofiber-preserving if $fi = i'$. Two cofiber-preserving maps $f, g : X \to X'$ are cofiber-homotopic if there is a homotopy $f_t : X \to X'$ such that $f_0 = f$, $f_1 = g$ and $f_t i = i'$, for all $t \in I$. If we take $i : A \to X$ to be an inclusion map and $i' : A \to X'$ any map, then the notion of cofiber-homotopy is precisely that of homotopy relative to A.

3.4 Examples of Fiber Bundles

In this section we present several classes of examples of bundle maps. By Proposition 3.3.9 all of these examples are weak fibrations. The first two classes of examples are only briefly discussed. To give a longer discussion would take us too far afield.

All topological spaces and continuous functions in this section are unbased, unless otherwise stated.

Covering Maps

Any covering map $p : E \to B$ is a bundle map. In fact, a *covering map* can be defined as a bundle map with the following additional property. There is a locally trivializing cover $\{U_\alpha\}_{\alpha \in A}$ such that $p^{-1}(U_\alpha)$ is a disjoint union of open sets of E each of which is mapped homeomorphically by p onto U_α, for every $\alpha \in A$ [61, Chap. 11]. For each $b \in B$, the inverse image $p^{-1}(b)$ is a discrete topological space.

Vector Bundles

If M is a differentiable m-manifold, then let $T(M)$ denote the space of all tangent vectors to M. A map $p : T(M) \to M$ is given by $p(v) = x$, where v is a tangent vector to M at $x \in M$. For any $x \in M$, the set of tangent vectors at x is $p^{-1}(x)$, which is an m-dimensional real vector space. Then p is a bundle map with fiber an m-dimensional vector space, and is called the *tangent bundle* of M (for details see [14, p. 88]). This is an important example of a real vector bundle over M. More generally, an m-dimensional real *vector bundle* consists of a bundle map $p : E \to B$ having fiber $\mathbb{R}^m$ with the following additional properties: (1) for every $b \in B$, the set $p^{-1}(b)$ is an m-dimensional real vector space; (2) there is a locally trivializing cover $\{U_a\}_{\alpha \in A}$ with local trivializations $\theta_\alpha : p^{-1}(U_\alpha) \to U_\alpha \times \mathbb{R}^n$ such that $\theta_\alpha | p^{-1}(b) : p^{-1}(b) \to \{b\} \times \mathbb{R}^m$ is a vector space isomorphism, for every $b \in U_\alpha$. In addition to the tangent bundle, there are many other examples of vector bundles, for instance, the normal bundle of the embedding of one manifold into another. Furthermore, many of the vector space operations such as direct sum, tensor product, exterior product, and so on can be extended to vector bundles, and thus provide methods to construct new vector bundles from given ones. Two m-dimensional vector bundles $p : E \to B$ and $p' : E' \to B$ are said to be *equivalent* if there is a homeomorphism $\phi : E \to E'$ such that $p'\phi = p$ and $\phi | p^{-1}(b) : p^{-1}(b) \to p'^{-1}(b)$ is a vector space isomorphism, for every $b \in B$. The set of equivalence classes of m-dimensional vector bundles over B is denoted $\mathrm{Vect}_m(B)$. We note that it is possible to define complex or quaternionic vector bundles by replacing the field of real numbers $\mathbb{R}$ by the complex numbers $\mathbb{C}$ or by the quaternions $\mathbb{H}$.

Hopf Fibrations

We describe in a unified way several classes of fiber bundles in which the fiber is a sphere. Let $\mathbb{F}$ be the real numbers $\mathbb{R}$, the complex numbers $\mathbb{C}$, or the quaternions $\mathbb{H}$, and let $d = 1$, 2, or 4, respectively. Then

$$\mathbb{F}^{n+1} = \begin{cases} \mathbb{R}^{n+1} & \text{if } \mathbb{F} = \mathbb{R} \\ \mathbb{C}^{n+1} \cong \mathbb{R}^{2(n+1)} & \text{if } \mathbb{F} = \mathbb{C} \\ \mathbb{H}^{n+1} \cong \mathbb{R}^{4(n+1)} & \text{if } \mathbb{F} = \mathbb{H}. \end{cases}$$

The $(d(n+1)-1)$-dimensional sphere $S^{d(n+1)-1} \subseteq \mathbb{F}^{n+1}$ is defined by

$$S^{d(n+1)-1} = \Big\{(u_0, u_1, \ldots, u_n) \,|\, u_i \in \mathbb{F} \text{ and } \sum_{k=0}^{n} |u_k|^2 = 1\Big\}.$$

By 1.5.10, the $\mathbb{F}$-projective n-space $\mathbb{F}\mathrm{P}^n$ is $\mathbb{F}^{n+1} - \{0\}/\sim$, where the equivalence relation is defined as follows: $(u_0, u_1, \ldots, u_n) \sim (v_0, v_1, \ldots, v_n)$ if and only if there exists $\lambda \in \mathbb{F} - \{0\}$ such that $v_i = \lambda u_i$, for $i = 0, 1, \ldots, n$. The equivalence class of $(u_0, u_1, \ldots, u_n) \in \mathbb{F}^{n+1} - \{0\}$ is denoted $[u_0, u_1, \ldots, u_n]$. There is then a continuous function

$$\phi : S^{d(n+1)-1} \to \mathbb{F}\mathrm{P}^n$$

defined by $\phi(u_0, u_1, \ldots, u_n) = [u_0, u_1, \ldots, u_n]$. We determine the set $F = \phi^{-1}[1, 0, \ldots, 0]$: $(u_0, u_1, \ldots, u_n) \in \phi^{-1}[1, 0, \ldots, 0]$ if and only if there exists $\lambda \in \mathbb{F} - \{0\}$ such that $u_0 = \lambda$ and $u_i = 0$ for $i = 1, \ldots, n$. Therefore

$$F = \{(\lambda, 0, \ldots, 0) \mid \lambda \in \mathbb{F}, \ |\lambda| = 1\} = S^{d-1}.$$

Proposition 3.4.1 *For* $\mathbb{F} = \mathbb{R}, \mathbb{C}$, *or* $\mathbb{H}$ *and* $d = 1$, 2, *or* 4, *respectively,*

$$S^{d-1} \longrightarrow S^{d(n+1)-1} \xrightarrow{\ \phi\ } \mathbb{F}\mathrm{P}^n$$

is a fiber bundle.

Proof. We construct a locally trivializing cover $\{W_0, W_1, \ldots, W_n\}$ of $\mathbb{F}\mathrm{P}^n$. For $i = 0, 1, \ldots, n$, set $W_i = \{[u_0, u_1, \ldots, u_n] \mid u_i \neq 0\} \subseteq \mathbb{F}\mathrm{P}^n$. Define free maps $\rho_i : W_i \times S^{d-1} \to \phi^{-1}(W_i)$ by

$$\rho_i([u_0, u_1, \ldots, u_n], \lambda) = \frac{|u_i|\lambda}{u_i\sqrt{\sum_k |u_k|^2}}(u_0, u_1, \ldots, u_n),$$

for $[u_0, u_1, \ldots, u_n] \in W_i$ and $\lambda \in S^{d-1}$, and define free maps $\theta_i : \phi^{-1}(W_i) \to W_i \times S^{d-1}$ by

$$\theta_i(u_0, u_1, \ldots, u_n) = \left([u_0, u_1, \ldots, u_n], \frac{u_i}{|u_i|}\right),$$

for $(u_0, u_1, \ldots, u_n) \in \phi^{-1}(W_i)$. Then $\rho_i\theta_i = \mathrm{id}$ and $\theta_i\rho_i = \mathrm{id}$. Thus the θ_i are homeomorphisms and so $\{W_0, W_1, \ldots, W_n\}$ is a locally trivializing cover. □

Definition 3.4.2 In Proposition 3.4.1, the maps ϕ are called *Hopf maps* and the fibrations are called *Hopf fibrations.*

Remark 3.4.3

1. More explicitly, the Hopf fibrations are
   - $S^0 \longrightarrow S^n \xrightarrow{\ \phi\ } \mathbb{R}\mathrm{P}^n$
   - $S^1 \longrightarrow S^{2n+1} \xrightarrow{\ \phi\ } \mathbb{C}\mathrm{P}^n$
   - $S^3 \longrightarrow S^{4n+3} \xrightarrow{\ \phi\ } \mathbb{H}\mathrm{P}^n$.

   These fibrations are the real, complex, and quaternionic Hopf fibrations.
2. The Hopf maps are the last attaching maps in the CW decomposition of $\mathbb{R}\mathrm{P}^{n+1}$, $\mathbb{C}\mathrm{P}^{n+1}$, and $\mathbb{H}\mathrm{P}^{n+1}$ given in Example 1.5.10. In Chapter 5 these maps will be shown to be generators of homotopy groups.
3. The real Hopf map $\phi : S^n \to \mathbb{R}\mathrm{P}^n$ is a two-sheeted covering map. For $n > 1$, ϕ is the universal covering map.

We next prove a lemma which identifies the 1-dimensional projective spaces.

Lemma 3.4.4 $\mathbb{F}\mathrm{P}^1 \cong S^d$.

Proof. For $d = 1$, we regard $\mathbb{R}\mathrm{P}^1$ as $E^1 = [-1, 1]$ with $\{-1\}$ and $\{1\}$ identified according to Example 1.5.10(6). This is homeomorphic to S^1. For $d = 2$, we have $\mathbb{C}\mathrm{P}^1 = \{[z_0, z_1] \mid (z_0, z_1) \in \mathbb{C}^2 - \{0\}\}$. We regard S^2 as the one point compactification of $\mathbb{C}$, that is, $S^2 = \mathbb{C} \cup \{\infty\}$ (the so-called Gaussian sphere). Then we define $\theta : \mathbb{C}\mathrm{P}^1 \to S^2$ by $\theta[z_0, z_1] = z_0/z_1$, where $z_0/z_1 = \infty$ if $z_1 = 0$. It follows that θ is a homeomorphism. The case $d = 4$ is similar, and hence omitted. □

Example 3.4.5 By Proposition 3.4.1 and Lemma 3.4.4, there are weak fibrations

$$S^1 \longrightarrow S^3 \xrightarrow{\phi} S^2 \quad \text{and} \quad S^3 \longrightarrow S^7 \xrightarrow{\phi} S^4.$$

The fiber map $\phi : S^3 \to S^2$ is often called the *Hopf map.* In addition, by using the *Octonions* (also called the *Cayley numbers*), $\mathbb{O} \cong \mathbb{R}^8$ [49, pp. 448–449], we can construct in a completely analogous way a fibration $S^7 \longrightarrow S^{15} \xrightarrow{\phi} \mathbb{O}P^1$. Since $\mathbb{O}P^1 \cong S^8$, we obtain a fiber sequence

$$S^7 \longrightarrow S^{15} \xrightarrow{\phi} S^8.$$

By forming the mapping cone of ϕ we obtain $\mathbb{O}$-projective 2-space. Because of the non-associativity of $\mathbb{O}$, an $\mathbb{O}$-projective n-space for $n > 2$ cannot be defined in analogy to $\mathbb{F}\mathrm{P}^n$.

Homogeneous Spaces

Let $p : E \to B$ be a continuous surjection and $U \subseteq B$ an open set. A *local section* of p on U is a continuous function $s : U \to E$ such that $ps = i$, where $i : U \to B$ is the inclusion function. If a point $x \in B$ has a neighborhood U and a local section s of p on U, then we say that s is a *local section at* x. Suppose now that

$$F \longrightarrow E \xrightarrow{p} B$$

is a fiber bundle with locally trivializing cover $\{U_\alpha\}_{\alpha \in A}$ of B and local trivialization $\theta_\alpha : p^{-1}(U_\alpha) \to U_\alpha \times F$. We choose $y \in F$ and define $i_y : U_\alpha \to U_\alpha \times F$ by $i_y(x) = (x, y)$, for $x \in U_\alpha$. Then a local section of p on U_α is defined as the composition

$$U_\alpha \xrightarrow{i_y} U_\alpha \times F \xrightarrow{\theta_\alpha^{-1}} p^{-1}(U_\alpha) \subseteq E.$$

Thus for any fiber bundle, there is a local section at each point of the base. We show next that the converse holds in certain cases.

Let G be a topological group with multiplication denoted by juxtaposition, and let $H \subseteq G$ be a closed subgroup. We can then form G/H, the set of all left cosets gH for $g \in G$, and give it the quotient topology obtained from the projection $p : G \to G/H$. Then G/H is called a *homogeneous space.* There is an action $G \times G/H \to G/H$ defined by $g \cdot (g'H) = (gg')H$, for $g, g' \in G$, such that $p(gg') = g \cdot p(g')$. Let e denote the coset H in G/H.

Lemma 3.4.6 *If $p : G \to G/H$ has a local section at e, then p has a local section at every point of G/H.*

Proof. Let $U \subseteq G/H$ be an open set containing e and let $s : U \to G$ be a local section. Given any $x = gH \in G/H$, we consider the open set $g \cdot U$ of G/H containing x. We define $s_x : g \cdot U \to G$ by

$$s_x(gg'H) = gs(g'H),$$

where $g'H \in U$. Then clearly s_x is a local section at x. □

Remark 3.4.7 We mention in passing that if G is a Lie group and H a closed subgroup, then $p : G \to G/H$ has a local section at e [18, p.110].

Proposition 3.4.8 *Let G be a topological group and let $K \subseteq H \subseteq G$ be closed subgroups. If $p : G \to G/H$ has a local section at $e = H$, then*

$$H/K \longrightarrow G/K \xrightarrow{p'} G/H$$

is a fiber bundle, where $p'(gK) = gH$, for every $g \in G$.

Proof. By Lemma 3.4.6, there is an open cover $\{U_\alpha\}_{\alpha \in A}$ of G/H and a local section $s_\alpha : U_\alpha \to G$, for every $\alpha \in A$. Define $\psi_\alpha : U_\alpha \times H/K \to G/K$ by

$$\psi_\alpha(gH, hK) = s_\alpha(gH)hK,$$

for $g \in G$, $h \in H$, and $gH \in U_\alpha$. Clearly, $p'\psi_\alpha(gH, hK) = gH$ and $\psi_\alpha(gH, hK) \subseteq p'^{-1}(U_\alpha)$. Now define $\theta_\alpha : p'^{-1}(U_\alpha) \to U_\alpha \times H/K$ by

$$\theta_\alpha(gK) = (gH, (s_\alpha(gH))^{-1}gK),$$

for $gK \in p'^{-1}(U_\alpha)$. A simple calculation shows that $\theta_\alpha\psi_\alpha = \mathrm{id}$ and $\psi_\alpha\theta_\alpha = \mathrm{id}$. Therefore $\{U_\alpha\}$ is a locally trivializing cover of G/H. □

As a special case we consider the *orthogonal group* $O(n)$. This consists of all $n \times n$ matrices A with real entries such that $AA^T = I$, where A^T is the transpose of A and $I = I_n$ is the $n \times n$ identity matrix. The columns of A, considered as vectors in $\mathbb{R}^n$, are a set of n orthonormal vectors. By writing the rows of a matrix in $O(n)$ one after the other as an n^2-tuple, we have that $O(n) \subseteq \mathbb{R}^{n^2}$. Thus $O(n)$ becomes a topological space by giving it the subspace topology from $\mathbb{R}^{n^2}$. It can be shown that $O(n)$ is a compact space

(Exercise 3.22). If $k \leqslant n$, then $O(k)$ can be embedded into $O(n)$ by assigning to $A \in O(k)$, the matrix

$$\begin{pmatrix} A & 0 \\ 0 & I_{n-k} \end{pmatrix}$$

in $O(n)$. Then $O(k)$, regarded as a subset of $O(n)$, is a closed subgroup of $O(n)$. We wish to show that the quotient map $p : O(n) \to O(n)/O(k)$ has a local section at $e = O(k)$, for each $k \leqslant n$. For this we introduce the Stiefel manifolds, which are interesting in their own right.

Stiefel Manifolds

Definition 3.4.9 If $0 \leqslant k \leqslant n$, then any ordered set of k orthonormal vectors in $\mathbb{R}^n$ is called a *k-frame* in $\mathbb{R}^n$. The *real Stiefel manifold* $V_k(\mathbb{R}^n)$ is the set of all k-frames. The basepoint of $V_k(\mathbb{R}^n)$ is the last k vectors $e_{n-k+1}, \ldots, e_n$ of the standard basis $e_1, \ldots, e_n$ of $\mathbb{R}^n$. A k-frame $v_1, \ldots, v_k$ can be regarded as an $n \times k$ matrix by taking the ith column to be the vector v_i. In this way we identify a k-frame with a point in $\mathbb{R}^{nk}$ and give $V_k(\mathbb{R}^n)$ the subspace topology. We note that if $l' \leqslant l$, then there is a map $q : V_l(\mathbb{R}^n) \to V_{l'}(\mathbb{R}^n)$ which assigns to an l-frame the last l' vectors. By replacing the reals $\mathbb{R}$ with the complex numbers $\mathbb{C}$ or with the quaternions $\mathbb{H}$, we obtain the *complex Stiefel manifolds* $V_k(\mathbb{C}^n)$ or the *quaternionic Stiefel manifolds* $V_k(\mathbb{H}^n)$.

We first consider real Stiefel manifolds, written $V_{k,n} = V_k(\mathbb{R}^n)$, and relate them to the orthogonal groups.

Lemma 3.4.10 *There is a homeomorphism $\phi_{l,n} : O(n)/O(n-l) \to V_{l,n}$ such that the following diagram commutes*

$$\begin{array}{ccc} O(n)/O(n-l) & \xrightarrow{\phi_{l,n}} & V_{l,n} \\ \downarrow{\scriptstyle p'} & & \downarrow{\scriptstyle q} \\ O(n)/O(n-k) & \xrightarrow{\phi_{k,n}} & V_{k,n}, \end{array}$$

for $k \leqslant l$.

Proof. Define $\phi : O(n) \to V_{l,n}$ by $\phi(A) = Ae_{n-l+1}, \ldots, Ae_n$ for $A \in O(n)$. Thus ϕ assigns to a matrix A the l-frame consisting of the last l columns of A. If C is an $(n-l) \times (n-l)$ matrix, then multiplication of A on the right with a matrix of the form

$$\begin{pmatrix} C & 0 \\ 0 & I_l \end{pmatrix},$$

where I_l is the $l \times l$ identity matrix, does not change the last l columns of A, and so it follows that ϕ is constant on cosets $AO(n-l)$ in $O(n)/O(n-l)$. Thus

ϕ induces a continuous surjection $\phi_{l,n} : O(n)/O(n-l) \to V_{l,n}$. It is easily seen that if $A, B \in O(n)$ each have the same last l columns, then $A^{-1}B = A^T B$ has the form

$$\begin{pmatrix} C & 0 \\ 0 & I_l \end{pmatrix},$$

where $C^T = C^{-1}$. This shows that $\phi_{l,n}$ is one–one. By Exercise 3.22, $O(n)$ is compact. Thus $\phi_{l,n}$ is a continuous bijection from a compact space $O(n)/O(n-l)$ to a Hausdorff space $V_{l,n}$, and hence is a homeomorphism. The commutativity of the diagram in Lemma 3.4.10 follows easily. □

We wish to show that $p' : O(n)/O(n-l) \to O(n)/O(n-k)$ is a bundle map with fiber $O(n-k)/O(n-l)$. By Proposition 3.4.8 we need to show that $p : O(n) \to O(n)/O(n-k)$ has a local section at e. By Lemma 3.4.10, it suffices to show that $q : V_{n,n} \to V_{k,n}$ has a local section at $e_{n-k+1}, \ldots, e_n$.

Lemma 3.4.11 *The map $q : V_{n,n} \to V_{k,n}$ that assigns to an n-frame the last k vectors has a local section at $e_{n-k+1}, \ldots, e_n$.*

Proof. We seek a continuous function s that assigns to a k-frame $v_1, \ldots, v_k$ in $\mathbb{R}^n$ an n-frame $u_1, \ldots, u_{n-k}, v_1, \ldots, v_k$ in $\mathbb{R}^n$. Let $M_{n,l}$ be the space of $n \times l$ matrices with real entries. By considering the columns of $M_{n,l}$ as vectors, we identify elements of $M_{n,l}$ with l-tuples of vectors in $\mathbb{R}^n$. If r is the function that assigns to any k vectors $w_1, \ldots, w_k$ in $\mathbb{R}^n$, the n vectors $e_1, \ldots, e_{n-k}, w_1, \ldots, w_k$, then $r(e_{n-k+1}, \ldots, e_n) = e_1, \ldots, e_n$. We regard r as a continuous function $r : M_{n,k} \to M_{n,n}$. The elements of $M_{n,n}$ that correspond to n linearly independent vectors in $\mathbb{R}^n$ form an open set (because the determinant of such matrices is non-zero) containing $e_1, \ldots, e_n$. By continuity of r, there is an open set $U \subseteq V_{k,n}$ containing $e_{n-k+1}, \ldots, e_n$ such that $v_1, \ldots, v_k \in U$ implies that $e_1, \ldots, e_{n-k}, v_1, \ldots, v_k$ is linearly independent. We now use the Gram–Schmidt process from linear algebra. This assigns an n-frame to n linearly independent vectors. It produces a continuous function that, when applied to $v_k, \ldots, v_1, e_{n-k}, \ldots e_1$ for $v_1, \ldots, v_k \in V_{k,n}$, yields an n-frame $v_k, \ldots, v_1, e'_{n-k}, \ldots e'_1$, for some vectors $e'_{n-k}, \ldots, e'_1$. Therefore $s : U \to V_{n,n}$ defined by $s(v_1, \ldots, v_k) = e'_1, \ldots, e'_{n-k}, v_1, \ldots, v_k$ is a local section of p at $e_{n-k+1}, \ldots, e_n$. □

Therefore we have the following proposition.

Proposition 3.4.12 *The sequence*

$$O(n-k)/O(n-l) \longrightarrow O(n)/O(n-l) \xrightarrow{p'} O(n)/O(n-k)$$

is a fiber bundle and the sequence

$$V_{l-k,n-k} \longrightarrow V_{l,n} \xrightarrow{q} V_{k,n}$$

is a fiber bundle, where $0 \leqslant k \leqslant l \leqslant n$.

We comment on some special cases of the proposition next. First note that $V_{1,n+1}$ is the space of unit vectors in $\mathbb{R}^{n+1}$ and so $O(n+1)/O(n) \cong V_{1,n+1} \cong S^n$. Therefore the orthogonal group fiber bundle in Proposition 3.4.12 with $l = n = m+1$ and $k = 1$ is

$$O(m) \longrightarrow O(m+1) \xrightarrow{p} S^m.$$

The corresponding Stiefel fiber bundle is

$$V_{m,m} \longrightarrow V_{m+1,m+1} \xrightarrow{q} S^m.$$

There is another Stiefel fiber bundle which is derived from Proposition 3.4.12 where the fiber is a sphere. It is obtained by setting $k = l-1$ and getting

$$S^{n-l} \cong V_{1,n-l+1} \longrightarrow V_{l,n} \xrightarrow{q} V_{l-1,n}.$$

These fibrations are used in Section 5.6 to calculate some homotopy groups of the orthogonal group and of Stiefel manifolds.

We next discuss a class of fiber bundles that involve Stiefel manifolds in which the base is a Grassmann manifold.

Grassmann Manifolds

Definition 3.4.13 If $0 \leqslant k \leqslant n$, then the *real Grassmann manifold* or *real Grassmannian* $G_k(\mathbb{R}^n)$ consists of all k-dimensional subspaces of $\mathbb{R}^n$ (also called k-planes in $\mathbb{R}^n$). There is a surjection $\pi : V_k(\mathbb{R}^n) \to G_k(\mathbb{R}^n)$ defined by $\pi(v_1, \ldots, v_k) = \langle v_1, \ldots, v_k \rangle$, where $\langle v_1, \ldots, v_k \rangle$ is the subspace of $\mathbb{R}^n$ spanned by $v_1, \ldots, v_k$. We give $G_k(\mathbb{R}^n)$ the quotient topology induced by π. By replacing $\mathbb{R}$ with the complex numbers $\mathbb{C}$ or with the quaternions $\mathbb{H}$, we obtain the *complex Grassmann manifold* $G_k(\mathbb{C}^n)$ or the *quaternionic Grassmann manifold* $G_k(\mathbb{H}^n)$. We note that when $k = 1$, the Grassmannians $G_k(\mathbb{R}^n)$, $G_k(\mathbb{C}^n)$ and $G_k(\mathbb{H}^n)$ are essentially real projective $(n-1)$-space $\mathbb{R}\mathrm{P}^{n-1}$, complex projective $(n-1)$-space $\mathbb{C}\mathrm{P}^{n-1}$, or quaternionic projective $(n-1)$-space $\mathbb{H}\mathrm{P}^{n-1}$, respectively.

We frequently write $G_{k,n}$ for the real Grassmann manifolds $G_k(\mathbb{R}^n)$. The notation $V_{k,n}$ and $G_{k,n}$ for the real Stiefel manifolds and the real Grassmann manifolds is not standard. Some authors denote these as $V_{n,k}$ and $G_{n,k}$. It seems that the most unambiguous notation is $V_k(\mathbb{R}^n)$ and $G_k(\mathbb{R}^n)$. Similar remarks hold for $\mathbb{C}$ and $\mathbb{H}$.

Lemma 3.4.14 *The function* $\pi : V_{k,n} \to G_{k,n}$ *is an open mapping.*

Proof. We identify elements of $V_{k,n}$ with $n \times k$ matrices of rank k, by assigning to each $v_1, \ldots, v_k \in V_{k,n}$, the matrix A whose columns are $v_1, \ldots, v_k$. If P

is a nonsingular $k \times k$ matrix, then $\theta_P : V_{k,n} \to V_{k,n}$ defined by $\theta_P(A) = AP$ is a homeomorphism since $\theta_{P^{-1}}$ is its inverse. If $x, y \in V_{k,n}$ with corresponding matrices A and B, then $\pi(x) = \pi(y)$ if and only if $B = AP$, for some nonsingular $k \times k$ matrix P. Hence if $U \subseteq V_{k,n}$, then

$$\pi^{-1}(\pi(U)) = \bigcup_P \theta_P(U),$$

for all nonsingular $k \times k$ matrices P. Therefore if U is open in $V_{k,n}$, then $\pi^{-1}(\pi(U))$ is open in $V_{k,n}$. Hence $\pi(U)$ is open in $G_{k,n}$, and so π is an open map. □

Next consider the orthogonal group $O(n)$ and let $O'(k)$ be the subgroup of all matrices

$$\begin{pmatrix} I_{n-k} & 0 \\ 0 & D \end{pmatrix},$$

for $D \in O(k)$. We form the subgroup $O(n-k) \cdot O'(k)$ of $O(n)$ and this subgroup consists of all matrices of the form

$$\begin{pmatrix} C & 0 \\ 0 & D \end{pmatrix},$$

where $C \in O(n-k)$ and $D \in O(k)$. Then $O(n-k) \subseteq O(n-k) \cdot O'(k) \subseteq O(n)$.

Lemma 3.4.15 *There is a homeomorphism $\psi_{k,n} : O(n)/(O(n-k) \cdot O'(k)) \to G_{k,n}$ such that the following diagram is commutative*

$$\begin{array}{ccc} O(n)/O(n-k) & \xrightarrow{\phi_{k,n}} & V_{k,n} \\ \downarrow{\scriptstyle p'} & & \downarrow{\scriptstyle \pi} \\ O(n)/(O(n-k) \cdot O'(k)) & \xrightarrow{\psi_{k,n}} & G_{k,n}. \end{array}$$

Proof. We define $\psi : O(n) \to G_{k,n}$ by $\psi(A) = \langle Ae_{n-k+1}, \ldots, Ae_n \rangle$, for $A \in O(n)$. If $B \in O(n-k) \cdot O'(k)$, then $\psi(AB) = \langle ABe_{n-k+1}, \ldots, ABe_n \rangle = \langle Ae_{n-k+1}, \ldots, Ae_n \rangle = \psi(A)$. Thus ψ is constant on cosets of $O(n-k) \cdot O'(k)$, and so ψ induces a continuous surjection $\psi_{k,n} : O(n)/(O(n-k) \cdot O'(k)) \to G_{k,n}$. It follows immediately that the above diagram is commutative. To show that $\psi_{k,n}$ is one–one, suppose that $\psi_{k,n}(A) = \psi_{k,n}(B) = W$, for $A, B \in O(n)$. Then $Ae_{n-k+1}, \ldots, Ae_n$ and $Be_{n-k+1}, \ldots, Be_n$ are bases for W. Let P be the change of basis matrix from the first basis to the second. Also, $\langle Ae_1, \ldots, Ae_{n-k} \rangle = \langle Be_1, \ldots, Be_{n-k} \rangle$ with bases $Ae_1, \ldots, Ae_{n-k}$ and $Be_1, \ldots, Be_{n-k}$. Let Q be the change of basis matrix from the first basis to the second one. Then

$$B = A \begin{pmatrix} Q & 0 \\ 0 & P \end{pmatrix} \quad \text{and} \quad \begin{pmatrix} Q & 0 \\ 0 & P \end{pmatrix} \in O(n-k) \cdot O'(k).$$

This shows that $\psi_{k,n}$ is one–one. Finally, $\psi_{k,n}$ is an open mapping by Lemma 3.4.14 and the commutative diagram in 3.4.15. Therefore $\psi_{k,n}$ is a homeomorphism. □

Next we show that π is a bundle map.

Proposition 3.4.16 *The sequence*

$$V_{k,k} \longrightarrow V_{k,n} \xrightarrow{\pi} G_{k,n}$$

is a fiber bundle.

Proof. By Lemma 3.4.15 the given sequence is essentially

$$(O(n-k)\cdot O'(k))/O(n-k) \longrightarrow O(n)/(O(n-k) \longrightarrow O(n)/(O(n-k)\cdot O'(k)),$$

therefore it suffices to show that π has a local section at $\langle e_{n-k+1}, \ldots, e_n \rangle$ by Proposition 3.4.8. Let $U \subseteq G_{k,n}$ be the set of all k-dimensional subspaces W of $\mathbb{R}^n$ such that $W \cap \langle e_1, \ldots, e_{n-k} \rangle = \{0\}$. We show that U is open in $G_{k,n}$ by showing that $\pi^{-1}(U)$ is open in $V_{k,n}$. Let $M_{n,k}$ be the space of $n \times k$ matrices and define a continuous function $F : M_{n,k} \to \mathbb{R}$ by $F(A) = \det(A')$, where A' is the $n \times n$ matrix obtained from A by putting the columns $e_1, \ldots, e_{n-k}$ on the left. Then $F^{-1}(\mathbb{R} - \{0\})$ is open in $M_{n,k}$. Regarding $V_{k,n}$ as a subset of $M_{n,k}$, we have $V_{k,n} \cap F^{-1}(\mathbb{R} - \{0\}) = \pi^{-1}(U)$, and so $\pi^{-1}(U)$ is open in $V_{k,n}$. Therefore U is open and we next define a local section $t : U \to V_{k,n}$. Given $W \in U$, let $\bar{e}_i$ be the orthogonal projection of e_i onto W, for $i = n-k+1, \ldots, n$. Then $\bar{e}_{n-k+1}, \ldots, \bar{e}_n$ is a linearly independent set in W. We apply the Gram–Schmidt process to these vectors and obtain a k-frame $\bar{e}'_{n-k+1}, \ldots, \bar{e}'_n$ in W. Then $t(W) = \bar{e}'_{n-k+1}, \ldots, \bar{e}'_n$ is a local section. □

We have discussed in detail several fibrations in which the real Stiefel manifolds and the real Grassmann manifolds appear. We have commented that complex Stiefel and Grassmann manifolds and quaternionic Stiefel and Grassmann manifolds exist. We now remark that there are results for these manifolds which are parallel to Lemmas 3.4.10 and 3.4.15 and Propositions 3.4.12 and 3.4.16. More precisely, let $U(n)$ be the *group of unitary matrices*, that is, $U(n)$ consists of $n \times n$ matrices A with complex entries such that the transpose of the matrix of complex conjugates of the entries of A equals A^{-1}. Then there are homeomorphisms

$$U(n)/U(n-l) \cong V_l(\mathbb{C}^n) \quad \text{and} \quad U(n)/(U(n-k)\cdot U'(k)) \cong G_k(\mathbb{C}^n).$$

Furthermore, for $k \leqslant l$, there are fiber bundles

- $U(n-k)/U(n-l) \longrightarrow U(n)/U(n-l) \xrightarrow{p'} U(n)/U(n-k)$, equivalently,
- $V_{l-k}(\mathbb{C}^{n-k}) \longrightarrow V_l(\mathbb{C}^n) \xrightarrow{q} V_k(\mathbb{C}^n)$ and

- $V_k(\mathbb{C}^k) \longrightarrow V_k(\mathbb{C}^n) \xrightarrow{\pi} G_k(\mathbb{C}^n)$.

In addition, similar results hold for the quaternions. The symplectic group $Sp(n)$ of $n \times n$ quaternion matrices A such that $\bar{A}A^T = I$ takes the role that the orthogonal group $O(n)$ plays in the real case. Moreover, for $k = 1$, the fibrations $V_k(\mathbb{F}^k) \longrightarrow V_k(\mathbb{F}^n) \xrightarrow{\pi} G_k(\mathbb{F}^n)$ reduce to the real, complex, or quaternionic Hopf fibrations when $\mathbb{F} = \mathbb{R}$, $\mathbb{C}$, or $\mathbb{H}$. The proofs of all of the above results for the complex numbers $\mathbb{C}$ and for the quaternions $\mathbb{H}$ are completely analogous to those that we have given for the real numbers $\mathbb{R}$. For more details, see [47, Chap. 7].

Remark 3.4.17 Apart from providing nontrivial examples of fibrations, the Stiefel manifolds and Grassmann manifolds have applications to other areas of topology. We first consider the real Stiefel manifolds $V_{k,n}$. Elements of $V_{2,n+1}$ are pairs of orthonormal vectors v_1, v_2 in $\mathbb{R}^{n+1}$. We can identify v_1 with a point of S^n and, by transferring v_2 to the endpoint of v_1, we can identify v_2 with a unit tangent vector to $v_1 \in S^n$. Thus $V_{2,n+1}$ can be regarded as the space of unit tangent vectors to S^n. A section of $q : V_{2,n+1} \to V_{1,n+1} \cong S^n$ is a continuous unit tangent vector field on S^n. Similarly a section of $q : V_{k+1,n+1} \to V_{1,n+1}$ is essentially k continuous orthonormal tangent vector fields on S^n. It is an interesting and difficult problem to determine the maximum number $\rho(n)$ of continuous orthonormal tangent vector fields on S^n or, what is equivalent by the Gram–Schmidt process, the maximum number of continuous linearly independent tangent vector fields on S^n. It is a classical result in algebraic topology that $\rho(n) = 0$ if and only if n is even [61, p. 190]. The problem of determining the exact number $\rho(n)$ for odd n has been investigated extensively and a complete solution has been given (see [2], [37, pp. 336–339] and [39, pp. 493–495]).

For the Grassmann manifolds $G_k(\mathbb{F}^n)$, where $\mathbb{F} = \mathbb{R}, \mathbb{C}$, or $\mathbb{H}$, we have $G_k(\mathbb{F}^n) \subseteq G_k(\mathbb{F}^{n+1})$ and so we can form

$$G_k(\mathbb{F}^\infty) = \bigcup_{n \geqslant 0} G_k(\mathbb{F}^{n+k}),$$

with the weak topology. Let $\mathrm{Vect}_k(B)$ be the set of equivalence classes of k-dimensional $\mathbb{F}$-vector bundles over B, where equivalence was defined at the beginning of this section. Then, if B is a CW complex, it can be proved that there is a natural bijection $\theta : [B, G_k(\mathbb{F}^\infty)]_{\mathsf{free}} \to \mathrm{Vect}_k(B)$. The function θ assigns to a free homotopy class $[f]$ in $[B, G_k(\mathbb{F}^\infty)]_{\mathsf{free}}$ the pullback of f and the fiber map of a canonical k-dimensional vector bundle over $G_k(\mathbb{F}^\infty)$. It follows that a homotopy class of maps $B \to G_k(\mathbb{F}^\infty)$ uniquely determines a k-dimensional equivalence class of $\mathbb{F}$-vector bundles over B. The space $G_k(\mathbb{F}^\infty)$ is called the *classifying space* for k-dimensional $\mathbb{F}$-vector bundles. For more details, see [47, p. 84].

3.5 Replacing a Map by a Cofiber or Fiber Map

Maps that are cofibrations or fibrations have important special properties in homotopy theory. We show many of these in the next chapter. Therefore it is very useful to be able to factor any map into the composition of a homotopy equivalence and a cofibration (or fibration). We do that in this section. We begin with cofibrations.

Definition 3.5.1 Given a map $f : X \to Y$, we take the disjoint union $X \times I \sqcup Y$ and introduce the equivalence relation $(x, 0) \sim f(x)$, for all $x \in X$, and $(*_X, t) \sim *_Y$, for all $t \in I$. This is a special case of the adjunction space (Definition 1.5.1(2)). The quotient space is denoted M_f and called the *mapping cylinder* of f. Equivalently, we define $X \ltimes I = X \times I/\{(*, t) \mid t \in I\}$ (see Remark 1.3.5) and then form $(X \ltimes I) \vee Y$. In the latter space we define the equivalence relation $(\langle x, 0\rangle, *) \sim (*, f(x))$. The resulting quotient space is M_f. There are maps $f' : X \to M_f$ and $f'' : M_f \to Y$ defined by $f'(x) = \langle x, 1\rangle$, $f''\langle x, t\rangle = f(x)$ and $f''\langle y\rangle = y$, for all $x \in X$, $y \in Y$ and $t \in I$. Clearly $f = f''f'$.

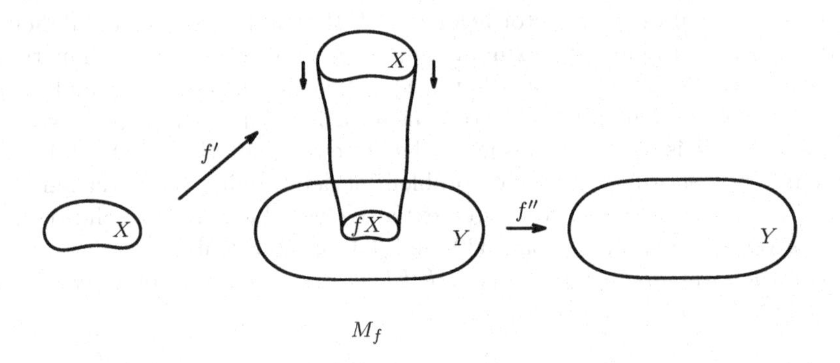

Figure 3.4

Proposition 3.5.2

1. *$f' : X \to M_f$ is a cofiber map.*
2. *$f'' : M_f \to Y$ is a homotopy equivalence.*

Proof. (1) Let $h_0 : M_f \to Z$ be a map and $g_s : X \to Z$ a homotopy such that $h_0 f' = g_0$. We set $S = (I \times \{0\}) \cup (\partial I \times I) \subseteq I \times I$ and define $H'' : X \times S \to Z$ by

$$H''(x, t, 0) = h_0\langle x, t\rangle, \quad H''(x, 0, s) = h_0\langle f(x)\rangle, \quad \text{and} \quad H''(x, 1, s) = g_s(x),$$

for every $x \in X$ and $s, t \in I$. It is easily checked that H'' is a well-defined, continuous function. We let $r : I \times I \to S$ be the retraction of Example 1.4.2(3) and define $H' : X \times I \times I \to Z$ by $H'(x,t,s) = H''(x, r(t,s))$. Then $H : M_f \times I \to Z$ is defined by

$$H(\langle x,t\rangle, s) = H'(x,t,s) \text{ and } H(\langle y\rangle, s) = h_0\langle y\rangle.$$

Therefore H is a well-defined continuous function that is a homotopy of h_0 such that $H(f'(x), s) = g_s(x)$.

(2) Define $l : Y \to M_f$ by $l(y) = \langle y\rangle$, for every $y \in Y$. Then $f''l(y) = f''\langle y\rangle = y$. Therefore $f''l = \mathrm{id}$, and so f'' is a retraction. Now

$$lf''\langle x,t\rangle = lf(x) = \langle f(x)\rangle \quad \text{and} \quad lf''\langle y\rangle = l(y) = \langle y\rangle.$$

We define a homotopy F such that $\mathrm{id} \simeq_F lf'' : M_f \to M_f$ by

$$F(\langle x,t\rangle, s) = \langle x, t(1-s)\rangle \quad \text{and} \quad F(\langle y\rangle, s) = \langle y\rangle. \qquad \square$$

This proposition was foreshadowed by Corollary 2.4.10. This result asserts that any map $f : X \to Y$ can be factored as

$$X \xrightarrow{i} K \xrightarrow{\overline{f}} Y$$

for a space K such that (K, X) is a relative CW complex (and hence the inclusion i is a cofiber map) and $\overline{f}$ is a weak homotopy equivalence.

Remark 3.5.3

1. Let X be a space of the homotopy type of a CW complex K, let Y be a space of the homotopy type of a CW complex L, and let $f : X \to Y$ be a map. If $g : K \to L$ is a cellular map homotopic to f, then it can be shown that the mapping cylinder M_g and the mapping cone C_g are CW complexes. It then follows from Exercise 3.27 and Proposition 3.2.15 that M_f and C_f have the homotopy type of CW complexes.
2. The homotopy F in the preceding proof can be described as "sliding down the cylinder". Note that F is a homotopy rel Y, so that the inclusion $l : Y \to M_f$ is a strong deformation retraction.
3. The cofiber of $f' : X \to M_f$ is just the mapping cone C_f of f. Thus we have a cofiber sequence

$$X \xrightarrow{f'} M_f \xrightarrow{p} C_f$$

(see Figure 3.5). We observe that this is one of two cofiber sequences in which the mapping cone C_f appears. The other is the principal cofibration

$$Y \xrightarrow{l} C_f \xrightarrow{q} \Sigma X$$

described in Definition 3.2.11. The space C_f, which is the cofiber of f', is also the homotopy cofiber of f.

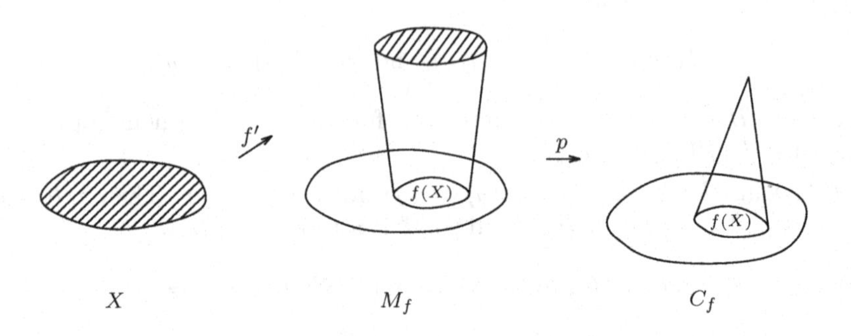

Figure 3.5

If $f : X \to Y$ is a cofiber map, then the cofiber of f is $Z = Y/f(X)$ and the homotopy cofiber of f is C_f. We show next that these two spaces have the same homotopy type. We may then refer to either of them as the cofiber of f or the homotopy cofiber of f.

Let

$$X \xrightarrow{f} Y \xrightarrow{p} Z$$

be a cofibration and let $f'' : M_f \to Y$ the homotopy equivalence of Proposition 3.5.2. The map f'' induces a map $\alpha : M_f/f'(X) = C_f \to Z = Y/f(X)$.

Proposition 3.5.4 *With the notation above, α is a homotopy equivalence such that the following diagram is commutative*

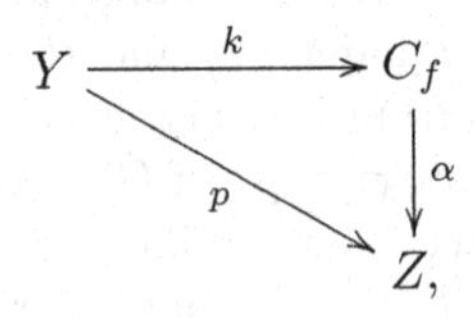

where k is the inclusion.

Proof. Define a map $l : Y \to M_f$ and a homotopy $g_s : X \to M_f$ by

$$l(y) = \langle y \rangle \text{ and } g_s(x) = \langle x, s \rangle,$$

for $y \in Y$ and $x \in X$. Then $lf = g_0$. Since f is a cofiber map, there is a homotopy $l_s : Y \to M_f$ such that $l_s f = g_s$ and $l_0 = l$. We set $l' = l_1 : Y \to M_f$ and note that $l'f(x) = \langle x, 1 \rangle = f'(x)$ for every $x \in X$. Thus l' induces $\widetilde{l'} : Z = Y/f(X) \to C_f = M_f/f'(X)$, and we show that $\widetilde{l'}$ is the homotopy inverse of α. The homotopy

$$Y \xrightarrow{l_s} M_f \xrightarrow{f''} Y$$

induces a homotopy $\widetilde{l}_s : Z \to Z$ with $\widetilde{l}_0 = \text{id}$ and $\widetilde{l}_1 = \alpha\widetilde{l}'$. Therefore $\text{id} \simeq \alpha\widetilde{l}'$. Next define a homotopy $h_t : M_f \to M_f$ by

$$h_t\langle y\rangle = l_t(y) \text{ and } h_t\langle x, s\rangle = \langle x, (1-s)t + s\rangle,$$

where $y \in Y$, $x \in X$, and $s, t \in I$. Clearly h_t is well-defined and $h_t\langle x, 1\rangle = \langle x, 1\rangle$. Then h_t induces a homotopy $H : C_f \times I \to C_f$ with $\text{id} \simeq_H \widetilde{l}'\alpha$. Thus α is a homotopy equivalence. Finally, $\alpha k = p$ follows from the definition of α, and so the proof is complete. □

Remark 3.5.5 We note for future use that $\alpha : C_f \to Z$ is given by $\alpha\langle x, t\rangle = *$ and $\alpha\langle y\rangle = \langle y\rangle$, for $x \in X$, $y \in Y$, and $t \in I$.

In order to obtain the coexact sequence of a map in Section 4.2, we need to iterate the mapping cone construction. To fix our notation, let $f : X \to Y$ be a cofiber map with cofiber Z, let $p_f : Y \to Z$ be the projection, let $k_f : Y \to C_f$ be the inclusion and let $\alpha_f : C_f \to Z$ be the homotopy equivalence of Proposition 3.5.4. Now we apply Proposition 3.5.4 and this notation to the cofiber map $k_f : Y \to C_f$. We write $k_f = l$ and obtain the diagram

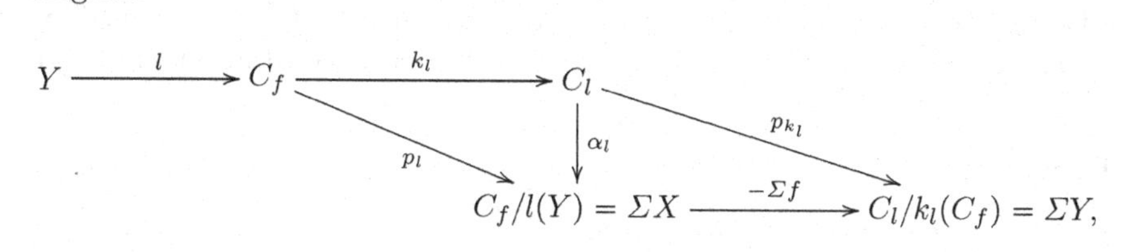

where the map $-\Sigma f : \Sigma X \to \Sigma Y$ is defined by $-\Sigma f\langle x, t\rangle = \langle f(x), 1-t\rangle$. Clearly $C_f/l(Y) = \Sigma X$ and $C_l/k_l(C_f) = \Sigma Y$ as illustrated in Figure 3.6.

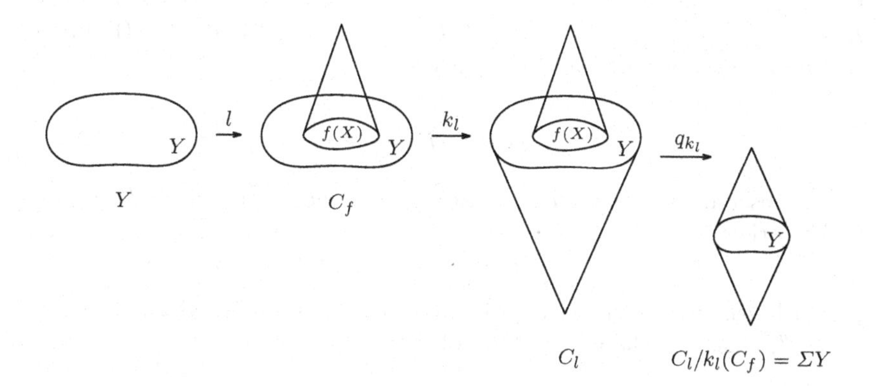

Figure 3.6

Lemma 3.5.6 *In the diagram above, the left triangle is commutative and the right triangle is commutative up to homotopy.*

Proof. Since the left triangle is commutative by 3.5.4, we need only to prove that there is a homotopy $F : C_l \times I \to \Sigma Y$ of p_{k_l} to $(-\Sigma f)\,\alpha_l$. Define $F(\langle x, s\rangle, t) = \langle f(x), t(1-s)\rangle$ for $\langle x, s\rangle \in C_l$, where $x \in X$ and $s, t \in I$, and $F(\langle y, s\rangle, t) = \langle y, s(1-t)+t\rangle$, for $\langle y, s\rangle \in CY \subseteq C_l$ and $t \in I$. Then F is the desired homotopy. □

Next we consider fibrations. All statements and proofs are dual to those for cofibrations, and so we omit the proofs.

Definition 3.5.7 Given a map $f : X \to Y$, let $E_f \subseteq X \times Y^I$ be defined by

$$E_f = \{(x, l) \mid x \in X,\ l \in Y^I,\ f(x) = l(0)\}.$$

Then E_f is called the *mapping path* of f. Clearly E_f is defined by the pullback square

$$\begin{array}{ccc} E_f & \longrightarrow & Y^I \\ \downarrow & & \downarrow{\scriptstyle p_0} \\ X & \xrightarrow{f} & Y, \end{array}$$

where $p_0(l) = l(0)$. There are maps $f' : X \to E_f$ and $f'' : E_f \to Y$ defined by $f'(x) = (x, l_{f(x)})$, where $l_{f(x)}$ is the path in Y that is constant at $f(x)$ and $f''(x, l) = l(1)$. Clearly $f = f''f'$.

Proposition 3.5.8

1. *$f'' : E_f \to Y$ is a fiber map.*
2. *$f' : X \to E_f$ is a homotopy equivalence.*

Remark 3.5.9

1. The homotopy inverse $k : E_f \to X$ of f' is given by $k(x, l) = x$.
2. The fiber of $f'' : E_f \to Y$ is just I_f the homotopy fiber of f (Definition 3.3.13). Thus we have a fiber sequence

$$I_f \longrightarrow E_f \xrightarrow{f''} Y.$$

We observe that this is one of two fiber sequences in which I_f appears. The other is

$$\Omega Y \longrightarrow I_f \xrightarrow{v} X$$

which was described in Definition 3.3.13. The space I_f, which is the fiber of f'', is also the homotopy fiber of f.

If $f : X \to Y$ is a fiber map, then the fiber of f is $F = f^{-1}(*)$ and the homotopy fiber of f is I_f. The next proposition states that these two spaces have the same homotopy type. We may then refer to either of them as the fiber of f or the homotopy fiber of f.

Let $F \xrightarrow{i} X \xrightarrow{f} Y$ be a fiber sequence and let $f' : X \to E_f$ be the map defined in Definition 3.5.7. Clearly f' induces a map $\beta : F \to I_f$ of fibers.

Proposition 3.5.10 *With the notation above, β is a homotopy equivalence such that the following diagram is commutative*

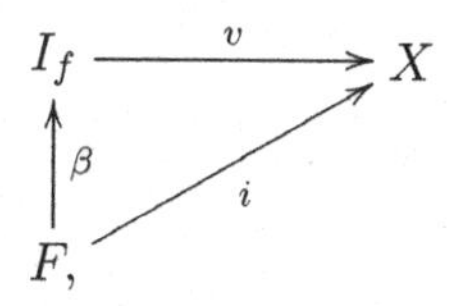

where v is the projection.

Remark 3.5.11 We note for future use that $\beta : F \to I_f$ is given by $\beta(x) = (i(x), *)$, for $x \in F$.

In order to obtain the exact sequence of a map in Section 4.2, we need to iterate the homotopy fiber construction. To fix our notation, let $f : X \to Y$ be a fiber map , $i_f : F \to X$ the inclusion of the fiber, $v_f : I_f \to X$ the fiber map, and $\beta_f : F \to I_f$ the homotopy equivalence of Proposition 3.5.10. Now we apply Proposition 3.5.10 to the fiber map $v_f : I_f \to X$. We write $v_f = \lambda$ and obtain a diagram

$$\begin{array}{ccccc} & & I_\lambda & \xrightarrow{v_\lambda} & I_f \xrightarrow{\lambda} X \\ & \nearrow^{i_{v_\lambda}} & \uparrow{\scriptstyle \beta_\lambda} & \nearrow_{i_\lambda} & \\ v_\lambda^{-1}(*) = \Omega X & \xrightarrow{-\Omega f} & \Omega Y = \lambda^{-1}(*), & & \end{array}$$

where the map $-\Omega f : \Omega X \to \Omega Y$ is defined by $((-\Omega f)(l))(t) = f(l(1-t))$, for $l \in \Omega X$ and $t \in I$.

Lemma 3.5.12 *In the diagram above, the left triangle is commutative up to homotopy and the right triangle is commutative.*

Exercises

Exercises marked with $(*)$ may be more difficult than the others. Exercises marked with (†) are used in the text.

3.1. (†) Consider the following pushout square

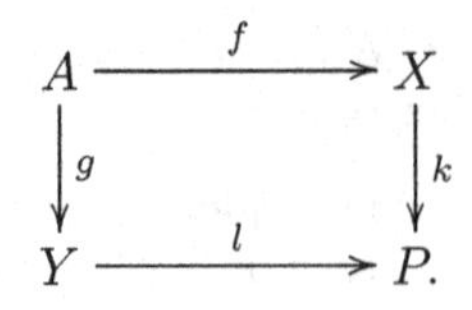

Prove that

$$\begin{array}{ccc} \Sigma A & \xrightarrow{\Sigma f} & \Sigma X \\ \downarrow{\scriptstyle \Sigma g} & & \downarrow{\scriptstyle \Sigma k} \\ \Sigma Y & \xrightarrow{\Sigma l} & \Sigma P \end{array}$$

is a pushout square.

3.2. If $f : X \to Y$ is a cofiber map and $f \simeq g : X \to Y$, then is g a cofiber map?

3.3. Let $i : A \to X$ be a map. Prove that if i is a free cofiber map, then i is a based cofiber map.

3.4. (†) Let W be a locally compact space. Prove that if

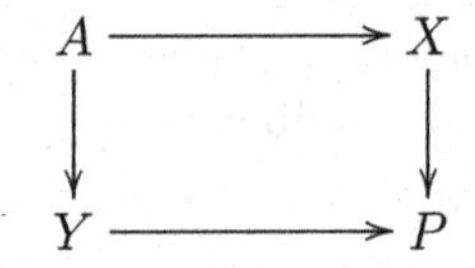

is a pushout square, then

$$\begin{array}{ccc} A \times W & \longrightarrow & X \times W \\ \downarrow & & \downarrow \\ Y \times W & \longrightarrow & P \times W \end{array}$$

is a pushout square. Can this result be dualized?

3.5. Consider the homotopy-commutative diagram

$$\begin{array}{ccccc} A & \xrightarrow{\alpha} & A' & \xrightarrow{\alpha'} & A'' \\ \downarrow{\scriptstyle f} & & \downarrow{\scriptstyle f'} & & \downarrow{\scriptstyle f''} \\ X & \xrightarrow{\beta} & X' & \xrightarrow{\beta'} & X''. \end{array}$$

Let F and G be homotopies such that $\beta f \simeq_F f'\alpha$ and $\beta' f' \simeq_G f''\alpha'$. Define a homotopy $F*G : A\times I \to X''$ by $F*G = \beta' F + G(\alpha \times \mathrm{id})$ of $\beta'\beta f$ with $f''\alpha'\alpha$. Then there are maps $\Phi_F : C_f \to C_{f'}$, $\Phi_G : C_{f'} \to C_{f''}$ and $\Phi_{F*G} : C_f \to C_{f''}$ as defined in Proposition 3.2.13. Prove that $\Phi_G \Phi_F \simeq \Phi_{F*G}$.

3.6. Prove the following.

1. The composition of fiber maps is a fiber map.
2. The product of fiber maps is a fiber map.
3. The loop of a fiber map is a fiber map.

What is the fiber in each case? What are the duals of these results?

3.7. If $p : E \to B$ is a fibration and B is path-connected, prove that p is onto.

3.8. Prove that $E(X \times Y; X, Y)$ is contractible, for any space X and Y.

3.9. (†) Prove that if $p : E \to B$ is a based weak fibration, then p is an unbased weak fibration.

3.10. (†) Given a map $p : E \to B$, we define $S \subseteq E \times B^I$ to be the space consisting of all (e, l) such that $p(e) = l(0)$, where $e \in E$ and $l \in B^I$. Then p is said to have the *path lifting property* if there is a map $\lambda : S \to E^I$ such that $\lambda(e, l)(0) = e$ and $p\lambda(e, l) = l$ (where $p\lambda(e, l)$ is the composition of $\lambda(e, l) : I \to E$ and $p : E \to B$). The map λ is called a *lifting map* for p. Prove that the following are equivalent.

1. $p : E \to B$ is a fiber map.
2. p has the path lifting property.
3. p has the covering homotopy property with respect to the mapping path space E_p.

3.11. Using reduced cylinders, dualize Exercise 3.10 by obtaining equivalent conditions for a map to be a cofibration.

3.12. Prove Proposition 3.3.12 using the characterization of fibrations via lifting maps in Exercise 3.10.

3.13. Consider the pullback square

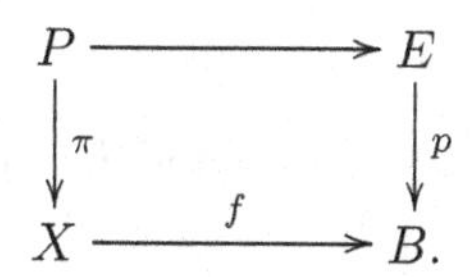

Show that f lifts to $E \iff \pi$ has a section. What conditions will yield a homotopy lift?

3.14. $(*)$ (†) Consider the commutative diagram

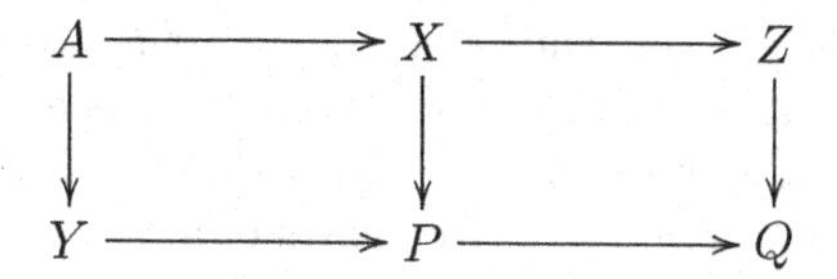

and denote the three commutative squares by $A-X-P-Y$, $X-Z-Q-P$, and $A-Z-Q-Y$. Prove the following.

1. If $A-X-P-Y$ and $X-Z-Q-P$ are pullback squares, then $A-Z-Q-Y$ is a pullback square.
2. If $X-Z-Q-P$ and $A-Z-Q-Y$ are pullback squares, then $A-X-P-Y$ is a pullback square.

3.15. (†) Let $f : X \to Y$ and $g : Y \to Z$ be maps and let I_g be the homotopy fiber of g with projection $v : I_g \to Y$. Consider the pullback square

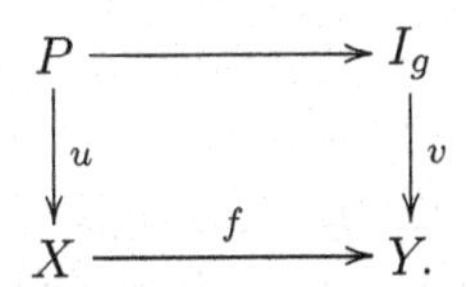

Show that there is a homeomorphism $\theta : P \to I_{gf}$ such that $w\theta = u$, where $w : I_{gf} \to X$ is the projection. Formulate and prove the dual result.

3.16. Let $f : X \to A$ and $g : X \to B$ be maps and $(f, g) : X \to A \times B$. Let I_f be the homotopy fiber of f with projection $v : I_f \to X$ and form the composition $gv : I_f \to B$. Prove $I_{gv} \cong I_{(f,g)}$. Formulate and prove the dual result.

3.17. Define the map Ψ_F in Proposition 3.3.15 and prove Proposition 3.3.15.

3.18. Recall that for $U, V \subseteq X$, we denote by $E_{U,V}$ the space of paths in X that begin in U and end in V. Now let $i : A \to X$ be an inclusion map. Show that the mapping path $E_i \cong E_{A,X}$ and that the homotopy fiber $I_i \cong E_{A,\{*\}}$.

3.19. Let $p : E \to B$ be a bundle map and $f : A \to B$ any map. If $Q(u, v)$ is the pullback of

$$A \xrightarrow{f} B \xleftarrow{p} E,$$

then prove that $u : Q \to A$ is a bundle map.

3.20. ($*$) Let G be a topological group and H a closed subgroup. Prove

1. $p : G \to G/H$ is an open map.
2. G/H is Hausdorff.
3. If H and G/H are connected, then G is connected.

3.21. (†) Let G be a topological group and X a space. We call X a *G-space* if there is a continuous function $G \times X \to X$ denoted $(g, x) \to g \cdot x$, for $g \in G$ and $x \in X$ such that $e \cdot x = x$ and $g' \cdot (g \cdot x) = (g'g) \cdot x$, for all $g, g' \in G$ and $x \in X$, where e is the identity of G. We say that X is a *transitive G-space* if for every $x, x' \in X$, there exists a $g \in G$ such that $x' = g \cdot x$. For an element $x \in X$, the *isotropy subgroup* G_x of G consists of all $g \in G$ such that $g \cdot x = x$.

For a transitive G-space X, define a continuous surjection $\theta : G \to X$ by $\theta(g) = g \cdot *$. Prove that θ induces a continuous bijection $\widetilde{\theta} : G/G_* \to X$. Prove that if θ is an open map or if G is compact and X is Hausdorff, then $\widetilde{\theta}$ is a homeomorphism. Apply this result to $V_k(\mathbb{F}^n)$, where $\mathbb{F} = \mathbb{R}, \mathbb{C}$, or $\mathbb{H}$ to obtain Lemma 3.4.10 and its analogues for $\mathbb{C}$ and $\mathbb{H}$.

3.22. (†) Consider the determinant function $\det : O(n) \to \{\pm 1\}$. Define the *special orthogonal group* $SO(n)$ to be the subgroup of $O(n)$ consisting of all matrices of determinant 1.

1. Prove that the orthogonal group $O(n)$ is compact.
2. Prove that the special orthogonal group $SO(n)$ is compact.

3.23. (†) Prove that $SO(n)/SO(n-k)$ is homeomorphic to the real Stiefel manifold $V_{k,n}$.

3.24. (*) (†) Prove

1. $O(n)$ is not connected.
2. $SO(n)$ is connected.
3. $O(n)$ is the union of two connected components, one of which is $SO(n)$ and the other of which is homeomorphic to $SO(n)$.

3.25. (*) Prove that $G_k(\mathbb{F}^n)$ is Hausdorff, where $\mathbb{F} = \mathbb{R}, \mathbb{C}$, or $\mathbb{H}$.

3.26. Prove that X and Y have the same homotopy type if and only if X and Y are deformation retracts of the same space.

3.27. Prove the following result about mapping cylinders. If $f_0 \simeq f_1 : X \to Y$, then $M_{f_0} \simeq M_{f_1}$.

3.28. Let $\Delta : X \to X \times X$ be the diagonal map and let $\Delta'' : E_\Delta \to X \times X$ be the mapping path fibration. Consider the fibration $p : X^I \to X \times X$ given by $p(l) = (l(0), l(1))$, with $X = A = B$ and $l \in X^I$, which is defined just before Proposition 3.3.2.

1. Prove that there is a homeomorphism $\theta : E_\Delta \to X^I$ such that $p\theta = \Delta''$.
2. Show that the homotopy fiber of Δ has homotopy type of ΩX.
3. If $j_1 : X \to X \times X$ is the inclusion into the first factor, prove that the pullback of j_1 and p gives the fiber sequence $\Omega X \to EX \to X$.

Chapter 4
Exact Sequences

4.1 Introduction

In the first section we use the ideas of Chapter 3 to derive several basic exact sequences. The main sequences that we consider are two long exact sequences of homotopy sets. One is associated to a fiber sequence $F \to E \to B$. The terms are the homotopy sets $[X, Y]$, where the X's are the iterated suspensions of some fixed space and the Y's are the successive spaces of the fiber sequence. The other sequence is associated to a cofiber sequence $A \to X \to Q$. The terms are the homotopy sets $[X, Y]$, where the Y's are the iterated loop spaces of some fixed space and the X's are the successive spaces of the cofiber sequence. As special cases we obtain the exact homotopy sequence of a fibration and the exact cohomology sequence of a cofibration. We next discuss the action of an H-space on a space and the coaction of a co-H-space on a space. The former is illustrated in a fiber sequence by the action of the loops of the base ΩB on the fiber F, and the latter in a cofiber sequence by the coaction of the suspension ΣA on the cofiber Q. These then yield operations of one homotopy set on another and give sharper exactness results at the end terms of the two exact sequences of homotopy sets mentioned above. In the final section we return to homotopy groups. We give equivalent characterizations of homotopy groups, define the homotopy groups of a pair, establish the exact homotopy sequence of a pair, and discuss the relative Hurewicz homomorphism. We conclude by considering certain maps, called excision maps, which are associated to maps $X \to Y \to Z$ whose composition is trivial. These maps are heavily used in Chapter 6.

Unless otherwise stated, all topological spaces are based and all maps and homotopies preserve the basepoint. We begin with some simple definitions.

4.2 The Coexact and Exact Sequence of a Map

Definition 4.2.1

1. A set S with a fixed element denoted 0 is called a *based set*. If S and T are based sets, then a function $\phi : S \to T$ such that $\phi(0) = 0$ is called a *based function*. The *kernel* of ϕ, denoted $\operatorname{Ker}\phi$, is the set $\{x \mid x \in S,\ \phi(x) = 0\}$ and the *image* of ϕ, denoted $\operatorname{Im}\phi$, is just the image of the function ϕ. Let S_i be based sets and $\phi_i : S_i \to S_{i+1}$ based functions for $i = 1, 2$. Then the sequence
$$S_1 \xrightarrow{\phi_1} S_2 \xrightarrow{\phi_2} S_3$$
is called *exact at* S_2 if $\operatorname{Im}\phi_1 = \operatorname{Ker}\phi_2$. The sequence (finite or infinite) of based sets and functions
$$\cdots \longrightarrow S_{i-1} \xrightarrow{\phi_{i-1}} S_i \xrightarrow{\phi_i} S_{i+1} \longrightarrow \cdots$$
is *exact* if it is exact at each S_i. If
$$\cdots \longrightarrow S'_{i-1} \xrightarrow{\phi'_{i-1}} S'_i \xrightarrow{\phi'_i} S'_{i+1} \longrightarrow \cdots$$
is another exact sequence, then a map of the first exact sequence into the second exact sequence consists of based functions $h_i : S_i \to S'_i$ such that the following diagram commutes
$$\begin{array}{ccccccccc} \cdots \longrightarrow & S_{i-1} & \xrightarrow{\phi_{i-1}} & S_i & \xrightarrow{\phi_i} & S_{i+1} & \longrightarrow \cdots \\ & \downarrow h_{i-1} & & \downarrow h_i & & \downarrow h_{i+1} & \\ \cdots \longrightarrow & S'_{i-1} & \xrightarrow{\phi'_{i-1}} & S'_i & \xrightarrow{\phi'_i} & S'_{i+1} & \longrightarrow \cdots . \end{array}$$

2. Now consider a sequence (finite or infinite) of spaces and maps
$$X_1 \xrightarrow{f_1} X_2 \xrightarrow{f_2} X_3 \xrightarrow{f_3} \cdots .$$
This sequence is called *coexact* if for every space Z, the sequence of based sets and functions
$$\cdots \xrightarrow{f_3^*} [X_3, Z] \xrightarrow{f_2^*} [X_2, Z] \xrightarrow{f_1^*} [X_1, Z]$$
is exact.

3. A sequence (finite or infinite) of spaces and maps
$$\cdots \xrightarrow{g_4} Y_3 \xrightarrow{g_3} Y_2 \xrightarrow{g_2} Y_1$$

is called *exact* if for every space W, the sequence of based sets and functions

$$\cdots \xrightarrow{g_{4*}} [W, Y_3] \xrightarrow{g_{3*}} [W, Y_2] \xrightarrow{g_{2*}} [W, Y_1]$$

is exact.

If the S_i are groups with identity 0 and the ϕ_i are homomorphisms, then Ker and Im are groups, and this notion of exactness agrees with the more familiar algebraic notion of exactness. In this context, a map of one exact sequence of groups into another has the additional requirement that each h_i is a homomorphism.

We refer to a sequence of based sets and functions as a sequence of sets, to a sequence of groups and homomorphisms as a sequence of groups, and to a sequence of spaces and maps as a sequence of spaces.

Definition 4.2.2 Let

$$\cdots \longrightarrow X_{i-1} \xrightarrow{f_{i-1}} X_i \xrightarrow{f_i} X_{i+1} \longrightarrow \cdots$$

and

$$\cdots \longrightarrow Y_{i-1} \xrightarrow{g_{i-1}} Y_i \xrightarrow{g_i} Y_{i+1} \longrightarrow \cdots$$

be two sequences of spaces. Suppose they are both finite on the left ($M \leqslant i$, for some M), finite on the right ($i \leqslant N$, for some N), finite ($M \leqslant i \leqslant N$), or infinite (no restriction on i). Then the sequences are called *equivalent* if there are homotopy equivalences $\alpha_i : X_i \to Y_i$ such that each square in the following diagram

$$\begin{array}{ccccccccc}
\cdots \longrightarrow & X_{i-1} & \xrightarrow{f_{i-1}} & X_i & \xrightarrow{f_i} & X_{i+1} & \longrightarrow \cdots \\
& \downarrow{\scriptstyle \alpha_{i-1}} & & \downarrow{\scriptstyle \alpha_i} & & \downarrow{\scriptstyle \alpha_{i+1}} & \\
\cdots \longrightarrow & Y_{i-1} & \xrightarrow{g_{i-1}} & Y_i & \xrightarrow{g_i} & Y_{i+1} & \longrightarrow \cdots
\end{array}$$

is homotopy-commutative.

We begin with a lemma.

Lemma 4.2.3 *Let* $X_1 \xrightarrow{f_1} X_2 \xrightarrow{f_2} X_3 \xrightarrow{f_3} \cdots$ *be a coexact sequence.*

1. *Then* $\Sigma X_1 \xrightarrow{\Sigma f_1} \Sigma X_2 \xrightarrow{\Sigma f_2} \Sigma X_3 \xrightarrow{\Sigma f_3} \cdots$ *is coexact.*
2. *If the sequence* $Y_1 \longrightarrow Y_2 \longrightarrow Y_3 \longrightarrow \cdots$ *is equivalent to the given sequence, then it is coexact.*

Proof. The only part of the proof requiring comment is (1), and this follows from the adjoint isomorphism $\kappa_* : [\Sigma X, Z] \cong [X, \Omega Z]$ and the fact that $f^* \kappa_* = \kappa_* (\Sigma f)^*$ for any $f : X \to X'$. □

The next lemma deals with the mapping cone.

Lemma 4.2.4

1. *If $f : X \to Y$ is any map, then* $X \xrightarrow{f} Y \xrightarrow{k} C_f$ *is coexact, where k is the inclusion.*
2. *Any cofiber sequence is coexact.*

Proof. (1) This is a consequence of Proposition 3.2.12.
(2) Since any cofiber sequence $X \to Y \to Z$ is equivalent to $X \to Y \to C_f$ by Proposition 3.5.4, the second assertion follows. □

We will show that the infinite sequence of spaces and maps obtained from the mapping cone construction is coexact. The main step in the proof is the following proposition.

Proposition 4.2.5 *Let $f : X \to Y$ be any map, let $l : Y \to C_f$ be the inclusion, and let $p : C_f \to C_f/l(Y) = \Sigma X$ be the projection. Then the sequence of spaces*

$$X \xrightarrow{f} Y \xrightarrow{l} C_f \xrightarrow{p} \Sigma X \xrightarrow{-\Sigma f} \Sigma Y,$$

is coexact.

Proof. For any space Z, we must show that

$$[\Sigma Y, Z] \xrightarrow{(-\Sigma f)^*} [\Sigma X, Z] \xrightarrow{p^*} [C_f, Z] \xrightarrow{l^*} [Y, Z] \xrightarrow{f^*} [X, Z]$$

is exact. Exactness at $[Y, Z]$ follows from Lemma 4.2.4. We also have that

$$Y \xrightarrow{l} C_f \xrightarrow{p} \Sigma X$$

is a cofiber sequence, and so by Lemma 4.2.4, there is exactness at $[C_f, Z]$. Next recall from Proposition 3.5.4 and Lemma 3.5.6 that there is a homotopy-commutative diagram

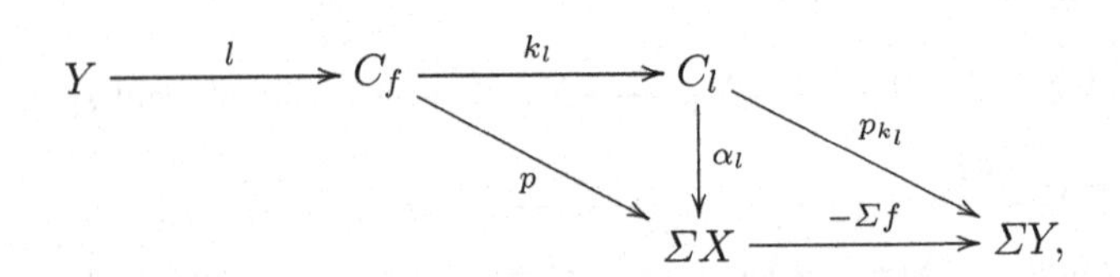

where $k_l : C_f \to C_f \cup_l CY = C_l$ is the inclusion, $\Sigma Y = C_l/k_l(C_f)$, $p = p_l$ and p_{k_l} are projections onto cofibers, and α_l is a homotopy equivalence. The sequence

$$C_f \xrightarrow{p} \Sigma X \xrightarrow{-\Sigma f} \Sigma Y$$

is therefore equivalent to the sequence

$$C_f \xrightarrow{k_l} C_l \xrightarrow{p_{k_l}} \Sigma Y,$$

and hence there is exactness at $[\Sigma X, Z]$. □

Remark 4.2.6 Because $(\Sigma f)^* : [\Sigma Y, Z] \to [\Sigma X, Z]$ is a group homomorphism and $-(\Sigma f)^* = (-\Sigma f)^*$, we obtain from Proposition 4.2.5 that the following sequence of spaces

$$X \xrightarrow{f} Y \xrightarrow{l} C_f \xrightarrow{p} \Sigma X \xrightarrow{\Sigma f} \Sigma Y$$

is coexact.

Theorem 4.2.7 *If $f : X \to Y$ is any map, then the sequence of spaces*

$$X \xrightarrow{f} Y \xrightarrow{l} C_f \xrightarrow{p} \Sigma X \longrightarrow \cdots$$

$$\cdots \longrightarrow \Sigma^n X \xrightarrow{\Sigma^n f} \Sigma^n Y \xrightarrow{\Sigma^n l} \Sigma^n C_f \xrightarrow{\Sigma^n p} \Sigma^{n+1} X \longrightarrow \cdots$$

is coexact.

Proof. We apply Lemma 4.2.3 to the coexact sequence in Remark 4.2.6 and obtain the coexact sequence

$$\Sigma X \xrightarrow{\Sigma f} \Sigma Y \xrightarrow{\Sigma l} \Sigma C_f \xrightarrow{\Sigma p} \Sigma^2 X \xrightarrow{\Sigma^2 f} \Sigma^2 Y.$$

We then apply Lemma 4.2.3 to this sequence. We continue in this way to complete the proof. □

As an immediate consequence of Theorem 4.2.7 we have the following corollary.

Corollary 4.2.8 *For any map $f : X \to Y$ and space Z, the following sequence of groups and sets is exact*

$$\longrightarrow [\Sigma^n C_f, Z] \xrightarrow{(\Sigma^n l)^*} [\Sigma^n Y, Z] \xrightarrow{(\Sigma^n f)^*} [\Sigma^n X, Z] \xrightarrow{(\Sigma^{n-1} p)^*} [\Sigma^{n-1} C_f, Z] \longrightarrow$$

$$\cdots \xrightarrow{p^*} [C_f, Z] \xrightarrow{l^*} [Y, Z] \xrightarrow{f^*} [X, Z].$$

Remark 4.2.9 The exact sequence in Corollary 4.2.8 consists of sets, groups, and abelian groups. In general, the last three terms are sets, the next three are groups, and all others are abelian groups.

By taking Z to be an Eilenberg–Mac Lane space in Corollary 4.2.8 we obtain the exact sequence in the following corollary.

Corollary 4.2.10 *If $f : X \to Y$ is a map, G an abelian group, and $n \geqslant 0$, then there is a homomorphism $\delta^n : H^n(X;G) \to H^{n+1}(C_f;G)$ such that the following sequence is exact*

$$H^0(C_f;G) \xrightarrow{l^*} H^0(Y;G) \xrightarrow{f^*} H^0(X;G) \xrightarrow{\delta^0} H^1(C_f;G) \longrightarrow \cdots$$

$$\longrightarrow H^n(C_f;G) \xrightarrow{l^*} H^n(Y;G) \xrightarrow{f^*} H^n(X;G) \xrightarrow{\delta^n} H^{n+1}(C_f;G) \longrightarrow \cdots$$

Proof. Define δ^n to be the composition

$$H^n(X;G) \xrightarrow{\cong} H^{n+1}(\Sigma X;G) \xrightarrow{p^*} H^{n+1}(C_f;G). \qquad \square$$

We introduce some terminology for the sequences that we have discussed.

Definition 4.2.11 The coexact sequence of spaces in Theorem 4.2.7 is called the *coexact sequence of a map*. The exact sequence of Corollary 4.2.8 is sometimes referred to as the *Puppe sequence* or the *Barratt–Puppe sequence.* We call the exact sequence in Corollary 4.2.10 the *exact cohomology sequence of a map.*

There are similar sequences for a cofibration. These are obtained as a consequence of the results above. We proceed as follows. Let

$$A \xrightarrow{i} X \xrightarrow{q} Q$$

be a cofiber sequence. In Proposition 3.5.4 we proved that there is a homotopy equivalence α from the mapping cone C_i of i into Q such that the following diagram commutes

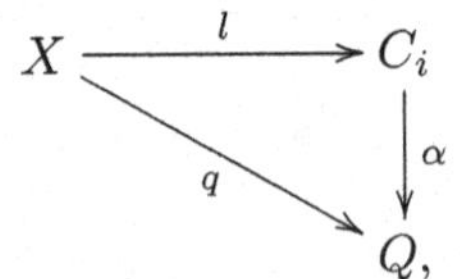

where l is the inclusion. Thus we have the following equivalence of sequences

$$\begin{array}{ccccccccccc} A & \xrightarrow{i} & X & \xrightarrow{l} & C_i & \xrightarrow{p} & \Sigma A & \xrightarrow{\Sigma i} & \Sigma X & \longrightarrow & \cdots \\ \downarrow{\scriptstyle \mathrm{id}} & & \downarrow{\scriptstyle \mathrm{id}} & & \downarrow{\scriptstyle \alpha} & & \downarrow{\scriptstyle \mathrm{id}} & & \downarrow{\scriptstyle \mathrm{id}} & & \\ A & \xrightarrow{i} & X & \xrightarrow{q} & Q & \xrightarrow{\partial} & \Sigma A & \xrightarrow{\Sigma i} & \Sigma X & \longrightarrow & \cdots, \end{array}$$

where $\partial = p\alpha^{-1}$. By Lemma 4.2.3, the lower sequence is coexact. This allows us to transfer the preceding results for the map $i : A \to X$ to the cofiber sequence $A \longrightarrow X \longrightarrow Q$. As a consequence of Corollaries 4.2.8 and 4.2.10, we have the following corollary.

Corollary 4.2.12 *Let $A \xrightarrow{i} X \xrightarrow{q} Q$ be a cofiber sequence.*
1. For any space Z, the following sequence is exact

$$\cdots \longrightarrow [\Sigma^n Q, Z] \xrightarrow{(\Sigma^n q)^*} [\Sigma^n X, Z] \xrightarrow{(\Sigma^n i)^*} [\Sigma^n A, Z] \xrightarrow{\sigma^n} [\Sigma^{n-1} Q, Z] \longrightarrow$$

$$\cdots \xrightarrow{\sigma^1} [Q, Z] \xrightarrow{q^*} [X, Z] \xrightarrow{i^*} [A, Z],$$

where $\sigma^n = (\Sigma^{n-1}\partial)^* : [\Sigma^n A, Z] \to [\Sigma^{n-1} Q, Z]$.

2. If G is an abelian group and $n \geqslant 0$, then there is a homomorphism $\Delta^n : H^n(A; G) \to H^{n+1}(Q; G)$ such that the following sequence is exact

$$H^0(Q; G) \xrightarrow{q^*} H^0(X; G) \xrightarrow{i^*} H^0(A; G) \xrightarrow{\Delta^0} H^1(Q; G) \longrightarrow \cdots$$

$$\longrightarrow H^n(Q; G) \xrightarrow{q^*} H^n(X; G) \xrightarrow{i^*} H^n(A; G) \xrightarrow{\Delta^n} H^{n+1}(Q; G) \longrightarrow \cdots .$$

Proof. Define Δ^n as the composition

$$H^n(A; G) \xrightarrow{\cong} H^{n+1}(\Sigma A; G) \xrightarrow{\partial^*} H^{n+1}(Q; G). \qquad \square$$

Definition 4.2.13 Let $A \xrightarrow{i} X \xrightarrow{q} Q$ be a cofiber sequence. The map $\partial = p\alpha^{-1} : Q \to \Sigma A$ is called the *connecting map of the cofibration.* The exact sequence (1) of Corollary 4.2.12 is called the *exact sequence of a cofibration* and the exact sequence (2) of Corollary 4.2.12 is called the *exact cohomology sequence of a cofibration.*

Remark 4.2.14

1. The exact cohomology sequence of a cofibration in Corollary 4.2.12 is isomorphic to the exact CW or singular cohomology sequence of the pair (Y, X) with coefficients in G, when the CW or singular cohomology groups $H^*(Y, X; G)$ are substituted for $H^*(Q; G)$ (see [32, p. 14]).
2. We note for later use that there is an exact homology sequence of a map and an exact homology sequence of a cofibration. These exact sequences are similar to the ones above for cohomology in Corollaries 4.2.10 and 4.2.12, but the homology groups replace the cohomology groups and the arrows are reversed (see [14, p. 434]).

We next discuss the dual results for exact sequences of spaces. We state the main results without proof and leave the statement of intermediate results and all the proofs as exercises. These are straightforward translations via duality of the corresponding results for coexact sequences.

For a map $f : X \to Y$, we denote by I_f the homotopy fiber of f (Definition 3.3.13). We consider the sequence of spaces

$$\Omega Y \xrightarrow{i} I_f \xrightarrow{v} X \xrightarrow{f} Y,$$

where v is the fiber map and i is the inclusion of the fiber ΩY into the total space I_f.

Theorem 4.2.15 *If $f : X \to Y$ is a map, then the following sequence of spaces is exact*

$$\longrightarrow \Omega^{n+1} Y \xrightarrow{\Omega^n i} \Omega^n I_f \xrightarrow{\Omega^n v} \Omega^n X \xrightarrow{\Omega^n f} \Omega^n Y \longrightarrow$$

$$\cdots \longrightarrow I_f \xrightarrow{v} X \xrightarrow{f} Y.$$

The next corollary is an immediate consequence of the theorem.

Corollary 4.2.16 *If $f : X \to Y$ is a map and W is any space, then the following sequence of groups and sets is exact*

$$\longrightarrow [W, \Omega^{n+1} Y] \xrightarrow{(\Omega^n i)_*} [W, \Omega^n I_f] \xrightarrow{(\Omega^n v)_*} [W, \Omega^n X] \xrightarrow{(\Omega^n f)_*} [W, \Omega^n Y] \longrightarrow$$

$$\cdots \longrightarrow [W, \Omega Y] \xrightarrow{i_*} [W, I_f] \xrightarrow{v_*} [W, X] \xrightarrow{f_*} [W, Y].$$

By taking W to be the Moore space $M(G, n)$ in Corollary 4.2.16, we obtain an exact sequence of homotopy groups.

Corollary 4.2.17 *If $f : X \to Y$ is a map, G an abelian group, and $n \geqslant 2$, then there is a homomorphism $d_{n+1} : \pi_{n+1}(Y; G) \to \pi_n(I_f; G)$ such that the following sequence of homotopy groups with coefficients is exact*

$$\longrightarrow \pi_{n+1}(Y; G) \xrightarrow{d_{n+1}} \pi_n(I_f; G) \xrightarrow{v_*} \pi_n(X; G) \xrightarrow{f_*} \pi_n(Y; G) \longrightarrow$$

$$\cdots \longrightarrow \pi_3(Y; G) \xrightarrow{d_3} \pi_2(I_f; G) \xrightarrow{v_*} \pi_2(X; G) \xrightarrow{f_*} \pi_2(Y; G).$$

If $G = \mathbb{Z}$, then this sequence can be continued to

$$\pi_1(I_f) \xrightarrow{v_*} \pi_1(X) \xrightarrow{f_*} \pi_1(Y) \xrightarrow{d_1} \pi_0(I_f) \xrightarrow{v_*} \pi_0(X) \xrightarrow{f_*} \pi_0(Y).$$

Proof. Define d_{n+1} to be the composition

$$\pi_{n+1}(Y; G) \xrightarrow{\cong} \pi_n(\Omega Y; G) \xrightarrow{i_*} \pi_n(I_f; G). \qquad \square$$

Next we show how to obtain exact sequences for fibrations. Let

$$F \xrightarrow{j} E \xrightarrow{p} B$$

be a fiber sequence. By Proposition 3.5.10, there is a homotopy equivalence $\beta : F \to I_p$, where I_p is the homotopy fiber of p.

Definition 4.2.18 The *connecting map* $\partial : \Omega B \to F$ of the above fibration is the composition

$$\Omega B \xrightarrow{i} I_p \xrightarrow{\beta^{-1}} F,$$

where i is the inclusion and β^{-1} is a homotopy inverse of β.

It now follows, analogous to the dual situation, that the sequence

$$\cdots \longrightarrow \Omega^{n+1}B \xrightarrow{\Omega^n \partial} \Omega^n F \xrightarrow{\Omega^n j} \Omega^n E \xrightarrow{\Omega^n p} \Omega^n B \longrightarrow$$

$$\cdots \longrightarrow \Omega B \xrightarrow{\partial} F \xrightarrow{j} E \xrightarrow{p} B.$$

is equivalent to the sequence

$$\cdots \longrightarrow \Omega^{n+1}B \xrightarrow{\Omega^n i} \Omega^n I_p \xrightarrow{\Omega^n v} \Omega^n E \xrightarrow{\Omega^n p} \Omega^n B \longrightarrow$$

$$\cdots \longrightarrow \Omega B \xrightarrow{i} I_p \xrightarrow{v} E \xrightarrow{p} B.$$

Thus we obtain the following corollary.

Corollary 4.2.19 *Let* $F \xrightarrow{j} E \xrightarrow{p} B$ *be a fiber sequence.*

1. If W *is any space, then the sequence*

$$\cdots \longrightarrow [W, \Omega^{n+1}B] \xrightarrow{(\Omega^n \partial)_*} [W, \Omega^n F] \xrightarrow{(\Omega^n j)_*} [W, \Omega^n E] \xrightarrow{(\Omega^n p)_*} [W, \Omega^n B] \longrightarrow$$

$$\cdots \longrightarrow [W, \Omega B] \xrightarrow{\partial_*} [W, F] \xrightarrow{j_*} [W, E] \xrightarrow{p_*} [W, B]$$

is exact.

2. If G *is an abelian group and* $n \geqslant 2$*, then there exists a homomorphism* $\partial_{n+1} : \pi_{n+1}(B; G) \to \pi_n(F; G)$ *such that the following sequence is exact*

$$\cdots \longrightarrow \pi_{n+1}(B;G) \xrightarrow{\partial_{n+1}} \pi_n(F;G) \xrightarrow{j_*} \pi_n(E;G) \xrightarrow{p_*} \pi_n(B;G) \longrightarrow$$

$$\cdots \longrightarrow \pi_3(B;G) \xrightarrow{\partial_3} \pi_2(F;G) \xrightarrow{j_*} \pi_2(E;G) \xrightarrow{p_*} \pi_2(B;G).$$

If $G = \mathbb{Z}$*, then this sequence can be continued to*

$$\pi_1(F) \xrightarrow{j_*} \pi_1(E) \xrightarrow{p_*} \pi_1(B) \xrightarrow{\partial_1} \pi_0(F) \xrightarrow{j_*} \pi_0(E) \xrightarrow{p_*} \pi_0(B).$$

Proof. Define ∂_{n+1} to be the composition

$$\pi_{n+1}(B;G) \xrightarrow{\cong} \pi_n(\Omega B;G) \xrightarrow{\partial_*} \pi_n(F;G). \qquad \square$$

We now introduce some terminology for the previous exact sequences.

Definition 4.2.20 The exact sequence of spaces in Theorem 4.2.15 is called the *exact sequence of a map*. The exact sequence in Corollary 4.2.17 is called the *exact homotopy sequence of a map*. The exact sequence in Corollary 4.2.19(2) is called the *exact homotopy sequence of a fibration*.

Remark 4.2.21 We observe that the exact sequences of sets and groups in Corollaries 4.2.8, 4.2.10, 4.2.12, 4.2.16, 4.2.17, and 4.2.19 are all natural with respect to the appropriate maps of spaces. This means that these maps induce maps of the exact sequences. Rather than stating this result for all of the sequences mentioned above, we just do it for one of the corollaries to illustrate the general principle.

We consider Corollary 4.2.8. Let $f : X \to Y$ and $f' : X' \to Y'$ be maps and assume that there is a homotopy-commutative diagram

$$\begin{array}{ccc} X & \xrightarrow{f} & Y \\ \downarrow a & & \downarrow b \\ X' & \xrightarrow{f'} & Y'. \end{array}$$

By Proposition 3.2.13, there is a map $\Phi : C_f \to C_{f'}$ such that the following diagram homotopy-commutes

$$\begin{array}{ccccc} Y & \xrightarrow{l} & C_f & \xrightarrow{p} & \Sigma X \\ \downarrow b & & \downarrow \Phi & & \downarrow \Sigma a \\ Y' & \xrightarrow{l'} & C_{f'} & \xrightarrow{p'} & \Sigma X'. \end{array}$$

From this we obtain a map of the exact sequence of f' in Corollary 4.2.8 into the analogous exact sequence of f

$$\begin{array}{ccccccccc} \longrightarrow & [\Sigma^n C_{f'}, Z] & \longrightarrow & [\Sigma^n Y', Z] & \longrightarrow & [\Sigma^n X', Z] & \longrightarrow & [\Sigma^{n-1} C_{f'}, Z] & \longrightarrow \\ & \downarrow (\Sigma^n \Phi)^* & & \downarrow (\Sigma^n b)^* & & \downarrow (\Sigma^n a)^* & & \downarrow (\Sigma^{n-1}\Phi)^* & \\ \longrightarrow & [\Sigma^n C_f, Z] & \longrightarrow & [\Sigma^n Y, Z] & \longrightarrow & [\Sigma^n X, Z] & \longrightarrow & [\Sigma^{n-1} C_f, Z] & \longrightarrow, \end{array}$$

where the top line is induced by the suspensions of l', f', and p' and the bottom line is induced by the suspensions of l, f, and p. Thus the sequence in Corollary 4.2.8 is natural with respect to the maps a and b.

4.3 Actions and Coactions

In this section we introduce the notion of an H-space acting on a space and the dual notion. These are used to refine exactness of the end terms of the some of the exact sequences of Section 4.2. We begin by giving the homotopy version of a group acting on a space.

Definition 4.3.1 Given a space X and an H-space (A, m). A *(right) action of A on X* is a map $\phi : X \times A \to X$ such that

1. $\phi j_1 \simeq \mathrm{id} : X \to X$, where $j_1 : X \to X \times A$ is the inclusion.
2. $\phi(\phi \times \mathrm{id}) \simeq \phi(\mathrm{id} \times m) : X \times A \times A \to X$:

$$\begin{array}{ccc} X \times A \times A & \xrightarrow{\phi \times \mathrm{id}} & X \times A \\ \downarrow{\scriptstyle \mathrm{id} \times m} & & \downarrow{\scriptstyle \phi} \\ X \times A & \xrightarrow{\phi} & X. \end{array}$$

We say that *A acts on X by ϕ*.

We now give a few simple examples. We give others later in this section.

Example 4.3.2

- If G is a group and if there is an action of G on X with equality instead of homotopy in Definition 4.3.1, then X is a right G-space (Exercise 3.21). This concept appears frequently in the literature ([14, p. 54], [49, Chap. 1, Sect. 12]).
- If $X = A$ and (X, m) is a homotopy-associative H-space, then X acts on itself. Definition 4.3.1(2) is just the homotopy-associativity of m.

We next show that in a fibration, the loops on the base act on the fiber. We begin with principal fibrations. Let $f : X \to Y$ be a map and let $I_f = \{(x, \omega) \mid x \in X,\ \omega \in EY,\ f(x) = \omega(0)\}$ be the homotopy fiber of f. Then

$$\Omega Y \xrightarrow{i} I_f \xrightarrow{v} X,$$

is a fiber sequence, where $i(\omega) = (*, \omega)$ and $v(x, \omega) = x$ for $\omega \in \Omega Y$ and $x \in X$. Recall that for any space Y, the loop space ΩY is a grouplike space with multiplication $m : \Omega Y \times \Omega Y \to \Omega Y$, and we write $m(\omega, \nu) = \omega + \nu$, for $\omega, \nu \in \Omega Y$. More generally, if $\omega \in EY$ and $\nu \in \Omega Y$, then $\omega + \nu \in EY$.

We define a map $\phi_0 : I_f \times \Omega Y \to I_f$ by $\phi_0((x, \omega), \nu) = (x, \omega + \nu)$, for $(x, \omega) \in I_f$ and $\nu \in \Omega Y$.

Proposition 4.3.3

1. *The map $\phi_0 : I_f \times \Omega Y \to I_f$ is an action of ΩY on I_f.*

2. *The following diagram commutes*

$$\begin{array}{ccc} I_f \times \Omega Y & \xrightarrow{\phi_0} & I_f \\ \Big\downarrow{\scriptstyle p_1} & & \Big\downarrow{\scriptstyle v} \\ I_f & \xrightarrow{v} & X, \end{array}$$

where $p_1 : I_f \times \Omega Y \to I_f$ *is the projection.*

3. *The following square diagram is commutative and the following triangular diagram is homotopy-commutative,*

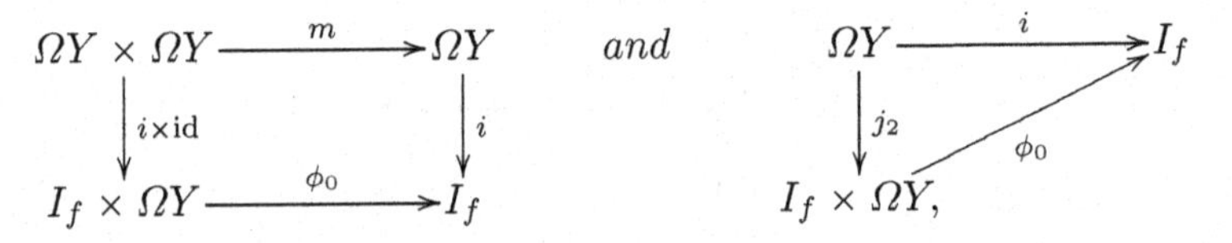

where j_2 *is inclusion into the second factor.*

4. *If the square*

$$\begin{array}{ccc} X & \xrightarrow{f} & Y \\ \Big\downarrow{\scriptstyle a} & & \Big\downarrow{\scriptstyle b} \\ X' & \xrightarrow{f'} & Y' \end{array}$$

is commutative and $\Psi : I_f \to I_{f'}$ *is defined by* $\Psi(x, \omega) = (a(x), b\,\omega)$, *for* $(x, \omega) \in I_f$, *then the square*

$$\begin{array}{ccc} I_f \times \Omega Y & \xrightarrow{\phi_0} & I_f \\ \Big\downarrow{\scriptstyle \Psi \times \Omega b} & & \Big\downarrow{\scriptstyle \Psi} \\ I_{f'} \times \Omega Y' & \xrightarrow{\phi'_0} & I_{f'}, \end{array}$$

is commutative, where ϕ_0 *and* ϕ'_0 *are the actions.*

Proof. (1) Let $j_1 : I_f \to I_f \times \Omega Y$ be the inclusion and let $(x, \omega) \in I_f$. Then

$$\phi_0 j_1(x, \omega) = \phi_0((x, \omega), *) = (x, \omega + *),$$

and so $\phi_0 j_1 \simeq \mathrm{id}$. Next if $(x, \omega) \in I_f$ and $\lambda, \mu \in \Omega Y$, then

$$\phi_0(\phi_0 \times \mathrm{id})((x, \omega), \lambda, \mu) = (x, (\omega + \lambda) + \mu)$$

and

$$\phi_0(\mathrm{id} \times m)((x, \omega), \lambda, \mu) = (x, \omega + (\lambda + \mu)).$$

Hence $\phi_0(\phi_0 \times \mathrm{id}) \simeq \phi_0(\mathrm{id} \times m)$.

(2) and (3) The proofs are straightforward and hence omitted.

(4)

$$\begin{aligned}\Psi\phi_0((x,\omega),\nu) &= \Psi(x,\omega+\nu)\\ &= (a(x), b\,(\omega+\nu))\\ &= \phi_0'((a(x), b\,\omega), b\,\nu)\\ &= \phi_0'(\Psi\times\Omega b)((x,\omega),\nu),\end{aligned}$$

for $(x,\omega)\in I_f$ and $\nu\in\Omega Y$. □

We next show how to define an action for any fibration $F\xrightarrow{j}E\xrightarrow{p}B$. Recall from Proposition 3.5.10 and Remark 3.5.11 that the map $\beta : F\to I_p$ which is given by $\beta(x) = (j(x), *)$ for $x\in F$ is a homotopy equivalence. Then we define a map $\phi : F\times\Omega B\to F$ as the composition

$$F\times\Omega B\xrightarrow{\beta\times\mathrm{id}} I_p\times\Omega B\xrightarrow{\phi_0} I_p\xrightarrow{\beta^{-1}} F,$$

where ϕ_0 is the action above and β^{-1} is a homotopy inverse of β.

Theorem 4.3.4

1. *The map $\phi : F\times\Omega B\to F$ is an action of ΩB on F.*
2. *The following diagrams are homotopy-commutative*

$$\begin{array}{ccc}\Omega B\times\Omega B & \xrightarrow{m} & \Omega B\\ \downarrow{\scriptstyle\partial\times\mathrm{id}} & & \downarrow{\scriptstyle\partial}\\ F\times\Omega B & \xrightarrow{\phi} & F\end{array}\qquad\text{and}\qquad\begin{array}{ccc}\Omega B & \xrightarrow{\partial} & F\\ \downarrow{\scriptstyle j_2} & \nearrow{\scriptstyle\phi} & \\ F\times\Omega B, & & \end{array}$$

where $\partial : \Omega B\to F$ is the connecting map (Definition 4.2.18) and j_2 is inclusion into the second factor.

3. *If $F'\xrightarrow{j'}E'\xrightarrow{p'}B'$ is another fibration and there is a commutative diagram*

$$\begin{array}{ccccc}F & \xrightarrow{j} & E & \xrightarrow{p} & B\\ \downarrow{\scriptstyle d} & & \downarrow{\scriptstyle a} & & \downarrow{\scriptstyle b}\\ F' & \xrightarrow{j'} & E' & \xrightarrow{p'} & B',\end{array}$$

then the following diagram is homotopy-commutative

$$\begin{array}{ccc}F\times\Omega B & \xrightarrow{\phi} & F\\ \downarrow{\scriptstyle d\times\Omega b} & & \downarrow{\scriptstyle d}\\ F'\times\Omega B' & \xrightarrow{\phi'} & F',\end{array}$$

where ϕ and ϕ' are the actions of the two fibrations.

Proof. Since $\beta : F \to I_p$ is a homotopy equivalence and $\phi = \beta^{-1}\phi_0(\beta \times \mathrm{id})$ and $\partial = \beta^{-1}i$, the proofs of (1) and (2) follow easily from Proposition 4.3.3. For (3) we first note that the maps a and b determine a map $\Psi : I_p \to I_{p'}$ defined by $\Psi(x,\omega) = (a(x), b\omega)$, for $x \in E$ and $\omega \in EB$. Then it is easily checked that $\Psi\beta = \beta' d$, and so the squares in the following diagram

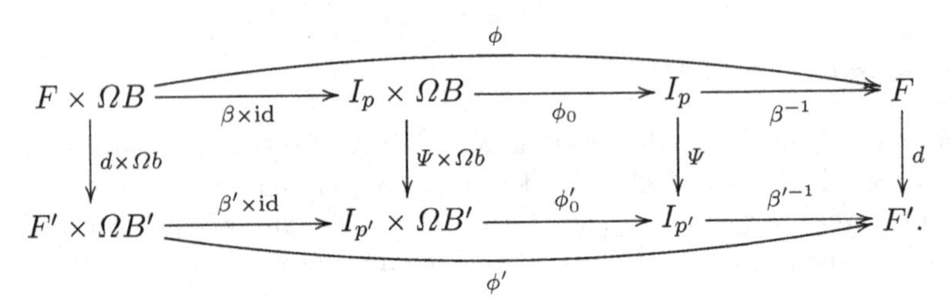

are homotopy-commutative. Therefore

$$d\phi = d\beta^{-1}\phi_0(\beta \times \mathrm{id}) \simeq \beta'^{-1}\phi_0'(\beta' \times \mathrm{id})(d \times \Omega b) = \phi'(d \times \Omega b).$$

This completes the proof. □

The map ϕ in Theorem 4.3.4 is sometimes referred to as the *holonomy map* or the *holonomy action.*

For the rest of this section we give the dual theory of coactions. These results follow easily by straightforward dualization, and so we state only the main results and omit their proofs.

Definition 4.3.5 Given a space Y and a co-H-space (B, c). A *(right) coaction of B on Y* is a map $\psi : Y \to Y \vee B$ such that

1. $q_1\psi \simeq \mathrm{id}$, where $q_1 : Y \vee B \to Y$ is the projection.
2. The following diagram is homotopy-commutative

$$\begin{array}{ccc} Y & \xrightarrow{\ \psi\ } & Y \vee B \\ \downarrow{\scriptstyle \psi} & & \downarrow{\scriptstyle \mathrm{id} \vee c} \\ Y \vee B & \xrightarrow{\ \psi \vee \mathrm{id}\ } & Y \vee B \vee B. \end{array}$$

We say that *B coacts on Y*.

We begin by defining a coaction for a principal cofibration. Let $f : X \to Y$ be a map and consider the cofiber sequence

$$Y \xrightarrow{\ l\ } C_f \xrightarrow{\ q\ } \Sigma X,$$

where l is the inclusion and q is the projection. We define a map $\psi_0 : C_f \to C_f \vee \Sigma X$ by $\psi_0\langle y\rangle = (\langle y\rangle, *)$ and

$$\psi_0\langle x,t\rangle = \begin{cases} (\langle x,2t\rangle,*) & \text{if } 0 \leqslant t \leqslant \frac{1}{2} \\ (*,\langle x,2t-1\rangle) & \text{if } \frac{1}{2} \leqslant t \leqslant 1, \end{cases}$$

for $x \in X$, $y \in Y$, and $t \in I$.

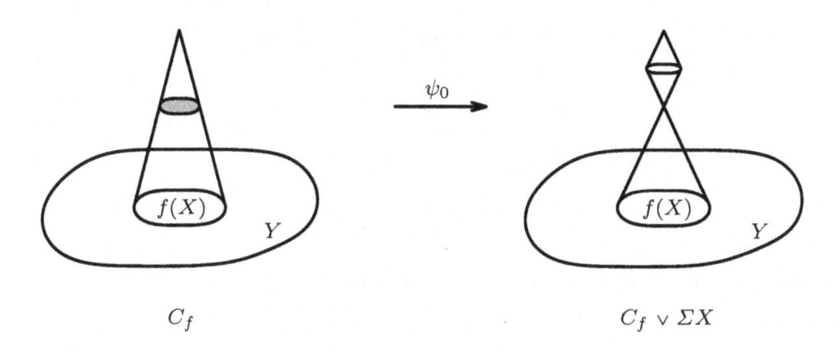

Figure 4.1

Proposition 4.3.6 *For any map $f : X \to Y$, the map $\psi_0 : C_f \to C_f \vee \Sigma X$ defined above is a coaction.*

For an arbitrary cofibration $A \xrightarrow{i} X \xrightarrow{p} Q$, the map $\alpha : C_i \to Q$ which is given by $\alpha\langle a,t\rangle = *$ and $\alpha\langle x\rangle = p(x)$, for $a \in A$, $x \in X$ and $t \in I$, is a homotopy equivalence (Proposition 3.5.4 and Remark 3.5.5). Then we define a map $\psi : Q \to Q \vee \Sigma A$ as the composition

$$Q \xrightarrow{\alpha^{-1}} C_i \xrightarrow{\psi_0} C_i \vee \Sigma A \xrightarrow{\alpha \vee \mathrm{id}} Q \vee \Sigma A.$$

Theorem 4.3.7

1. *The map $\psi : Q \to Q \vee \Sigma A$ is a coaction of ΣA on Q.*
2. *The following diagrams are homotopy-commutative*

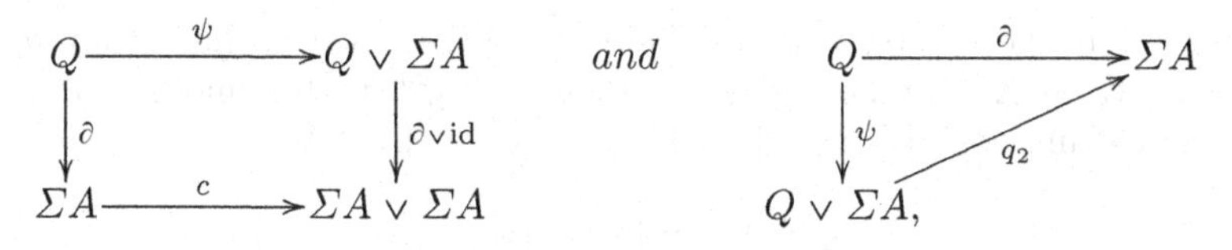

where $\partial : Q \to \Sigma A$ is the connecting map of the cofibration (Definition 4.2.13) and q_2 is the projection.

3. *If $A' \xrightarrow{i'} X' \xrightarrow{p'} Q'$ is a cofibration and there is a commutative diagram*

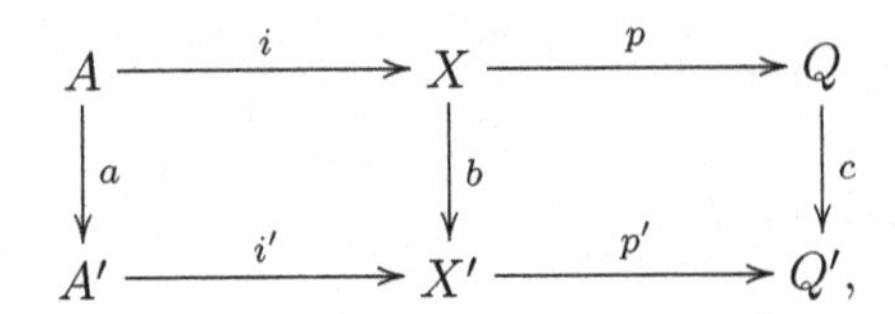

then the following diagram is homotopy-commutative

$$\begin{array}{ccc} Q & \xrightarrow{\psi} & Q \vee \Sigma A \\ \downarrow{\scriptstyle c} & & \downarrow{\scriptstyle c \vee \Sigma a} \\ Q' & \xrightarrow{\psi'} & Q' \vee \Sigma A', \end{array}$$

where ψ and ψ' are the coactions of the two cofibrations.

4.4 Operations

In this section we show how the actions and coactions of the previous section lead to operations of one homotopy set on another. We note that several authors use the terms "action" and "operation" interchangeably. However we attempt to use the word "action" to apply only to spaces.

Definition 4.4.1 Let S be a based set with a binary operation (denoted $+$) and two-sided identity 0. We say that S *operates* on a set V if for every $s \in S$ and $x \in V$, there is an element $x^s \in V$ such that (1) $x^0 = x$ and (2) $(x^s)^t = x^{s+t}$, for all $x \in V$ and $s, t \in S$.

Suppose that (A, m) is an H-space, X is a space, and $\phi : X \times A \to X$ is an action of A on X. If W is any space and $g : W \to X$ and $a : W \to A$ are any maps, then define $g^a : W \to X$ to be the composition

$$W \xrightarrow{\Delta} W \times W \xrightarrow{g \times a} X \times A \xrightarrow{\phi} X,$$

where Δ is the diagonal map. If $a \simeq a' : W \to A$ with homotopy a_t and $g \simeq g' : W \to X$ with homotopy g_t, then $g^a \simeq g'^{a'}$ with homotopy $g_t^{a_t}$. Thus there is a function $[W, X] \times [W, A] \to [W, X]$ defined by

$$[g]^{[a]} = [g^a] \in [W, X],$$

for $[g] \in [W, X]$ and $[a] \in [W, A]$.

Proposition 4.4.2 *The function $[W, X] \times [W, A] \to [W, X]$ defined above is an operation of $[W, A]$ on $[W, X]$.*

Proof. (1) Let $\rho = [f] \in [W, X]$ and $0 = [*] \in [W, A]$, let $p_1 : W \times W \to W$ be the projection onto the first factor, and let $j_1 : X \to X \times A$ be the inclusion. Then in the diagram

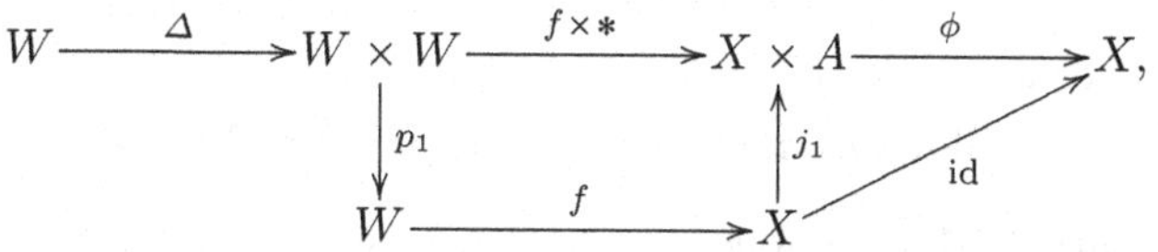

$$\begin{array}{ccccccc} W & \xrightarrow{\Delta} & W \times W & \xrightarrow{f \times *} & X \times A & \xrightarrow{\phi} & X, \\ & & \downarrow p_1 & & \uparrow j_1 & \nearrow \mathrm{id} & \\ & & W & \xrightarrow{f} & X & & \end{array}$$

the top line is $f^* : W \to X$ (i.e., $*$ operating on f), the square is commutative and the triangle is homotopy-commutative by Definition 4.3.1. Thus $f^* \simeq f p_1 \Delta = f$, and so $\rho^0 = \rho$.

(2) Let $\rho = [f] \in [W, X]$ and $\alpha = [a]$, $\beta = [b] \in [W, A]$. Then in the following diagram

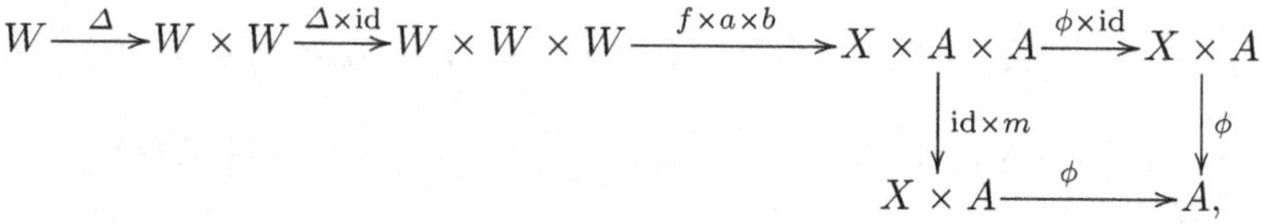

$$\begin{array}{ccccccccc} W & \xrightarrow{\Delta} & W \times W & \xrightarrow{\Delta \times \mathrm{id}} & W \times W \times W & \xrightarrow{f \times a \times b} & X \times A \times A & \xrightarrow{\phi \times \mathrm{id}} & X \times A \\ & & & & & & \downarrow \mathrm{id} \times m & & \downarrow \phi \\ & & & & & & X \times A & \xrightarrow{\phi} & A, \end{array}$$

the top line is $(f^a)^b$ and the triangle is homotopy-commutative by Definition 4.3.1. Therefore

$$\begin{aligned} (f^a)^b &\simeq \phi(f \times m(a \times b))(\Delta \times \mathrm{id})\Delta \\ &= \phi(f \times m(a \times b))(\mathrm{id} \times \Delta)\Delta \\ &= \phi(f \times (a + b))\Delta \\ &= f^{a+b}, \end{aligned}$$

and so $(\rho^\alpha)^\beta = \rho^{\alpha+\beta}$. □

Next we let $f : X \to Y$ be any map and consider the principal fiber sequence

$$\Omega Y \xrightarrow{i} I_f \xrightarrow{v} X.$$

By Proposition 4.3.3, there is an action $\phi_0 : I_f \times \Omega Y \to I_f$. This yields an operation of $[W, \Omega Y]$ on $[W, I_f]$, for any space W.

Lemma 4.4.3 *If* $i_* : [W, \Omega Y] \to [W, I_f]$ *and* $\alpha, \gamma \in [W, \Omega Y]$, *then*

1. $i_*(\alpha + \gamma) = (i_*(\alpha))^\gamma$.
2. $i_*(\gamma) = 0^\gamma$.

Proof. (1) Let $\alpha = [a]$ and $\gamma = [c]$ and consider the diagram

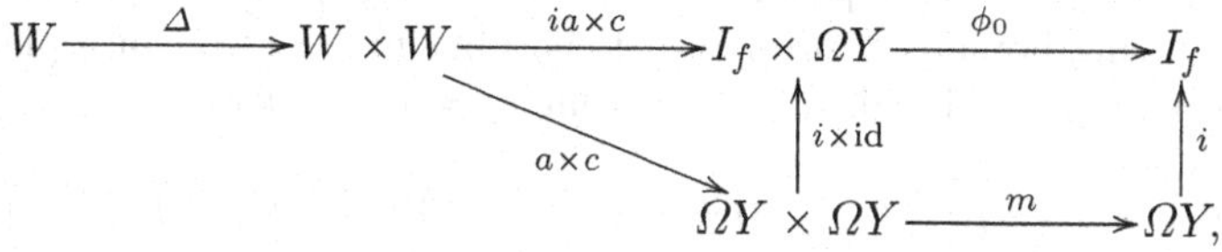

$$\begin{array}{ccccccc} W & \xrightarrow{\Delta} & W \times W & \xrightarrow{ia \times c} & I_f \times \Omega Y & \xrightarrow{\phi_0} & I_f \\ & & & \searrow a \times c & \uparrow i \times \mathrm{id} & & \uparrow i \\ & & & & \Omega Y \times \Omega Y & \xrightarrow{m} & \Omega Y, \end{array}$$

where the top line equals $(ia)^c$. The triangle is clearly commutative and the square is commutative by Proposition 4.3.3. Therefore

$$(ia)^c = i(a+c),$$

and this proves (1).

(2) The proof follows from Part (1) by setting $\alpha = 0$. □

By Corollary 4.2.16, the following sequence is exact

$$\cdots \longrightarrow [W, \Omega X] \xrightarrow{(\Omega f)_*} [W, \Omega Y] \xrightarrow{i_*} [W, I_f] \xrightarrow{v_*} [W, X],$$

for every space W. We use the operation of $[W, \Omega Y]$ on $[W, I_f]$ to obtain more information about the exactness at $[W, I_f]$ and at $[W, \Omega Y]$.

Theorem 4.4.4

1. *Let $\rho, \sigma \in [W, I_f]$. Then $v_*(\rho) = v_*(\sigma) \iff$ there exists $\gamma \in [W, \Omega Y]$ with $\sigma = \rho^\gamma$.*
2. *If $\gamma, \delta \in [W, \Omega Y]$, then $i_*(\gamma) = i_*(\delta) \iff$ there exists $\epsilon \in [W, \Omega X]$ with $\gamma = (\Omega f)_*(\epsilon) + \delta$.*

Proof. (1) Let $\rho = [g]$ and $\gamma = [c]$. We first show $vg^c \simeq vg$. We have that vg^c is the top line of the diagram

$$\begin{array}{ccccccc} W & \xrightarrow{\Delta} & W \times W & \xrightarrow{g \times c} & I_f \times \Omega Y & \xrightarrow{\phi_0} & I_f \\ & & & & \downarrow p_1 & & \downarrow v \\ & & & & I_f & \xrightarrow{v} & X, \end{array}$$

where Δ is the diagonal map. By Proposition 4.3.3(2), $v\phi_0 = vp_1$. Therefore

$$vg^c = vp_1(g \times c)\Delta = vg.$$

This shows that $v_*(\rho^\gamma) = v_*(\rho)$.

We prove the other implication. Recall that I_f is the following pullback

$$\begin{array}{ccc} I_f & \xrightarrow{u} & EY \\ \downarrow v & & \downarrow p_0 \\ X & \xrightarrow{f} & Y, \end{array}$$

where $p_0(l) = l(0)$. Now suppose that $\rho = [g]$, $\sigma = [h] \in [W, I_f]$ and $vg \simeq vh$. By the covering homotopy property of the fiber map v, we can replace g by a homotopic map (still called g) such that $vg = vh$. Thus

$$(ug)(w)(0) = (p_0 ug)(w) = (fvg)(w) = (fvh)(w) = (p_0 uh)(w) = (uh)(w)(0),$$

for every $w \in W$. We then define $c : W \to \Omega Y$ by

$$c(w)(t) = \begin{cases} (ug)(w)(1-2t) & \text{if } 0 \leqslant t \leqslant \frac{1}{2} \\ (uh)(w)(2t-1) & \text{if } \frac{1}{2} \leqslant t \leqslant 1, \end{cases}$$

for $w \in W$ and $t \in I$, so $c(w) = -(ug)(w) + (uh)(w)$. We show $g^c \simeq h$. We have

$$\begin{aligned}(g^c)(w) &= \Big((vg)(w), (ug)(w) + c(w)\Big) \\ &= \Big((vg)(w), (ug)(w) + \big(-(ug)(w) + (uh)(w)\big)\Big),\end{aligned}$$

and so $ug + (-ug + uh) \simeq_F uh : W \to EY$ for some homotopy F such that $F(w,t)(0) = (fvh)(w)$. Since $h(w) = ((vh)(w), (uh)(w))$, it follows that $g^c \simeq h : W \to I_f$, and so $\rho^{[c]} = \sigma$.

(2) We first prove the implication "$\Longrightarrow$". Suppose $i_*(\gamma) = i_*(\delta)$. Then for every $\nu \in [W, \Omega Y]$,

$$i_*(\gamma + \nu) = (i_*(\gamma))^\nu = (i_*(\delta))^\nu = i_*(\delta + \nu),$$

by Lemma 4.4.3. Therefore with $\nu = -\delta$, we have $i_*(\gamma - \delta) = 0$. By exactness, there exists $\epsilon \in [W, \Omega Y]$ such that $\gamma = (\Omega f)_*(\epsilon) + \delta$.

Now we prove "$\Longleftarrow$". Suppose $\gamma = (\Omega f)_*(\epsilon) + \delta$, for some $\epsilon \in [W, \Omega X]$. Then

$$i_*(\gamma) = i_*((\Omega f)_*(\epsilon) + \delta) = (i_*(\Omega f)_*(\epsilon))^\delta = 0^\delta = i_*(\delta),$$

by Lemma 4.4.3. □

We extend Theorem 4.4.4 to arbitrary fibrations. Let $F \xrightarrow{j} E \xrightarrow{p} B$ be a fiber sequence. By Proposition 3.5.10 and Definition 4.2.18, the map $\beta : F \to I_p$ is a homotopy equivalence such that the following diagram homotopy-commutes

$$\begin{array}{ccccccc} \Omega E & \xrightarrow{\Omega p} & \Omega B & \xrightarrow{\partial} & F & \xrightarrow{j} & E \\ \| & & \| & & \downarrow{\scriptstyle \beta} & & \| \\ \Omega E & \xrightarrow{\Omega p} & \Omega B & \xrightarrow{i} & I_p & \xrightarrow{v} & E, \end{array}$$

where $\partial : \Omega B \to F$ is the connecting map. By Section 4.3 there is an action $\phi : F \times \Omega B \to F$ defined as the composition

$$F \times \Omega B \xrightarrow{\beta \times \mathrm{id}} I_p \times \Omega B \xrightarrow{\phi_0} I_p \xrightarrow{\beta^{-1}} F.$$

Then ϕ determines an operation of $[W, \Omega B]$ on $[W, F]$, for any space W. Thus we obtain the following commutative diagram

$$\begin{array}{ccccccc}
[W,\Omega E] & \xrightarrow{(\Omega p)_*} & [W,\Omega B] & \xrightarrow{\partial_*} & [W,F] & \xrightarrow{j_*} & [W,E] \\
\| & & \| & & \cong \downarrow \beta_* & & \| \\
[W,\Omega E] & \xrightarrow{(\Omega p)_*} & [W,\Omega B] & \xrightarrow{i_*} & [W,I_p] & \xrightarrow{v_*} & [W,E]
\end{array}$$

such that the operation of $[W, \Omega B]$ on $[W, F]$ in the top line is compatible with the operation of $[W, \Omega B]$ on $[W, I_p]$ in the bottom line, that is, $\beta_*(\chi^\gamma) = (\beta_*(\chi))^\gamma$, for $\chi \in [W, F]$ and $\gamma \in [W, \Omega B]$.

These remarks together with Theorem 4.4.4 give the following theorem.

Theorem 4.4.5 *Let $F \xrightarrow{j} E \xrightarrow{p} B$ be a fiber sequence with connecting map $\partial : \Omega B \to F$. Let $\phi : F \times \Omega B \to F$ be the action with corresponding induced operation of $[W, \Omega B]$ on $[W, F]$, for every space W.*

1. *If $\chi, \xi \in [W, F]$, then $j_*(\chi) = j_*(\xi) \iff \xi = \chi^\gamma$ for some $\gamma \in [W, \Omega B]$.*
2. *If $\gamma, \delta \in [W, \Omega B]$, then $\partial_*(\gamma) = \partial_*(\delta) \iff \gamma = (\Omega p)_*(\epsilon) + \delta$, for some $\epsilon \in [W, \Omega E]$.*

We conclude this section by briefly considering operations obtained from coactions.

Definition 4.4.6 Let (B, c) be a co-H-space and $\psi : Y \to Y \vee B$ a coaction of B on Y. If Z is any space and $g : Y \to Z$ and $a : B \to Z$ are any maps, then define $g^a : Y \to Z$ as the composition

$$Y \xrightarrow{\psi} Y \vee B \xrightarrow{g \vee a} Z \vee Z \xrightarrow{\nabla} Z,$$

where ∇ is the folding map. This defines an *operation of* $[B, Z]$ *on* $[Y, Z]$ by

$$[g]^{[a]} = [g^a] \in [Y, Z].$$

For any map $f : X \to Y$, we have seen in Proposition 4.3.6 that there is a coaction $\psi_0 : C_f \to C_f \vee \Sigma X$. This yields an operation of $[\Sigma X, Z]$ on $[C_f, Z]$, for every space Z. By Corollary 4.2.8, the map $f : X \to Y$ gives rise to the following exact sequence

$$\cdots \longrightarrow [\Sigma Y, Z] \xrightarrow{(\Sigma f)^*} [\Sigma X, Z] \xrightarrow{q^*} [C_f, Z] \xrightarrow{l^*} [Y, Z].$$

Theorem 4.4.7

1. *Let $\rho, \sigma \in [C_f, Z]$. Then $l^*(\rho) = l^*(\sigma) \iff \sigma = \rho^\gamma$, for some $\gamma \in [\Sigma X, Z]$.*
2. *Let $\gamma, \delta \in [\Sigma X, Z]$. Then $q^*(\gamma) = q^*(\delta) \iff \gamma = (\Sigma f)^*(\epsilon) + \delta$, for some $\epsilon \in [\Sigma Y, Z]$.*

If $A \xrightarrow{i} X \xrightarrow{p} Q$ is any cofibration, then there is a homotopy equivalence $\alpha : C_i \to Q$ by Proposition 3.5.4 and a coaction $\psi : Q \to Q \vee \Sigma A$ which is defined as the composition

$$Q \xrightarrow{\alpha^{-1}} C_i \xrightarrow{\psi_0} C_i \vee \Sigma A \xrightarrow{\alpha \vee \mathrm{id}} Q \vee \Sigma A.$$

This gives an operation of $[\Sigma A, Z]$ on $[Q, Z]$, for any space Z. We have the following exact sequence

$$\cdots \longrightarrow [\Sigma X, Z] \xrightarrow{(\Sigma i)^*} [\Sigma A, Z] \xrightarrow{\partial^*} [Q, Z] \xrightarrow{p^*} [X, Z]$$

by Corollary 4.2.12.

Using α, we map the exact sequence of the cofibration into the exact sequence of the map $i : A \to X$ and obtain the following result from Theorem 4.4.7.

Theorem 4.4.8 *Let* $A \xrightarrow{i} X \xrightarrow{p} Q$ *be a cofibration and consider the operation of* $[\Sigma A, Z]$ *on* $[Q, Z]$.

1. *Let* $\chi, \xi \in [Q, Z]$. *Then* $p^*(\chi) = p^*(\xi) \iff \xi = \chi^\gamma$, *for some* $\gamma \in [\Sigma A, Z]$.
2. *Let* $\gamma, \delta \in [\Sigma A, Z]$. *Then* $\partial^*(\gamma) = \partial^*(\delta) \iff \gamma = (\Sigma i)^*(\epsilon) + \delta$, *for some* $\epsilon \in [\Sigma X, Z]$.

4.5 Homotopy Groups II

In this section we discuss the homotopy groups in more detail. We begin with an alternative characterization of the homotopy groups of a space. We then introduce and study the homotopy groups of a pair of spaces. After that we derive exact homotopy sequences which are obtained from two maps whose composition is trivial.

We adopt the following notation. If $f : X \to Y$ is a map and $f(A) \subseteq B$ for $A \subseteq X$ and $B \subseteq Y$, then we write $f' : (X, A) \to (Y, B)$ for the map of pairs determined by f. The pairs (X, A) and (Y, B) are homeomorphic if there is a homeomorphism $\theta : X \to Y$ such that $\theta|A : A \to B$ is a homeomorphism.

Now let I^n be the *n-cube* $I \times I \times \cdots \times I$ (n-factors) and let ∂I^n be the boundary of I^n, the set of all $(t_1, t_2, \ldots, t_n)$ in I^n such that some $t_i = 0$ or 1 and let the basepoint $*$ of ∂I^n be $(1, 0, \ldots, 0)$. For a space X we consider maps of pairs $f : (I^n, \partial I^n) \to (X, \{*\})$ and denote by $\pi'_n(X)$ the set of homotopy classes of these maps. Two maps $f, g : (I^n, \partial I^n) \to (X, \{*\})$ are homotopic if and only if $f \simeq g \operatorname{rel} \partial I^n$. Clearly there is a homeomorphism from I^n to E^n that carries ∂I^n onto S^{n-1}, and so the pair $(I^n, \partial I^n)$ is homeomorphic to the pair (E^n, S^{n-1}). Thus we could regard $\pi'_n(X)$ as the set of homotopy classes of maps $(E^n, S^{n-1}) \to (X, \{*\})$. The pair (E^n, S^{n-1}) is homeomorphic to the pair (E^n_+, S^{n-1}), where E^n_+ is the upper cap of S^n, therefore we could also regard $\pi'_n(X)$ as the set of homotopy classes of maps $(E^n_+, S^{n-1}) \to (X, \{*\})$.

We set $S = (0, 0, \ldots, -1) \in S^n$, called the *South Pole*, and $N = (0, 0, \ldots, 1) \in S^n$, called the *North Pole*, and recall that $* = (1, 0, \ldots, 0)$

is the basepoint of S^n (as well as of E_+^n and E_-^n). We define a continuous function $k : E_+^n \to S^n$ as follows. Represent the points $p \in S^n \subseteq \mathbb{R}^{n+1}$ as $[x, u]$, where $x \in S^{n-1}$ and $-1 \leqslant u \leqslant 1$ is the last coordinate of p, as indicated in Figure 4.2.

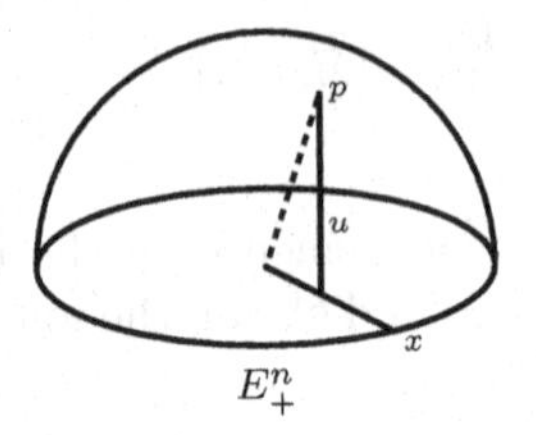

Figure 4.2

Thus $x = (x_1, \ldots, x_n)$ is the point in S^{n-1} determined by the ray from the origin through the projection of p onto $\mathbb{R}^n$. Therefore $[x, u]$ corresponds to $(p_1, \ldots, p_n, u)$ with $p_i = x_i\sqrt{1-u^2}$ for $i = 1, \ldots, n$. Define $k : E_+^n \to S^n$ by $k[x, u] = [x, 2u - 1]$, for $0 \leqslant u \leqslant 1$ and $x \in S^{n-1}$. Then $k(S^{n-1}) = \{S\}$ and $k|(E_+^n - S^{n-1})$ is a free map that carries $E_+^n - S^{n-1}$ homeomorphically onto $S^n - \{S\}$. It can be described as follows. The upper cap $E_+^n \subseteq S^n$ is stretched on S^n with N fixed and $S^{n-1} \subseteq E_+^n$ going to S.

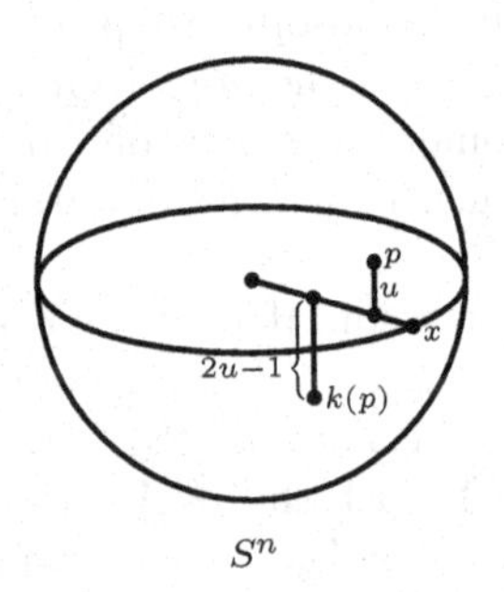

Figure 4.3

In order to obtain a map $E_+^n \to S^n$ that carries S^{n-1} to $\{*\}$ and is a homeomorphism $E_+^n - S^{n-1} \cong S^n - \{*\}$, we compose k with a rotation ρ of S^n. If we define ρ by $\rho(t_1, t_2, \ldots, t_n, t_{n+1}) = (-t_{n+1}, t_2, \ldots, t_n, t_1)$, then ρ carries S to $*$. Thus we have the map $h = \rho k : E_+^n \to S^n$ that carries S^{n-1} onto $\{*\}$ and maps $E_+^n - S^{n-1}$ homeomorphically onto $S^n - \{*\}$. Now h determines a map of pairs $h' : (E_+^n, S^{n-1}) \to (S^n, \{*\})$.

Proposition 4.5.1 *For all spaces X and $n \geqslant 1$, the function $\theta : \pi_n(X) \to \pi_n'(X)$ defined by $\theta[f] = [fh']$ is a bijection.*

Proof. We first show that θ is one–one. Suppose $f, g : S^n \to X$ and $fh' \simeq_F gh' : (E^n_+, S^{n-1}) \to (X, \{*\})$. Since h is an identification map, so is $h \times \mathrm{id} : E^n_+ \times I \to S^n \times I$ (Appendix A). Therefore $F : E^n_+ \times I \to X$ induces $G : S^n \times I \to X$. Hence $f \simeq_G g$, and so $[f] = [g]$.

To show θ is onto, let $l : (E^n_+, S^{n-1}) \to (X, \{*\})$ be a map. Then l induces a map $\bar{l} : S^n \to X$ such that $\bar{l}h' = l$ Thus $\theta[\bar{l}] = [l]$. □

Next we see that this alternative characterization of homotopy groups gives rise to an alternative definition of addition.

Remark 4.5.2 If $f, g : (I^n, \partial I^n) \to (X, \{*\})$, then define $f +' g : (I^n, \partial I^n) \to (X, \{*\})$ by

$$(f +' g)(t_1, t_2, \ldots, t_n) = \begin{cases} f(t_1, t_2, \ldots, 2t_n) & \text{if } 0 \leqslant t_n \leqslant \frac{1}{2} \\ g(t_1, t_2, \ldots, 2t_n - 1) & \text{if } \frac{1}{2} \leqslant t_n \leqslant 1, \end{cases}$$

for $(t_1, t_2, \ldots, t_n) \in I$. Clearly this induces a binary operation on $\pi'_n(X)$ denoted $+'$. Then $\theta(\alpha + \beta) = \theta(\alpha) +' \theta(\beta)$ by Exercise 4.7. Because of this and Proposition 4.5.1, we drop the prime in $\pi'_n(X)$ and $+'$ and regard the group $\pi_n(X)$ as homotopy classes of maps $S^n \to X$, as homotopy classes of maps $(I^n, \partial I^n) \to (X, \{*\})$, or as homotopy classes of maps $(E^n, S^{n-1}) \to (X, \{*\})$.

We next study the relative homotopy groups, that is, the homotopy groups of a pair of spaces. These are generalizations of the homotopy groups and are similar to the relative homology and cohomology groups. For a pair of spaces (X, A), the relative homotopy groups give information on the homotopy homomorphism induced by the inclusion map $A \to X$.

Now let X be a space and $A \subseteq X$ a subspace. We denote by $E(X; A, \{*\})$ the subspace of the path space EX consisting of paths that begin in A and end in $\{*\}$. Clearly $E(X; A, \{*\})$ is just the homotopy fiber of the inclusion map $A \to X$.

Definition 4.5.3 For $A \subseteq X$ and an abelian group G, we define the *nth relative homotopy group of the pair* (X, A) *with coefficients in* G by

$$\pi_n(X, A; G) = \pi_{n-1}\big(E(X; A, \{*\}); G\big),$$

for $n \geqslant 3$. When $G = \mathbb{Z}$, we have

$$\pi_n(X, A) = \pi_{n-1}(E(X; A, \{*\})),$$

for $n \geqslant 1$. We refer refer to the $\pi_n(X, A)$ for $n \geqslant 1$ as *relative homotopy groups* even though $\pi_1(X, A)$ is in general just a based set. A map $f : (X, A) \to (Y, B)$ of pairs induces a map $E(X; A, \{*\}) \to E(Y; B, \{*\})$ and hence a homomorphism $f_* : \pi_n(X, A; G) \to \pi_n(Y, B; G)$, for $n \geqslant 3$ ($n \geqslant 2$, if $G = \mathbb{Z}$). Furthermore, the induced function $f_* : \pi_1(X, A) \to \pi_1(Y, B)$ is a based function and is called a homomorphism.

We can interpret relative homotopy groups with coefficients as follows. If $\alpha = [a]$ is an element of $\pi_n(X, A; G)$, then $a : \Sigma M(G, n-1) = M(G, n) \to E(X; A, \{*\})$. By taking the adjoint of a we obtain a map of pairs $\tilde{a} : (CM(G, n-1), M(G, n-1)) \to (X, A)$. Clearly we may then regard $\pi_n(X, A; G)$ as homotopy classes of maps $(CM(G, n-1), M(G, n-1)) \to (X, A)$. In the case $G = \mathbb{Z}$, we have $M(\mathbb{Z}, n-1) = S^{n-1}$ and $CS^{n-1} \cong E^n$ (proof of Proposition 2.3.9), and so we may regard the relative group $\pi_n(X, A)$ as homotopy classes of maps $(E^n, S^{n-1}) \to (X, A)$.

We give another interpretation of the relative homotopy groups with $\mathbb{Z}$ coefficients. Regard $I^{n-1} \subseteq I^n$ by identifying I^{n-1} with $I^{n-1} \times \{0\} \subseteq I^n$. Let $J^{n-1} \subseteq I^n$ be the set of all $(t_1, t_2, \ldots, t_n) \in I^n$ such that either $t_i = 0$ or 1, for some $i = 1, 2, \ldots, n-1$, or $t_n = 1$.

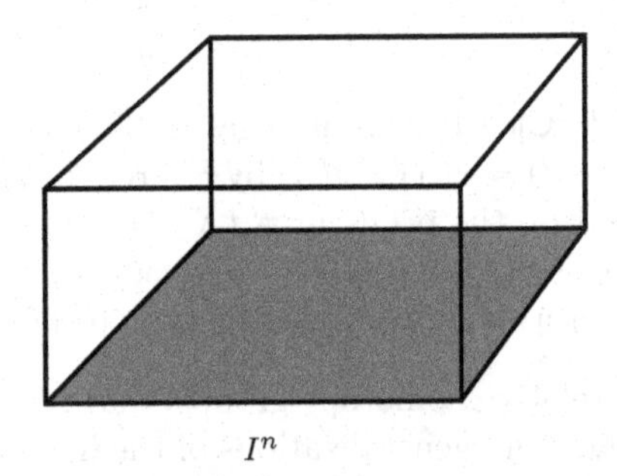

I^n

Shaded face is I^{n-1}, union of other faces is J^{n-1}

Figure 4.4

Note that $J^{n-1} = (\partial I^{n-1} \times I) \cup (I^{n-1} \times \{1\})$ and $I^{n-1} \cup J^{n-1} = \partial I^n$, where ∂I^{n-1} is the boundary of I^{n-1}. If $f, g : (I^n, I^{n-1}, J^{n-1}) \to (X, A, \{*\})$, then $f \simeq g$ means that there is a homotopy f_t between f and g such that $f_t(I^{n-1}) \subseteq A$ and $f_t(J^{n-1}) = \{*\}$.

Remark 4.5.4 If (X, A) is a pair of spaces and $n \geqslant 1$, then there is a bijection of $\pi_n(X, A)$ with the set of homotopy classes of maps $(I^n, I^{n-1}, J^{n-1}) \to (X, A, \{*\})$. The bijection is given as follows. Assume that $f : (I^{n-1}, \partial I^{n-1}) \to (E(X; A, \{*\}), \{*\})$. Then we consider the adjoint $\tilde{f} : I^n = I^{n-1} \times I \to X$ of f. Clearly $\tilde{f}(I^{n-1} \times \{0\}) \subseteq A$ and $\tilde{f}(I^{n-1} \times \{1\}) = \{*\} = \tilde{f}(\partial I^{n-1} \times I)$, and so $\tilde{f} : (I^n, I^{n-1}, J^{n-1}) \to (X, A, \{*\})$. Then the function μ defined by $\mu[f] = [\tilde{f}]$ is the desired bijection. For details, see [41, pp. 17-19].

We next obtain an exact sequence for relative homotopy groups. By Proposition 3.3.2, the sequence of spaces

$$\Omega X \longrightarrow E(X; A, \{*\}) \xrightarrow{q_0} A$$

is a fiber sequence, where $q_0(\omega) = \omega(0)$, for $\omega \in E(X; A, \{*\})$. By applying Corollary 4.2.19(2) to this fibration, we have the following *exact homotopy sequence of the pair* (X, A) *with coefficients in* G

$$\longrightarrow \pi_{n+1}(X, A; G) \longrightarrow \pi_n(A; G) \longrightarrow \pi_n(X; G) \longrightarrow \pi_n(X, A; G) \longrightarrow$$

$$\cdots \longrightarrow \pi_3(X, A; G) \longrightarrow \pi_2(A; G) \longrightarrow \pi_2(X; G).$$

For $G = \mathbb{Z}$ we can adjoin the following exact sequence to the preceding one

$$\pi_2(X, A) \longrightarrow \pi_1(A) \longrightarrow \pi_1(X) \longrightarrow \pi_1(X, A) \longrightarrow \pi_0(A) \longrightarrow \pi_0(X).$$

In general, the last three terms are not groups and the previous three terms are groups that are not necessarily abelian. Let the boundary map in this latter sequence be denoted by $\partial_n : \pi_n(X, A) \to \pi_{n-1}(A)$. If we regard $\pi_n(X, A)$ as homotopy classes of maps $f : (E^n, S^{n-1}) \to (X, A)$, then by Exercise 4.16, $\partial_n[f] = [f|S^{n-1}]$.

Furthermore, if $g : (X, A) \to (Y, B)$ is a map, then g gives rise to a commutative diagram of maps from one fiber sequence to another

$$\begin{array}{ccccc} \Omega X & \longrightarrow & E(X; A, \{*\}) & \longrightarrow & A \\ \downarrow{\scriptstyle \Omega g} & & \downarrow{\scriptstyle Eg} & & \downarrow{\scriptstyle g|A} \\ \Omega Y & \longrightarrow & E(Y; B, \{*\}) & \longrightarrow & B, \end{array}$$

and hence a map from the exact homotopy sequence of the pair (X, A) into the exact homotopy sequence of the pair (Y, B).

We study the relative group in more detail. This leads to a proof of Lemma 4.5.7, a generalization of Lemma 2.4.5, which was stated without proof.

Lemma 4.5.5 *Let* (Y, B) *be a pair of spaces and let* $f : (E^n, S^{n-1}) \to (Y, B)$ *represent an element* $\alpha \in \pi_n(Y, B)$. *Then* $\alpha = 0$ *if and only if there exists a map* $g : (E^n, S^{n-1}) \to (Y, B)$ *with* $g|S^{n-1} = f|S^{n-1}$, $f \simeq g$ *rel* S^{n-1}, *and* $g(E^n) \subseteq B$.

Proof. Suppose that $\alpha = 0$. We first define a function $F : E^n \times I \to E^n \times I$ by

$$F(x, t) = \begin{cases} (x/(1 - t/2), t) & \text{if } 0 \leqslant |x| \leqslant 1 - t/2 \\ (x/|x|, 2 - 2|x|) & \text{if } 1 - t/2 \leqslant |x| \leqslant 1, \end{cases}$$

for $x \in E^n$ and $t \in I$. If E^n_t denotes $E^n \times \{t\} \subseteq E^n \times I$, then Figure 4.5 illustrates the image of E^n_t under F.

Clearly $F|E^n_0 = \mathrm{id}$ and F maps $\{x \in E^n_1 \,|\, |x| \leqslant \frac{1}{2}\}$ to E^n_1 and maps $\{x \in E^n_1 \,|\, |x| \geqslant \frac{1}{2}\}$ to $S^{n-1} \times I$. If $f \simeq_G * : (E^n, S^{n-1}) \to (Y, B)$, then set $g(x) = GF(x, 1)$ for $x \in E^n$. Thus $g(E^n) \subseteq B$ and $f \simeq_{GF} g$ rel S^{n-1}.

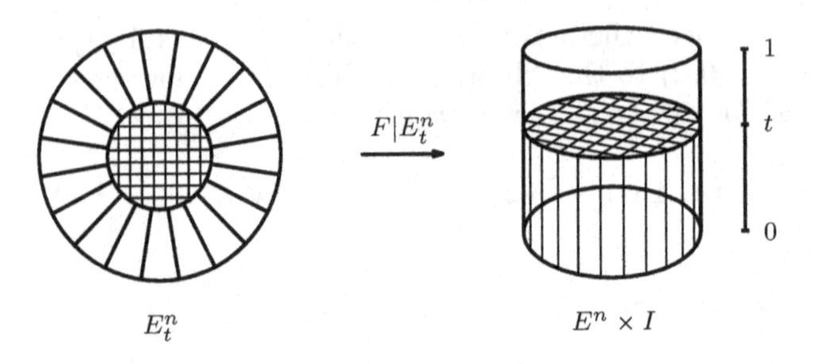

Figure 4.5

Conversely suppose $f \simeq_H g$ rel S^{n-1} and $g(E^n) \subseteq B$. Define a homotopy $G : E^n \times I \to Y$ by

$$G(x,t) = \begin{cases} H(x, 2t) & \text{if } 0 \leqslant t \leqslant 1/2 \\ g((2-2t)x + (2t-1)*) & \text{if } 1/2 \leqslant t \leqslant 1, \end{cases}$$

for $x \in E^n$ and $t \in I$. From $t = 0$ to $t = 1/2$ the homotopy G is just the homotopy H and from $t = 1/2$ to $t = 1$ the homotopy G is g on the line segment in E^n from x to $*$. Then $f \simeq_G * : (E^n, S^{n-1}) \to (Y, B)$. □

This lemma plays a key role in the following result.

Lemma 4.5.6 *Let $e : B \to Y$ be a map such that $e_* : \pi_{k-1}(B) \to \pi_{k-1}(Y)$ is a monomorphism and $e_* : \pi_k(B) \to \pi_k(Y)$ is an epimorphism, for some k. Let $i : S^{k-1} \to E^k$ be the inclusion map and let $a : E^k \to Y$ and $b : S^{k-1} \to B$ be maps with $ai \simeq_H eb$ for some homotopy $H : S^{k-1} \times I \to Y$,*

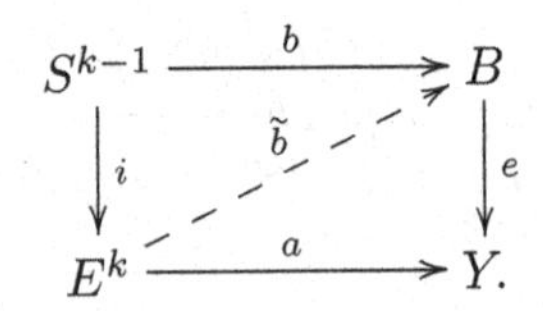

Then there exists a map $\tilde{b} : E^k \to B$ and a homotopy $J : E^k \times I \to Y$ such that $\tilde{b}i = b$, $a \simeq_J e\tilde{b}$, and $J|S^{k-1} \times I = H$.

Proof. We can replace e by the inclusion of B into the mapping cylinder M_e of e (Definition 3.5.1 and Proposition 3.5.2), and so can assume that e is an inclusion map. Then the hypothesis on e_* and the exact homotopy sequence of the pair (Y, B) imply that $\pi_k(Y, B) = 0$. Write $H(x,t) = h_t(x)$ and consider the homotopy-commutative square above. Because i is a cofibration, there is a homotopy $k_t : E^k \to Y$ such that $k_0 = a$ and $k_t\, i = h_t$. Thus the following square is commutative

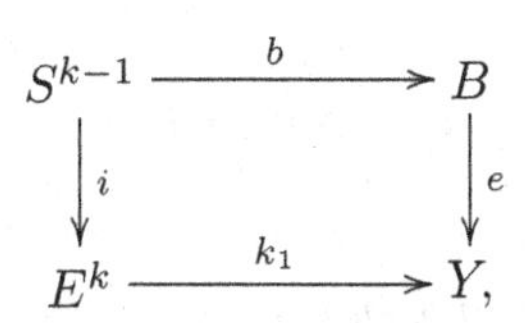

and a map of pairs $k_1' : (E^k, S^{k-1}) \to (Y, B)$ is determined by k_1. Since $[k_1'] \in \pi_k(Y, B)$, it follows that $[k_1'] = 0$. By Lemma 4.5.5, there exists a map $\widetilde{b} : E^k \to B$ such that $\widetilde{b}\, i = b$ and $k_1 \simeq_G e\widetilde{b}$ rel S^{k-1}, for some homotopy $G : E^k \times I \to Y$. If $K(x,t) = k_t(x)$, then

$$a = k_0 \simeq_K k_1 \simeq_G e\widetilde{b}.$$

We then define the homotopy $J' : E^k \times I \to Y$ to be the concatenation of K and G, that is, $J' = K + G$ (see the discussion before Definition 1.3.3). Then $a \simeq_{J'} e\widetilde{b}$ and $J'|S^{k-1} \times I = H + F_{eb}$, where $F_{eb} : S^{k-1} \times I \to Y$ is the stationary homotopy determined by $eb : S^{k-1} \to Y$. Clearly $H + F_{eb} \simeq H$ rel $S^{k-1} \times \partial I : S^{k-1} \times I \to Y$ and we denote this homotopy by $\Lambda_t : S^{k-1} \times I \to Y$. Define a homotopy $\Gamma_t : S^{k-1} \times I \cup E^k \times \partial I \to Y$ by

$$\begin{aligned} \Gamma_t(x,s) &= \Lambda_t(x,s) \\ \Gamma_t(y,0) &= a(y) \\ \Gamma_t(y,1) &= e\widetilde{b}(y), \end{aligned}$$

for $x \in S^{k-1}$, $y \in E^k$, and $s \in I$. Then we have the diagram

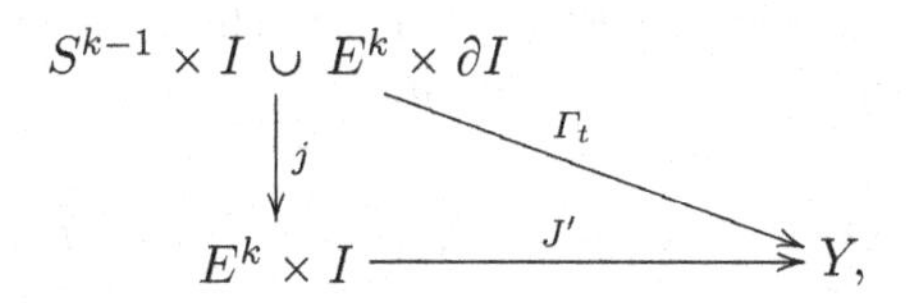

where j is the inclusion. But j is a cofiber map and $J'j = \Gamma_0$. Therefore J' is homotopic to a homotopy $J : E^k \times I \to Y$ such that $a \simeq_J e\widetilde{b}$ and $J|S^{k-1} \times I = H$. This completes the proof. □

We next prove the lemma which implies the result (2.4.5) that was used to prove Whitehead's first theorem 2.4.7. The property that appears in this lemma is often referred to as the homotopy extension lifting property and the lemma is called the HELP lemma.

For the proof we recall Zorn's lemma. Let $\mathcal{S}$ be a set with partial ordering "$\leqslant$". A subset of $\mathcal{S}$ is called a chain if for any two elements a and b in the subset, either $a \leqslant b$ or $b \leqslant a$. If every nonempty chain in $\mathcal{S}$ has an upper bound, (that is, an element greater than or equal to every element of the chain), then there is a maximal element in S, (that is, an element that is not strictly less than any element of $\mathcal{S}$) [26, pp. 30–32].

Lemma 4.5.7 *Let (X, A) be a relative CW complex and let $e : B \to Y$ be a map. Assume that for every k such that there exists one or more relative k-cells of (X, A), the homomorphism $e_* : \pi_{k-1}(B) \to \pi_{k-1}(Y)$ is a monomorphism and the homomorphism $e_* : \pi_k(B) \to \pi_k(Y)$ is an epimorphism. Let $j : A \to X$ be the inclusion and suppose that there are maps $f : X \to Y$ and $g : A \to B$ and a diagram*

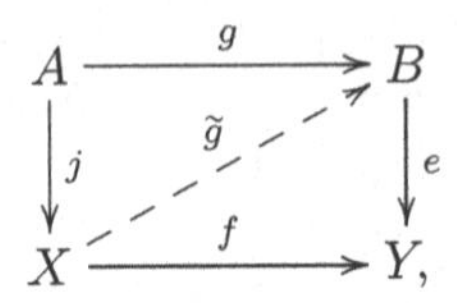

such that $eg \simeq_L fj$, for some homotopy $L : A \times I \to Y$. Then there exists a map $\widetilde{g} : X \to B$ and a homotopy $F : X \times I \to Y$ such that $\widetilde{g}j = g$ and $e\widetilde{g} \simeq_F f$, where $F|A \times I = L$.

Proof. We consider the set $\mathcal{S}$ of all triples $((X', A), \widetilde{g}', F')$, where (1) (X', A) is a subrelative complex of (X, A) (Remark 1.5.16), (2) $\widetilde{g}' : X' \to B$ is a map such that $\widetilde{g}'j = g$; and (3) $F' : X' \times I \to Y$ is a homotopy such that $e\widetilde{g}' \simeq_{F'} f|X'$ and $F'|A \times I = L$. Then $\mathcal{S}$ is a set that is partially ordered as follows: $((X_1, A), \widetilde{g}_1, F_1) \leqslant ((X_2, A), \widetilde{g}_2, F_2)$ if $X_1 \subseteq X_2$, $\widetilde{g}_2$ is an extension of $\widetilde{g}_1$, and F_2 is an extension of F_1. Since $\mathcal{S}$ satisfies the hypotheses of Zorn's lemma, there is a maximal element $((X_m, A), \widetilde{g}_m, F_m)$ in $\mathcal{S}$. We show that $X_m = X$. Suppose X_m is a proper subset of X. Then there is a relative cell in (X, A) that is not contained in X_m, and we assume that e^k is such an open cell of lowest dimension. If $\varPhi : E^k \to X$ is the characteristic map of this cell with $\varPhi|S^{k-1} = \phi$, then there is a diagram

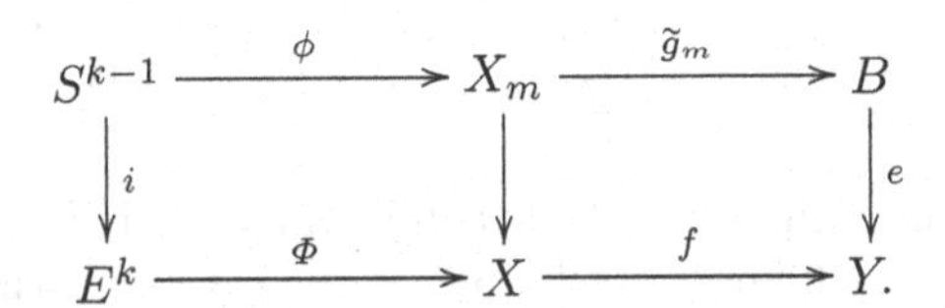

Because $e\,\widetilde{g}_m \simeq_{F_m} f|X_m$, it follows that $e\,\widetilde{g}_m\phi \simeq_{F_m(\phi \times \mathrm{id})} f\varPhi\, i$. By Lemma 4.5.6 there exists a map $\widetilde{b} : E^k \to B$ such that $\widetilde{b}\, i = \widetilde{g}_m\phi$ and a homotopy $J : E^k \times I \to Y$ such that $e\,\widetilde{b} \simeq_J f\varPhi$ and $J|S^{k-1} \times I = F_m(\phi \times \mathrm{id})$,

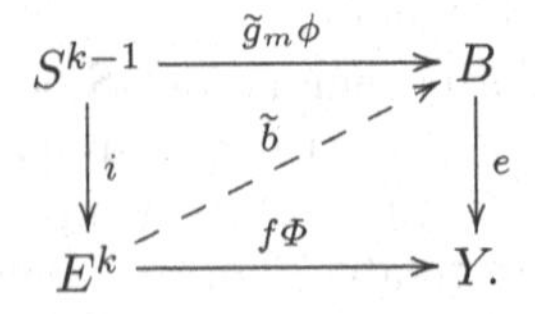

We define X' by the following pushout square

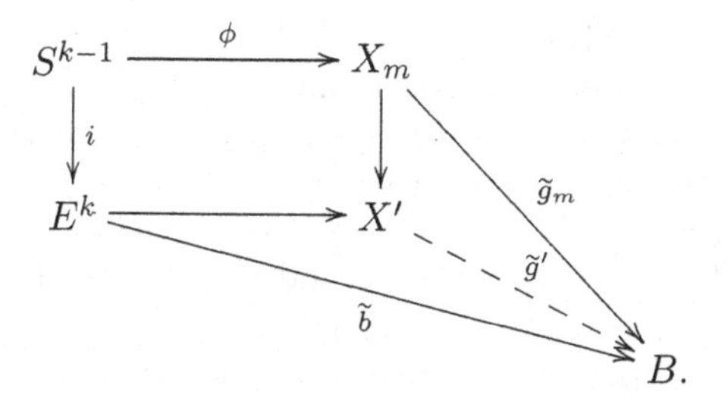

and define $\widetilde{g}' : X' \to B$ as the map of the pushout X' determined by $\widetilde{b}$ and $\widetilde{g}_m$. Similarly consider the diagram

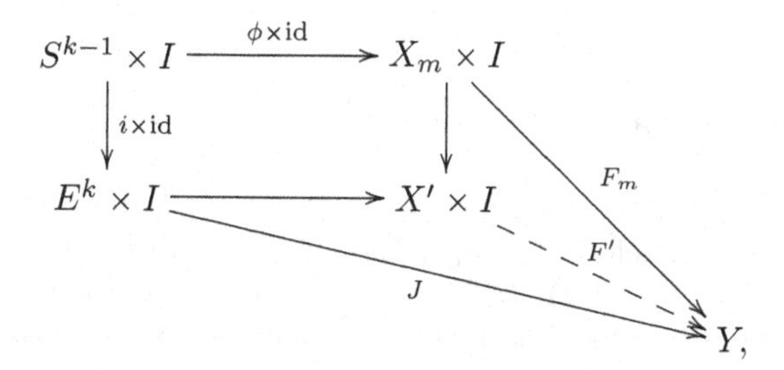

where the square is a pushout square (Exercise 3.4) and the homotopy F' is determined by J and F_m. It is then easily verified that F' is a homotopy between $e\widetilde{g}'$ and $f|X'$. Thus the triple $((X', A), \widetilde{g}', F')$ is in $\mathcal{S}$ and $((X_m, A), \widetilde{g}_m, F_m)$ is not less than or equal to $((X', A), \widetilde{g}', F')$. This contradicts the maximality of $((X_m, A), \widetilde{g}_m, F_m)$. Therefore $X_m = X$ and we set $\widetilde{g} = \widetilde{g}_m$ and $F = F_m$. □

Lemma 4.5.6 is a special case of the HELP lemma 4.5.7 which occurs when the relative CW complex $(X, A) = (E^k, S^{k-1})$. In the proof above, Lemma 4.5.6 was used to prove the HELP lemma.

We next make some additional comments on the HELP lemma 4.5.7. We first note that this lemma immediately yields Lemma 2.4.5. In the latter lemma the hypothesis is that $\dim(X, A) \leqslant n$ and $e : B \to Y$ is an n-equivalence. This clearly implies the hypothesis of the HELP lemma.

Remark 4.5.8 1. The hypothesis on e_* may appear somewhat unwieldy. However, as noted earlier, if e is an inclusion, then it is equivalent to the condition that $\pi_k(Y, B) = 0$, and if e is not an inclusion, then it is equivalent to $\pi_k(M_e, B) = 0$, where M_e is the mapping cylinder of e.
2. It is possible to prove the HELP lemma without invoking Zorn's lemma. This is done by defining the map $\widetilde{g}$ and the homotopy F inductively over the skeleta of (X, A). If this has been done on the $(k-1)$-skeleton $(X, A)^{k-1}$, then the extension to each relative k-cell is carried out exactly as in the proof of the HELP lemma above. This defines the map and the

homotopy on the k-skeleton $(X,A)^k$. These functions can then be put together to yield the desired map $\tilde{g}$ and homotopy F.

3. The HELP lemma can be used to give a proof of the cellular approximation theorem. We sketch the proof. We are given a map $f : (X,A) \to (Y,B)$ of relative CW complexes. We construct a cellular approximation $g : (X,A) \to (Y,B)$ of f by induction over the skeleta $(X,A)^n$. Assume that a map $g_{n-1} : (X,A)^{n-1} \to (Y,B)^{n-1}$ exists such that $f|(X,A)^{n-1} \simeq j_{n-1}g_{n-1} : (X,A)^{n-1} \to Y$, where $j_{n-1} : (Y,B)^{n-1} \to Y$ is the inclusion. We then have the diagram

$$\begin{array}{ccccc} (X,A)^{n-1} & \xrightarrow{g_{n-1}} & (Y,B)^{n-1} & \xrightarrow{i_{n-1}} & (Y,B)^n \\ \downarrow & & & & \downarrow{\scriptstyle j_n} \\ (X,A)^n & & \xrightarrow{f|(X,A)^n} & & Y, \end{array}$$

where i_{n-1} is the inclusion. Since $j_{n-1} = j_n i_{n-1}$, the diagram is homotopy-commutative. Therefore, to apply the HELP lemma to this diagram and obtain $g_n : (X,A)^n \to (Y,B)^n$, we need $\pi_n(Y,(Y,B)^n) = 0$. This is true but the proof is technical and we do not give it. We refer to [91, p. 73] or [14, p. 207].

Next we derive a useful corollary of the HELP lemma.

Corollary 4.5.9 *Let $j: A \to X$, $f: X \to Y$, $g: A \to B$, and $e: B \to Y$ be maps such that $eg = fj$. Assume that e is a fiber map with fiber F*

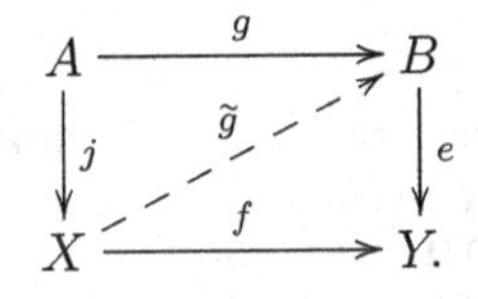

1. *If (X,A) is a relative CW complex with inclusion map j and $\pi_{k-1}(F) = 0$ for every k for which there exists one or more relative k-cells of (X,A), then there exists a map $\tilde{g} : X \to B$ such that $\tilde{g}j = g$ and $e\tilde{g} = f$.*
2. *If j is a k-equivalence and $\pi_i(F) = 0$, for $i \geqslant k$, then there exists a map $\tilde{g} : X \to B$ such that $\tilde{g}j \simeq g$ and $e\tilde{g} \simeq f$.*

Proof. (1) The condition $\pi_{k-1}(F) = 0$ is equivalent to the condition that $e_* : \pi_{k-1}(B) \to \pi_{k-1}(Y)$ is a monomorphism and $e_* : \pi_k(B) \to \pi_k(Y)$ is an epimorphism by the exact homotopy sequence of a fibration. Furthermore if we let $M : A \times I \to B$ be the stationary homotopy defined by $M(a,t) = g(a)$, for $a \in A$ and $t \in I$, then $L = eM : A \times I \to Y$ is a (stationary) homotopy between eg and fj. Therefore by the HELP lemma 4.5.7, there exists a map $\tilde{g}_0 : X \to B$ such that $\tilde{g}_0 j = g$ and a homotopy $F : X \times I \to Y$ such that $e\tilde{g}_0 \simeq_F f$, where $F|A \times I = L$. If we write $m_t(a) = M(a,t)$ and $f_t(x) = F(x,t)$

for $a \in A$ and $x \in X$, then $em_t = f_t j$. By the CHEP (Corollary 3.3.6) applied to the diagram

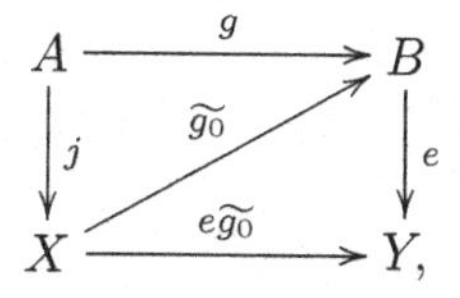

there is a homotopy $\widetilde{g}_t : X \to B$ of $\widetilde{g}_0$ such that $\widetilde{g}_t j = m_t = g$ and $e\widetilde{g}_t = f_t$. Then $\widetilde{g} = \widetilde{g}_1 : X \to B$ is the desired map.

(2) By Corollary 2.4.10, we replace $j : A \to X$ by an inclusion $i : A \to K$, where (K, A) is a relative CW complex with relative cells in dimensions $\geqslant k + 1$. We then apply (1). □

Corollary 4.5.9(1) also follows from the obstruction theory developed in Chapter 9 (Exercise 9.1).

We next discuss the Hurewicz homomorphism for relative homotopy groups. We choose a generator $\lambda_n \in H_n(I^n, \partial I^n) \cong \mathbb{Z}$. Then we define the function

$$h_n : \pi_n(X, A) \to H_n(X, A)$$

as follows. If $\alpha = [f]$ is an element of the relative homotopy group $\pi_n(X, A)$, then the map $f : (I^n, I^{n-1}, J^{n-1}) \to (X, A, \{*\})$ determines a map $f : (I^n, \partial I^n) \to (X, A)$. This latter map induces $f_* : H_n(I^n, \partial I^n) \to H_n(X, A)$, and we set $h_n(\alpha) = f_*(\lambda_n)$. Then $h_n : \pi_n(X, A) \to H_n(X, A)$ is a homomorphism for $n \geqslant 2$ (Exercise 4.14) called the *(relative) Hurewicz homomorphism.* Clearly if $g : (X, A) \to (Y, B)$ is a map of pairs, then the following diagram is commutative

$$\begin{array}{ccc} \pi_n(X, A) & \xrightarrow{g_*} & \pi_n(Y, B) \\ \downarrow{\scriptstyle h_n} & & \downarrow{\scriptstyle \bar{h}_n} \\ H_n(X, A) & \xrightarrow{g_*} & H_n(Y, B), \end{array}$$

where h_n and $\bar{h}_n$ are nth Hurewicz homomorphisms for (X, A) and (Y, B), respectively.

We relate the Hurewicz homomorphism of the pair (X, A) to the Hurewicz homomorphism of A. Recall that the definition of the Hurewicz homomorphism $h'_{n-1} : \pi_{n-1}(A) \to H_{n-1}(A)$ in Section 2.4 requires the choice of a generator $\gamma_{n-1} \in H_{n-1}(\partial I^n)$. We choose $\lambda_n \in H_n(I^n, \partial I^n)$ and $\gamma_{n-1} \in H_{n-1}(\partial I^n)$ with the property that $\Delta_n(\lambda_n) = \gamma_{n-1}$, where $\Delta_n : H_n(I^n, \partial I^n) \to H_{n-1}(\partial I^n)$ is the boundary homomorphism in the exact homology sequence of the pair $(I^n, \partial I^n)$. The proof of the next lemma is then straightforward, and hence omitted.

Lemma 4.5.10 *The following diagram commutes*

$$\begin{array}{ccc} \pi_n(X,A) & \xrightarrow{\partial_n} & \pi_{n-1}(A) \\ \downarrow h_n & & \downarrow h'_{n-1} \\ H_n(X,A) & \xrightarrow{\Delta_n} & H_{n-1}(A), \end{array}$$

where ∂_n and Δ_n are the boundary homomorphisms in the exact homotopy sequence of a pair and the exact homology sequence of a pair, respectively.

We now describe the adjoint isomorphism $\bar{\kappa}_* : \pi_{n-1}(\Omega X) \to \pi_n(X)$ using the path-space fibration.

Lemma 4.5.11 *Let*

$$\Omega X \longrightarrow EX \xrightarrow{p} X$$

be the path-space fibration and $p' : (EX, \Omega X) \to (X, \{*\})$ be the corresponding map of pairs. Then the following diagram is commutative

$$\begin{array}{ccc} \pi_n(EX,\Omega X) & \xrightarrow{\partial_n} & \pi_{n-1}(\Omega X) \\ \downarrow p'_* & \swarrow \bar{\kappa}_* & \\ \pi_n(X), & & \end{array}$$

where ∂_n is the boundary homomorphism.

Proof. The proof follows from Exercises 4.2 and 4.24. □

Then $\bar{\kappa}_* = p'_* \partial_n^{-1}$ since ∂_n is an isomorphism and this suggests that we study the homology analogue of the homomorphism $p'_* \partial_n^{-1}$.

Definition 4.5.12 Let $\Omega X \longrightarrow EX \xrightarrow{p} X$ be the path-space fibration and consider the homomorphisms

$$H_n(X;G) \xleftarrow{p'_*} H_n(EX,\Omega X;G) \xrightarrow{\Delta_n} H_{n-1}(\Omega X;G),$$

where Δ_n is the boundary homomorphism in the exact homology sequence of a pair and $p' : (EX, \Omega X) \to (X, \{*\})$ is the map of pairs obtained from p. Since EX is contractible, Δ_n is an isomorphism. Define the *homology suspension* to be the homomorphism $\sigma_{n-1} = p'_* \Delta_n^{-1} : H_{n-1}(\Omega X; G) \to H_n(X;G)$.

The term "homology suspension" is well established for this homomorphism, even though it is not directly related to the suspension. In Lemma 6.4.7 we give conditions for σ_{n-1} to be an isomorphism. For now we note its relation to the Hurewicz homomorphism. The proof follows from Lemma 4.5.10.

Lemma 4.5.13 *For any space X, the following diagram is commutative*

$$\begin{array}{ccc} \pi_{n-1}(\Omega X) & \xrightarrow{\bar{\kappa}_*} & \pi_n(X) \\ \downarrow{\scriptstyle h_n} & & \downarrow{\scriptstyle h'_n} \\ H_{n-1}(\Omega X) & \xrightarrow{\sigma_{n-1}} & H_n(X), \end{array}$$

where h_n and h'_n are Hurewicz homomorphisms.

We have seen in previous sections of this chapter that an important role was played by certain maps called α and β assigned to a sequence

$$X \xrightarrow{f} Y \xrightarrow{g} Z.$$

If this sequence is a fiber sequence, then $\beta : X \to I_g$ is a homotopy equivalence (Proposition 3.5.10 and Remark 3.5.11). If the sequence is a cofiber sequence, then $\alpha : C_f \to Z$ is a homotopy equivalence (Proposition 3.5.4 and Remark 3.5.5). We next define these maps more generally and introduce two other closely related maps. Let

$$X \xrightarrow{f} Y \xrightarrow{g} Z$$

be a sequence of spaces such that $gf = *$. Consider the homotopy fibers I_f and I_g of f and g

$$\begin{array}{ccccc} I_f & & I_g & & \\ \downarrow{\scriptstyle w} & & \downarrow{\scriptstyle v} & & \\ X & \xrightarrow{f} & Y & \xrightarrow{g} & Z, \end{array}$$

where v and w are projections.

Definition 4.5.14 We define two *excision maps* as follows: $\beta : X \to I_g$ is given by $\beta(x) = (f(x), *)$, for $x \in X$ and $\gamma : I_f \to \Omega Z$ is given by $\gamma(x, l) = g\,l$, for $(x, l) \in I_f$.

Next note that $v\beta = f$ and $\gamma j = \Omega g$,

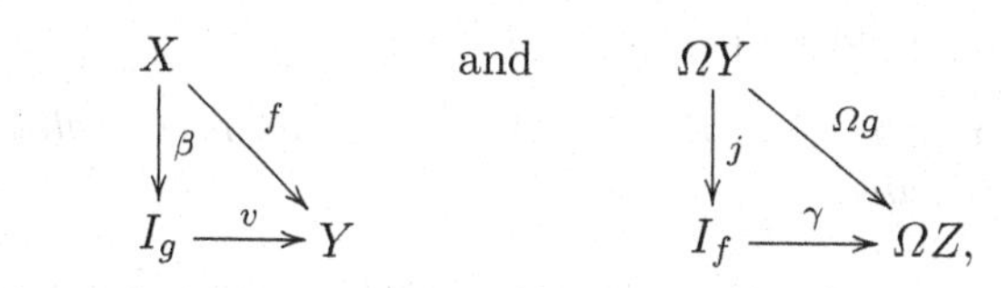

where j is the inclusion. By Corollary 4.2.17 we have the following diagram with exact rows

$$\begin{array}{ccccccccc}
\cdots \longrightarrow & \pi_{r+1}(Y) & \xrightarrow{j'_*} & \pi_r(I_f) & \xrightarrow{w_*} & \pi_r(X) & \xrightarrow{f_*} & \pi_r(Y) & \longrightarrow \cdots \\
& \downarrow \mathrm{id} & & \downarrow \gamma_* & & \downarrow \beta_* & & \downarrow \mathrm{id} & \\
\cdots \longrightarrow & \pi_{r+1}(Y) & \xrightarrow{g'_*} & \pi_r(\Omega Z) & \xrightarrow{k_*} & \pi_r(I_g) & \xrightarrow{v_*} & \pi_r(Y) & \longrightarrow \cdots,
\end{array}$$

where j'_* is the composition

$$\pi_{r+1}(Y) \xrightarrow{\cong} \pi_r(\Omega Y) \xrightarrow{j_*} \pi_r(I_f),$$

g'_* is the composition

$$\pi_{r+1}(Y) \xrightarrow{g_*} \pi_{r+1}(Z) \xrightarrow{\cong} \pi_r(\Omega Z),$$

and $k : \Omega Z \to I_g$ is the inclusion.

Proposition 4.5.15 *In the above diagram with exact rows, the first and third squares are commutative, and the second square is anticommutative, that is, $\beta_* w_* = -k_* \gamma_*$.*

Proof. Because $v\beta = f$ and $\gamma j = \Omega g$, the first and third squares are commutative. We sketch the proof that the second square is anticommutative. Let us write $a = \beta w$, $b = k\gamma : I_f \to I_g$. Then $a(x, l) = (f(x), *)$ and $b(x, l) = (*, g\, l)$, for $(x, l) \in I_f$. Let $v : I_g \to Y$ and $u : I_g \to EZ$ be the two projections of the pullback I_g and $w : I_f \to X$ and $t : I_f \to EY$ be the two projections of the pullback I_f. If $[\phi] \in \pi_r(I_f)$, then

$$u(a\phi + b\phi) = * + (Eg)t\phi \quad \text{and} \quad v(a\phi + b\phi) = fw\phi + *.$$

Now define $\theta : I_f \to I_g$ by $\theta(x, l) = (f(x), Eg(l))$, where $(x, l) \in I_f$. Then $u\theta\phi = (Eg)t\phi$ and $v\theta\phi = fw\phi$, and from this it follows that $a\phi + b\phi \simeq \theta\phi$. Next we define a homotopy $F : I_f \times I \to I_g$ such that $\theta \simeq_F *$. For a path $l \in EY$ and $s \in I$, let $l_{s,1}$ be the path l linearly reparametrized so as to start at $l(s)$ and end at $l(1) = *$, that is, $l_{s,1}(t) = l((1-t)s + t)$. Then we set $F((x, l), s) = (l(s), g\, l_{s,1})$, for $(x, l) \in I_f$ and $s \in I$, and so F is a nullhomotopy of θ. It follows that $a\phi + b\phi \simeq *$, and thus $a\phi \simeq -b\phi$. Therefore $\beta_* w_*[\phi] = -k_* \gamma_*[\phi]$, for all $[\phi] \in \pi_r(I_f)$. □

Corollary 4.5.16 *If $n \leqslant \infty$, then $\beta : X \to I_g$ is an n-equivalence $\iff \gamma : I_f \to \Omega Z$ is an n-equivalence.*

Proof. The proof is an immediate consequence of Proposition 4.5.15 and the five lemma (Appendix C). □

Next we consider $X \xrightarrow{f} Y \xrightarrow{g} Z$ with $gf = *$ in the case when f is an inclusion map. Then we identify I_f with $E(Y; X, \{*\})$, the space of paths in Y that begin in X and end at $\{*\}$, and so $\pi_n(I_f) \cong \pi_{n+1}(Y, X)$. Let $\gamma : I_f \to \Omega Z$ be the excision map and let $g' : (Y, X) \to (Z, \{*\})$ be the map of pairs determined by g.

Lemma 4.5.17 *The following diagram is commutative for all $n \geqslant 0$,*

$$\begin{array}{ccc} \pi_n(I_f) & \xrightarrow{\cong} & \pi_{n+1}(Y, X) \\ \downarrow{\scriptstyle \gamma_*} & & \downarrow{\scriptstyle g'_*} \\ \pi_n(\Omega Z) & \xrightarrow{\cong} & \pi_{n+1}(Z). \end{array}$$

Proof. This follows from the definition of γ. □

The next result is an immediate consequence of the previous lemma, Proposition 3.5.10, and Corollary 4.5.16.

Proposition 4.5.18 *If $F \xrightarrow{i} E \xrightarrow{p} B$ is a fiber sequence and if $p' : (E, F) \to (B, \{*\})$ is the map of pairs obtained from p, then $p'_* : \pi_r(E, F) \to \pi_r(B)$ is an isomorphism for all $r > 1$ and a bijection for $r = 1$.*

Remark 4.5.19

1. We observe that Proposition 4.5.15 holds for homotopy groups with coefficients. For the proof we simply replace the sphere by a Moore space.
2. Proposition 4.5.18 is often proved directly as a consequence of the definition of a fibration (e.g., see [37, pp. 83–84]).

We next briefly comment on the dual results. Let

$$X \xrightarrow{f} Y \xrightarrow{g} Z$$

be a sequence of spaces such that $gf = *$. We consider the mapping cones C_f and C_g of f and g

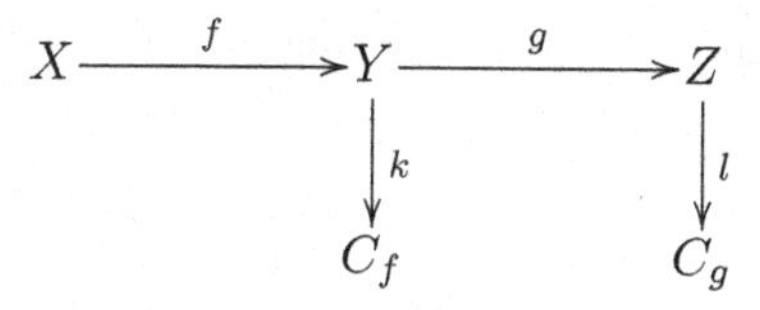

where k and l are inclusions.

Definition 4.5.20 We define *excision maps* as follows: $\alpha : C_f \to Z$ by $\alpha\langle y \rangle = g(y)$ and $\alpha\langle x, t \rangle = *$ and $\delta : \Sigma X \to C_g$ by $\delta\langle x, t \rangle = \langle f(x), t \rangle$, for $x \in X$, $y \in Y$, and $t \in I$.

All of these excision maps play an important role in the sequel. They are used to prove the classical theorems in Section 6.4. With some modifications, the results on the excision maps β and γ dualize to the maps α and δ. We state two of these without proof. The first is the dual of Proposition 4.5.15.

Consider the sequences of spaces

$$Y \xrightarrow{k} C_f \xrightarrow{p} \Sigma X \quad \text{and} \quad Z \xrightarrow{l} C_g \xrightarrow{q} \Sigma Y$$

and the corresponding diagram of cohomology groups

$$\begin{array}{ccccccccc} \cdots \longrightarrow & H^{r-1}(Y;G) & \xrightarrow{q'^*} & H^r(C_g;G) & \xrightarrow{l^*} & H^r(Z;G) & \xrightarrow{g^*} & H^r(Y;G) & \longrightarrow \cdots \\ & \downarrow \mathrm{id} & & \downarrow \delta^* & & \downarrow \alpha^* & & \downarrow \mathrm{id} & \\ \cdots \longrightarrow & H^{r-1}(Y;G) & \xrightarrow{f'^*} & H^r(\Sigma X;G) & \xrightarrow{p^*} & H^r(C_f;G) & \xrightarrow{k^*} & H^r(Y;G) & \longrightarrow \cdots, \end{array}$$

where q'^* is the composition

$$H^{r-1}(Y;G) \xrightarrow{\cong} H^r(\Sigma Y;G) \xrightarrow{q^*} H^r(C_g;G)$$

and f'^* is the composition

$$H^{r-1}(Y;G) \xrightarrow{f^*} H^{r-1}(X;G) \xrightarrow{\cong} H^r(\Sigma X;G).$$

Proposition 4.5.21 *In the preceding diagram the rows are exact, the first and third squares are commutative, and the second square is anticommutative.*

We give a definition analogous to n-equivalence.

Definition 4.5.22 A map $f : X \to Y$ is called a *cohomological n-equivalence* if for every abelian group G, the induced homomorphism $f^* : H^i(Y;G) \to H^i(X;G)$ is an isomorphism for $i < n$ and a monomorphism for $i = n$.

Corollary 4.5.23 *If $n \leqslant \infty$, then $\alpha : C_f \to Z$ is a cohomological n-equivalence $\iff \delta : \Sigma X \to C_g$ is a cohomological n-equivalence.*

Finally, we mention that we are following [40, p. 13] in calling these maps excision maps.

Exercises

Exercises marked with (∗) may be more difficult than the others. Exercises marked with (†) are used in the text.

4.1. (∗) Let $F \xrightarrow{j} E \xrightarrow{p} B$ be a fibration and let $\lambda : S \to E^I$ be a path lifting map (Exercise 3.10). Let $i_2 : \Omega B \to S$ be the inclusion into the second factor and let $p_1 : E^I \to E$ be defined by $p_1(l) = l(1)$. Define $\lambda_1 = p_1\lambda i_2 : \Omega B \to E$ and note that λ_1 induces $\widehat{\lambda}_1 : \Omega B \to F$. Prove that $\widehat{\lambda}_1$ is homotopic to the connecting map $\partial : \Omega B \to F$ (Definition 4.2.18).

4.2. (*) (†) For the path space fibration $\Omega B \longrightarrow EB \longrightarrow B$, prove that the connecting map $\partial : \Omega B \to \Omega B$ is homotopic to $-\mathrm{id} : \Omega B \to \Omega B$.

4.3. (*) Let K be a CW complex and $L \subseteq K$ a subcomplex. Consider the cofiber sequence $L \to K \to K/L$ and let $\partial : K/L \to \Sigma L$ be the connecting map. Let $i : K/L \to CK/L$ be the inclusion. Prove the following.

1. $CK/L \simeq \Sigma L$.
2. $\partial \simeq * \Longleftrightarrow i \simeq *$.
3. ΣL is a homotopy retract of $\Sigma K \Longleftrightarrow i \simeq *$.

4.4. (*) (†) Let $\Omega X \longrightarrow E_0 X \xrightarrow{p_1} X$ be the path space fibration where E_0X consists of paths l in X with $l(0) = *$ and $p_1(l) = l(1)$. Prove that the action $\phi : \Omega X \times \Omega X \to \Omega X$ of the loops of the base on the fiber is homotopic to the loop space multiplication m of ΩX.

4.5. (*) Let $F \xrightarrow{j} E \xrightarrow{p} B$ be a fiber sequence. Prove that if B is contractible, then j is a homotopy equivalence.

4.6. Let $\theta_{X,Y} : \Omega X \times \Omega Y \to \Omega(X \times Y)$ be the homeomorphism defined by $\theta_{X,Y}(\omega,\nu)(t) = (\omega(t),\nu(t))$, for $\omega \in \Omega X$, $\nu \in \Omega Y$, and $t \in I$. Furthermore, let (A, m) be an H-space and $\phi : X \times A \to X$ an action of A on X. Prove that $(\Omega\phi)\theta_{X,A}$ is an action of $(\Omega A, \mu)$ on ΩX, where $\mu : \Omega A \times \Omega A \to \Omega A$ is the loop-space multiplication. (Note that $\mu \simeq (\Omega m)\theta_{A,A}$.)

4.7. (*) (†) For each $i = 1, 2, \ldots, n$, define a binary operation on $\pi'_n(X)$ by setting $[f] +_i [g] = [f +_i g]$, where

$$(f +_i g)(t_1, \ldots, t_n) = \begin{cases} f(t_1, \ldots, 2t_i, \ldots, t_n) & \text{if } 0 \leqslant t_i \leqslant \frac{1}{2} \\ g(t_1, \ldots, 2t_i - 1, \ldots, t_n) & \text{if } \frac{1}{2} \leqslant t_i \leqslant 1, \end{cases}$$

for $f, g : (I^n, \partial I^n) \to (X, \{*\})$. These induce n binary operations in $\pi_n(X)$. Prove that each of these operations coincides with the standard binary operation in $\pi_n(X)$ (obtained by regarding S^n as a co-H-space).

4.8. $\pi_n(X, A)$ consists of homotopy classes of maps $(I^n, I^{n-1}, J^{n-1}) \to (X, A, \{*\})$. If f and g are two such maps, define $f +_i g : (I^n, I^{n-1}, J^{n-1}) \to (X, A, \{*\})$ by

$$(f +_i g)(t_1, \ldots, t_n) = \begin{cases} f(t_1, \ldots, 2t_i, \ldots, t_n) & \text{if } 0 \leqslant t_i \leqslant \frac{1}{2} \\ g(t_1, \ldots, 2t_i - 1, \ldots, t_n) & \text{if } \frac{1}{2} \leqslant t_i \leqslant 1, \end{cases}$$

for $i = 1, \ldots, n-1$. This induces $n-1$ binary operations on $\pi_n(X, A)$. Prove that these binary operations all agree with the given operation on $\pi_n(X, A)$.

4.9. Regard $\{*\}$ as a subspace of X. Show that the relative homotopy group $\pi_n(X, \{*\})$ is isomorphic to $\pi_n(X)$.

4.10. Let K be a CW complex and let $L \subseteq K$ be a subcomplex with inclusion map $i : L \to K$. Prove that if i induces isomorphisms of all homotopy groups, then L is a strong deformation retract of K.

4.11. Given maps $f : X \to Y$ and $f' : X' \to Y'$. A *map (of maps)* from f to f' is a pair (u, v), where $u : X \to X'$ and $v : Y \to Y'$ are maps such that $f'u = vf$. A homotopy of maps is a pair of homotopies (u_t, v_t), where $u_t : X \to X'$ and $v_t : Y \to Y'$

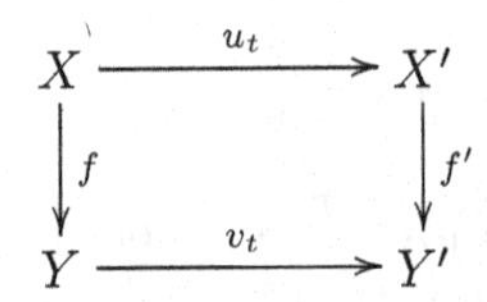

such that $f'u_t = v_t f$, for every $t \in I$. Define a set $H^n(X, A; G)$ using homotopies of maps, where A is a subspace of X and G is an abelian group. Prove that there is a bijection of $H^n(X, A; G)$ onto $H^n(C_i; G)$, where $i : A \to X$ is the inclusion map. Consider the dual result.

4.12. (†) Establish the claim made in the proof of Lemma 4.5.6 that it suffices to consider the case when e is an inclusion.

4.13. (†) Verify the assertion in Lemma 4.5.6 that J' is homotopic to J such that $a \simeq_J e\tilde{b}$ and $J|S^{k-1} \times I = H$.

4.14. (*) (†) Prove that $h_n : \pi_n(X, A) \to H_n(X, A)$ is a homomorphism for $n \geqslant 2$.

4.15. In the proof of Proposition 4.5.15 show that $a\phi + b\phi \simeq \theta\phi$.

4.16. (†) Using the characterization of the relative homotopy group $\pi_n(X, A)$ as homotopy classes of maps $(E^n, S^{n-1}) \to (X, A)$, show that the boundary homomorphism $\partial_n : \pi_n(X, A) \to \pi_{n-1}(A)$ satisfies $\partial_n[f] = [f|S^{n-1}]$.

4.17. Suppose that a map $f : X \to Y$ is factored as

$$X \xrightarrow{f_1} M \xrightarrow{f_2} Y,$$

where f_1 is an inclusion and f_2 is a homotopy equivalence. Prove that there exists a map $\theta : M_f \to M$ such that $\theta|X = \mathrm{id}_X$ and $\theta'_* : \pi_n(M_f, X) \to \pi_n(M, X)$ is an isomorphism for all n.

4.18. If G is a topological group and H is a subgroup, prove that the second relative homotopy group $\pi_2(G, H)$ is abelian.

4.19. (*) If A is a subspace of X and A is contractible in X (see Definition 1.4.3), then prove that

$$\pi_n(X, A) \cong \pi_n(X) \oplus \pi_{n-1}(A),$$

for $n \geqslant 2$. (For $n = 2$, you may use the fact that the image of $\pi_2(X) \to \pi_2(X, A)$ is contained in the center, which is proved in Corollary 5.4.3.)

4.20. Let the isomorphism $\tau_{n-1} : H_{n-1}(X) \to H_n(\Sigma X)$ be taken as the composition

$$H_{n-1}(X) \xrightarrow{\Delta_n^{-1}} H_n(CX, X) \xrightarrow{q'_*} H_n(\Sigma X),$$

where Δ_n is the boundary homomorphism in the exact homology sequence of a pair and $q : CX \to \Sigma X$ is the quotient map. Prove that for any space X, the following diagram is commutative

$$\begin{array}{ccc} \pi_{n-1}(X) & \xrightarrow{\Sigma_{n-1}} & \pi_n(\Sigma X) \\ \downarrow{h_n} & & \downarrow{h'_n} \\ H_{n-1}(X) & \xrightarrow{\tau_{n-1}} & H_n(\Sigma X), \end{array}$$

where h_n and h'_n are Hurewicz homomorphisms and Σ_{n-1} is defined by $\Sigma_{n-1}[f] = [\Sigma f]$. (For this problem, it is necessary to choose compatible generators $\gamma_{n-1} \in H_{n-1}(S^{n-1})$ and $\gamma_n \in H_n(S^n)$.)

4.21. For a space X, define $\tau_{n-1} : H_{n-1}(X) \to H_n(\Sigma X)$ as in Exercise 4.20. Let $f : \Sigma X \to Y$ and let $\overline{f} : X \to \Omega Y$ be the adjoint of f. Show that the following diagram is commutative

$$\begin{array}{ccccc} \pi_{n-1}(X) & \longrightarrow & H_{n-1}(X) & \xrightarrow{\tau_{n-1}} & H_n(\Sigma X) \\ \downarrow{\overline{f}_*} & & & & \downarrow{f_*} \\ \pi_{n-1}(\Omega Y) & \xrightarrow{\overline{\kappa}_*} & \pi_n(Y) & \longrightarrow & H_n(Y), \end{array}$$

where $\overline{\kappa}_*$ is the adjoint isomorphism and the unmarked arrows are Hurewicz homomorphisms.

4.22. (†) Let X be a space, let A and B be subspaces and set $C = A \cap B$. Then $(X; A, B)$ is called a *triad* and we define the *triad homotopy groups* by

$$\pi_n(X; A, B) = \pi_{n-1}(E(X; B, \{*\}), E(A; C, \{*\})),$$

for all $n \geqslant 2$ (when $n = 2$, this is a based set).

1. Prove that $\pi_n(X; A, B)$ consists of homotopy classes of maps $f : I^n \to X$ such that $f(t_1, \dots, t_n) \in A$ if $t_{n-1} = 0$, $f(t_1, \dots, t_n) \in B$ if $t_n = 0$, and $f(t_1, \dots, t_n) = *$ if $(t_1, \dots, t_n) \in \partial I^n$ with $t_{n-1} \neq 0$ or $t_n \neq 0$.
2. Prove that the following is an exact sequence.

$$\longrightarrow \pi_{n+1}(X; A, B) \longrightarrow \pi_n(A, C) \longrightarrow \pi_n(X, B) \longrightarrow \pi_n(X; A, B) \longrightarrow$$

   This is called the *exact homotopy sequence of a triad.*
3. Prove that $\pi_n(X; A, B) \cong \pi_n(X; B, A)$.
4. If $A \subseteq B \subseteq X$, then show that the following sequence is exact.

$$\longrightarrow \pi_n(B, A) \longrightarrow \pi_n(X, A) \longrightarrow \pi_n(X, B) \longrightarrow \pi_{n-1}(B, A) \longrightarrow$$

   This is called the *exact homotopy sequence of a triple.*

4.23. Let

$$X \xrightarrow{f} Y \xrightarrow{g} Z$$

be a sequence of spaces such that $gf = *$ and let $\alpha : C_f \to Z$ and $\gamma : I_f \to \Omega Z$ be excision maps. Define a map $\eta : I_f \to \Omega C_f$ such that the following diagram is anti-homotopy-commutative

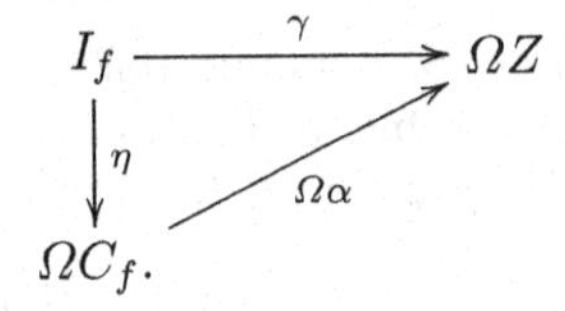

State and prove the dual result.

4.24. (*) (†) Let $F \xrightarrow{j} E \xrightarrow{p} B$ be a fiber sequence. Let $p' : (E, F) \to (B, *)$ be the map of pairs determined by p, let $\partial_n : \pi_n(E, F) \to \pi_{n-1}(F)$ be the boundary homomorphism in the exact homotopy sequence of the pair (E, F), and let $\partial_* : \pi_n(B) \to \pi_{n-1}(F)$ be the homomorphism induced by the connecting map $\partial : \Omega B \to F$ of the fibration. Prove that the following diagram is anticommutative

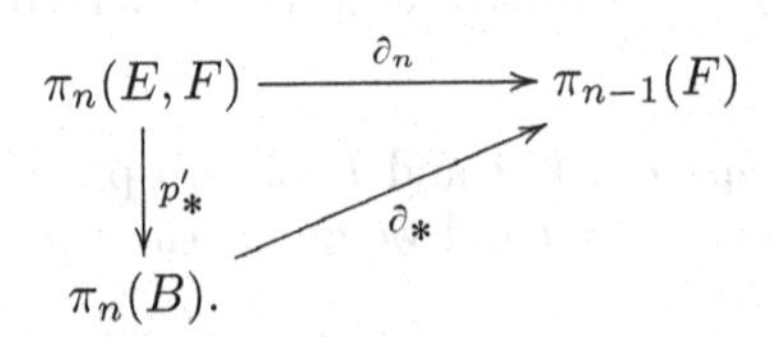

Chapter 5
Applications of Exactness

5.1 Introduction

In this chapter we give some diverse applications of the material in Chapter 4. In Definition 2.5.10 we defined the (homotopical) cohomology groups with coefficients as $H^n(X;G) = [X, K(G,n)]$. When $G = \mathbb{Z}$, we refer to these groups as *ordinary cohomology groups* or *integral cohomology groups.* Analogously, the homotopy groups with coefficients are $\pi_n(X;G) = [M(G,n), X]$ (Definition 2.5.6). When $G = \mathbb{Z}$, these are just the *homotopy groups* of X. In the first section we present two universal coefficient theorems. The first expresses cohomology with coefficients in terms of ordinary cohomology, the tensor product, and the torsion product. The second expresses homotopy groups with coefficients in terms of the homotopy groups, Hom, and Ext. In the next section we show that the cohomology groups of a CW complex are naturally isomorphic to the singular cohomology groups of the complex. This is done by a simple inductive argument over the dimension of the complex. In Section 5.4, we study the action of $[W, \Omega B]$ on $[W, F]$, when $F \to E \to B$ is a fiber sequence and W is a co-H-space. We also give conditions for certain fibrations and cofibrations to be trivial. For the Hopf fibrations this implies that the loops on an $\mathbb{F}$-projective n-space is equivalent to the product of S^{d-1} (the fiber) and the loops on $S^{d(n+1)-1}$ (the total space). In Section 5.5 we consider the space of free maps $M(X,Y)$ of X into Y and show that the evaluation fiber sequence $Y^X \to M(X,Y) \to Y$ yields an operation of $[W, \Omega Y]$ on $[W, Y^X]$, for any space W. This operation is needed for comparing based and free homotopy classes. By taking $W = S^0$ and $X = S^n$, we obtain an operation of $\pi_1(Y)$ on $\pi_n(Y)$. We discuss several characterizations of this operation, extend it to relative homotopy groups, and prove that the homomorphisms in the exact homotopy sequence of a pair are all π_1-homomorphisms. In the last section we calculate some homotopy groups of spheres, Moore spaces, topological groups, Stiefel manifolds, and Grassmannians. Our computational methods include the exact homotopy sequence of a fiber sequence, the truncated exact homotopy sequence of a cofiber sequence, and the Freudenthal suspension theorem.

5.2 Universal Coefficient Theorems

We assume some familiarity with the tensor product $A \otimes B$ and the group of homomorphisms $\mathrm{Hom}(A, B)$ of two abelian groups A and B. In addition, we refer to basic facts about the torsion product $\mathrm{Tor}(A, B) = A * B$ and the group of extensions $\mathrm{Ext}(A, B)$, that can be found in Appendix C.

We begin by noting the following extension of Proposition 2.5.9 which is an immediate consequence of Lemma 2.5.13.

Lemma 5.2.1 *If G and H are abelian groups and $f, g : K(G, n) \to K(H, n)$ are two maps, then $f_* = g_* : \pi_n(K(G, n)) = G \to \pi_n(K(H, n)) = H$, $n \geqslant 1$, if and only if $f \simeq g$.*

Next we define a homomorphism of the universal coefficient theorem.

Definition 5.2.2 For a space X and an abelian group G, we define a homomorphism $\alpha = \alpha_G : H^n(X) \otimes G \to H^n(X; G)$ as follows. Given $[f] \in H^n(X) = [X, K(\mathbb{Z}, n)]$ and $\gamma \in G$, then γ induces a homomorphism $\theta_\gamma : \mathbb{Z} \to G$ defined by $\theta_\gamma(1) = \gamma$. By Proposition 2.5.9 and Lemma 5.2.1, θ_γ determines a unique homotopy class $[\phi_\gamma] \in [K(\mathbb{Z}, n), K(G, n)]$ given by $\phi_{\gamma *} = \theta_\gamma : \pi_n(K(\mathbb{Z}, n)) \to \pi_n(K(G, n))$. We set $\alpha([f] \otimes \gamma) = [\phi_\gamma f]$,

$$X \xrightarrow{f} K(\mathbb{Z}, n) \xrightarrow{\phi_\gamma} K(G, n).$$

Then α is a well-defined homomorphism (Exercise 5.1).

We need the following elementary fact in our proof below. If $l : K(G, n) \to K(H, n)$ induces the homomorphism $\lambda : G \to H$ on n-dimensional homotopy groups, then the following diagram commutes

$$\begin{array}{ccc} H^n(X) \otimes G & \xrightarrow{\mathrm{id} \otimes \lambda} & H^n(X) \otimes H \\ \downarrow{\scriptstyle \alpha_G} & & \downarrow{\scriptstyle \alpha_H} \\ H^n(X; G) & \xrightarrow{l_*} & H^n(X; H), \end{array}$$

where l_* is the coefficient homomorphism induced by l.

Lemma 5.2.3 *If F is a finitely generated, free-abelian group, then $\alpha_F : H^n(X) \otimes F \to H^n(X; F)$ is an isomorphism.*

Proof. Let $F = \mathbb{Z}$ and consider $\alpha_{\mathbb{Z}} : H^n(X) \otimes \mathbb{Z} \to H^n(X)$. Thus $\alpha_{\mathbb{Z}}([f] \otimes 1) = [f]$ because $\phi_1 = \mathrm{id}$. Therefore $\alpha_{\mathbb{Z}}$ is an isomorphism. Now let F have rank k, so that F is the direct sum of k copies of $\mathbb{Z}$, and we write $F \cong \bigoplus_k \mathbb{Z}$. Then $H^n(X) \otimes F \cong \bigoplus_k (H^n(X) \otimes \mathbb{Z})$ and $K(F, n) = K(\mathbb{Z}, n) \times \cdots \times K(\mathbb{Z}, n)$ (k factors), and so $H^n(X; F) \cong \bigoplus_k H^n(X)$ by Corollary 1.3.7. The commutativity of the diagram

$$\begin{array}{ccc} H^n(X) \otimes F & \xrightarrow{\cong} & \bigoplus_k (H^n(X) \otimes \mathbb{Z}) \\ \downarrow{\scriptstyle \alpha_F} & & \cong \downarrow{\scriptstyle \bigoplus_k \alpha_{\mathbb{Z}}} \\ H^n(X; F) & \xrightarrow{\cong} & \bigoplus_k H^n(X) \end{array}$$

then completes the proof. □

Theorem 5.2.4 (universal coefficient theorem for cohomology) *If X is a space, G a finitely generated abelian group, and $n \geqslant 1$, then there is a homomorphism $\beta : H^n(X; G) \to H^{n+1}(X) * G$ and a short exact sequence of groups*

$$0 \longrightarrow H^n(X) \otimes G \xrightarrow{\alpha} H^n(X; G) \xrightarrow{\beta} H^{n+1}(X) * G \longrightarrow 0,$$

for $n \geqslant 1$.

Proof. Write $G = F/R$, where F is a finitely generated, free-abelian group and R is a subgroup. Let $\iota : R \to F$ be the inclusion and $\nu : F \to G$ the projection. By Proposition 2.5.9(2), there exists a map $q : K(F, n+1) \to K(G, n+1)$ such that $q_* = \nu : \pi_{n+1}(K(F, n+1)) \to \pi_{n+1}(K(G, n+1))$. By Proposition 3.5.8, we may assume that q is a fiber map. Thus we have a fiber sequence

$$W \xrightarrow{i} K(F, n+1) \xrightarrow{q} K(G, n+1),$$

for some space W. Applying Corollary 4.2.19 to this fibration, we see that

$$\pi_l(W) = \begin{cases} 0 & \text{if } l \neq n+1 \\ R & \text{if } l = n+1. \end{cases}$$

Therefore $K(R, n+1) \xrightarrow{i} K(F, n+1) \xrightarrow{q} K(G, n+1)$ is a fiber sequence and $i_* : \pi_{n+1}(K(R, n+1)) \to \pi_{n+1}(K(F, n+1))$ is the inclusion $\iota : R \to F$. By Corollary 4.2.19, for any space X, the following sequence of groups is exact

$$H^n(X; R) \xrightarrow{i_*} H^n(X; F) \xrightarrow{q_*} H^n(X; G) \xrightarrow{\partial_*} H^{n+1}(X; R) \xrightarrow{i_*} H^{n+1}(X; F).$$

This gives the short exact sequence

$$0 \longrightarrow \frac{H^n(X; F)}{i_* H^n(X; R)} \xrightarrow{q'_*} H^n(X; G) \xrightarrow{\partial'_*} \operatorname{Im} \partial_* \longrightarrow 0,$$

where q'_* and ∂'_* are determined by q_* and ∂_*, respectively. Now $\operatorname{Ker} i_* = \operatorname{Im} \partial'_*$, where $i_* : H^{n+1}(X;R) \to H^{n+1}(X;F)$, and we show that $\operatorname{Ker} i_* \cong H^{n+1}(X) * G$. For this consider the commutative diagram

$$\begin{array}{ccc} H^{n+1}(X) \otimes R & \xrightarrow{\mathrm{id} \otimes \iota} & H^{n+1}(X) \otimes F \\ \cong \downarrow \alpha_R & & \cong \downarrow \alpha_F \\ H^{n+1}(X;R) & \xrightarrow{i_*} & H^{n+1}(X;F). \end{array}$$

Since $0 \longrightarrow R \xrightarrow{\iota} F \xrightarrow{\nu} G \longrightarrow 0$ is a free resolution of G,

$$\operatorname{Ker} i_* \cong \operatorname{Ker} (\mathrm{id} \otimes \iota) = H^{n+1}(X) * G$$

by Appendix C. We define β to be the composition

$$H^n(X;G) \xrightarrow{\partial'_*} \operatorname{Ker} i_* \xrightarrow{\cong} H^{n+1}(X) * G.$$

We next show that there is an isomorphism

$$\theta : H^n(X) \otimes G \longrightarrow \frac{H^n(X;F)}{i_* H^n(X;R)}$$

such that the following diagram commutes

$$\begin{array}{ccc} H^n(X) \otimes G & \xrightarrow{\alpha} & H^n(X;G) \\ \cong \downarrow \theta & \nearrow q'_* & \\ \dfrac{H^n(X;F)}{i_* H^n(X;R)}. & & \end{array}$$

Consider the diagram with exact rows

$$\begin{array}{ccccccc} H^n(X) \otimes R & \xrightarrow{\mathrm{id} \otimes \iota} & H^n(X) \otimes F & \xrightarrow{\mathrm{id} \otimes \nu} & H^n(X) \otimes G & \longrightarrow & 0 \\ \cong \downarrow \alpha_R & & \cong \downarrow \alpha_F & & \downarrow \theta & & \\ H^n(X;R) & \xrightarrow{i_*} & H^n(X;F) & \xrightarrow{\pi} & \dfrac{H^n(X;F)}{i_* H^n(X;R)} & \longrightarrow & 0, \end{array}$$

where π is the projection onto the quotient group. Since the left square is commutative, there exists an isomorphism θ such that $\theta\,(\mathrm{id} \otimes \nu) = \pi\,\alpha_F$. Furthermore, it is easily checked that $\alpha = q'_*\,\theta$. Thus α is a monomorphism and

$$\operatorname{Im} \alpha = \operatorname{Im} q'_* = \operatorname{Ker} \partial'_* = \operatorname{Ker} \beta.$$

This completes the proof. □

Remark 5.2.5 Theorem 5.2.4 has been proved in [83, p. 246] and elsewhere by purely algebraic methods. In fact, the result in [83] is more general than Theorem 5.2.4 in that (1) the hypothesis is that either G or the homology groups of X are finitely generated; and (2) the exact sequence splits.

We next consider the dual theorem for homotopy groups with coefficients $\pi_n(X;G) = [M(G,n),X]$, where $M(G,n)$ is the Moore space defined in Lemma 2.5.2. Recall that $M(G,n)$ has been constructed relative to a choice of presentation of G.

Definition 5.2.6 Let X be a space, G an abelian group, and n an integer greater than 1. If $[f] \in \pi_n(X;G)$, then $f_* : G = \pi_n(M(G,n)) \to \pi_n(X)$. Thus there is a function $\eta = \eta_G : \pi_n(X;G) \to \mathrm{Hom}(G,\pi_n(X))$ defined by $\eta[f] = f_*$. It can be shown that η is a homomorphism (Exercise 5.3).

Lemma 5.2.7 *The homomorphism $\eta_F : \pi_n(X;F) \to \mathrm{Hom}(F,\pi_n(X))$ is an isomorphism if F is a free-abelian group.*

Proof. For some set A, we have $M(F,n) = \bigvee_{\alpha\in A} S^n_\alpha$, where S^n_α is the n-sphere. Let $i_\alpha : S^n_\alpha \to M(F,n)$ be the injection into the αth copy in the wedge. Then $\{[i_\alpha] \mid \alpha \in A\} \subseteq \pi_n(M(F,n)) = F$ is a basis for F by Lemma 2.4.17. If $f,g : M(F,n) \to X$ are maps such that $f_* = g_* : \pi_n(M(F,n)) \to \pi_n(X)$, then $f_*[i_\alpha] = g_*[i_\alpha]$. Hence $fi_\alpha \simeq gi_\alpha$, so $f \simeq g$. Therefore η_F is one–one.

If $\phi : F \to \pi_n(X)$ is a homomorphism, then $\phi[i_\alpha] = [f_\alpha] \in \pi_n(X)$, for some $f_\alpha : S^n_\alpha \to X$. Then the f_α determine $f : M(F,n) \to X$ such that $f_* = \phi : \pi_n(M(F,n)) \to \pi_n(X)$. Therefore η_F is onto. □

Corollary 5.2.8 *The homomorphism $\eta : [M(F,n),M(F',n)] \to \mathrm{Hom}(F,F')$ is an isomorphism if F and F' are free-abelian groups.*

Theorem 5.2.9 (universal coefficient theorem for homotopy) [40, p. 30] *For any space X, abelian group G, and integer $n \geqslant 2$, there exists a homomorphism $\xi : \mathrm{Ext}(G,\pi_{n+1}(X)) \to \pi_n(X;G)$ such that the following is a short exact sequence of groups*

$$0 \longrightarrow \mathrm{Ext}(G,\pi_{n+1}(X)) \xrightarrow{\xi} \pi_n(X;G) \xrightarrow{\eta} \mathrm{Hom}(G,\pi_n(X)) \longrightarrow 0.$$

Proof. We sketch the proof. The group G has a presentation given by a short exact sequence

$$0 \longrightarrow R \xrightarrow{\iota} F \xrightarrow{\nu} G \longrightarrow 0,$$

where R and F are free-abelian. By Corollary 5.2.8, ι determines a map $i : M(R,n) \to M(F,n)$ such that $i_* = \iota : \pi_n(M(R,n)) \to \pi_n(M(F,n))$. Then by Lemma 2.5.2, $M(G,n)$ is the mapping cone of i with inclusion $j : M(F,n) \to M(G,n)$. By Corollary 4.2.8, we have the exact sequence of groups

$$\pi_{n+1}(X;F) \xrightarrow{i^*} \pi_{n+1}(X;R) \xrightarrow{\partial^*} \pi_n(X;G) \xrightarrow{j^*} \pi_n(X;F) \xrightarrow{i^*} \pi_n(X;R)$$

which yields the short exact sequence

$$0 \longrightarrow \frac{\pi_{n+1}(X;R)}{i^*\pi_{n+1}(X;F)} \xrightarrow{\partial'^*} \pi_n(X;G) \xrightarrow{j'^*} \operatorname{Ker} i^* \longrightarrow 0,$$

where j'^* and ∂'^* are induced by j^* and ∂^*, respectively. Using Lemma 5.2.7, we see that there is an isomorphism $\theta : \operatorname{Ker} i^* \to \operatorname{Hom}(G, \pi_n(X))$ such that $\theta j'^* = \eta$. Also using Lemma 5.2.7 and the definition of Ext (Appendix C), we see that

$$\frac{\pi_{n+1}(X;R)}{i^*\pi_{n+1}(X;F)} \cong \operatorname{Ext}(G, \pi_{n+1}(X)).$$

This completes the sketch of the proof. □

Remark 5.2.10

1. In general the exact sequence in Theorem 5.2.9 does not split. See [39, p. 463] for an example. However, the exact sequence is natural with respect to homomorphisms induced by maps of spaces and to homomorphisms induced by a coefficient homomorphism.
2. There are other universal coefficient theorems besides the two given in this section. One gives homology with arbitrary coefficients in terms of integral homology and another gives cohomology with coefficients in terms of integral homology. These are discussed in Appendix C. In addition, there is a universal coefficient theorem for homotopy with coefficients which is different from Theorem 5.2.9. This gives homotopy groups with coefficients in terms of the ordinary homotopy groups and the functors tensor and tor (see [75]).

5.3 Homotopical Cohomology Groups

The purpose of this section is to prove the result stated in Remark 2.5.11 that the nth homotopical cohomology group of a CW complex X with coefficients in an abelian group G, namely, $H^n(X;G) = [X, K(G,n)]$, is naturally isomorphic to the n-th singular cohomology group $H^n_{\text{sing}}(X;G)$.

We first define the basic class. If $K = K(G,n)$ then $\mu : H^n_{\text{sing}}(K;G) \to \operatorname{Hom}(H_n(K), G)$, the epimorphism of the universal coefficient theorem for cohomology (Appendix C), is an isomorphism because $H_{n-1}(K) = 0$. By Exercise 2.34, the Hurewicz homomorphism $h : \pi_n(K) = G \to H_n(K)$ is an isomorphism.

Definition 5.3.1 The *nth basic class* in singular cohomology (or *nth fundamental class*) $b_n \in H^n_{\text{sing}}(K;G)$ is defined by $\mu(b_n) = h^{-1} \in \operatorname{Hom}(H_n(K), G)$.

Recall that in Section 2.5 there is a homomorphism $\rho = \rho_X : H^n(X;G) \to H^n_{\text{sing}}(X;G)$ defined by

$$\rho[f] = f^*(b_n) \in H^n_{\text{sing}}(X; G),$$

where $[f] \in [X, K]$.

Theorem 5.3.2 *If X is a CW complex, then $\rho = \rho_X : H^n(X; G) \to H^n_{\text{sing}}(X; G)$ is an isomorphism.*

We discuss some preliminaries before beginning the proof. As earlier, let $g' : (X, A) \to (Y, B)$ denote the map of pairs determined by $g : X \to Y$ such that $g(A) \subseteq B$. We also consider cones $C_0X = X \times I/X \times \{0\} \cup \{*\} \times I$ and path spaces $E_0X = \{\nu \mid \nu \in X^I,\ \nu(0) = *\}$. For any space A, the *cohomology suspension* $\omega : H^n_{\text{sing}}(A; G) \to H^{n-1}_{\text{sing}}(\Omega A; G)$ is defined as the following composition

$$H^n_{\text{sing}}(A; G) \xrightarrow{p'^*} H^n_{\text{sing}}(E_0A, \Omega A; G) \xrightarrow{\delta^{-1}_{\Omega A}} H^{n-1}_{\text{sing}}(\Omega A; G),$$

where $p : E_0A \to A$ with $p(l) = l(1)$ and $\delta_{\Omega A} : H^{n-1}_{\text{sing}}(\Omega A; G) \to H^n_{\text{sing}}(E_0A, \Omega A; G)$ is the coboundary homomorphism in the exact singular cohomology sequence of the pair $(E_0A, \Omega A)$. If $A = K = K(G, n)$ so that $\omega : H^n_{\text{sing}}(K; G) \to H^{n-1}_{\text{sing}}(\Omega K; G)$, then it follows that

$$\omega(b_n) = b_{n-1},$$

where $\pi_n(K) \cong \pi_{n-1}(\Omega K)$ via the adoint isomorphism (see Lemma 4.5.11).

We next present two lemmas for the proof of Theorem 5.3.2.

Lemma 5.3.3 *If $f : \Sigma X \to Y$ is a map, then there exists a map $\tilde{f} : C_0X \to E_0Y$ such that the following diagram commutes*

$$\begin{array}{ccccc}
X & \xrightarrow{\ i\ } & C_0X & \xrightarrow{\ q\ } & \Sigma X \\
\downarrow{\scriptstyle \kappa(f)} & & \downarrow{\scriptstyle \tilde{f}} & & \downarrow{\scriptstyle f} \\
\Omega Y & \xrightarrow{\ j\ } & E_0Y & \xrightarrow{\ p\ } & Y,
\end{array}$$

where p and q are projections, i and j are inclusions, and $\kappa(f)$ is the adjoint of f.

Proof. Let $\tilde{f}\langle x, s\rangle(t) = f\langle x, st\rangle$, for $x \in X$ and $s, t \in I$. □

Lemma 5.3.4 *If $K = K(G, n)$ and $q : C_0X \to \Sigma X$ is the projection, then the following diagram commutes*

$$\begin{array}{ccccc}
[\Sigma X, K] & & \xrightarrow[\cong]{\ \kappa_*\ } & & [X, \Omega K] \\
\downarrow{\scriptstyle \rho_{\Sigma X}} & & & & \downarrow{\scriptstyle \rho_X} \\
H^n_{\text{sing}}(\Sigma X; G) & \xrightarrow[\cong]{q'^*} & H^n_{\text{sing}}(C_0X, X; G) & \xleftarrow[\cong]{\delta_X} & H^{n-1}_{\text{sing}}(X; G),
\end{array}$$

where δ_X is the coboundary homomorphism in the exact singular cohomology sequence of the pair (C_0X, X).

Proof. If $[f] \in [\Sigma X, K]$, then

$$\begin{aligned}\delta_X \rho_X \kappa_*[f] &= \delta_X \kappa(f)^*(b_{n-1}) \\ &= \delta_X \kappa(f)^* \delta_{\Omega K}^{-1} p'^*(b_n) \\ &= \widetilde{f}'^* p'^*(b_n) \\ &= (p'\widetilde{f}')^*(b_n)\end{aligned}$$

since the diagram

$$\begin{array}{ccc} H^{n-1}_{\text{sing}}(\Omega K; G) & \xrightarrow{\delta_{\Omega K}} & H^n_{\text{sing}}(E_0 K, \Omega K; G) \\ \Big\downarrow{\scriptstyle \kappa(f)^*} & & \Big\downarrow{\scriptstyle \widetilde{f}'^*} \\ H^{n-1}_{\text{sing}}(X; G) & \xrightarrow{\delta_X} & H^n_{\text{sing}}(C_0 X, X; G) \end{array}$$

is commutative, where $\widetilde{f} : C_0X \to E_0K$ is the map of Lemma 5.3.3. But

$$q'^* \rho_{\Sigma X}[f] = q'^* f^*(b_n).$$

Since $fq = p\widetilde{f}$ by Lemma 5.3.3, $\delta_X \rho_X \kappa_*[f] = q'^* \rho_{\Sigma X}[f]$. □

Now we prove Theorem 5.3.2.

Proof. We first prove the result when X is finite-dimensional by showing by induction on $\dim X$ that ρ_X is an isomorphism. Let X^i be the ith skeleton of X and denote $\rho_{X^i} : H^n(X^i; G) = [X^i, K(G,n)] \to H^n_{\text{sing}}(X^i; G)$ by ρ_i. If X is one-dimensional, then $X = \bigvee_{\alpha \in A} S^1_\alpha$ and there is a commutative diagram

$$\begin{array}{ccc} [\bigvee_\alpha S^1_\alpha, K] & \xrightarrow[\cong]{(i^*_\alpha)} & \prod_\alpha [S^1_\alpha, K] \\ \Big\downarrow{\scriptstyle \rho_1} & & \Big\downarrow{\scriptstyle \prod \rho_\alpha} \\ H^n_{\text{sing}}(\bigvee_\alpha S^1_\alpha; G) & \xrightarrow[\cong]{(i^*_\alpha)} & \prod_\alpha H^n_{\text{sing}}(S^1_\alpha; G), \end{array}$$

where $K = K(G,n)$, $\rho_\alpha = \rho_{S^1_\alpha}$, $\rho_1 = \rho_{\vee_\alpha S^1_\alpha}$, $i_\alpha : S^1_\alpha \to \bigvee_\alpha S^1_\alpha$ is the inclusion and (i^*_α) is the map into the product determined by the i^*_α. Note that we are dealing here with direct products of groups (not direct sums). Both (i^*_α) are isomorphisms, therefore to show ρ_1 an isomorphism, it suffices to show that $\rho_\alpha : [S^1_\alpha, K] \to H^n_{\text{sing}}(S^1_\alpha; G)$ is an isomorphism. If $n > 1$ both groups are trivial, so we assume that $n = 1$. We consider the diagram

$$\begin{array}{ccc} [S^1_\alpha, K] & \xrightarrow{\rho_\alpha} & H^1_{\text{sing}}(S^1_\alpha; G) \\ \downarrow{\scriptstyle \eta_\pi} & & \downarrow{\scriptstyle \mu} \\ \text{Hom}(\pi_1(S^1_\alpha), G) & \xleftarrow{h^*} & \text{Hom}(H_1(S^1_\alpha), G), \end{array}$$

where η_π assigns to a homotopy class the induced homomorphism of fundamental groups, μ is the epimorphism of the universal coefficient theorem and h^* is obtained from the Hurewicz homomorphism $h : \pi_1(S^1_\alpha) \to H_1(S^1_\alpha)$. It is easily checked that the diagram is commutative. Since μ, h^*, and η_π are isomorphisms (for the latter, see Proposition 2.5.13), it follows that ρ_α is an isomorphism. This establishes the induction for $\dim X = 1$. Now suppose that the result holds for all CW complexes of dimension $\leqslant i - 1$. If X is i-dimensional, then $X = X^i$. We consider the coexact sequence

$$\bigvee_\beta S^{i-1}_\beta \xrightarrow{\phi} X^{i-1} \xrightarrow{j} X^i \xrightarrow{q} \Sigma(\bigvee_\beta S^{i-1}_\beta) \xrightarrow{\Sigma\phi} \Sigma X^{i-1},$$

where ϕ is determined by the attaching maps, j is the inclusion, and q is the projection. This gives rise to a map of exact sequences

$$\begin{array}{ccccccccc} [\Sigma X^{i-1}, K] & \xrightarrow{(\Sigma\phi)^*} & [\Sigma A, K] & \xrightarrow{q^*} & [X^i, K] & \xrightarrow{j^*} & [X^{i-1}, K] & \xrightarrow{\phi^*} & [A, K] \\ \downarrow{\scriptstyle \rho} & & \downarrow{\scriptstyle \rho'} & & \downarrow{\scriptstyle \rho_i} & & \downarrow{\scriptstyle \rho_{i-1}} & & \downarrow{\scriptstyle \rho''} \\ H^n_{\text{s}}(\Sigma X^{i-1}) & \xrightarrow{(\Sigma\phi)^*} & H^n_{\text{s}}(\Sigma A) & \xrightarrow{q^*} & H^n_{\text{s}}(X^i) & \xrightarrow{j^*} & H^n_{\text{s}}(X^{i-1}) & \xrightarrow{\phi^*} & H^n_{\text{s}}(A), \end{array}$$

where $A = \bigvee_\beta S^{i-1}_\beta$ and $H^n_{\text{s}}(Y)$ denotes $H^n_{\text{sing}}(Y; G)$, for any space Y. By the inductive hypothesis, ρ_{i-1} and ρ'' are isomorphisms. By Lemma 5.3.4 and the inductive hypothesis, ρ and ρ' are isomorphisms. By the five lemma, ρ_i is an isomorphism. This completes the induction and proves the theorem for finite-dimensional complexes. If X is an arbitrary CW complex, then the inclusion map $i : X^{n+1} \to X$ induces an isomorphism $i^* : H^n_{\text{sing}}(X; G) \to H^n_{\text{sing}}(X^{n+1}; G)$. Moreover, $i^* : [X, K(G,n)] \to [X^{n+1}, K(G,n)]$ is an isomorphism by Proposition 2.4.13. It now follows that $\rho : [X, K(G,n)] \to H^n_{\text{sing}}(X; G)$ is an isomorphism. □

5.4 Applications to Fiber and Cofiber Sequences

We begin this section by investigating the operation of $[W, \Omega B]$ on $[W, F]$, when W is a co-H-space and $F \to E \to B$ is a fiber sequence.

Proposition 5.4.1 *Let $F \xrightarrow{j} E \xrightarrow{p} B$ be a fibration with action $\phi : F \times \Omega B \to F$ and resulting operation of $[W, \Omega B]$ on $[W, F]$. If (W, c) is a co-H-space and $\alpha, \beta \in [W, \Omega B]$ and $u, v \in [W, F]$, then*

$$(u+v)^{\alpha+\beta} = u^\alpha + v^\beta,$$

where + denotes the binary operation obtained from the comultiplication c.

Proof. Let $\alpha = [a]$, $\beta = [b]$, $u = [f]$, and $v = [g]$. Then it is easily checked that the following diagram is commutative

$$\begin{array}{ccccc}
W & \xrightarrow{\Delta} & W \times W & \xrightarrow{c \times c} & (W \vee W) \times (W \vee W) \\
\downarrow{\scriptstyle c} & & & & \downarrow{\scriptstyle (f \vee g) \times (a \vee b)} \\
W \vee W & & & & (F \vee F) \times (\Omega B \vee \Omega B) \\
\downarrow{\scriptstyle \Delta \vee \Delta} & & & & \downarrow{\scriptstyle \nabla \times \nabla} \\
(W \times W) \vee (W \times W) & & & & F \times \Omega B \\
\downarrow{\scriptstyle (f \times a) \vee (g \times b)} & & & & \downarrow{\scriptstyle \phi} \\
(F \times \Omega B) \vee (F \times \Omega B) & \xrightarrow{\phi \vee \phi} & F \vee F & \xrightarrow{\nabla} & F.
\end{array}$$

Therefore

$$\begin{aligned}
(f+g)^{a+b} &= \phi(\nabla \times \nabla)((f \vee g) \times (a \vee b))(c \times c)\Delta \\
&= \nabla(\phi \vee \phi)((f \times a) \vee (g \times b))(\Delta \vee \Delta)c \\
&= f^a + g^b.
\end{aligned}$$

This completes the proof. □

Corollary 5.4.2 *Assume the hypothesis of Proposition 5.4.1 and let* $\partial_* : [W, \Omega B] \to [W, F]$ *be induced by the connecting map* $\partial : \Omega B \to F$*. Then*

1. $v^\alpha = \partial_*(\alpha) + v$*, for all* $\alpha \in [W, \Omega B]$ *and* $v \in [W, F]$*.*
2. $u^\beta = u + \partial_*(\beta)$*, for all* $\beta \in [W, \Omega B]$ *and* $u \in [W, F]$*.*

Consequently each element of $\operatorname{Im} \partial_*$ *commutes with each element of* $[W, F]$*.*

Proof. Let $\alpha = [a] \in [W, \Omega B]$ and $0 = [*] \in [W, F]$ and consider the diagram

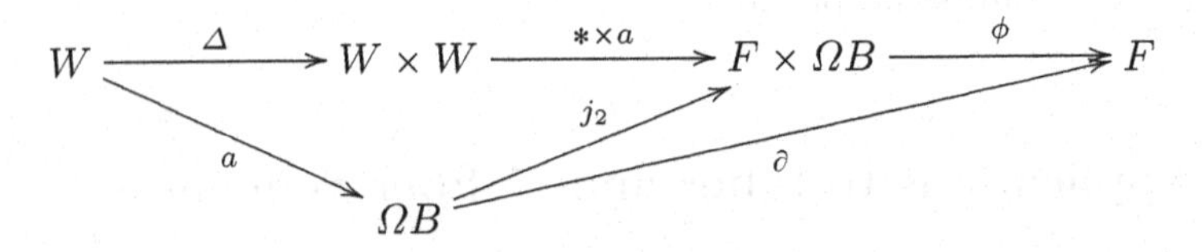

which is homotopy-commutative by Theorem 4.3.4(2). Therefore $0^\alpha = \partial_*(\alpha)$. We now prove (1) by setting $u = 0$ and $\beta = 0$ in Proposition 5.4.1. We prove (2) by setting $v = 0$ and $\alpha = 0$. The last assertion of the corollary follows from (1) and (2). □

This corollary has the following corollary.

Corollary 5.4.3

1. *If* $F \xrightarrow{j} E \xrightarrow{p} B$ *is a fibration, then the image of* $\partial_* : \pi_2(B) \to \pi_1(F)$ *is contained in the center* $Z(\pi_1(F))$ *of* $\pi_1(F)$.
2. *In the exact homotopy sequence of the pair* (X, A),

$$\cdots \longrightarrow \pi_2(X) \xrightarrow{j} \pi_2(X, A) \longrightarrow \pi_1(A) \longrightarrow \cdots,$$

   $\operatorname{Im} j$ *is contained in the center* $Z(\pi_2(X, A))$, *where the homomorphism* j *is induced by the inclusion* $(X, \{*\}) \to (X, A)$.

Proof. (1) is a special case of the previous corollary. For (2), let $i : A \to X$ be the inclusion and apply (1) to the mapping path fibration $I_i \to E_i \to X$, where $I_i = E(X; A, \{*\})$. □

We next consider when a principal fibration is trivial.

Proposition 5.4.4 *Let* $f : X \to Y$ *be a map and let* $\Omega Y \xrightarrow{i} I_f \xrightarrow{v} X$ *be the principal fiber sequence induced by* f. *Then the following three statements are equivalent.*

1. $f \simeq *$.
2. v *has a homotopy section.*
3. *There is a homotopy equivalence* $\theta : X \times \Omega Y \to I_f$ *such that the following diagram is homotopy-commutative*

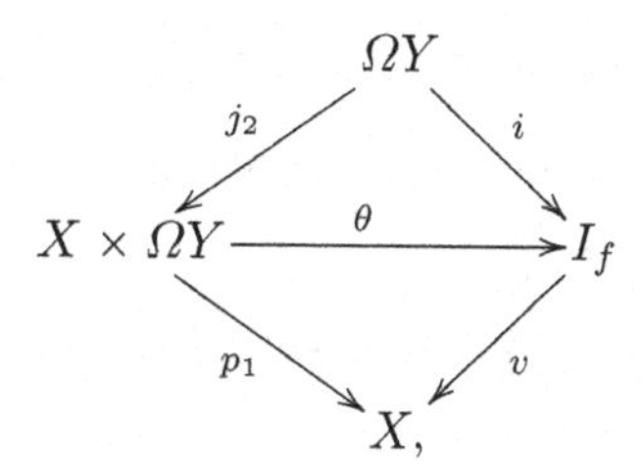

where j_2 *is the injection and* p_1 *is the projection.*

Proof. Consider the exact sequence

$$[X, I_f] \xrightarrow{v_*} [X, X] \xrightarrow{f_*} [X, Y].$$

If $f \simeq *$, then v_* is onto. Thus there exists $s : X \to I_f$ such that $vs \simeq \mathrm{id}$. Therefore (1) $\Longrightarrow$ (2). If v has a homotopy section s, then $vs \simeq \mathrm{id}$. Hence $* \simeq fvs \simeq f$. This shows that (1) and (2) are equivalent.

Next we show that (2) $\Longrightarrow$ (3). Assume that v has a homotopy section $s : X \to I_f$ and define θ to be the following composition

$$X \times \Omega Y \xrightarrow{s \times \mathrm{id}} I_f \times \Omega Y \xrightarrow{\phi_0} I_f,$$

where ϕ_0 is the action of ΩY on I_f defined in Proposition 4.3.3. Then it follows from Proposition 4.3.3(2) and (3) that the diagram of Proposition 5.4.4 is homotopy-commutative. We next show that θ is a homotopy equivalence. If A and B are any spaces, then $\mu : \pi_r(A) \oplus \pi_r(B) \to \pi_r(A \times B)$ defined by $\mu(\alpha, \beta) = j_{1*}(\alpha) + j_{2*}(\beta)$, where $j_1 : A \to A \times B$ and $j_2 : B \to A \times B$ are the injections, is an isomorphism by Section 2.4. Hence there is a diagram

$$\begin{array}{ccccc}
\pi_r(X) \oplus \pi_r(\Omega Y) & \xrightarrow{s_* \oplus \mathrm{id}} & \pi_r(I_f) \oplus \pi_r(\Omega Y) & & \\
\cong \downarrow \mu & & \cong \downarrow \mu & \searrow^{\lambda} & \\
\pi_r(X \times \Omega Y) & \xrightarrow{(s \times \mathrm{id})_*} & \pi_r(I_f \times \Omega Y) & \xrightarrow{\phi_{0*}} & \pi_r(I_f),
\end{array}$$

with vertical arrows isomorphisms, where $\lambda(\alpha, \beta) = \alpha + i_*(\beta)$. The square is commutative and, because $\phi_0 j_1 \simeq \mathrm{id}$ and $\phi_0 j_2 \simeq i$, the triangle is also commutative. Thus $\theta_* \mu(\alpha, \beta) = s_*(\alpha) + i_*(\beta)$ for $\alpha \in \pi_r(X)$ and $\beta \in \pi_r(\Omega Y)$. Since $v_* s_* = \mathrm{id}$, a straightforward argument shows that $\theta_* \mu$ is an isomorphism. Therefore θ induces isomorphisms of homotopy groups. It now follows from Whitehead's first theorem 2.4.7 that θ is a homotopy equivalence. This proves that (2) $\Longrightarrow$ (3).

For the opposite implication, we assume that θ exists and define s as the composition

$$X \xrightarrow{j_1} X \times \Omega Y \xrightarrow{\theta} I_f.$$

Then s is a homotopy section of v. □

As a consequence, we obtain a result about fibrations whose fiber is contractible in the total space.

Corollary 5.4.5 *Let $F \xrightarrow{i} E \xrightarrow{p} B$ be a fibration with $i \simeq *$. Then*

1. *$\Omega B \simeq F \times \Omega E$.*
2. *F is an H-space.*

Proof. (1) Consider the principal fibration $\Omega E \longrightarrow I_i \longrightarrow F$. By Proposition 5.4.4, $I_i \simeq F \times \Omega E$. By Proposition 3.5.10, the excision map $\beta : F \to I_p$ of the given fibration is a homotopy equivalence. By Corollary 4.5.16, $\gamma : I_i \to \Omega B$ is then a homotopy equivalence. Thus $\Omega B \simeq F \times \Omega E$.

(2) By Part (1), F is a homotopy retract of the H-space ΩB. By Exercise 2.2, F is an H-space. □

In Chapter 8 we need the dual of Corollary 5.4.2. We therefore state it and leave the proof as an exercise (Exercise 5.5).

Proposition 5.4.6 *Let $A \xrightarrow{i} X \xrightarrow{q} Q$ be a cofibration with coaction $\psi : Q \to Q \vee \Sigma A$ and resulting operation of $[\Sigma A, Z]$ on $[Q, Z]$. Let (Z, m) be an H-space and let $\partial^* : [\Sigma A, Z] \to [Q, Z]$ be induced by the connecting map $\partial : Q \to \Sigma A$. Then*

1. *$v^\alpha = \partial^*(\alpha) + v$, for all $\alpha \in [\Sigma A, Z]$ and $v \in [Q, Z]$, and*
2. *$u^\beta = u + \partial^*(\beta)$, for all $\beta \in [\Sigma A, Z]$ and $u \in [Q, Z]$,*

where + denotes the binary operation obtained from the multiplication m. Consequently each element of $\operatorname{Im} \partial^*$ commutes with each element of $[Q, Z]$.

We next note that there are obvious duals to Proposition 5.4.4 and Corollary 5.4.5, and we state without proof the result which is dual to Corollary 5.4.5.

Corollary 5.4.7 *Let $A \xrightarrow{j} X \xrightarrow{q} Q$ be a cofibration with $q \simeq *$ and let Q be simply connected. Then*

1. *$\Sigma A \simeq Q \vee \Sigma X$.*
2. *Q is a co-H-space.*

We conclude this section by illustrating Corollaries 5.4.5 and 5.4.7. We first apply Corollary 5.4.5 to the Hopf fibrations.

Example 5.4.8 Consider the Hopf fibrations of Proposition 3.4.1

$$S^{d-1} \xrightarrow{j} S^{d(n+1)-1} \xrightarrow{p} \mathbb{F}\mathrm{P}^n,$$

where $\mathbb{F}$ is the real numbers, the complex numbers, or the quaternions, $d = 1$, 2, or 4, respectively, and $\mathbb{F}\mathrm{P}^n$ is the $\mathbb{F}$-projective n-space. Since $j \simeq *$ by Proposition 2.4.18, we apply Corollary 5.4.5 to conclude that

$$\Omega \mathbb{F}\mathrm{P}^n \simeq S^{d-1} \times \Omega S^{d(n+1)-1}$$

in each of these three cases. In addition, the result holds for the fibration

$$S^7 \longrightarrow S^{15} \longrightarrow S^8$$

obtained from the Cayley numbers (see Example 3.4.5), and so $\Omega S^8 \simeq S^7 \times \Omega S^{15}$.

Remark 5.4.9 If $F \xrightarrow{i} E \xrightarrow{p} B$ is a fiber sequence with $i \simeq *$, then by Corollary 5.4.5, $\Omega B \simeq F \times \Omega E$, and F is an H-space. Thus ΩB and $F \times \Omega E$ are H-spaces, and it is reasonable to ask if there is a homotopy equivalence between them which is an H-map (such an equivalence is called an *H-equivalence*). For the fibrations considered in Example 5.4.8, the answer has been given by Ganea [34].

We next give some definitions.

Definition 5.4.10 Let $j : X \vee Y \to X \times Y$ be the inclusion map. The *smash product* $X \wedge Y$ of X and Y is defined to be C_j, the mapping cone of j and the inclusion is denoted by $l : X \times Y \to X \wedge Y$. If X and Y are CW complexes, j is a cofiber map. Then $X \wedge Y$ can be identified with $X \times Y/X \vee Y$ and l can be identified with the quotient map $q : X \times Y \to X \times Y/X \vee Y$. The *flat product* of X and Y denoted by $X \flat Y$ is defined to be I_j, the homotopy fiber of j. Then $v : X \flat Y \to X \vee Y$ denotes the projection. Clearly $X \flat Y = E(X \times Y; X \vee Y, \{*\})$, and so $\pi_n(X \flat Y) = \pi_{n+1}(X \times Y, X \vee Y)$.

Lemma 5.4.11

1. *The map* $\Sigma j : \Sigma(X \vee Y) \to \Sigma(X \times Y)$ *has a homotopy retraction.*
2. *The map* $\Omega j : \Omega(X \vee Y) \to \Omega(X \times Y)$ *has a homotopy section.*

Proof. The proofs of the two statements are dual to each other, so we only prove (1). We begin by fixing notation. Let $p_1 : X \times Y \to X$, $p_2 : X \times Y \to Y$, $q_1 : X \vee Y \to X$, and $q_2 : X \vee Y \to Y$ be the projections and let $i_1 : X \to X \vee Y$, $i_2 : Y \to X \vee Y$, $\iota_1 : \Sigma X \to \Sigma X \vee \Sigma Y$ and $\iota_2 : \Sigma Y \to \Sigma X \vee \Sigma Y$ be the injections. If $\mu : \Sigma X \vee \Sigma Y \to \Sigma(X \vee Y)$ is the homeomorphism given by $\mu(\langle x, t\rangle, *) = \langle (x, *), t\rangle$ and $\mu(*, \langle y, t\rangle) = \langle (*, y), t\rangle$, for all $x \in X$, $y \in Y$, and $t \in I$, then $\mu\iota_1 = \Sigma i_1$ and $\mu\iota_2 = \Sigma i_2$ (Exercise 2.11). Now define $r : \Sigma(X \times Y) \to \Sigma(X \vee Y)$ by $r = \Sigma i_1 \Sigma p_1 + \Sigma i_2 \Sigma p_2$. Then

$$r\,\Sigma j = \Sigma(i_1 p_1 j) + \Sigma(i_2 p_2 j).$$

Therefore to show $r\Sigma j \simeq \mathrm{id}$, it suffices to prove that $r(\Sigma j)\mu \simeq \mu$, that is, $r\Sigma j \Sigma i_k \simeq \Sigma i_k$, for $k = 1, 2$. But

$$\begin{aligned} r\Sigma j \Sigma i_k &= \Sigma(i_1 q_1 i_k) + \Sigma(i_2 q_2 i_k) \\ &\simeq \begin{cases} \Sigma i_1 \text{ if } k = 1 \\ \Sigma i_2 \text{ if } k = 2, \end{cases} \end{aligned}$$

because $q_k i_k = \mathrm{id}$ and $q_k i_l = *$, for $k \neq l$. □

This leads to a splitting of the suspension of a product and of the loop space of a wedge.

Proposition 5.4.12

1. *For any spaces* X *and* Y*, there is a homotopy equivalence*

$$\Sigma(X \times Y) \simeq \Sigma(X \wedge Y) \vee \Sigma(X \vee Y)$$

which is given as the composition

$$\Sigma(X \times Y) \xrightarrow{c} \Sigma(X \times Y) \vee \Sigma(X \times Y) \xrightarrow{\Sigma l \vee r} \Sigma(X \wedge Y) \vee \Sigma(X \vee Y),$$

where c is the comultiplication of $\Sigma(X \times Y)$ and r is the homotopy retraction of Σj.

2. *For any spaces X and Y, there is a homotopy equivalence*

$$\Omega(X \flat Y) \times \Omega(X \times Y) \simeq \Omega(X \vee Y)$$

which is given as the composition

$$\Omega(X \flat Y) \times \Omega(X \times Y) \xrightarrow{\Omega v \times s} \Omega(X \vee Y) \times \Omega(X \vee Y) \xrightarrow{m} \Omega(X \vee Y),$$

where m is the multiplication of $\Omega(X \vee Y)$ and s is the homotopy section of Ωj.

Proof. We prove (1). Because $X \wedge Y = C_j$, there is a cofiber sequence

$$X \times Y \xrightarrow{l} X \wedge Y \xrightarrow{p} \Sigma(X \vee Y)$$

and a resulting exact sequence

$$[\Sigma(X \times Y), \Sigma(X \vee Y)] \xrightarrow{(\Sigma j)^*} [\Sigma(X \vee Y), \Sigma(X \vee Y)] \xrightarrow{p^*} [(X \wedge Y), \Sigma(X \vee Y)].$$

Since Σj has a homotopy retraction, $(\Sigma j)^*$ is onto. Thus $[p] = p^*[\text{id}] = 0$, and so $p \simeq *$. By Corollary 5.4.7, $\Sigma(X \times Y)$ and $\Sigma(X \wedge Y) \vee \Sigma(X \vee Y)$ have the same homotopy type. It is easily verified that the homotopy equivalence is as stated in the proposition. Assertion (2) of the proposition is proved similarly. □

5.5 The Operation of the Fundamental Group

In this section we study the operation of the fundamental group on homotopy groups and on homotopy sets. The operation of the fundamental group also appears in the statement of the Hurewicz theorem (see Theorem 6.4.21) and in the construction of the Postnikov decomposition. We obtain the operation from the action of the loops on the base on the fiber in a certain fibration which we next consider.

If X and Y are spaces, then we denote by Y^X the space of based maps from X to Y with the compact–open topology. We now need to consider free maps between based spaces, and so we introduce the following notation.

NOTATION *The space of all free maps from the space X to the space Y with the compact–open topology is denoted by $M(X,Y)$. The set of all free homotopy classes $[X,Y]_{\text{free}}$ is denoted in this section by $\langle X,Y\rangle$. The basepoint of $M(X,Y)$ is the constant map $* : X \to Y$.*

We show next that there is a function $\pi : M(X,Y) \to Y$ which is a fibration. This is a consequence of a more general result. Before stating this proposition, we discuss some preliminaries. If $f : W \to M(X,Y)$ and $g : X \times W \to Y$ are functions such that $f(w)(x) = g(x,w)$, for all $w \in W$ and $x \in X$, then f and g are adjoint. If W is locally compact, then f is continuous if and only if g is continuous (Appendix A). However, as we indicate in Appendix A, it is possible to get rid of this hypothesis, and so in this section we will ignore this assumption. Now suppose $f : W \to M(X,Y)$ and $f' : X \to M(W,Y)$ are such that $f(w)(x) = f'(x)(w)$. Therefore f and f' are essentially double adjoints of each other, and so f is continuous if and only if f' is continuous for spaces X, Y and W.

Proposition 5.5.1 *If $i : A \to X$ is a free cofiber map, then the map $\pi : M(X,Y) \to M(A,Y)$ defined by $\pi(f) = fi$ for $f \in M(X,Y)$ is a fiber map.*

Proof. Let W be a space, let $g : W \to M(X,Y)$ be a map, and let $F : W \times I \to M(A,Y)$ be a (based) homotopy such that $F(w,0) = \pi(g(w))$ for $w \in W$. Then $g' : X \to M(W,Y)$ is continuous and $g'(X) \subseteq Y^W$, so we regard g' as a continuous function from X to Y^W. Similarly F determines a continuous function $F' : A \times I \to Y^W$ with $F'(a,0) = g'(i(a))$, for $a \in A$. But $i : A \to X$ is a free cofiber map, and so there is a free homotopy $G' : X \times I \to Y^W$ of g' such that $G'(i \times \mathrm{id}) = F'$. Then the function $G : W \times I \to M(X,Y)$ defined by $G(w,t)(x) = G'(x,t)(w)$ is the desired based homotopy of g that covers F. □

We apply this to the case when $A = \{*\}$ is the base point of the CW complex X and Y is a based space. Since $M(\{*\},Y) \cong Y$, we have a fiber sequence

$$Y^X \xrightarrow{i} M(X,Y) \xrightarrow{\pi} Y,$$

where $\pi(f) = f(*)$. By Theorem 4.3.4, there is an action $\phi : Y^X \times \Omega Y \to Y^X$. For any space W, this determines an operation

$$[W, Y^X] \times [W, \Omega Y] \to [W, Y^X]. \qquad (*)$$

Lemma 5.5.2 *There are bijections*

1. $\mu : \pi_0(Y^X) \to [X,Y]$.
2. $\nu : \pi_0(M(X,Y)) \to \langle X,Y \rangle$.

Proof. The proofs of the two assertions are similar, therefore we only prove (1). Let $S^0 = \{-1,1\}$ with basepoint 1. If $f : S^0 \to Y^X$, then $f(-1) : X \to Y$. If $f \simeq_F g : S^0 \to Y^X$, then $G : X \times I \to Y$ defined by $G(x,t) = F(-1,t)(x)$, for $x \in X$ and $t \in I$, is a homotopy between $f(-1)$ and $g(-1)$. Thus there is a well-defined function $\mu : \pi_0(Y^X) \to [X,Y]$ given by $\mu[f] = [f(-1)]$. It is easily checked that μ is a bijection. □

We take $W = S^0$ in $(*)$ and recall that as groups $\pi_0(\Omega Y) \cong \pi_1(Y)$ (Exercise 2.24). Thus there is a function $[X,Y] \times \pi_1(Y) \to [X,Y]$. We then have the following definition.

Definition 5.5.3 For any spaces X and Y, the above operation of $\pi_1(Y)$ on $[X,Y]$ is called the *operation of the fundamental group* $\pi_1(Y)$ *on the homotopy set* $[X,Y]$. For $u \in [X,Y]$ and $\alpha \in \pi_1(Y)$, the operation of α on u is denoted $u \cdot \alpha \in [X,Y]$. If we set $X = S^n$ for $n \geqslant 1$, we obtain the *operation of the fundamental group* $\pi_1(Y)$ *on the nth homotopy group* $\pi_n(Y)$.

The following properties hold (see Definition 4.4.1):

- $u \cdot 0 = u$, for $u \in [X,Y]$.
- $(u \cdot \alpha) \cdot \beta = u \cdot (\alpha + \beta)$, for $u \in [X,Y]$ and $\alpha, \beta \in \pi_1(Y)$.
- If $h : Y \to Y'$ is a map, then $h_*(u \cdot \alpha) = h_*(u) \cdot h_*(\alpha)$, for $u \in [X,Y]$ and $\alpha \in \pi_1(Y)$.
- The diagram

$$\begin{array}{ccc} \pi_0(Y^X) & \xrightarrow{i_*} & \pi_0(M(X,Y)) \\ \downarrow \mu & & \downarrow \nu \\ [X,Y] & \xrightarrow{\theta} & \langle X,Y \rangle \end{array}$$

  is commutative, where $i : Y^X \to M(X,Y)$ is the inclusion function and θ is the function that assigns to the homotopy class of a map, its free homotopy class. Since μ and ν are bijections, we have the following consequence of Theorem 4.4.5.
- $\theta(u) = \theta(v)$ if and only if there is an $\alpha \in \pi_1(Y)$ such that $v = u \cdot \alpha$. Let X be a CW complex, let Y be a path-connected space, and let $[X,Y]'$ be the orbit space of the operation of $\pi_1(Y)$ on $[X,Y]$. Then θ induces a function $\theta' : [X,Y]' \to \langle X,Y \rangle$. Because θ is onto (Exercise 2.25), it follows that
- $\theta' : [X,Y]' \to \langle X,Y \rangle$ is a bijection. In particular, if the operation of $\pi_1(Y)$ on $[X,Y]$ is trivial, then $\theta : [X,Y] \to \langle X,Y \rangle$ is a bijection. In this case, two based maps $X \to Y$ which are freely homotopic are based homotopic. This holds, for example, if Y is simply connected or if Y is an H-space (Lemma 5.5.6). We note that if the operation of $\pi_1(Y)$ on $\pi_n(Y)$ is trivial, then $\theta : \pi_n(Y) \to \langle S^n, Y \rangle$ is a bijection.

An alternative definition of the operation of the fundamental group on the homotopy set in terms of covering spaces is given in [23, p. 157]. Another definition of this operation using loop spaces and conjugation in groups appears in [67, p. 330].

We next give a more concrete description of the operation of $\pi_1(Y)$ on $[X,Y]$. Let $g : X \to Y$ be a map, let $a : (I, \partial I) \to (Y, \{*\})$, and let $g \cup a : X \times \{0\} \cup \{*\} \times I \to Y$ be the function that is g on $X \times \{0\}$ and a on $\{*\} \times I$. Since $\{*\} \subseteq X$ is a subcomplex, $X \times \{0\} \cup \{*\} \times I$ is a retract of $X \times I$ by Proposition 1.5.13. Thus $g \cup a$ extends to a function $G : X \times I \to Y$ and we define the map $\widetilde{g}_a : X \to Y$ by $\widetilde{g}_a(x) = G(x,1)$. If $g \simeq_L g' : X \to Y$ and $a \simeq_A$

$a' : (I, \partial I) \to (Y, \{*\})$, we show that $\widetilde{g}_a \simeq \widetilde{g}'_{a'} : X \to Y$. Let $G' : X \times I \to Y$ be the function used to define $\widetilde{g}'_{a'}$. We assume that X is a CW complex and so the inclusion $X \times \partial I \cup \{*\} \times I \to X \times I$ is a cofibration. By Proposition 1.5.13, there is a retraction $X \times I \times I \to (X \times I \times \{0\}) \cup (X \times \partial I \times I) \cup (\{*\} \times I \times I)$. Now define $H' : (X \times I \times \{0\}) \cup (X \times \partial I \times I) \cup (\{*\} \times I \times I) \to Y$ by

$$H'(x, s, t) = \begin{cases} L(x, s) & \text{if } t = 0 \text{ and } x \in X \\ G(x, t) & \text{if } s = 0 \text{ and } x \in X \\ G'(x, t) & \text{if } s = 1 \text{ and } x \in X \\ A(t, s) & \text{if } x = * \text{ and } s, t \in I. \end{cases}$$

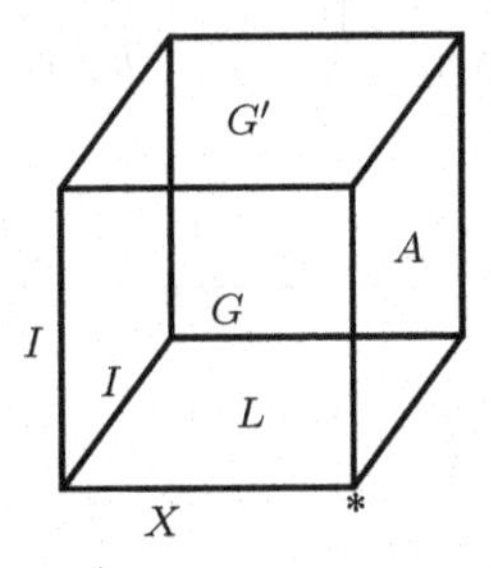

$X \times I \times I$

Homotopy H: L on bottom face, G on front face, G' on back face, A on right face

Figure 5.1

Let $H : X \times I \times I \to Y$ be an extension of H' (see Figure 5.1). Then $K : X \times I \to Y$ defined by $K(x, s) = H(x, s, 1)$ is a homotopy between $\widetilde{g}_a$ and $\widetilde{g}'_{a'}$. Thus if $\alpha = [a] \in \pi_1(Y)$ and $u = [g] \in [X, Y]$, then there is a well-defined element $u * \alpha = [\widetilde{g}_a] \in [X, Y]$.

We show that this operation agrees with the one in Definition 5.5.3.

Proposition 5.5.4 *For all $\alpha \in \pi_1(Y)$ and $u \in [X, Y]$,*

$$u \cdot \alpha = u * \alpha.$$

Proof. According to Theorem 4.3.4, the action $\psi : Y^X \times \Omega Y \to Y^X$ of the fibration

$$Y^X \longrightarrow M(X, Y) \xrightarrow{\pi} Y$$

is obtained as follows. Let I_π be the homotopy fiber of π, let $\beta : Y^X \to I_\pi$ be the excision map which is a homotopy equivalence, and let $\phi : I_\pi \times \Omega Y \to I_\pi$ be the action of Proposition 4.3.3. Recall that $\beta(g) = (g, *)$, for $g \in Y^X$, and $\phi((g, \nu), \omega) = (g, \nu + \omega)$, for $(g, \nu) \in I_\pi$ and $\omega \in \Omega Y$. Then the action ψ is defined by the homotopy-commutativity of the diagram

$$\begin{array}{ccc} Y^X \times \Omega Y & \xrightarrow{\psi} & Y^X \\ \downarrow{\scriptstyle \beta \times \mathrm{id}} & & \downarrow{\scriptstyle \beta} \\ I_\pi \times \Omega Y & \xrightarrow{\phi} & I_\pi . \end{array}$$

Then we define maps $\rho, \sigma, \tau : Y^X \times \Omega Y \to I_\pi$ by

$$\begin{aligned} \rho(g, \omega) &= (g, * + \omega), \\ \sigma(g, \omega) &= (\widetilde{g}_\omega, *), \\ \tau(g, \omega) &= (g, \omega), \end{aligned}$$

for $g \in Y^X$ and $\omega \in \Omega Y$. If $r : X \times I \to X \times \{0\} \cup \{*\} \times I$ is a fixed retraction, we set $G = (g \cup \omega)r : X \times I \to Y$, and so $\widetilde{g}_\omega(x) = G(x, 1)$, for $x \in X$. It suffices to prove that $\rho \simeq \sigma$ because $\rho \simeq \beta\psi$. Clearly $\rho \simeq \tau$, so we show that $\tau \simeq \sigma$. For this, let $\widetilde{G} : I \to M(X, Y)$ be the adjoint of $G : X \times I \to Y$. Then, with $g \in Y^X$, $\omega \in \Omega Y$ and $t \in I$, we define $H : Y^X \times \Omega Y \times I \to I_\pi$ by $H(g, \omega, t) = (\widetilde{G}(t), \omega_{t,1})$, where $\omega_{t,1}$ is the path ω reparametrized so that it starts at $\omega(t)$ and ends at $\omega(1) = *$. It follows that $\tau \simeq_H \sigma$. □

Thus we have two descriptions of the operation of $\pi_1(Y)$ on $\pi_n(Y)$, the one given in Definition 5.5.3 and the one just given with $X = S^n$. We now give a third description which is frequently used. In this characterization, maps representing homotopy classes are regarded as defined on $(I^n, \partial I^n)$.

Let $u = [f] \in \pi_n(Y)$ and $\alpha = [a] \in \pi_1(Y)$, where $f : (I^n, \partial I^n) \to (Y, \{*\})$ and $a : (I, \partial I) \to (Y, \{*\})$. These maps determine a function $f \cup a : I^n \times \{0\} \cup \partial I^n \times I \to Y$ defined by

$$(f \cup a)(x, 0) = f(x) \quad \text{and} \quad (f \cup a)(x', t) = a(t),$$

for $x \in I^n$, $x' \in \partial I^n$, and $t \in I$. Since $I^n \times \{0\} \cup \partial I^n \times I$ is a retract of $I^n \times I$, the function $f \cup a$ can be extended to a function $F : I^n \times I \to Y$. We define $\bar{f}_a : (I^n, \partial I^n) \to (Y, \{*\})$ by $\bar{f}_a(x) = F(x, 1)$.

We next show that the homotopy class of $\bar{f}_a$ depends only on the homotopy classes of f and a. The proof is similar to the proof of the analogous statement for the second definition, but we sketch it for completeness. Suppose $f \simeq_G f' : (I^n, \partial I^n) \to (Y, \{*\})$ and $a \simeq_A a' : (I, \partial I) \to (Y, \{*\})$. Furthermore, let $F : I^n \times I \to Y$ be any function that is an extension of $f \cup a$ and let $F' : I^n \times I \to Y$ be any function that is an extension of $f' \cup a'$. Set $\bar{f}_a(x) = F(x, 1)$ and $\bar{f}'_{a'}(x) = F'(x, 1)$. Because $I^n \times \partial I \cup \partial I^n \times I$ is a subcomplex of $I^n \times I$, it follows that $(I^n \times I \times \{0\}) \cup (I^n \times \partial I \times I) \cup (\partial I^n \times I \times I)$ is a retract of $I^n \times I \times I$. We define $H : (I^n \times I \times \{0\}) \cup (I^n \times \partial I \times I) \cup (\partial I^n \times I \times I) \to Y$ by

$$H(x, s, t) = \begin{cases} G(x, s) & \text{if } t = 0 \text{ and } x \in I^n \\ F(x, t) & \text{if } s = 0 \text{ and } x \in I^n \\ F'(x, t) & \text{if } s = 1 \text{ and } x \in I^n \\ A(t, s) & \text{if } x \in \partial I^n, \end{cases}$$

where $s, t \in I$. Therefore there exists a function $K : I^n \times I \times I \to Y$ that extends H. Thus if $K' : I^n \times I \to Y$ is defined by $K'(x, s) = K(x, s, 1)$, then $\bar{f}_a \simeq_{K'} \bar{f}'_{a'} : (I^n, \partial I^n) \to (Y, \{*\})$. Consequently $u \circ \alpha = [\bar{f}_a] \in \pi_n(Y)$ is well-defined for $\alpha = [a] \in \pi_1(Y)$ and $u = [f] \in \pi_n(Y)$. We note for later use that we have shown that if $F : I^n \times I \to Y$ is *any* homotopy which is an extension of $f \cup a$, then the map $k : (I^n, \partial I^n) \to (Y, \{*\})$ defined by $k(x) = F(x, 1)$ is homotopic to $\bar{f}_a$.

We now show that this operation agrees with the earlier ones. We represent $u \in \pi_n(Y)$ by a map $g : S^n \to Y$ and represent $\alpha \in \pi_1(Y)$ by $a : (I, \partial I) \to (Y, \{*\})$. Then $g \cup a : S^n \times \{0\} \cup \{*\} \times I \to Y$ is the function defined above. Thus $g \cup a$ extends to a function $G : S^n \times I \to Y$ and $\tilde{g}_a : S^n \to Y$ is given by $\tilde{g}_a(y) = G(y, 1)$. If $h : I^n \to S^n$ is the continuous function that carries ∂I^n to $\{*\}$ and is a homeomorphism from $I^n - \partial I^n$ to $S^n - \{*\}$ (see the beginning of Section 4.5) and $f = gh$, then $f : (I^n, \partial I^n) \to (Y, \{*\})$ also represents u. Furthermore, if the extension F of $f \cup a : I^n \times \{0\} \cup \partial I^n \times I \to Y$ is taken to be the composition

$$I^n \times I \xrightarrow{h \times \mathrm{id}} S^n \times I \xrightarrow{G} X,$$

then $\bar{f}_a(x) = G(h(x), 1) = \tilde{g}_a(h(x))$. Therefore $\bar{f}_a : (I^n, \partial I^n) \to (Y, \{*\})$ and $\tilde{g}_a : S^n \to Y$ represent the same element in $\pi_n(Y)$, and so $u \cdot \alpha = u * \alpha = u \circ \alpha$.

Let us consider the case $n = 1$. Then $\alpha = [a], u = [f]$ and $u \cdot \alpha = [\bar{f}_a]$ are all in $\pi_1(Y)$ and we have

$$u \cdot \alpha = -\alpha + u + \alpha.$$

This follows at once from Figure 5.2.

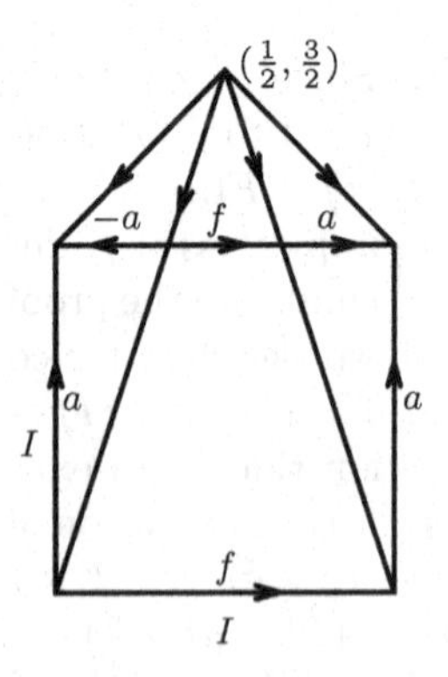

Figure 5.2

Thus the operation of $\alpha \in \pi_1(Y)$ on $u \in \pi_1(Y)$ is conjugation of u by $-\alpha$. In particular, if $\alpha \in \pi_1(Y)$ and $u, v \in \pi_1(Y)$, then $(u+v) \cdot \alpha = u \cdot \alpha + v \cdot \alpha$. We next see that this holds more generally.

Proposition 5.5.5 *If $\alpha \in \pi_1(Y)$ and $u, v \in \pi_n(Y)$, then $(u+v) \cdot \alpha = u \cdot \alpha + v \cdot \alpha$.*

Proof. Let $\alpha = [a]$, $u = [f]$ and $v = [g]$, where $a : (I, \partial I) \to (Y, \{*\})$ and $f, g : (I^n, \partial I^n) \to (Y, \{*\})$. Let $F, G : I^n \times I \to Y$ be free homotopies such that for $x \in I^n$, we have $\bar{f}_a(x) = F(x, 1)$, $\bar{g}_a(x) = G(x, 1)$, $f \simeq_F \bar{f}_a$, and $g \simeq_G \bar{g}_a$. Let $x = (t_1, t_2, \dots, t_n) \in I^n$ and define $F + G : I^n \times I \to Y$ by

$$(F+G)(x,t) = \begin{cases} F(2t_1, t_2, \dots, t_n, t) & \text{if } 0 \leqslant t_1 \leqslant \frac{1}{2} \\ G(2t_1 - 1, t_2, \dots, t_n, t) & \text{if } \frac{1}{2} \leqslant t_1 \leqslant 1. \end{cases}$$

Then $(F+G)(x, 0) = (f+g)(x)$ and $(F+G)(x', t) = a(t)$, for $x' \in \partial I^n$ and $t \in I$. Therefore

$$(\overline{f+g})_a(x) = (F+G)(x,1) = (\bar{f}_a + \bar{g}_a)(x).$$

Hence $(u+v) \cdot \alpha = u \cdot \alpha + v \cdot \alpha$. □

Next we consider the operation when Y is an H-space.

Lemma 5.5.6 *If Y is an H-complex, then the operation of $\pi_1(Y)$ on $[X, Y]$ is trivial.*

Proof. We assume that the multiplication m of Y has the property that $m(y, *) = y = m(*, y)$, for all $y \in Y$ by Exercise 2.1. If $\alpha = [a] \in \pi_1(Y)$ and $u = [f] \in [X, Y]$, then define $F : X \times I \to Y$ by $F(x, t) = m(f(x), a(t))$, for $x \in X$ and $t \in I$. Then $F(x, 0) = f(x)$ and $F(x', t) = a(t)$, for $x \in X$, $x' \in \partial I^n$, and $t \in I$, and so F is an extension of $f \cup a$. Therefore $\bar{f}_a(x) = F(x, 1) = f(x)$, and so $u \cdot \alpha = u$. □

Definition 5.5.7 A space Y is called *n-simple* if the operation of $\pi_1(Y)$ on the homotopy group $\pi_n(Y)$ is trivial, that is, if for every $\alpha \in \pi_1(Y)$ and $u \in \pi_n(Y)$, we have $u \cdot \alpha = u$. The space Y is called *simple* if Y is n-simple for every $n \geqslant 1$.

Clearly a space is 1-simple if and only if its fundamental group is abelian. In addition, every 1-connected space is simple and every H-complex is simple. Hatcher calls simple spaces *abelian spaces* [39, p. 342].

For a pair of spaces (X, A), we consider the operation of $\pi_1(A)$ on $\pi_n(X, A)$. Let $\alpha = [a] \in \pi_1(A)$ and let $w = [h] \in \pi_n(X, A)$, where $h : (I^n, I^{n-1}, J^{n-1}) \to (X, A, \{*\})$, and let $f = h|I^{n-1} : (I^{n-1}, \partial I^{n-1}) \to (A, \{*\})$. Since $I^{n-1} \times \{0\} \cup \partial I^{n-1} \times I$ is a retract of $I^{n-1} \times I$, the map $f \cup a : I^{n-1} \times \{0\} \cup \partial I^{n-1} \times I \to A$ extends to $F'' : I^{n-1} \times I \to A$. Now define $F' : I^n \times \{0\} \cup \partial I^n \times I \to X$ by

$$F'(x,t) = \begin{cases} h(x) & \text{if } x \in I^n \text{ and } t = 0 \\ F''(x,t) & \text{if } x \in I^{n-1} \text{ and } t \in I \\ a(t) & \text{if } x \in J^{n-1} \text{ and } t \in I. \end{cases}$$

Because $I^n \times \{0\} \cup \partial I^n \times I$ is a retract of $I^n \times I$, we extend F' to a function $F : I^n \times I \to X$ and define $\bar{h}_a(x) = F(x,1)$, for $x \in I^n$. Then $\bar{h}_a : (I^n, I^{n-1}, J^{n-1}) \to (X, A, \{*\})$. We note that $\bar{h}_a|I^{n-1} = \bar{f}_a : (I^{n-1}, \partial I^{n-1}) \to (A, \{*\})$, where $h|I^{n-1} = f$.

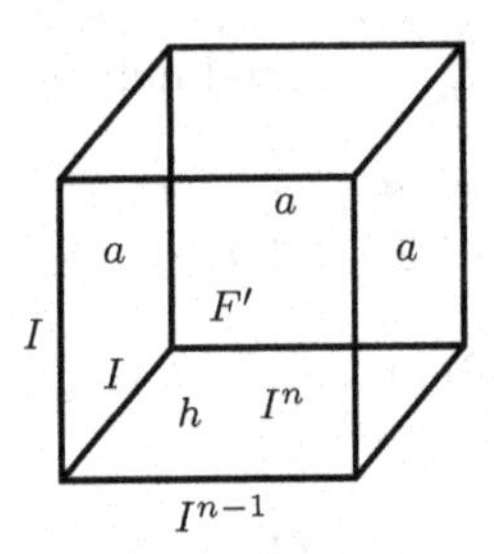

Homotopy F: h on bottom face, F' on front face, a on all other side faces

Figure 5.3

Lemma 5.5.8 *If $h \simeq h' : (I^n, I^{n-1}, J^{n-1}) \to (X, A, \{*\})$ and $a \simeq a' : (I, \partial I) \to (A, \{*\})$, then $\bar{h}_a \simeq \bar{h}'_{a'} : (I^n, I^{n-1}, J^{n-1}) \to (X, A, \{*\})$.*

Proof. This is Exercise 5.16. □

Definition 5.5.9 If (X, A) is a pair of spaces, then there is an operation of $\pi_1(A)$ on $\pi_n(X, A)$ which is called the *operation of the fundamental group $\pi_1(A)$ on the relative homotopy group $\pi_n(X, A)$*. For $w = [h] \in \pi_n(X, A)$ and $\alpha = [a] \in \pi_1(A)$, the operation of α on w is $w \cdot \alpha = [\bar{h}_a] \in \pi_n(X, A)$.

Let π be a group that operates (on the right) on groups G and H and let $\theta : G \to H$ be a homomorphism. If $\theta(u \cdot \alpha) = \theta(u) \cdot \alpha$, then θ is called a *π-homomorphism*. We consider the case $\pi = \pi_1(A)$. If (X, A) is a pair of spaces, then $\pi_1(A)$ operates on $\pi_n(A)$ and $\pi_n(X, A)$. In addition, $\pi_1(A)$ can be made to operate on $\pi_n(X)$ by setting

$$v \cdot \alpha = v \cdot i_*(\alpha),$$

where $v \in \pi_n(X)$, $\alpha \in \pi_1(A)$, and $i : A \to X$ is the inclusion map. If we consider the exact homotopy sequence of the pair (X, A),

$$\cdots \longrightarrow \pi_{n+1}(X, A) \xrightarrow{\partial_{n+1}} \pi_n(A) \xrightarrow{i_*} \pi_n(X) \xrightarrow{j_*} \pi_n(X, A) \longrightarrow \cdots,$$

where $j : (X, \{*\}) \to (X, A)$ is the inclusion map of pairs and ∂_{n+1} is the connecting homomorphism, then each of the groups that appears in the exact sequence is operated on by $\pi_1(A)$.

Proposition 5.5.10 *i_*, j_*, and ∂_{n+1} are $\pi_1(A)$-homomorphisms.*

Proof. From the definition of the operation of $\pi_1(A)$ on $\pi_n(X)$, it follows that i_* is a $\pi_1(A)$-homomorphism. To show ∂_n is a $\pi_1(A)$-homomorphism, we use the notation in the definition of the operation of $\pi_1(A)$ on $\pi_n(X, A)$. Since $\bar{h}_a|I^{n-1} = \bar{f}_a$, where $f = h|I^{n-1}$, we have $\partial_n(w \cdot \alpha) = (\partial_n w) \cdot \alpha$.

Finally we prove that j_* is a $\pi_1(A)$-homomorphism. Let $\alpha = [a] \in \pi_1(A)$ and $v = [k] \in \pi_n(X)$. Consider $jk : (I^n, I^{n-1}, J^{n-1}) \to (X, A, \{*\})$ and observe that $jk|I^{n-1} = *$. Following the definition of the operation of $\pi_1(A)$ on $\pi_n(X, A)$, define the extension $F'' : I^{n-1} \times I \to A$ of $* \cup a : I^{n-1} \times \{0\} \cup \partial I^{n-1} \times I \to A$ by $F''(x, t) = a(t)$. We then proceed as above, extending F'' to F' and F' to $F : I^n \times I \to X$. It follows that $F(x, 1) = (\overline{jk})_a(x)$ and $(\overline{jk})_a : (I^n, I^{n-1}, J^{n-1}) \to (X, A, \{*\})$ represents $j_*(v) \cdot \alpha$. On the other hand, the free homotopy F of k satisfies $F(\partial x, t) = ia(t)$, for $\partial x \in \partial I^n$. Therefore $F(x, 1) = \bar{k}_{ia}(x)$, for $x \in I^n$, where $\bar{k}_{ia} : (I^n, \partial I^n) \to (X, \{*\})$. Because $\bar{k}_{ia}$ represents $v \cdot \alpha$, we have that j_* is a $\pi_1(A)$-homomorphism. □

5.6 Calculation of Homotopy Groups

In this section we compute some homotopy groups of specific spaces. We consider a number of different spaces, but in most cases only compute the first few nontrivial homotopy groups. It is possible to make further computations with more work, but our purpose here is to illustrate some elementary methods for determining homotopy groups. Homotopy groups are difficult to calculate, even for familiar spaces. Part of the difficulty is that the homotopy groups $\pi_i(X)$ of any finite, 1-connected, noncontractible CW complex X are nontrivial for infinitely many values of i (see the discussion after Proposition 5.6.1).

We begin with spheres.

Spheres

We first consider the 1-sphere S^1, that is, the unit circle in $\mathbb{R}^2$. We have seen in Definition 2.4.14 that there is a function deg: $\pi_1(S^1) \to \mathbb{Z}$ which is an isomorphism by Proposition 2.4.16. Therefore $\pi_1(S^1) \cong \mathbb{Z}$. For $\pi_n(S^1)$, when $n > 1$, we consider the exact fiber homotopy sequence of the universal covering map $p : \mathbb{R} \to S^1$ [61, p. 118],

$$\cdots \longrightarrow \pi_n(\mathbb{Z}) \longrightarrow \pi_n(\mathbb{R}) \xrightarrow{p_*} \pi_n(S^1) \longrightarrow \pi_{n-1}(\mathbb{Z}) \longrightarrow \cdots,$$

where the fiber $\mathbb{Z}$ is a discrete topological space. Then $\pi_i(\mathbb{Z}) = 0$, for all $i > 0$, and $\pi_i(\mathbb{R}) = 0$, for all $i > 0$, since $\mathbb{R}$ is contractible. Therefore $\pi_n(S^1) = 0$, for $n > 1$.

Next we consider the n-sphere S^n for $n > 1$. We have seen in Proposition 2.4.18 that $\pi_i(S^n) = 0$ for $i < n$ and that $\pi_n(S^n) \cong \mathbb{Z}$. The latter isomorphism is either given by the degree function $\deg : \pi_n(S^n) \to \mathbb{Z}$ or equivalently by the Hurewicz homomorphism $h_n : \pi_n(S^n) \to H_n(S^n)$.

Additional results on the homotopy groups of spheres are obtained by considering Example 5.4.8 for $n = 1$ which yields $\Omega S^d = \Omega \mathbb{F}\mathrm{P}^1 \simeq \Omega S^{2d-1} \times S^{d-1}$. Here $\mathbb{F}$ is the complex numbers, the quaternions, or the octonions, when d =2, 4, or 8, respectively. We obtain the following proposition.

Proposition 5.6.1 *For $i \geqslant 3$ and $d = 2, 4,$ or 8,*

$$\pi_i(S^d) \cong \pi_i(S^{2d-1}) \oplus \pi_{i-1}(S^{d-1}).$$

The case $d = 2$ is particularly interesting. Then $\pi_i(S^2) \cong \pi_i(S^3)$, for all $i \geqslant 3$. Thus

$$\pi_3(S^2) \cong \pi_3(S^3) \cong \mathbb{Z}$$

and a generator of the infinite cyclic group $\pi_3(S^2)$ is the homotopy class of the Hopf map $\phi : S^3 \to S^2$. This surprising and important result, which is due to Hopf [44], was the first example to illustrate that, unlike homology groups, the nth homotopy group of an m-dimensional complex could be nonzero for $n > m$. Later work showed that the homotopy groups $\pi_i(S^n)$ $n > 1$, are nontrivial for infinitely many values of i. There is the result of McGibbon–Neisendorfer [68] which states that if X is a 1-connected complex and p is a prime such that $H_j(X; \mathbb{Z}_p) \neq 0$ for some j, then $\pi_i(X)$ contains a subgroup of order p for infinitely many values i.

We next show that the isomorphism $\pi_i(S^2) \cong \pi_i(S^3)$, for all $i \geqslant 3$ leads to an example of two spaces with isomorphic homotopy groups which are not of the same homotopy type.

Example 5.6.2 Let $X = S^2$ and $Y = S^3 \times K(\mathbb{Z}, 2)$. Then it is clear that X and Y have isomorphic homotopy groups. But they do not have the same homotopy type because they have nonisomorphic homology groups.

For the group $\pi_3(S^2)$, the difference between the degree of the homotopy group and the dimension of the sphere is one. This raises the question of what is the $(n+1)$st homotopy group of the n-sphere for $n > 2$. To answer this we introduce some ideas that bear on this question, but also have wider application.

Consider the function $\Sigma : [X, Y] \to [\Sigma X, \Sigma Y]$ defined by $\Sigma[f] = [\Sigma f]$. It can be shown directly that Σ is a homomorphism if X is a cogroup. However, we prove this differently. Define $e = e_Y : Y \to \Omega\Sigma Y$ by $e(y)(t) = \langle y, t \rangle$, for

$y \in Y$ and $t \in I$. Let κ be the adjoint function that assigns to a map $g : \Sigma Y \to Z$, the map $\kappa(g) : Y \to \Omega Z$. By taking $Z = \Sigma Y$ and $g = \mathrm{id} : \Sigma Y \to \Sigma Y$, we see that $\kappa(\mathrm{id}) = e : Y \to \Omega \Sigma Y$.

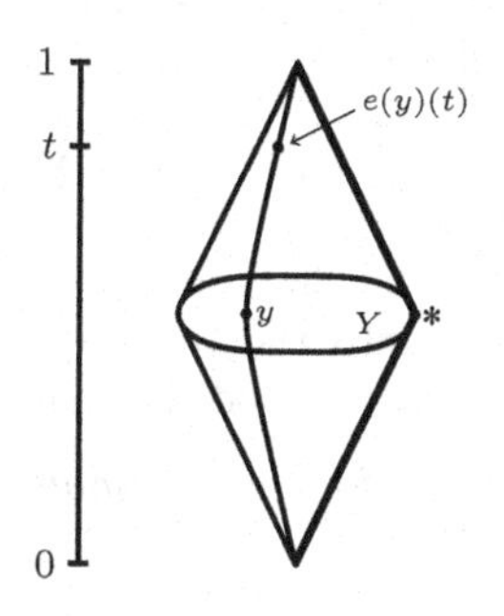

Figure 5.4

Lemma 5.6.3 *For any spaces X and Y, the following diagram commutes*

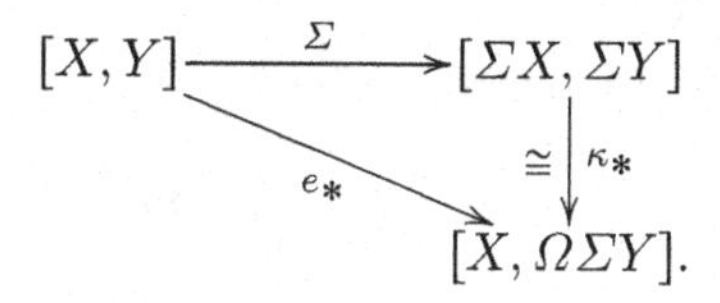

Consequently, if X is a cogroup, then $\Sigma : [X, Y] \to [\Sigma X, \Sigma Y]$ is a homomorphism called the suspension homomorphism.

Proof. This follows since $\kappa(\Sigma f) = ef$ and e_* is a homomorphism. □

Next we consider the sequence of groups and homomorphisms

$$\pi_3(S^2) \xrightarrow{\Sigma} \pi_4(S^3) \xrightarrow{\Sigma} \pi_5(S^4) \xrightarrow{\Sigma} \cdots .$$

We show that the first homomorphism is an epimorphism and the others are isomorphisms. For this we prove the Freudenthal suspension theorem which is a consequence of the Blakers–Massey theorem. We begin with the latter theorem [10] (also called the homotopy excision theorem). Recall that a map is an n-equivalence if it induces an isomorphism of homotopy groups in degrees $< n$ and an epimorphism in degree n.

Theorem 5.6.4 *Let*

$$A \xrightarrow{i} X \xrightarrow{q} Q$$

be a cofiber sequence with A r-connected and Q s-connected, $r, s \geqslant 1$. Let I_q be the homotopy fiber of q and let $\beta : A \to I_q$ be the excision map defined by $\beta(a) = (i(a), *)$, for $a \in A$. Then β is an $(r + s)$-equivalence.

This statement is one version of several that have been called the Blakers–Massey theorem (see [86]). The proof is given in Section 6.4. This theorem implies the generalized Freudenthal suspension theorem. Before proving this theorem, we give a simple lemma.

We first recall some notation. Let

$$X \xrightarrow{i} CX \xrightarrow{q} \Sigma X$$

be the cone cofibration, let I_q be the homotopy fiber of q, let $j : \Omega\Sigma X \to I_q$ be the inclusion of the fiber into the total space, let $\beta : X \to I_q$ be the excision map, and let $e : X \to \Omega\Sigma X$ be the map defined above.

Lemma 5.6.5 *The following triangle is homotopy-commutative*

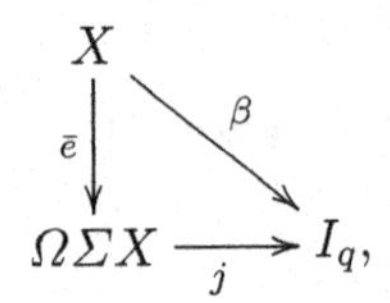

where $\bar{e} = -e : X \to \Omega\Sigma X$.

Proof. For $x \in X$, $s, t \in I$, define $e_t : X \to E\Sigma X$ by

$$e_t(x)(s) = \langle x, t(1-s)\rangle.$$

Then define $F : X \times I \to I_q$ by

$$F(x,t) = (\langle x,t\rangle, e_t(x)).$$

Hence $\beta \simeq_F j\bar{e}$. □

With the notation above we next prove the generalized Freudenthal theorem.

Theorem 5.6.6 *Let A be a finite-dimensional CW complex and let X be an $(n-1)$-connected space, $n \geqslant 2$. Then the suspension function $\Sigma : [A, X] \to [\Sigma A, \Sigma X]$ is bijective if $\dim A \leqslant 2n-2$ and surjective if $\dim A = 2n-1$.*

Proof. By the exactness of the homotopy sequence of the fibration

$$\Omega\Sigma X \xrightarrow{j} I_q \longrightarrow CX,$$

it follows that j is a weak equivalence. Since X is $(n-1)$-connected, ΣX is n-connected, and, by the Blakers–Massey theorem applied to

$$X \xrightarrow{i} CX \xrightarrow{q} \Sigma X\ ,$$

we see that $\beta : X \to I_q$ is a $(2n-1)$-equivalence. By Lemma 5.6.5, $\bar{e}$ is a $(2n-1)$-equivalence and thus so is e by Exercise 5.20. Therefore e_* is a bijection if $\dim A \leqslant 2n-2$ and a surjection if $\dim A = 2n-1$ by Proposition 2.4.6. The theorem now follows by Lemma 5.6.3. □

One consequence is the classical Freudenthal suspension theorem.

Theorem 5.6.7 *If X is an $(n-1)$-connected space, $n \geqslant 2$, then the suspension homomorphism $\Sigma : \pi_r(X) \to \pi_{r+1}(\Sigma X)$ is an isomorphism for $r < 2n-1$ and an epimorphism for $r = 2n-1$.*

The Blakers–Massey theorem 5.6.4 also yields a truncated exact homotopy sequence for a cofibration. We call this result the Blakers–Massey exact sequence of a cofibration.

Corollary 5.6.8 [40] *Let*

$$A \xrightarrow{i} X \xrightarrow{q} Q$$

be a cofiber sequence with A r-connected and Q s-connected, $r, s \geqslant 1$. Then there is a homomorphism $\partial_k : \pi_k(Q) \to \pi_{k-1}(A)$ for $k = 3, \ldots, r+s$, called the connecting homomorphism, such that the following sequence of groups is exact

$$\pi_{r+s}(A) \xrightarrow{i_*} \pi_{r+s}(X) \xrightarrow{q_*} \pi_{r+s}(Q) \xrightarrow{\partial_{r+s}} \pi_{r+s-1}(A) \longrightarrow \cdots$$

$$\cdots \longrightarrow \pi_2(A) \xrightarrow{i_*} \pi_2(X) \xrightarrow{q_*} \pi_2(Q) \longrightarrow 0.$$

Proof. The proof consists of taking the exact homotopy sequence of the map q (Corollary 4.2.17) and then using the Blakers–Massey theorem 5.6.4 to replace $\pi_k(I_q)$ by $\pi_k(A)$ in the appropriate degrees k. The connecting homomorphism ∂_k is the composition

$$\pi_k(Q) \xrightarrow{d_k} \pi_{k-1}(I_q) \xrightarrow{\beta_*^{-1}} \pi_{k-1}(A),$$

where d_k is the homomorphism of Corollary 4.2.17. □

Remark 5.6.9 We observe that there is a Blakers–Massey exact sequence for homotopy groups with coefficients. If the coefficient group is not free-abelian, then the truncated exact sequence starts in degree $r+s-1$. This can be proved just like Corollary 5.6.8.

We return to the sequence of groups and homomorphisms

$$\pi_3(S^2) \xrightarrow{\Sigma} \pi_4(S^3) \xrightarrow{\Sigma} \pi_5(S^4) \xrightarrow{\Sigma} \cdots .$$

By the Freudenthal theorem, the first homomorphism is an epimorphism and all subsequent ones are isomorphisms. Since $\pi_3(S^2) \cong \mathbb{Z}$, it follows that

$\pi_{n+1}(S^n)$ is a cyclic group, for all $n \geqslant 2$. In fact, there is the following result on the $(n+1)$st and $(n+2)$nd homotopy group of S^n.

Theorem 5.6.10 *Let* $n \geqslant 3$. *Then*

1. $\pi_{n+1}(S^n) \cong \mathbb{Z}_2$, *with generator* $[\Sigma^{n-2}\phi]$, *where* $\phi : S^3 \to S^2$ *is the Hopf map.*
2. $\pi_{n+2}(S^n) \cong \mathbb{Z}_2$, *with generator* $[(\Sigma^{n-2}\phi)\,(\Sigma^{n-1}\phi)]$.

The proof of (1) is given in Appendix D. For (2), see [45, p. 328] or [39, pp. 475, 478].

For a CW complex X (not necessarily path-connected), the iterated suspension $\Sigma^n X$ is $(n-1)$-connected. Therefore if r is an integer $\geqslant 0$, then the groups $\pi_{n+r}(\Sigma^n X)$ are all isomorphic by the Freudenthal theorem for n sufficiently large $(n > r+1)$. This group is denoted $\pi_r^S(X)$ and called the *rth stable homotopy group* of X. If $X = S^0$, then $\pi_r^S(S^0)$ equals $\pi_{n+r}(S^n)$ for $n > r+1$. This group is called the *stable r-stem* and written π_r^S. Thus $\pi_0^S = \mathbb{Z}$ and, by Theorem 5.6.10, $\pi_1^S = \mathbb{Z}_2$ and $\pi_2^S = \mathbb{Z}_2$. The determination of the stable stems π_r^S is a difficult problem and some deep results have been obtained [78].

Moore Spaces

Let m and n be integers $\geqslant 2$. We first consider the Moore space $M(\mathbb{Z}_m, n)$. Let $\mathrm{m} : S^n \to S^n$ be a map of degree m, let X be the mapping cylinder of m, and let $M = S^n \cup_{\mathrm{m}} e^{n+1}$ be the mapping cone of m. Then $M = M(\mathbb{Z}_m, n)$ and we have a cofiber sequence

$$S^n \xrightarrow{\mathrm{m}'} X \longrightarrow M,$$

where S^n and M are each $(n-1)$-connected and m' is the map into the mapping cylinder induced by m. We take the Blakers–Massey exact sequence of this cofibration, replace X by S^n, and m' by m, and obtain the exact sequence

$$\pi_i(S^n) \xrightarrow{\mathrm{m}_*} \pi_i(S^n) \xrightarrow{j_*} \pi_i(M) \xrightarrow{d_i} \pi_{i-1}(S^n) \xrightarrow{\mathrm{m}_*} \pi_{i-1}(S^n) \quad (*),$$

where $j : S^n \to M$ is the inclusion, d_i is the boundary homomorphism and $i \leqslant 2n-2$. It follows from the Freudenthal theorem that $\Sigma : \pi_{i-1}(S^{n-1}) \to \pi_i(S^n)$ is an epimorphism. Therefore if $\alpha \in \pi_i(S^n)$, then $\alpha = \Sigma\beta$, for some $\beta = [g] \in \pi_{i-1}(S^{n-1})$. Hence $\mathrm{m}_*(\alpha) = [\mathrm{m}\,\Sigma g] = (\Sigma g)^*[\mathrm{m}]$. But $\mathrm{m} = m\,(\mathrm{id})$ and $(\Sigma g)^*$ is a homomorphism by Proposition 2.3.4. Therefore

$$\mathrm{m}_*(\alpha) = m[\Sigma g] = m\alpha.$$

Similarly $m_*(\gamma) = m\gamma$, for $\gamma \in \pi_{i-1}(S^n)$. Thus the two homomorphisms m_* in the exact sequence $(*)$ are both multiplication by m, denoted $\times m$.

Now suppose $i = n+1$ in $(*)$, where $n \geqslant 3$. Then, by Theorem 5.6.10, we have the exact sequence

$$\mathbb{Z}_2 \xrightarrow{\times m} \mathbb{Z}_2 \xrightarrow{j_*} \pi_{n+1}(M) \xrightarrow{d_{n+1}} \mathbb{Z} \xrightarrow{\times m} \mathbb{Z}.$$

Therefore if m is even, then $\pi_{n+1}(M) = \mathbb{Z}_2$, and if m is odd, $\pi_{n+1}(M) = 0$.

Next suppose $i = n+2$ in $(*)$, where $n \geqslant 4$. Then we have the exact sequence

$$\mathbb{Z}_2 \xrightarrow{\times m} \mathbb{Z}_2 \xrightarrow{j_*} \pi_{n+2}(M) \xrightarrow{d_{n+2}} \mathbb{Z}_2 \xrightarrow{\times m} \mathbb{Z}_2,$$

by Theorem 5.6.10. If m is even we obtain $\pi_{n+2}(M)$ as an extension of $\mathbb{Z}_2$ by $\mathbb{Z}_2$. If m is odd, then $\pi_{n+2}(M) = 0$. We summarize these calculations in the next proposition.

Proposition 5.6.11 *Let $m \geqslant 2$ and $n \geqslant 3$. Then*

1.

$$\pi_{n+1}(M(\mathbb{Z}_m, n)) = \begin{cases} \mathbb{Z}_2 & \text{if } m \text{ is even} \\ 0 & \text{if } m \text{ is odd} \end{cases}$$

2. If m is even and $n \geqslant 4$, then there is a short exact sequence

$$0 \longrightarrow \mathbb{Z}_2 \longrightarrow \pi_{n+2}(M(\mathbb{Z}_m, n)) \longrightarrow \mathbb{Z}_2 \longrightarrow 0.$$

3. If m is odd and $n \geqslant 4$, then $\pi_{n+2}(M(\mathbb{Z}_m, n)) = 0$.

We next discuss the relation between the homotopy groups of a Moore space of type $(\mathbb{Z}_m, n)$ and those of a Moore space of type (G, n), for a finitely generated group G.

Let G be a finitely generated abelian group and let P be the set of primes that divide the order of the torsion subgroup of G. We write G as a direct sum

$$G = \underbrace{\mathbb{Z} \oplus \cdots \oplus \mathbb{Z}}_{r \text{ summands}} \oplus \bigoplus_{p \in P} A_p,$$

where r is the rank of G and A_p is the p-primary component of G, that is, the direct sum of cyclic groups of order a power of p. Then for $n \geqslant 2$ we form the wedge

$$X = \underbrace{S^n \vee \cdots \vee S^n}_{r \text{ terms}} \vee \bigvee_{p \in P} M(A_p, n),$$

where $M(A_p, n)$ is a wedge of Moore spaces of the form $M(\mathbb{Z}_{p^a}, n)$. Clearly X is a Moore space $M(G, n)$. Furthermore, the spaces S^n and $M(A_p, n)$ can be assumed to be CW complexes whose $(n-1)$-skeleton is the basepoint by Corollary 2.4.10.

Proposition 5.6.12 *If $i \leqslant 2n-2$ and $n \geqslant 2$, then*

$$\pi_i(X) \cong \underbrace{\pi_i(S^n) \oplus \cdots \oplus \pi_i(S^n)}_{r \text{ summands}} \oplus \bigoplus_{p \in P} \pi_i(M(A_p, n)).$$

Proof. Let

$$Y = \underbrace{S^n \times \cdots \times S^n}_{r \text{ factors}} \times \prod_{p \in P} M(A_p, n)$$

and let $j : X \to Y$ be the inclusion. Then $Y^{2n-1} = X$ and so $j_* : \pi_i(X) \to \pi_i(Y)$ is an isomorphism by Proposition 1.5.24. Since

$$\pi_i(Y) \cong \pi_i(S^n) \oplus \cdots \oplus \pi_i(S^n) \oplus \bigoplus_{p \in P} \pi_i(M(A_p, n)),$$

the proof is complete. □

Corollary 5.6.13 *Let G be a finitely generated abelian group of rank r whose 2-primary component is a direct sum of s cyclic groups of order a power of 2. Then*

1. *If $n \geqslant 3$, then* $\pi_{n+1}(M(G,n)) \cong \underbrace{\mathbb{Z}_2 \oplus \cdots \oplus \mathbb{Z}_2}_{r+s \text{ summands}}$.
2. *If $n \geqslant 4$, then* $\pi_{n+2}(M(G,n)) \cong H \oplus \underbrace{\mathbb{Z}_2 \oplus \cdots \oplus \mathbb{Z}_2}_{r \text{ summands}}$, *and there is a short exact sequence*

$$0 \longrightarrow \mathbb{Z}_2 \oplus \cdots \oplus \mathbb{Z}_2 \longrightarrow H \longrightarrow \mathbb{Z}_2 \oplus \cdots \oplus \mathbb{Z}_2 \longrightarrow 0,$$

where each of the two direct sums in the preceding exact sequence is a direct sum of s copies of $\mathbb{Z}_2$.

Remark 5.6.14

1. The methods of Proposition 5.6.11 and Corollary 5.6.13 can be used to obtain additional results on $\pi_i(M(G,n))$ for $i \leqslant 2n-2$. If we know the homotopy groups of spheres $\pi_i(S^n)$, then we can determine $\pi_i(M(G,n))$ up to an extension.
2. In [8, p. 269], Baues has shown that $\pi_{n+2}(M(G,n))$ is an extension of $G * \mathbb{Z}_2$ by $G \otimes \mathbb{Z}_2$. In addition, there is a great deal of information about the spaces $M(\mathbb{Z}_m, n)$ in [75].

Topological Groups

We begin with the orthogonal group $O(n)$ of $n \times n$ real matrices A such that $AA^T = I$. Then $\det A = \pm 1$ and the subgroup $SO(n)$ of $O(n)$ consisting of those matrices of determinant 1 is the special orthogonal group (Exercise

3.22). The group $SO(n)$ acts on $\mathbb{R}^n$ and can be shown to be the group of all rotations of $\mathbb{R}^n$. It is easily seen (Exercise 3.24) that $O(n)$ is the union of two path-connected components one of which is $SO(n)$. The other is also homeomorphic to $SO(n)$ and is obtained by multiplying $SO(n)$ by a fixed matrix of $O(n)$ of determinant -1. Thus $O(n) \cong SO(n) \times \mathbb{Z}_2$, where $\mathbb{Z}_2$ denotes the discrete space with two elements. Therefore $\pi_0(O(n))$ is the two element set and $\pi_0(SO(n)) = 0$. Furthermore,

$$\pi_i(O(n)) = \pi_i(SO(n)),$$

for $i \geqslant 1$. By Proposition 3.4.12 there are fiber sequences

$$O(n-1) \xrightarrow{j_{n-1}} O(n) \xrightarrow{p_n} S^{n-1}$$

and so it follows from the exact homotopy sequence of a fibration that

$$j_{n-1*} : \pi_i(O(n-1)) \to \pi_i(O(n)) \text{ is } \begin{cases} \text{an isomorphism if } i < n-2 \\ \text{an epimorphism if } i = n-2. \end{cases}$$

Therefore for fixed i, the groups $\pi_i(O(n))$ are all isomorphic if $n > i+1$. We denote this group by $\pi_i(O)$. Now $SO(2)$ acts on S^1 by rotation and this action is transitive, that is, for any two elements of S^1, there is a matrix in $SO(2)$ (a rotation) which carries one to the other. The isotropy group of this action is trivial, therefore it follows from Exercise 3.21 that $SO(2) \cong S^1$. Thus for $n \geqslant 1$,

$$\pi_n(O(2)) \cong \pi_n(SO(2)) \cong \pi_n(S^1) = \begin{cases} \mathbb{Z} & \text{if } n = 1 \\ 0 & \text{if } n > 1. \end{cases}$$

Next we sketch a proof that $SO(3)$ is homeomorphic to $\mathbb{R}\mathrm{P}^3$. We identify $S^3 \subseteq \mathbb{R}^4$ with the quaternions of unit length and identify $\mathbb{R}^3$ with the pure quaternions, that is, those of the form $b\mathbf{i}+c\mathbf{j}+d\mathbf{k}$, for $b, c, d \in \mathbb{R}$. For every $x \in S^3$, let ρ_x be conjugation of the pure quaternions by x, that is, $\rho_x(y) = xyx^{-1}$, where $y \in \mathbb{R}^3$. One then shows the following.

1. $\rho_x(y)$ is a pure quaternion. Therefore $\rho_x : \mathbb{R}^3 \to \mathbb{R}^3$ is a linear transformation. With the standard $\mathbb{R}^3$ basis, ρ_x can be regarded as a 3×3 matrix.
2. $\rho_x \in SO(3)$ and so a continuous function $\rho : S^3 \to SO(3)$ is defined by $\rho(x) = \rho_x$.
3. $\rho(x) = I \iff x = \pm 1$.
4. ρ induces a continuous injection $\tilde{\rho} : S^3/\!\sim \to SO(3)$, where $x \sim x'$ if and only if $x' = \pm x$, for $x, x' \in S^3$. Thus $\tilde{\rho} : \mathbb{R}\mathrm{P}^3 \to SO(3)$ is a continuous injection.
5. $\tilde{\rho}$ is a homeomorphism: Since $\mathbb{R}\mathrm{P}^3$ is compact and $SO(3)$ is Hausdorff, it suffices to show that $\tilde{\rho}$ is onto: Given $A \in SO(3)$ with columns A_1, A_2, A_3 (regarded as pure quaternions), let x' be a quaternion of unit length defined

by $x' = x - A_1x\mathbf{i} - A_2x\mathbf{j} - A_3x\mathbf{k}$, for an arbitrary quaternion x. Then a long calculation shows that $\widetilde{\rho}(x') = A$. Alternatively, the following is a shorter and more sophisticated argument: $\mathbb{R}\mathrm{P}^3$ and $SO(3)$ are compact, path-connected three-dimensional topological manifolds. Then $\mathbb{R}\mathrm{P}^3$ and $\widetilde{\rho}(\mathbb{R}\mathrm{P}^3)$ are homeomorphic spaces. By the invariance of domain theorem [61, p. 217, 6.5], $\widetilde{\rho}(\mathbb{R}\mathrm{P}^3)$ is open in $SO(3)$. Since $\widetilde{\rho}(\mathbb{R}\mathrm{P}^3)$ is closed in $SO(3)$, we obtain $\widetilde{\rho}(\mathbb{R}\mathrm{P}^3) = SO(3)$ because $SO(3)$ is connected.

This completes the sketch of the proof.

Hence for $n \geqslant 1$,

$$\pi_n(O(3)) \cong \pi_n(SO(3)) \cong \pi_n(\mathbb{R}\mathrm{P}^3).$$

Because S^3 is the universal cover of $\mathbb{R}\mathrm{P}^3$, we have by Theorem 5.6.10,

$$\pi_n(O(3)) \cong \pi_n(SO(3)) \cong \begin{cases} \mathbb{Z}_2 & \text{if } n = 1, 4, 5 \\ 0 & \text{if } n = 2 \\ \mathbb{Z} & \text{if } n = 3. \end{cases}$$

Thus we see that $\pi_1(O) = \mathbb{Z}_2$ and, since $j_{2*} : \pi_2(O(3)) \to \pi_2(O(4))$ is an epimorphism, $\pi_2(O) = 0$. An interesting discussion of the fundamental group of $SO(3)$ appears in [14, pp. 164–167] including an illustration with photographs of the fact that a loop in $SO(3)$ added to itself is homotopic to the identity.

For $n \geqslant 1$, the unitary group $U(n)$ consists of $n \times n$ complex matrices A such that $\bar{A}A^T = I$, where $\bar{A}$ is the matrix of complex conjugates of the entries of A. Then there are the fiber sequences

$$U(n-1) \xrightarrow{k_{n-1}} U(n) \xrightarrow{p_n} S^{2n-1},$$

and so it follows from the exact homotopy sequence of a fibration that

$$k_{n-1*} : \pi_i(U(n-1)) \to \pi_i(U(n)) \text{ is } \begin{cases} \text{an isomorphism if } i < 2n-2 \\ \text{an epimorphism if } i = 2n-2. \end{cases}$$

Therefore for fixed i, the groups $\pi_i(U(n))$ are all isomorphic if $2n > i$. We denote this group by $\pi_i(U)$.

Now $U(1) = S^1$ and so $\pi_1(U(1)) = \mathbb{Z}$ and $\pi_i(U(1)) = 0$ for $i \geqslant 2$. Thus

$$\pi_1(U) = \pi_1(U(n)) = \mathbb{Z} \quad \text{for } n \geqslant 1.$$

For $U(2)$ we have the fibration $S^1 \to U(2) \to S^3$. From its exact homotopy sequence we see that $\pi_i(U(2)) \cong \pi_i(S^3)$ for $i \geqslant 2$. Therefore

$$\pi_2(U) = \pi_2(U(n)) = 0 \quad \text{for } n \geqslant 2.$$

Consider the fibration $U(2) \to U(3) \to S^5$. We obtain $\pi_3(U(3)) = \mathbb{Z}$ and so

$$\pi_3(U) = \pi_3(U(n)) = \mathbb{Z} \quad \text{for } n \geqslant 3.$$

Part of the exact homotopy sequence of this fibration is

$$\pi_4(U(2)) \longrightarrow \pi_4(U(3)) \longrightarrow 0.$$

Thus if $n \geqslant 3$,

$$\pi_4(U) = \pi_4(U(n)) = 0 \text{ or } \mathbb{Z}_2,$$

because $\pi_4(U(2)) = \mathbb{Z}_2$. It can be shown that the group is 0.

The symplectic group $Sp(n)$ of $n \times n$ quaternion matrices A such that $\bar{A}A^T = I$ can be treated in complete analogy to the orthogonal group and the unitary group. We can determine the homotopy groups of $Sp(1)$ and $Sp(2)$ and calculate $\pi_i(Sp)$ for $i = 1, 2, 3, 4$. We omit details.

Going back to the groups $\pi_i(U)$, we see that $\pi_i(U) \cong \pi_{i+2}(U)$ for $i = 1, 2$. This is a small part of a more general, well-known, and deep result due to Bott [11]. We state this theorem which is called the Bott periodicity theorem.

Theorem 5.6.15 *For all $i \geqslant 0$,*

- $\pi_i(O) \cong \pi_{i+8}(O)$.
- $\pi_i(U) \cong \pi_{i+2}(U)$.
- $\pi_i(Sp) \cong \pi_{i+8}(Sp)$.

Thus $\pi_i(O)$ and $\pi_i(Sp)$ are each periodic in i with period 8 and $\pi_i(U)$ is periodic with period 2. All of the groups above have been calculated.

Stiefel Manifolds

We consider the real Stiefel manifolds $V_k(\mathbb{R}^n)$ which we write as $V_{k,n}$. For $0 \leqslant k \leqslant l$, we have by Proposition 3.4.12 the general Stiefel fibration

$$V_{l-k,n-k} \xrightarrow{j} V_{l+1,n+1} \xrightarrow{q} V_{k+1,n+1}.$$

We consider special cases. For $k = 0$, we obtain

$$V_{l,n} \xrightarrow{j} V_{l+1,n+1} \xrightarrow{q} V_{1,n+1} = S^n.$$

From the exact homotopy sequence of a fibration we have

$$j_* : \pi_i(V_{l,n}) \to \pi_i(V_{l+1,n+1}) \text{ is } \begin{cases} \text{an isomorphism if } i < n-1 \\ \text{an epimorphism if } i = n-1. \end{cases}$$

Corollary 5.6.16 *For $k < n$, $V_{k,n}$ is $(n-k-1)$-connected.*

Proof. Suppose $i < n-k$. Then we have the following string of isomorphisms

$$\pi_i(V_{k,n}) \cong \pi_i(V_{k-1,n-1}) \cong \cdots \cong \pi_i(V_{1,n-k+1}) \cong \pi_i(S^{n-k}).$$

But the latter group is 0 because $i < n-k$. □

A special case of the fibration $V_{l,n} \xrightarrow{j} V_{l+1,n+1} \xrightarrow{q} S^n$ occurs when $l = 1$ and we obtain

$$S^{n-1} \xrightarrow{j} V_{2,n+1} \xrightarrow{q} S^n .$$

As we noted in Remark 3.4.17 it is a classical result in algebraic topology that this bundle has a section if and only if n is odd. Therefore for odd n, there is a section and so

$$\pi_i(V_{2,n+1}) \cong \pi_i(S^n) \oplus \pi_i(S^{n-1}).$$

It can be shown [41, p. 88] that for odd $n \geqslant 5$,

$$\pi_{n-2}(V_{2,n}) \cong \pi_{n-1}(V_{2,n}) \cong \mathbb{Z}_2.$$

Another special case of the general Stiefel fibration occurs when $l = k+1$:

$$S^{n-l} = V_{1,n-l+1} \xrightarrow{j} V_{l+1,n+1} \xrightarrow{q} V_{l,n+1} .$$

Then

$$q_* : \pi_i(V_{l+1,n+1}) \to \pi_i(V_{l,n+1}) \text{ is } \begin{cases} \text{an isomorphism if } i < n-l \\ \text{an epimorphism if } i = n-l. \end{cases}$$

By taking $l = n-1$ in the previous fibration, we obtain the fibration

$$S^1 \xrightarrow{j} V_{n,n+1} \xrightarrow{q} V_{n-1,n+1} .$$

But $V_{n,n+1} \cong SO(n+1)$ (Exercise 3.23), and so for $i \geqslant 3$,

$$\pi_i(V_{n-1,n+1}) \cong \pi_i(SO(n+1)) \cong \pi_i(O(n+1)).$$

We omit a discussion of the complex and quaternionic Stiefel manifold. For more information about the Stiefel manifolds and their homotopy groups, see [53, 91].

Grassmann Manifolds

Since projective spaces are special Grassmann manifolds, we discuss these first. From the splitting of the homotopy groups of $\Omega\mathbb{F}\mathrm{P}^n$ in Example 5.4.8, we have for all $i \geqslant 1$,

$$\pi_i(\mathbb{F}\mathrm{P}^n) \cong \pi_i(S^{d(n+1)-1}) \oplus \pi_{i-1}(S^{d-1}),$$

where $d = 1, 2$ or 4 according as $\mathbb{F} = \mathbb{R}, \mathbb{C}$, or $\mathbb{H}$. More explicitly,

$$\pi_i(\mathbb{R}\mathrm{P}^n) \cong \pi_i(S^n) \text{ for } i \geqslant 2$$

$$\pi_i(\mathbb{C}\mathrm{P}^n) \cong \begin{cases} \mathbb{Z} & \text{if } i = 2 \\ \pi_i(S^{2n+1}) & \text{if } i > 2 \end{cases}$$

$$\pi_i(\mathbb{H}\mathrm{P}^n) \cong \pi_i(S^{4n+3}) \oplus \pi_{i-1}(S^3) \text{ for } i \geqslant 1.$$

The first of these isomorphisms also follows from the fact that S^n is the universal cover of $\mathbb{R}\mathrm{P}^n$, $n \geqslant 2$.

We next briefly consider general Grassmann manifolds, but only in the real case. By Proposition 3.4.16 there is a fiber sequence

$$O(k) = V_{k,k} \longrightarrow V_{k,n} \longrightarrow G_{k,n}.$$

From Corollary 5.6.16 and the exact homotopy sequence of a fibration, we conclude that

$$\pi_i(G_{k,n}) \cong \pi_{i-1}(O(k)) \text{ if } i < n - k.$$

In addition, it follows easily that if $0 < k < n$,

$$\pi_1(G_{k,n}) \cong \begin{cases} \mathbb{Z} & \text{if } n = 2 \\ \mathbb{Z}_2 & \text{if } n > 2. \end{cases}$$

Additional results on the homotopy groups of Grassmannians can be found in [91, pp. 202-209].

As we have seen, it is difficult to calculate homotopy groups. We have used elementary methods to determine the first few nontrivial homotopy groups of certain spaces. But even with more advanced methods, which can be used to obtain many more homotopy groups, a great deal is unknown. To end this section on a positive note, we mention a few known, general results.

In [82], Serre defined a *class* $\mathcal{C}$ *of abelian groups* by certain properties. Rather than give the definition, let us just say that the following collections of abelian groups are classes: the collection consisting only of the zero group, all finitely generated groups, all finite groups, and all finite groups of order not divisible by a given prime p. A homomorphism $f : G \to H$ of groups is called a $\mathcal{C}$-isomorphism if $\operatorname{Ker} f$ and $\operatorname{Coker} f = G/\operatorname{Im} f$ are in $\mathcal{C}$. A consequence of the work of Serre is the following theorem which is a $\mathcal{C}$-theory analogue of the Hurewicz theorem 6.4.8 (for details, see [83, pp. 504-512]).

Theorem 5.6.17 *Let* $\mathcal{C}$ *be a class of abelian groups, let* X *be a 1-connected space, and let* $n \geqslant 2$ *be an integer. If* $H_m(X) \in \mathcal{C}$ *for* $1 < m < n$, *then*

$\pi_m(X) \in \mathcal{C}$ for $1 < m < n$. Furthermore, the Hurewicz homomorphism $h_n : \pi_n(X) \to H_n(X)$ is a $\mathcal{C}$-isomorphism.

Thus if X is a 1-connected finite CW complex, then $\pi_m(X)$ is finitely generated for all m. Moreover, if $H_m(X)$ is finite or is finite with no p-torsion for all $m < n$, then the same holds for $\pi_m(X)$.

By using this and other work of Serre [81], the following results on the homotopy groups of spheres have been obtained.

- The homotopy groups $\pi_m(S^n)$ are all finite except in the following cases: (a) $\pi_n(S^n) = \mathbb{Z}$, for all n; and (b) $\pi_{2n-1}(S^n)$ is a group of rank 1, for all even n.
- If $n \geqslant 3$ is odd and p is a prime, then $\pi_m(S^n)_{(p)}$, the p-primary component of $\pi_m(S^n)$, is zero if $m < n + 2p - 3$ and is $\mathbb{Z}_p$ if $m = n + 2p - 3$.
- If $n \geqslant 4$ is even and p is an odd prime, then

$$\pi_m(S^n)_{(p)} \cong \pi_m(S^{2n-1})_{(p)} \oplus \pi_{m-1}(S^{n-1})_{(p)}.$$

We refer the reader to [19, 78, 88] for more on the homotopy groups of spheres.

Exercises

Exercises marked with (∗) may be more difficult than the others. Exercises marked with (†) are used in the text.

5.1. (†) In Definition 5.2.2 a homomorphism $\alpha : H^n(X) \otimes G \to H^n(X; G)$ was defined. Show that α is well-defined by showing that $\phi_\gamma(f + f') \simeq \phi_\gamma(f) + \phi_\gamma(f')$ and $\phi_{\gamma+\gamma'}(f) \simeq \phi_\gamma(f) + \phi_{\gamma'}(f)$, for $f, f' : X \to K(\mathbb{Z}, n)$ and $\gamma, \gamma' \in G$.

5.2. Let $0 \longrightarrow G_1 \longrightarrow G_2 \longrightarrow G_3 \longrightarrow 0$ be a short exact sequence of abelian groups and let X be a space. Show that there is a homomorphism $\delta^n : H^n(X; G_3) \to H^{n+1}(X; G_1)$ such that the following sequence of cohomology groups is exact

$$H^0(X;G_1) \longrightarrow H^0(X;G_2) \longrightarrow H^0(X;G_3) \xrightarrow{\delta^0} H^1(X;G_1) \longrightarrow \cdots$$

$$\cdots \longrightarrow H^n(X;G_1) \longrightarrow H^n(X;G_2) \longrightarrow H^n(X;G_3) \xrightarrow{\delta^n} H^{n+1}(X;G_1) \longrightarrow .$$

5.3. (†) Prove that the function $\eta : \pi_n(X; G) \to \operatorname{Hom}(G, \pi_n(X))$ given in Definition 5.2.6 is a homomorphism.

5.4. Let G be a finitely generated abelian group of rank r whose 2-primary component is a direct sum of s cyclic groups of order a power of 2. Prove

$$\pi_n(S^n; G) = \underbrace{\mathbb{Z} \oplus \cdots \oplus \mathbb{Z}}_{r \text{ summands}} \oplus \underbrace{\mathbb{Z}_2 \oplus \cdots \oplus \mathbb{Z}_2}_{s \text{ summands}},$$

where $n \geqslant 3$. You may assume that $\pi_{n+1}(S^n) = \mathbb{Z}_2$ (Theorem 5.6.10).

5.5. (†) Under the hypotheses of Proposition 5.4.6, prove that

$$(u + v)^{\alpha+\beta} = u^\alpha + v^\beta$$

by showing that the following diagram is commutative

$$\begin{array}{ccccc}
Q & \xrightarrow{\Delta} & Q \times Q & \xrightarrow{\psi \times \psi} & (Q \vee \Sigma A) \times (Q \vee \Sigma A) \\
\downarrow{\scriptstyle \psi} & & & & \downarrow{\scriptstyle (f \vee a) \times (g \vee b)} \\
Q \vee \Sigma A & & & & (Z \vee Z) \times (Z \vee Z) \\
\downarrow{\scriptstyle \Delta \vee \Delta} & & & & \downarrow{\scriptstyle \nabla \times \nabla} \\
(Q \times Q) \vee (\Sigma A \times \Sigma A) & & & & Z \times Z \\
\downarrow{\scriptstyle (f \times g) \vee (a \times b)} & & & & \downarrow{\scriptstyle m} \\
(Z \times Z) \vee (Z \times Z) & \xrightarrow{m \vee m} & Z \vee Z & \xrightarrow{\nabla} & Z,
\end{array}$$

where $\alpha = [a], \beta = [b] \in [\Sigma A, Z]$ and $u = [f]$, $v = [g] \in [Q, Z]$. As a consequence prove Proposition 5.4.6.

5.6. $(*)$ Let $A \longrightarrow X \longrightarrow Q$ be a cofiber sequence and let $r : X \to A$ be a homotopy retraction. It is well known that $H_j(X) \cong H_j(A) \oplus H_j(Q)$, for all $j \geqslant 0$.

1. Give an example for which X does not have the homotopy type of $A \vee Q$.
2. What additional hypothesis would ensure that $X \simeq A \vee Q$?
3. Prove that $\Sigma X \simeq \Sigma A \vee \Sigma Q$.

5.7. (†) Prove the following for the smash product of CW complexes.

1. $X \wedge Y \cong Y \wedge X$.
2. $(X \vee Y) \wedge Z \cong (X \wedge Z) \vee (Y \wedge Z)$.
3. $(\Sigma X) \wedge Y \cong \Sigma(X \wedge Y) \cong X \wedge (\Sigma Y)$.
4. If $1 \in I$ is the basepoint, then $CX = X \wedge I$.
5. $S^1 \wedge X \cong \Sigma X$.
6. $S^m \wedge S^n \cong S^{m+n}$.

5.8. Let X be a co-H-space.

1. If Y is any space, prove that $X \wedge Y$ is a co-H-space.
2. If X is a cogroup, prove that $X \wedge Y$ is also a cogroup.
3. If Y is a co-H-space, prove that $X \wedge Y$ is a commutative co-H-space.

5.9. In this problem we deal only with unbased spaces, functions, and homotopies. We also assume for convenience that all spaces under consideration are locally compact (but see Appendix A). Let $i : A \to X$ be a cofiber map and let W be any space. Define $i^\# : M(X, W) \to M(A, W)$ by $i^\#(f) = fi$, for $f \in M(X, W)$. Prove that $i^\#$ is a fiber map. You may assume that $M(X \times Y, Z) \cong M(Y, M(X, Z))$ (Proposition A.9).

5.10. Let $f : X \to Y$ be a map of path-connected CW complexes and let $* : X \to Y$ be the constant map. Prove: $f \simeq * \iff f \simeq_{free} *$. Does this contradict the fact that the comb space C is not contractible to $*_1$ (see Example 1.4.5)?

5.11. ($*$) If X is a free space and $x \in X$ is any point, then let $\pi_n(X, x)$ denote the nth homotopy group of the based space (X, x). Let a be a path in X with initial point $x_0 = a(0)$ and terminal point $x_1 = a(1)$. We define a function $a^\# : \pi_n(X, x_0) \to \pi_n(X, x_1)$ as follows. Let $f : (I^n, \partial I^n) \to (X, x_0)$ and consider $f \cup a : I^n \times \{0\} \cup \partial I^n \times I \to X$ (see Section 5.5). Since $I^n \times \{0\} \cup \partial I^n \times I$ is a retract of $I^n \times I$, there is an extension $G : I^n \times I \to X$ of $f \cup a$. If $g(x) = G(x, 1)$, for $x \in I^n$, then $g : (I^n, \partial I^n) \to (X, x_1)$ and we define $a^\#[f] = [g]$.

1. Show that $a^\# : \pi_n(X, x_0) \to \pi_n(X, x_1)$ is a well-defined function.
2. Prove that if a and b are paths in X from x_0 to x_1 and $a \simeq b : I \to X \text{ rel } \partial I$, then $a^\# = b^\# : \pi_n(X, x_0) \to \pi_n(X, x_1)$. Thus if $\alpha = [a]$ is the homotopy class rel ∂I of a path a from x_0 to x_1, define $\alpha^\# = a^\# : \pi_n(X, x_0) \to \pi_n(X, x_1)$.
3. Show that $\alpha^\#$ is a homomorphism.
4. Consider $\alpha = [a]$ and $\beta = [b]$, homotopy classes rel ∂I, where a is a path from x_0 to x_1 and b is a path from x_1 to x_2. Let $\alpha + \beta = [a + b]$. Prove $(\alpha + \beta)^\# = \beta^\# \alpha^\#$.
5. Prove that $\alpha^\#$ is an isomorphism.

5.12. ($*$) Let X and Y be free spaces and let $f : X \to Y$ be a free map which is a free homotopy equivalence. For any point $x_0 \in X$ and any $n \geqslant 1$, prove that $f_* : \pi_n(X, x_0) \to \pi_n(Y, f(x_0))$ is an isomorphism.

5.13. Show that the comb space C with basepoint $*_1$ of Example 1.4.5 is simply connected and that there are two (based) maps $C \to C$ which are freely homotopic but not (based) homotopic. Why does this not contradict the result in Section 5.5 which asserts that θ is a bijection?

5.14. ($*$) Find spaces X and Y (not both CW complexes) and weak equivalences $X \to Y$ and $Y \to X$ such that $X \not\simeq Y$.

5.15. (†) Let $\alpha \in \pi_1(Y)$ and $u \in \pi_n(Y)$. Represent u by a map $f : S^n \to Y$ and represent $u \cdot \alpha$ by a map $g : S^n \to Y$. Prove $f \simeq_{\text{free}} g$.

5.16. (†) Prove Lemma 5.5.8.

5.17. If $F \longrightarrow E \xrightarrow{p} B$ is a fiber sequence and M_p is the mapping cylinder of p, then show that $F \simeq E(M_p; E, \{*\})$. Use the action of the fundamental group on relative homotopy to define an action of $\pi_1(E)$ on $\pi_n(F)$.

5.18. $(*)$ Prove that a space Y is dominated by a loop space (Definition 1.4.1) $\Longleftrightarrow$ the map $e : Y \to \Omega\Sigma Y$ has a left homotopy inverse.

5.19. (†) Show that $\mathbb{R}\mathrm{P}^\infty$ is the Eilenberg–Mac Lane space $K(\mathbb{Z}_2, 1)$ and $\mathbb{C}\mathrm{P}^\infty$ is $K(\mathbb{Z}, 2)$.

5.20. (†) Let $f : X \to \Omega Y$ be any map and let $j : \Omega Y \to \Omega Y$ be the inverse map. Show that the composition $j^2 \simeq \mathrm{id} : \Omega Y \to \Omega Y$. Now define $-f = jf : X \to \Omega Y$ and consider the functions $f_* : [A, X] \to [A, \Omega Y]$ and $(-f)_* : [A, X] \to [A, \Omega Y]$, for any space A. Prove that f_* is injective $\Longleftrightarrow$ $(-f)_*$ is injective and that f_* is surjective $\Longleftrightarrow$ $(-f)_*$ is surjective.

5.21. Show that

$$\pi_k(\mathbb{R}\mathrm{P}^n, \mathbb{R}\mathrm{P}^{n-1}) = \begin{cases} 0 & k < n \\ \mathbb{Z} \oplus \mathbb{Z} & k = n, \end{cases}$$

where $n \geqslant 2$ and $k \geqslant 1$.

Chapter 6
Homotopy Pushouts and Pullbacks

6.1 Introduction

We begin the chapter by discussing homotopy pushouts and homotopy pullbacks in the first two sections. These are analogues in the homotopy category of pushouts and pullbacks. They have many applications in homotopy theory and are necessary for the proofs in Section 6.4. That section contains the statement and proof of many of the major theorems of classical homotopy theory.

Sections 6.2 and 6.3 establish several basic properties of homotopy pushouts and pullbacks and state two more advanced properties whose proofs appear in Appendix E. In Section 6.3 we prove that a certain homotopy-commutative square associated to a sequence

$$X \xrightarrow{f} Y \xrightarrow{g} Z,$$

such that $gf = *$, is a homotopy pushout square (Proposition 6.3.5). This proposition applied when the sequence above is a fiber sequence is an essential ingredient in the proof of Serre's theorem. This theorem asserts that there is an N-equivalence between the mapping cone of f and the base Z, where N depends on the connectivities of X and Z. This gives a truncated, long exact sequence whose terms are the cohomology groups of the spaces of the fiber sequence. The homotopy pushout proposition applied when the sequence above is a cofiber sequence is essential for the proof of the Blakers–Massey theorem. This result, which is dual to the Serre theorem, asserts that there is an M-equivalence between the homotopy fiber of g and X, where M depends on the connectivities of X and Z. This leads to a truncated, long exact sequence for the homotopy groups of a cofiber sequence.

From these theorems we derive the Hurewicz theorem and Whitehead's second theorem. The former states that the degree of the first nontrivial homotopy group of a space is the same as the degree of the first nontrivial homology group of the space, and that these groups are isomorphic via the Hurewicz homomorphism. The latter asserts that a map between simply connected spaces induces a homology isomorphism in dimensions $< n$ and an epimorphism in dimension n if and only if the same is true for the induced homotopy homomorphism. In addition, we prove a relative Hurewicz theorem and show that the homotopy type of a Moore space depends only on the group and the degree. We conclude the chapter with a section that continues our discussion of Eckmann–Hilton duality which was begun in Section 2.5.

6.2 Homotopy Pushouts and Pullbacks I

We begin with homotopy pushouts. For motivation we recall the defining property of pushouts (Definition 3.2.8). Let

$$Y \xleftarrow{g} A \xrightarrow{f} X$$

be spaces and maps, let $a : X \to Z$ and $b : Y \to Z$ be maps such that $af = bg$ and let P together with maps $r : X \to P$ and $s : Y \to P$, be the pushout of f and g. Then there is a unique map $c : P \to Z$ such that $cr = a$ and $cs = b$,

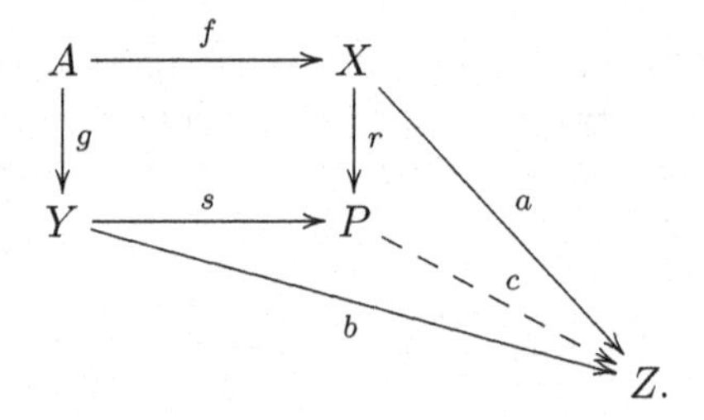

We seek a replacement P', r', s' for P, r, s such that if $af \simeq bg$, then there is a map $c' : P' \to Z$ with $c'r' \simeq a$ and $c's' \simeq b$. The homotopy pushout does this.

For any spaces X, Y, and Z, we regard $X \vee Y \vee Z$ as the subset of $X \times Y \times Z$ consisting of all triples (x, y, z) such that at least two coordinates are the basepoint. Recall that the reduced cylinder $X \ltimes I$ of X is the space $X \times I/\{(*, t) \,|\, t \in I\}$.

Definition 6.2.1 Given spaces and maps

$$Y \xleftarrow{g} A \xrightarrow{f} X \qquad (*)$$

the *standard homotopy pushout* or *double mapping cylinder* of $(*)$ consists of a space O and maps $u : X \to O$ and $v : Y \to O$ defined as follows: (1) $O = (X \vee (A \ltimes I) \vee Y)/\sim$, where $(*, \langle a, 0\rangle, *) \sim (f(a), *, *)$ and $(*, \langle a, 1\rangle, *) \sim$

$(*, *, g(a))$, for $a \in A$ and (2) $u(x) = \langle x, *, * \rangle$, for $x \in X$ and $v(y) = \langle *, *, y \rangle$, for $y \in Y$. The square

$$\begin{array}{ccc} A & \xrightarrow{f} & X \\ \downarrow g & & \downarrow u \\ Y & \xrightarrow{v} & O \end{array}$$

is then called the *standard homotopy pushout square*. Clearly $uf \simeq_F vg$, where $F : A \times I \to O$ is defined by $F(a,t) = \langle *, \langle a, t \rangle, * \rangle$, for $a \in A$ and $t \in I$.

Remark 6.2.2 The standard homotopy pushout can be realized as a pushout in the following way. Recall that any map $h : W \to Z$ can be factored as

$$W \xrightarrow{h'} M_h \xrightarrow{h''} Z,$$

where M_h is the mapping cylinder of h, the map h' is a cofibration, and the map h'' is a homotopy equivalence with homotopy inverse k_h. Given the diagram $(*)$ above, we consider

$$M_g \xleftarrow{g'} A \xrightarrow{f'} M_f$$

and take its pushout

$$\begin{array}{ccc} A & \xrightarrow{f'} & M_f \\ \downarrow g' & & \downarrow r \\ M_g & \xrightarrow{s} & P. \end{array}$$

Then P together with the maps $rk_f : X \to P$ and $sk_g : Y \to P$ is the standard homotopy pushout of $(*)$. See Figure 6.1.

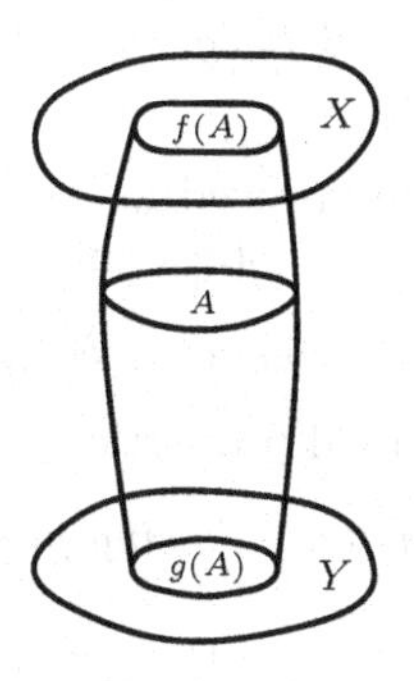

Figure 6.1

Remark 6.2.3 The standard homotopy pushout of the following diagram is the mapping cone C_f.

$$\{*\} \longleftarrow A \xrightarrow{f} X$$

Definition 6.2.4 *Let*

$$\begin{array}{ccc} A & \xrightarrow{f} & X \\ {\scriptstyle g}\downarrow & & \downarrow{\scriptstyle r} \\ Y & \xrightarrow{s} & P \end{array}$$

be a homotopy-commutative square. Suppose that there is a homotopy equivalence $\theta : O \to P$ defined on the standard homotopy pushout O of f and g such that $\theta\, u \simeq r$ and $\theta\, v \simeq s$:

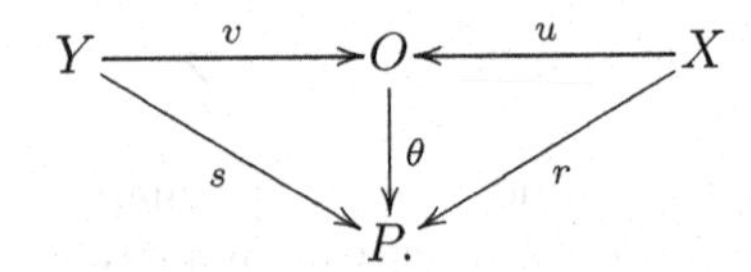

Then we call P together with r and s a *homotopy pushout* and the given square a *homotopy pushout square.*

Proposition 6.2.5 *Let the square in the following diagram*

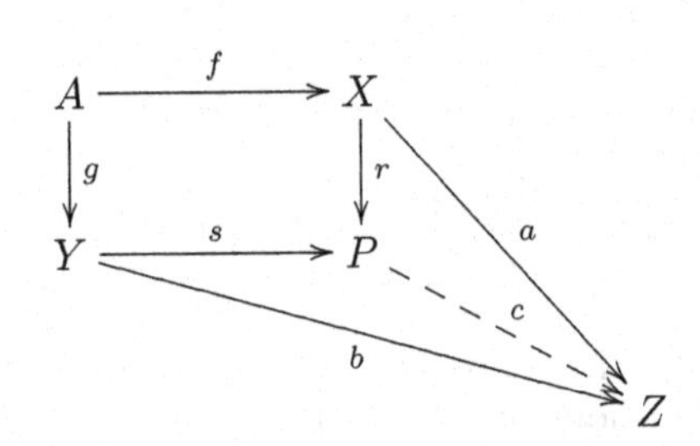

be a homotopy pushout square and let Z be any space. If $a : X \to Z$ and $b : Y \to Z$ are maps such that $af \simeq bg$, then there is a map $c : P \to Z$ such that $cr \simeq a$ and $cs \simeq b$.

Proof. The proposition is easily proved for the standard homotopy pushout square. It then follows for any homotopy pushout square. □

The homotopy class of the map c is not necessarily unique.

We next consider some general properties of homotopy pushouts.

Proposition 6.2.6 *Consider the following pushout square*

$$\begin{array}{ccc} A & \xrightarrow{f} & X \\ {\scriptstyle g}\downarrow & & \downarrow{\scriptstyle r} \\ Y & \xrightarrow{s} & P. \end{array}$$

If f or g is a cofiber map, then this square is a homotopy pushout square.

Proof. We assume that f is a cofiber map and consider the standard homotopy pushout O of f and g and the diagram

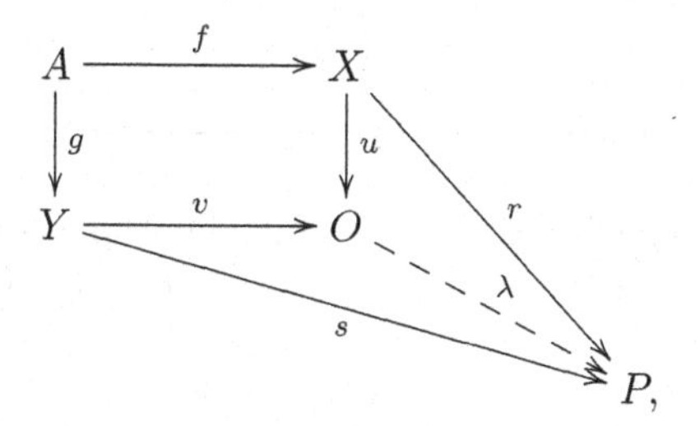

where λ is the map induced by r and s. It suffices to show that λ is a homotopy equivalence. Note that $O = M_f \vee Y/\sim$, where M_f is the mapping cylinder of f and $(\langle a, 1\rangle, *) \sim (*, g(a))$ for $a \in A$. Let us use $\langle\ ,\ \rangle$ for equivalence classes in P and M_f and $[\ ,\ ]$ for equivalence classes in O. Then $\lambda[x] = \langle x\rangle$, $\lambda[y] = \langle y\rangle$ and $\lambda[a, t] = rf(a) = sg(a)$, for $x \in X$ $y \in Y$, $a \in A$, and $t \in I$. We have the diagram

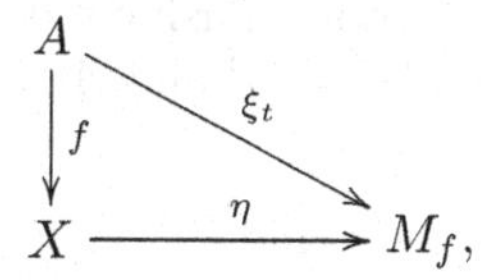

where $\eta(x) = \langle x\rangle$ and $\xi_t(a) = \langle a, t\rangle$. Since f is a cofiber map and $\eta f = \xi_0$, there exists a homotopy $\sigma : X \times I \to M_f$ such that $\sigma(x, 0) = \langle x\rangle$ and $\sigma(f(a), t) = \langle a, t\rangle$. Now define $\mu : P \to O$ by

$$\mu\langle x\rangle = [\sigma(x, 1)] \quad \text{and} \quad \mu\langle y\rangle = [y].$$

We show μ is a homotopy inverse to λ. Define a homotopy $A : P \times I \to P$ by

$$A(\langle x\rangle, t) = \lambda[\sigma(x, t)] \quad \text{and} \quad A(\langle y\rangle, t) = \langle y\rangle.$$

Then $\mathrm{id} \simeq_A \lambda\mu$. Next define $B : O \times I \to O$ by

$$B([x], t) = [\sigma(x, t)], \quad B([y], t) = [y], \quad \text{and} \quad B([a, s], t) = [a, (1 - s)t + s],$$

for $a \in A$, $x \in X$ and $s, t \in I$. Then $\mathrm{id} \simeq_B \mu\lambda$. This completes the proof. □

Definition 6.2.7 Let

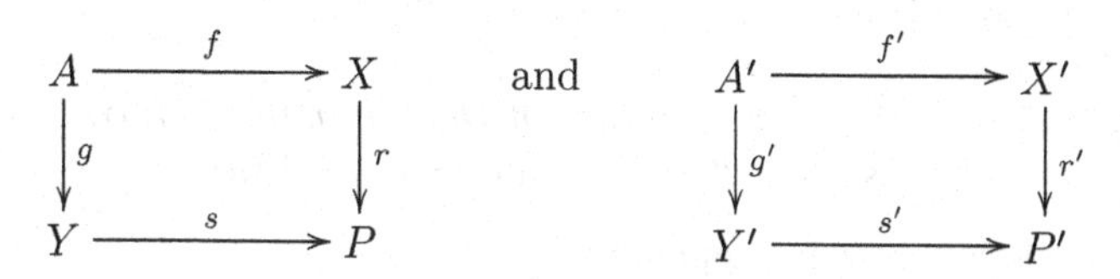

be two homotopy pushout squares and let $\alpha : A \to A'$, $\beta : X \to X'$ and $\gamma : Y \to Y'$ be maps such that the following diagram is homotopy-commutative

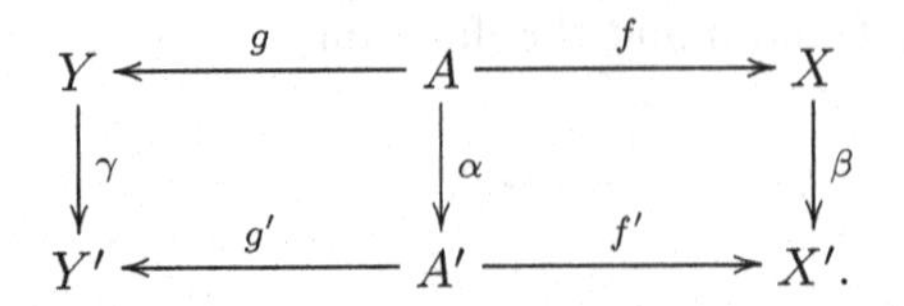

Let $F : A \times I \to X'$ and $G : A \times I \to Y'$ be homotopies such that $f'\alpha \simeq_F \beta f$ and $g'\alpha \simeq_G \gamma g$. We define a map $\Psi : P \to P'$ such that $\Psi r \simeq r'\beta$ and $\Psi s \simeq s'\gamma$,

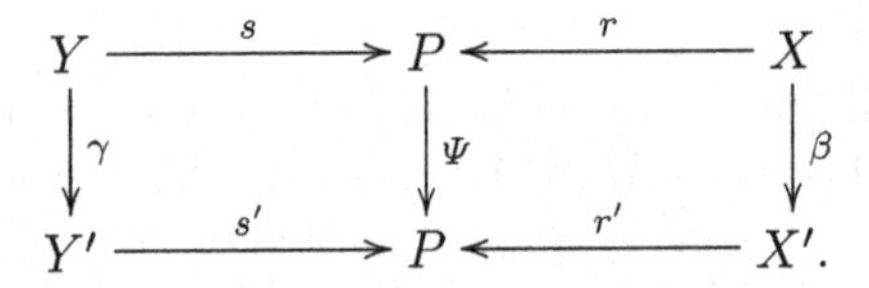

First note that by Definition 6.2.4, there are homotopy equivalences $\theta : O \to P$ and $\theta' : O' \to P'$, where O is the standard homotopy pushout of f and g and O' is the standard homotopy pushout of f' and g'. We first define $\Lambda = \Lambda_{F,G} : O \to O'$ such that $\Lambda u = u'\beta$ and $\Lambda v = v'\gamma$, where

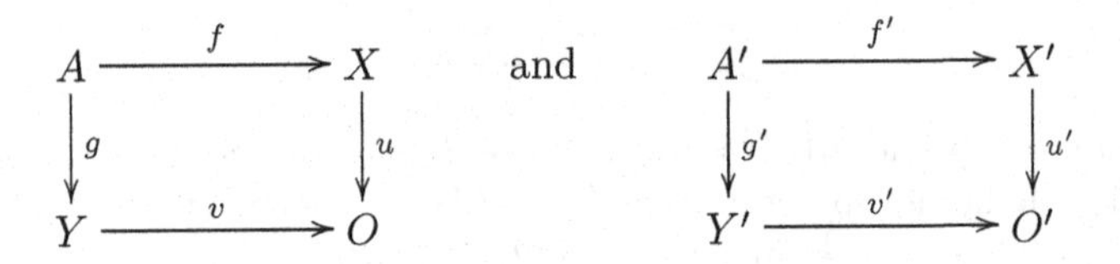

are standard homotopy pushout squares. Then $\Lambda : O \to O'$ is given by

$$\Lambda\langle x, *, *\rangle = \langle \beta(x), *, *\rangle$$
$$\Lambda\langle *, *, y\rangle = \langle *, *, \gamma(y)\rangle$$

and

$$\Lambda\langle *, (a,t), *\rangle = \begin{cases} \langle F(a, 1-3t), *, *\rangle & \text{if } 0 \leqslant t \leqslant \frac{1}{3} \\ \langle *, (\alpha(a), 3t-1), *\rangle & \text{if } \frac{1}{3} \leqslant t \leqslant \frac{2}{3} \\ \langle *, *, G(a, 3t-2)\rangle & \text{if } \frac{2}{3} \leqslant t \leqslant 1, \end{cases}$$

for $x \in X$, $y \in Y$, $a \in A$, and $t \in I$. See Figure 6.2. The map $\Psi : P \to P'$ is then defined by

$$\Psi = \theta' \Lambda \theta^{-1}.$$

Clearly Λ and Ψ depends on the choice of homotopies F and G.

Theorem 6.2.8 *Assume the notation of Definition* 6.2.7. *If* α, β *and* γ *are homotopy equivalences, then* Ψ *is a homotopy equivalence.*

The proof is be given in Appendix E.

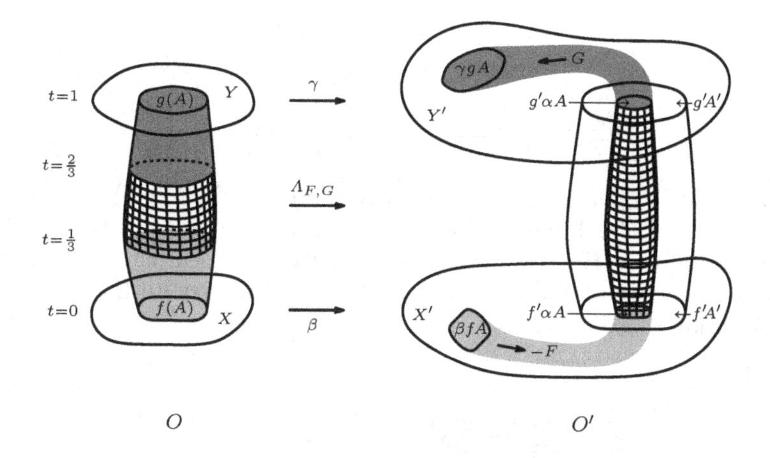

Figure 6.2

Corollary 6.2.9 *If*

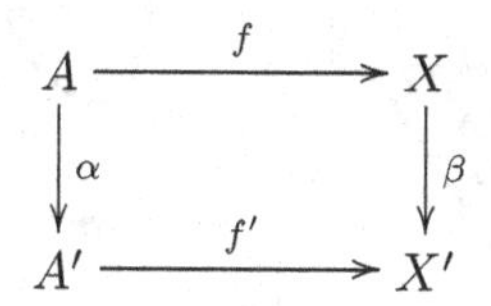

is a homotopy-commutative square with homotopy $F : A \times I \to X'$ *such that* $f'\alpha \simeq_F \beta f$, *then there is a map* $\Lambda : C_f \to C_{f'}$ *such that the following diagram commutes*

$$\begin{array}{ccc} X & \xrightarrow{\;l\;} & C_f \\ \downarrow{\scriptstyle \beta} & & \downarrow{\scriptstyle \Lambda} \\ X' & \xrightarrow{\;l'\;} & C_{f'}, \end{array}$$

where l *and* l' *are inclusions. If* α *and* β *are homotopy equivalences, then* Λ *is a homotopy equivalence.*

Proof. Represent C_f and $C_{f'}$ as standard homotopy pushouts (Remark 6.2.3) and let $S : A \times I \to \{*\}$ be the constant homotopy. Then $\Lambda = \Lambda_{F,S}$. The last assertion of the corollary follows from Theorem 6.2.8. □

By Exercise 6.1, Λ is homotopic to the map $\Phi_{\bar{F}}$ defined in Proposition 3.2.13 where $\bar{F}(a,t) = F(a, 1-t)$. There is a particular homotopy pushout that has interesting properties and which we consider in subsequent sections.

Definition 6.2.10 The standard homotopy pushout of the two projections $p_1 : X \times Y \to X$ and $p_2 : X \times Y \to Y$ is denoted by $X * Y$,

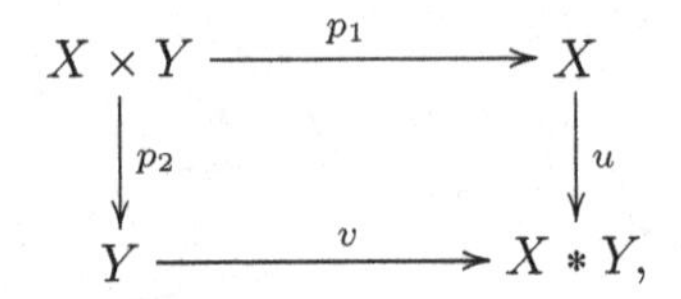

and is called the *join* of X and Y. Clearly $X * Y \cong (X \times Y \times I)/\sim$, where for all $x, x' \in X$, $y, y' \in Y$, and $t, t' \in I$, we have $(x, y, 0) \sim (x, y', 0)$, $(x, y, 1) \sim (x', y, 1)$, and $(*, *, t) \sim (*, *, t')$, with the latter representing the basepoint. Then $u(x) = \langle x, *, 0\rangle$ and $v(y) = \langle *, y, 1\rangle$. We usually regard the join as this identification space.

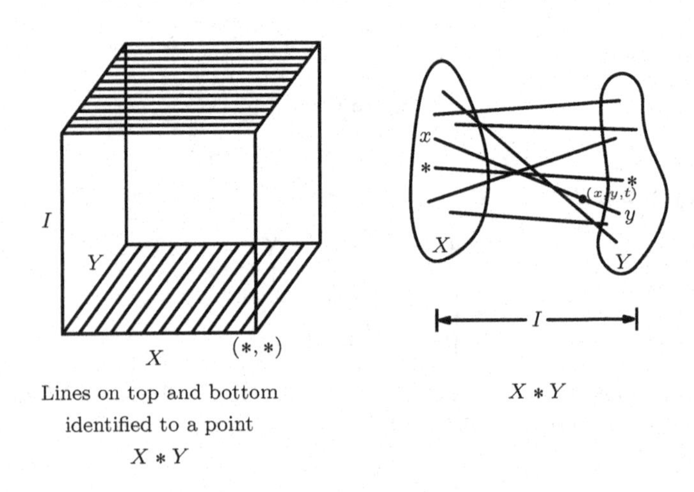

Figure 6.3

The following proposition gives some properties of the join.

Proposition 6.2.11 *Let X and Y be CW complexes. Then*

1. *The maps $u : X \to X * Y$ and $v : Y \to X * Y$ are nullhomotopic.*
2. *$X * Y \simeq \Sigma(X \wedge Y)$, where $X \wedge Y = X \times Y/X \vee Y$.*
3. *If X is $(p-1)$-connected and Y is $(q-1)$-connected, $p, q \geqslant 1$, then $X * Y$ is $(p+q)$-connected.*
4. *The projection $p_2 : X \times Y \to Y$ induces $\widetilde{p_2} : X \times Y/X \times \{*\} \to Y$ and $\widehat{p_2} : C_{j_1} \to Y$, where $j_1 : X \to X \times Y$ is the inclusion. Then $X * Y \simeq C_{\widetilde{p_2}} \simeq C_{\widehat{p_2}}$.*

Proof. (1) If we define $F : X \times I \to X * Y$ by $F(x, t) = \langle x, *, t\rangle$, then $u \simeq_F *$. In a similar way, $v \simeq *$.

(2) The spaces $X * Y$ and $\Sigma(X \wedge Y)$ are both identification spaces of $X \times Y \times I$ and there is a map $q : X * Y \to \Sigma(X \wedge Y)$ given by $q\langle x, y, t\rangle =$

$\langle\langle x, y\rangle, t\rangle$. If $C_1 = \{\langle x, *, t\rangle \,|\, x \in X,\ t \in I\} \subseteq X * Y$ and $C_0 = \{\langle *, y, t\rangle \,|\, y \in Y,\ t \in I\} \subseteq X * Y$, then C_1 and C_0 are cones on X and Y, respectively, that is, $C_1 \cong C_1X$ and $C_0 \cong C_0Y$. Furthermore, $X * Y/C_1 \vee C_0 \cong \Sigma(X \wedge Y)$, and the result follows from Corollary 1.5.19 since $C_1 \vee C_0$ is contractible.

(3) By Corollary 2.4.10, we may assume that X and Y are CW complexes with $X^{p-1} = \{*\}$ and $Y^{q-1} = \{*\}$. Therefore any cell of $X \times Y$ of dimension $< p+q$ is in $X \vee Y$. Hence $X \wedge Y$ is $(p+q-1)$-connected and so $\Sigma(X \wedge Y)$ is $(p+q)$-connected.

(4) In forming mapping cones, we consider cones $C_0A = A \times I/A \times \{0\} \cup \{*\} \times I$, for any space A. We first show that $X * Y \simeq C_{\widetilde{p_2}}$. We can regard $C_{\widetilde{p_2}}$ as the quotient space $X \times Y \times I/\sim$, where

$$(x, y, 0) \sim *, \quad (x, y, 1) \sim (x', y, 1), \quad \text{and} \quad (x, *, t) \sim *,$$

for $x, x' \in X$, $y \in Y$, and $t \in I$. Thus $X * Y$ and $C_{\widetilde{p_2}}$ are each a quotient space of $X \times Y \times I$. Then $C_{\widetilde{p_2}}$ is obtained from $X * Y$ by identifying $\{\langle x, *, t\rangle \,|\, x \in X,\ t \in I\} \subseteq X * Y$ to the basepoint. Since this subspace is homeomorphic to the cone C_1X, it is contractible, and so it follows from Corollary 1.5.19 that $C_{\widetilde{p_2}} \simeq X * Y$. To show $C_{\widetilde{p_2}} \simeq C_{\widehat{p_2}}$, consider the cofibration $X \to X \times Y \to X \times Y/X \times \{*\}$. Then the excision map $\alpha : C_{j_1} \to X \times Y/X \times \{*\}$ is a homotopy equivalence by Proposition 3.5.4. Then the following diagram is commutative

$$\begin{array}{ccc} C_{j_1} & \xrightarrow{\ \alpha\ } & X \times Y/X \times \{*\} \\ \Big\downarrow{\scriptstyle \widehat{p_2}} & & \Big\downarrow{\scriptstyle \widetilde{p_2}} \\ Y & \xrightarrow{\ \mathrm{id}\ } & Y. \end{array}$$

By Corollary 6.2.9, $C_{\widetilde{p_2}} \simeq C_{\widehat{p_2}}$. □

We conclude the discussion of elementary properties of homotopy pushouts by showing that that mapping cones of the vertical (or horizontal) maps in a homotopy pushout square have the same homotopy type.

Proposition 6.2.12 *If*

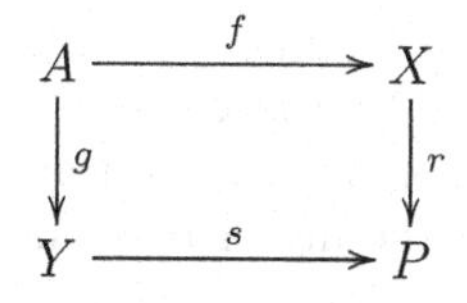

is a homotopy pushout square, then the mapping cone C_g has the same homotopy type as the mapping cone C_r.

Proof. We first establish the proposition when the given square is the following standard homotopy pushout square

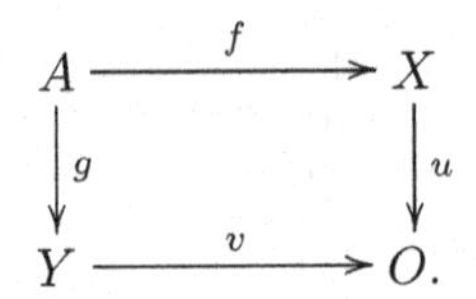

Using mapping cylinders, we replace X and Y by M_f and M_g, respectively, and consider the pushout square

$$\begin{array}{ccc} A & \xrightarrow{f'} & M_f \\ \downarrow{\scriptstyle g'} & & \downarrow{\scriptstyle \widetilde{u}} \\ M_g & \xrightarrow{\widetilde{v}} & O'. \end{array}$$

Since g' is a cofiber map, this is a homotopy pushout square (Proposition 6.2.6). Furthermore, $\widetilde{u}$ is a cofiber map and

$$C_g = M_g/g'(A) \simeq O'/\widetilde{u}(M_f) \simeq C_{\widetilde{u}},$$

by Propositions 3.2.10 and 3.5.4. But the commutative diagram

$$\begin{array}{ccccc} M_g & \xleftarrow{g'} & A & \xrightarrow{f'} & M_f \\ \downarrow{\scriptstyle g''} & & \downarrow{\scriptstyle \mathrm{id}} & & \downarrow{\scriptstyle f''} \\ Y & \xleftarrow{g} & A & \xrightarrow{f} & X \end{array}$$

induces a homotopy equivalence $\Psi : O' \to O$ such that $\Psi\widetilde{u} = uf''$ by Theorem 6.2.8. Then $C_{\widetilde{u}} \simeq C_u$ by Corollary 6.2.9. This proves the result for the standard homotopy pushout square. For an arbitrary homotopy pushout of f and g, as in the statement of Proposition 6.2.12, there is a homotopy equivalence $\theta : O \to P$ such that $\theta u \simeq r$. By Corollary 6.2.9, $C_u \simeq C_r$. This completes the proof. □

It is possible to state and prove results for homotopy pullbacks that are dual to those above given for homotopy pushouts. In particular, there are duals of Remarks 6.2.2 and 6.2.3, Propositions 6.2.5 and 6.2.6, Theorem 6.2.8, Corollary 6.2.9, and Proposition 6.2.12, and the proofs are dual. However, we just state the definitions and the major results.

Definition 6.2.13 Given spaces and maps

$$X \xrightarrow{f} A \xleftarrow{g} Y \qquad (*)$$

the *standard homotopy pullback* of $(*)$ consists of a space Q and maps $u : Q \to X$ and $v : Q \to Y$ defined as follows. (1) $Q \subseteq X \times A^I \times Y$ consists of all triples (x, ω, y) with $x \in X$, $y \in Y$, and $\omega \in A^I$ such that $\omega(0) = f(x)$

and $\omega(1) = g(y)$; (2) $u(x,\omega,y) = x$ and $v(x,\omega,y) = y$. The homotopy-commutative square

$$\begin{array}{ccc} Q & \xrightarrow{v} & Y \\ \downarrow{\scriptstyle u} & & \downarrow{\scriptstyle g} \\ X & \xrightarrow{f} & A \end{array}$$

is called the *standard homotopy pullback square.* Next let

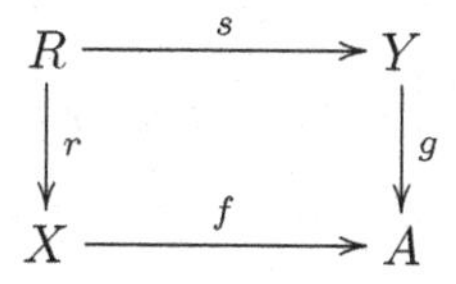

be a homotopy-commutative square. If there is a map $\lambda : R \to Q$ that is a homotopy equivalence such that $u\,\lambda \simeq r$ and $v\,\lambda \simeq s$, then the given square is called a *homotopy pullback square* and R together with r and s is called a *homotopy pullback.*

Clearly the homotopy fiber I_f of a map $f : X \to A$ can be represented as a standard homotopy pullback. Also, it is clear that if the square in the following diagram

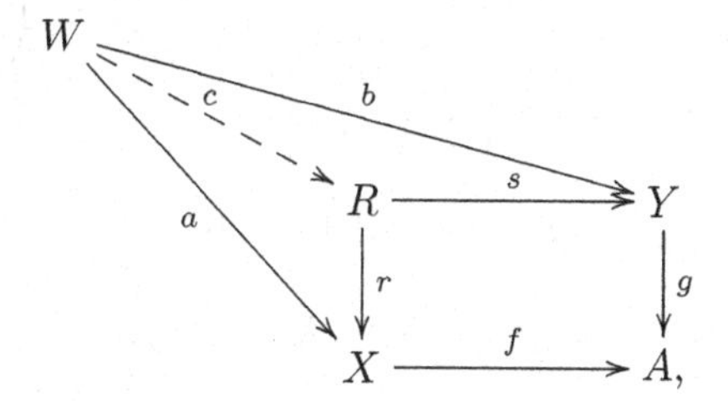

is a homotopy pullback square and $a : W \to X$ and $b : W \to Y$ are maps such that $fa \simeq gb$, then there is a map $c : W \to R$ such that $rc \simeq a$ and $sc \simeq b$.

We next state the dual of Proposition 6.2.6.

Proposition 6.2.14 *If*

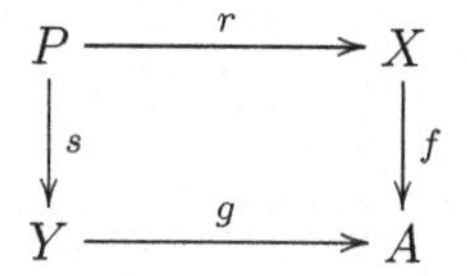

is a pullback square and if the map f or the map g is a fiber map, then this square is a homotopy pullback square.

We next indicate how to induce a natural map of homotopy pullbacks. Clearly it suffices to do this for standard homotopy pullbacks.

Definition 6.2.15 Given two standard homotopy pullback squares

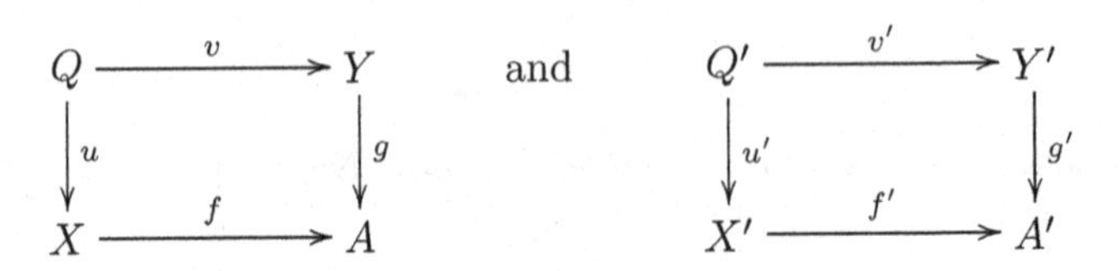

and maps $\alpha : A \to A'$, $\beta : X \to X'$, and $\gamma : Y \to Y'$ such that the following diagram is homotopy-commutative

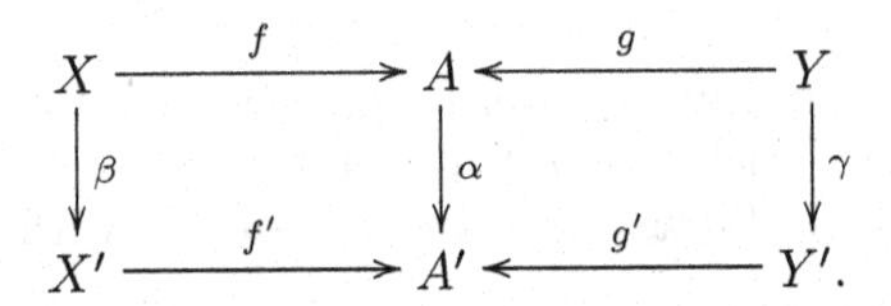

Let $F : X \times I \to A'$ and $G : Y \times I \to A'$ be homotopies such that $\alpha f \simeq_F f'\beta$ and $\alpha g \simeq_G g'\gamma$. We define a map $\Lambda = \Lambda_{F,G} : Q \to Q'$ such that $u'\Lambda = \beta u$ and $v'\Lambda = \gamma v$, If $(x, \omega, y) \in Q$, first define $\lambda_{x,y}(\omega) \in A'^I$ by

$$\lambda_{x,y}(\omega)(t) = \begin{cases} F(x, 1-3t) \text{ if } 0 \leqslant t \leqslant \frac{1}{3} \\ \alpha\omega(3t-1) \text{ if } \frac{1}{3} \leqslant t \leqslant \frac{2}{3} \\ G(y, 3t-2) \text{ if } \frac{2}{3} \leqslant t \leqslant 1, \end{cases}$$

for $t \in I$. Then $\Lambda = \Lambda_{F,G} : Q \to Q'$ is defined by

$$\Lambda(x, \omega, y) = (\beta(x), \lambda_{x,y}(\omega), \gamma(y)).$$

Theorem 6.2.16 *Assume the notation of Definition* 6.2.15. *If* α, β, *and* γ *are homotopy equivalences, then* Λ *is a homotopy equivalence.*

See Appendix E. We conclude the section by stating the following dual of Proposition 6.2.12.

Proposition 6.2.17 *If the square*

$$\begin{array}{ccc} Q & \xrightarrow{v} & Y \\ \downarrow{\scriptstyle u} & & \downarrow{\scriptstyle g} \\ X & \xrightarrow{f} & A \end{array}$$

is a homotopy pullback square, then it follows that the homotopy fibers I_u *and* I_g *have the same homotopy type.*

6.3 Homotopy Pushouts and Pullbacks II

In this section we present further results on homotopy pushouts and homotopy pullbacks. Many of these results are due to Mather [63]. We begin by defining equivalence of homotopy-commutative squares.

Definition 6.3.1 Let

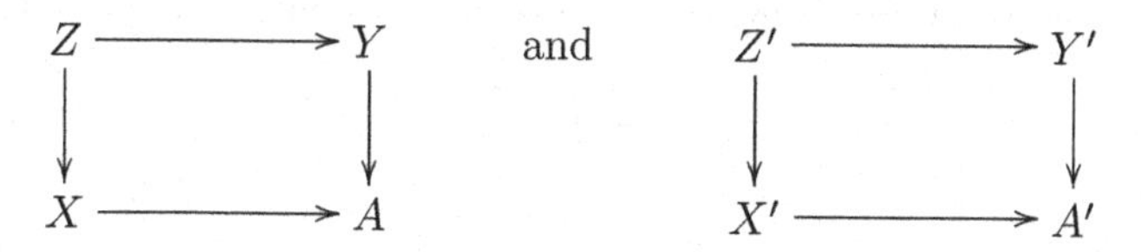

be two homotopy-commutative squares. If there exist homotopy equivalences $\alpha : A \to A'$, $\beta : X \to X'$, $\gamma : Y \to Y'$, and $\delta : Z \to Z'$ such that each of the square faces of the following diagram

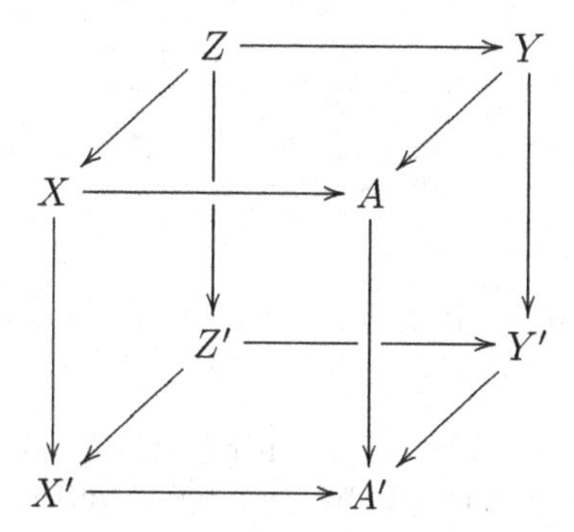

is homotopy-commutative, where the vertical arrows are β, δ, α, and γ, then we say that the two given squares are *equivalent.* This is clearly an equivalence relation.

Proposition 6.3.2 *If a homotopy-commutative square is equivalent to a homotopy pullback square, then it is a homotopy pullback square. If a homotopy-commutative square is equivalent to a homotopy pushout square, then it is a homotopy pushout square.*

Proof. We only prove the first assertion. A homotopy pullback square is equivalent to a standard homotopy pullback square, therefore we assume that the given homotopy-commutative square

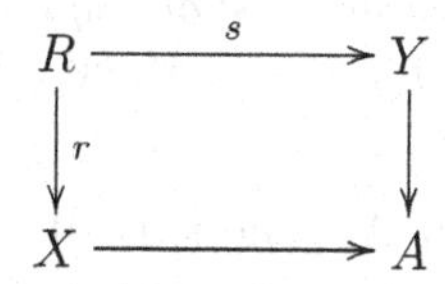

is equivalent to a standard homotopy pullback square

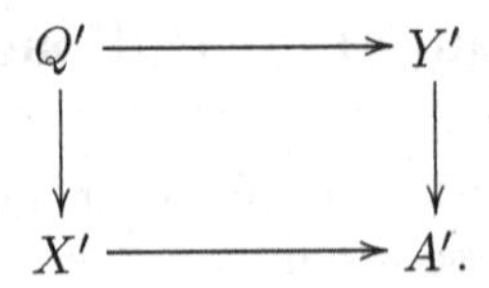

Thus there are homotopy equivalences $\alpha : A \to A'$, $\beta : X \to X'$, $\gamma : Y \to Y'$, and $\delta : R \to Q'$ that yield a cube diagram with each of the square faces homotopy-commutative. Now form the standard homotopy pullback

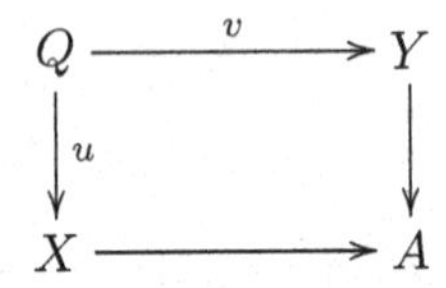

of $X \longrightarrow A \longleftarrow Y$. By Theorem 6.2.16, there is a homotopy equivalence $\Lambda : Q \to Q'$. If $\lambda = \Lambda^{-1}\delta : R \to Q$, then the diagram

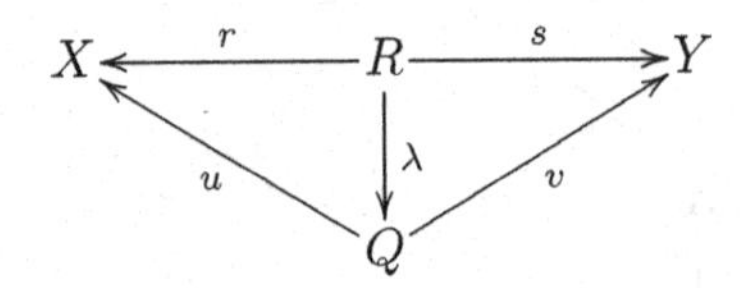

is homotopy-commutative. It follows from Definition 6.2.13 that the given square is a homotopy pullback square. □

We next state the prism theorem. In this result we sometimes refer to a homotopy-commutative rectangle by its labeled vertices.

Theorem 6.3.3 [63, pp. 231–234] *Given a homotopy-commutative diagram*

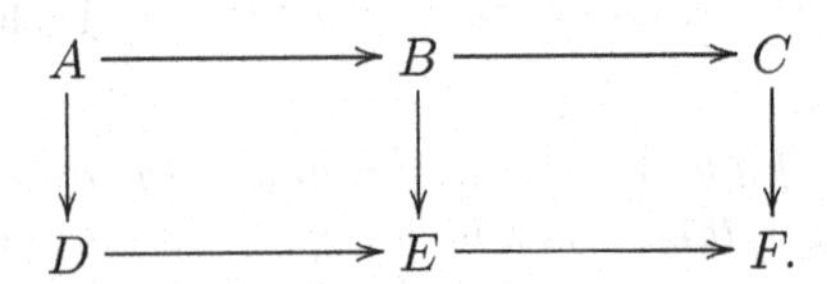

Then

1. *If the right square is a homotopy pullback square, then the left square is a homotopy pullback square if and only if the rectangle A-C-F-D is a homotopy pullback square.*
2. *If the left square is a homotopy pushout square, then the right square is a homotopy pushout square if and only if the rectangle A-C-F-D is a homotopy pushout square.*

This theorem is proved in Appendix E. It is called the prism theorem because the diagram in the statement of the theorem can be written as

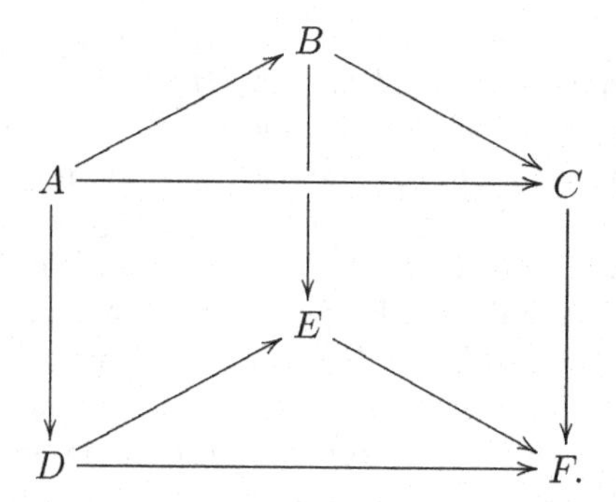

We next turn to several more specialized results on homotopy pushouts and pullbacks. We first introduce some necessary notation.

Let

$$X \xrightarrow{f} Y \xrightarrow{g} Z$$

be a sequence of spaces such that $gf = *$. We set $Q = C_f$, the mapping cone of f, and recall that the excision map $\alpha : Q \to Z$ is defined by $\alpha|Y = g$ and $\alpha|CX = *$ (Section 4.5). We let $H = I_\alpha$ be the homotopy fiber of α and let $P = I_g$ be the homotopy fiber of g. The excision map $\beta : X \to P$ is defined by $\beta(x) = (f(x), *)$, for $x \in X$ (Section 4.5), and we set $K = C_\beta$, the mapping cone of β. We have the following diagram with commutative triangles

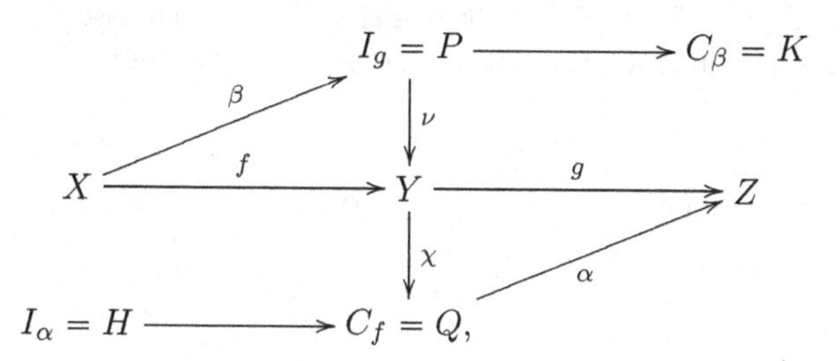

where ν is the projection and χ is the inclusion.

We next indicate our approach to the Blakers–Massey theorem in Section 6.4. If the three term sequence above is a cofiber sequence with X r-connected, and Z s-connected, then we must show that the excision map $\beta : X \to P$ is $(r+s)$-connected for the Blakers-Massey theorem. Let C_{j_1} be the mapping cone of the injection $j_1 : X \to X \times \Omega Z$. We prove that there is a homotopy pushout diagram

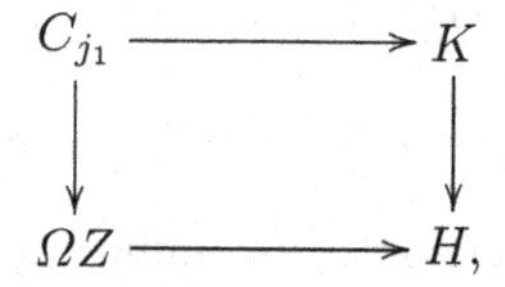

where the map $C_{j_1} \to \Omega Z$ is $\widehat{p}_2$, the map induced by the projection $p_2 : X \times \Omega Z \to \Omega Z$. We sketch the proof of how the Blakers–Massey theorem follows from this; details are given in the next section. We have seen that α is a homotopy equivalence and thus $\pi_i(H) = 0$, for all i. Therefore H is contractible, and so the mapping cone of $K \to H$ is ΣK. Because the two vertical maps have mapping cones of the same homotopy type and the mapping cone of $C_{j_1} \to \Omega Z$ has the homotopy type of the join $X * \Omega Z$ by Proposition 6.2.11, we conclude that $\Sigma K \simeq X * \Omega Z$. But $X * \Omega Z$ is $(r+s+1)$-connected by Proposition 6.2.11. From this it follows that $H_i(K) = 0$, for all $i \leqslant r+s$ and hence $\beta_* : H_i(X) \to H_i(P)$ is an isomorphism for $i < r+s$ and an epimorphism for $i = r+s$. This implies that β is an $(r+s)$-equivalence.

The Serre theorem (6.4.2) is similar to the Blakers–Massey theorem. It asserts that if the three term sequence of spaces above is a fiber sequence with X r-connected and Z s-connected, then the excision map $\alpha : Q \to Z$ is an $(r+s+2)$-equivalence. It is somewhat surprising that the same homotopy pushout square that was used for the Blakers–Massey theorem can be used to prove the Serre theorem. We sketch the proof. We have that $\beta : X \to P$ is a homotopy equivalence and so $H_i(K) = 0$ for all i. Consequently K is contractible and, from the homotopy pushout square, $H \simeq X * \Omega Z$. Therefore H is $(r+s+1)$-connected, and thus α is an $(r+s+2)$-equivalence.

We now return to the general case of a sequence of spaces

$$X \xrightarrow{f} Y \xrightarrow{g} Z$$

with $gf = *$. Our goal is to prove Proposition 6.3.5 which asserts the existence of the homotopy pushout previously mentioned. We define $\tau : X \times \Omega Z \to P = I_g$ by

$$\tau(x, \omega) = (f(x), \omega),$$

for $x \in X$ and $\omega \in \Omega Z$, and we define $\sigma : P \to H$ by

$$\sigma(y, \omega) = (\chi(y), \omega),$$

for $y \in Y$ and $\omega \in EZ$.

The following lemma is the main step in proving Proposition 6.3.5.

Lemma 6.3.4 *The square*

$$\begin{array}{ccc} X \times \Omega Z & \xrightarrow{\tau} & P \\ \big\downarrow{\scriptstyle p_2} & & \big\downarrow{\scriptstyle \sigma} \\ \Omega Z & \xrightarrow{i} & H, \end{array}$$

is a homotopy pushout square, where i is inclusion.

Proof. The given square is homotopy-commutative by a homotopy $F : X \times \Omega Z \times I \to H$ defined by

$$F(x, \omega, t) = (\langle x, t \rangle, \omega),$$

for $x \in X$, $\omega \in \Omega Z$, $t \in I$, and $\langle x, t\rangle \in C_f = Q$. Let O be the standard homotopy pushout of

$$\Omega Z \xleftarrow{p_2} X \times \Omega Z \xrightarrow{\tau} P.$$

Then i, σ and F determine a map $\theta_F : O \to H$ defined by

$$\begin{aligned} \theta_F\langle p, *, *\rangle &= \sigma(p) \\ \theta_F\langle *, ((x,\omega),t), *\rangle &= F(x,\omega,t) \\ \theta_F\langle *, *, \omega\rangle &= i(\omega), \end{aligned}$$

for all $p \in P$, $x \in X$, $\omega \in \Omega Z$, and $t \in I$. Then $\theta_F\, v = i$ and $\theta_F\, u = \sigma$,

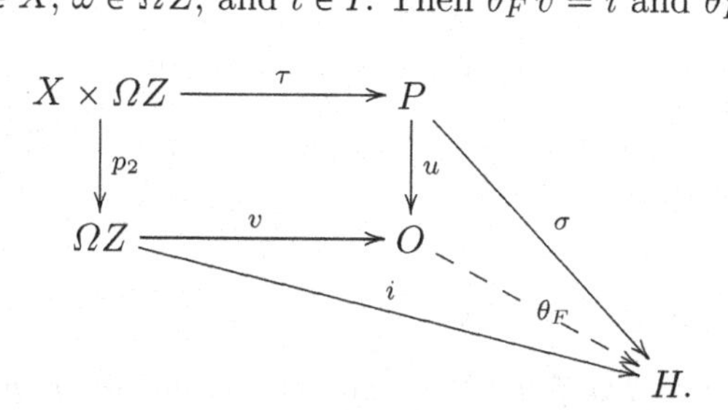

We prove the lemma by showing that θ_F is a homeomorphism. We regard O as the quotient space of $P \sqcup (X \times \Omega Z \times I)$ with equivalence relation given by (1) $(x,\omega,1) \sim (x',\omega,1)$, for $x, x' \in X$; (2) $(*,*,t) \sim (*,*)$, where $(*,*) \in P$; and (3) $(x,\omega,0) \sim \tau(x,\omega) = (f(x),\omega)$. On the other hand, H can be regarded as a quotient space of $(X \times I \times \Omega Z) \sqcup P$ with equivalence relation given by (1) $(x,1,\omega) \sim (x',1,\omega)$; (2) $(*,t,*) \sim (*,*)$; and (3) $(x,0,\omega) \sim (f(x),\omega)$. Then the function $\theta : P \sqcup (X \times \Omega Z \times I) \to (X \times I \times \Omega Z) \sqcup P$ which is the identity on P and carries $(x,\omega,t) \in X \times \Omega Z \times I$ to $(x,t,\omega) \in X \times I \times \Omega Z$ is a homeomorphism that induces the homeomorphism $\theta_F : O \to H$. □

There is another proof of Lemma 6.3.4 that we discuss in Appendix E. The proof depends on the cube theorem which is stated in Section 6.5. The proof of this theorem is long and difficult, and we do not give it. However, in Appendix E we use the cube theorem to derive Lemma 6.3.4.

We introduce some additional notation for the next result. The mapping cone of the injection $j_1 : X \to X \times \Omega Z$ is C_{j_1} and $p_2 : X \times \Omega Z \to \Omega Z$ induces $\widehat{p}_2 : C_{j_1} \to \Omega Z$. Let $w : X \times \Omega Z \to C_{j_1}$ and $\lambda : P = I_g \to K = C_\beta$ be inclusions. We denote the equivalence classes in C_{j_1}, K, and $Q = C_f$ by $\langle - \rangle$, $\langle - \rangle'$, and $\langle - \rangle''$, respectively. Since the diagram

$$\begin{array}{ccc} X & \xrightarrow{\text{id}} & X \\ \downarrow{\scriptstyle j_1} & & \downarrow{\scriptstyle \beta} \\ X \times \Omega Z & \xrightarrow{\tau} & P \end{array}$$

is commutative, there is a map $\tilde{\tau} : C_{j_1} \to K$ such that

$$\tilde{\tau}\langle x, \omega\rangle = \langle f(x), \omega\rangle'$$
$$\tilde{\tau}\langle x, t\rangle = \langle x, t\rangle',$$

for $x \in X$, $\omega \in \Omega Z$, and $t \in I$.

Also $\sigma : P \to H = I_\alpha$ induces $\tilde{\sigma} : K \to H$ such that

$$\tilde{\sigma}\langle p\rangle' = \sigma(p)$$
$$\tilde{\sigma}\langle x, t\rangle' = (\langle x, t\rangle'', *),$$

for $p \in P$, $x \in X$, and $t \in I$. Then we have the commutative diagram

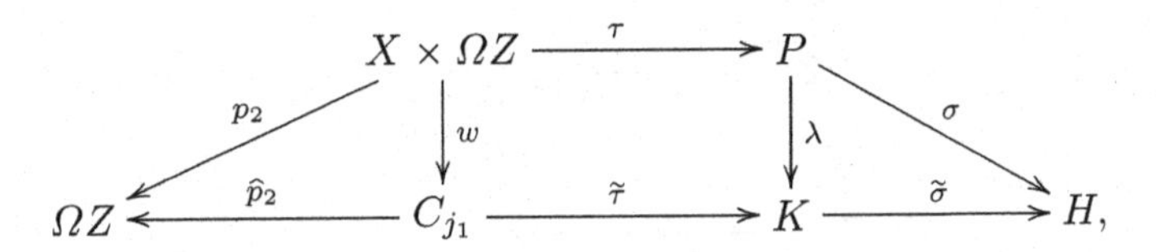

by the definitions of $\widehat{p}_2$, $\tilde{\tau}$, and $\tilde{\sigma}$.

Proposition 6.3.5 *The following diagram is a homotopy pushout square*

$$\begin{array}{ccc} C_{j_1} & \xrightarrow{\tilde{\tau}} & K \\ \downarrow{\scriptstyle \widehat{p}_2} & & \downarrow{\scriptstyle \tilde{\sigma}} \\ \Omega Z & \xrightarrow{i} & H. \end{array}$$

Proof. We first prove that the square is homotopy-commutative. Define $G : C_{j_1} \times I \to H$ by

$$G(\langle x, \omega\rangle, t) = (\langle x, t\rangle'', \omega)$$
$$G(\langle x, s\rangle, t) = (\langle x, (1-t)s + t\rangle'', *),$$

for $x \in X$, $\omega \in \Omega Z$, and $s, t \in I$. Then $\tilde{\sigma}\tilde{\tau} \simeq_G i\widehat{p}_2$. Next consider the diagram

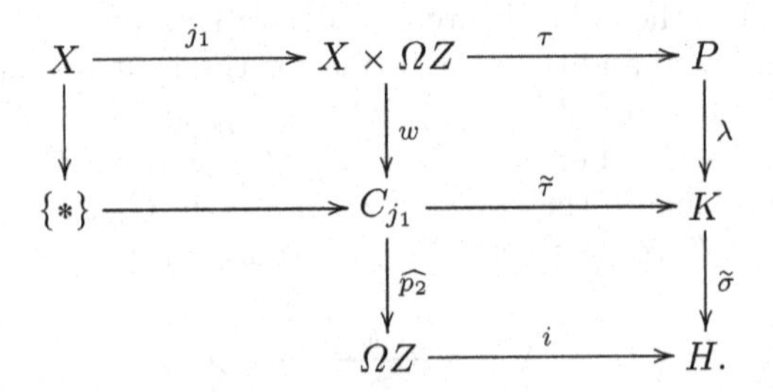

The top square on the left is a homotopy pushout square and, since $\tau j_1 = \beta$, it follows that $X - P - K - \{*\}$ is a homotopy pushout square. Therefore by the prism theorem, the top square on the right is a homotopy pushout

square. But by Proposition 6.3.4, $X \times \Omega Z - P - H - \Omega Z$ is a homotopy pushout square. Therefore by the prism theorem, the bottom square on the right is a homotopy pushout square. □

The following corollary of Proposition 6.3.5, which appears in [62], is used in the next section.

Corollary 6.3.6 *The mapping cone of $\widetilde{\sigma} : K \to H$ has the same homotopy type as the join $X * \Omega Z$.*

Proof. By the previous proposition and Proposition 6.2.12, the mapping cone of $\widetilde{\sigma}$ has the same homotopy type as the mapping cone of $\widehat{p}_2$. But by Proposition 6.2.11, the latter space has the homotopy type of $X * \Omega Z$. □

To illustrate one of the applications of Corollary 6.3.6, we show how to obtain a result of James [50]. The cofiber sequence

$$X \xrightarrow{i} CX \xrightarrow{q} \Sigma X$$

yields the homotopy-commutative diagram

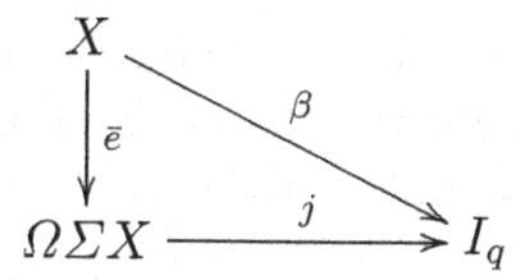

by Lemma 5.6.5, where $\bar{e} = -e$, β is an excision map, and j is inclusion. Since

$$\Omega\Sigma X \xrightarrow{j} I_q \longrightarrow CX$$

is a fibration with contractible base, j is a homotopy equivalence (Exercise 4.5). Therefore $C_\beta \simeq C_{\bar{e}} \simeq C_e$. Since $i : X \to CX$ is a cofiber map, the excision map $\alpha : C_i \to \Sigma X$ is a homotopy equivalence, and so $H = I_\alpha$ is contractible. It now follows from Corollary 6.3.6, that $\Sigma K \simeq X * \Omega\Sigma X$, where $K = C_\beta$. Therefore $\Sigma C_e \simeq X * \Omega\Sigma X \simeq \Sigma(X \wedge \Omega\Sigma X) \simeq X \wedge (\Sigma\Omega\Sigma X)$ by Proposition 6.2.11 and Exercise 5.7. Next we consider the cofiber sequence

$$\Omega\Sigma X \longrightarrow C_e \xrightarrow{r} \Sigma X\ ,$$

where r is the projection onto the cofiber. Let $\pi_A : \Sigma\Omega A \to A$ be defined by $\pi_A\langle\omega, t\rangle = \omega(t)$, for any A, where $\omega \in \Omega A$ and $t \in I$. Then it easily follows that $\pi_{\Sigma X}\, \Sigma e = \mathrm{id} : \Sigma X \to \Sigma X$. Thus Σe has a left homotopy inverse, and so it is a consequence of the exact sequence of Corollary 4.2.8 that $r \simeq * : C_e \to \Sigma X$. We now apply Corollary 5.4.7 to conclude that $\Sigma\Omega\Sigma X \simeq \Sigma X \vee \Sigma C_e$. Thus we have proved the following.

Proposition 6.3.7 *For any space X, $\Sigma\Omega\Sigma X \simeq \Sigma X \vee (X \wedge \Sigma\Omega\Sigma X)$.*

Remark 6.3.8 The previous proposition provides a recursive expression for $\Sigma\Omega\Sigma X$. By applying this formula inductively, it is possible to show that

$$\Sigma\Omega\Sigma X \simeq \bigvee_{i=1}^{\infty} \Sigma(X^{(i)}),$$

where $X^{(i)}$ is the smash product $X \wedge X \wedge \cdots \wedge X$ of i copies of X. This result is only part of the work of James in this area. In particular, James constructs a space, called the reduced product space, which is a homotopy type model for the space $\Omega\Sigma X$.

6.4 Theorems of Serre, Hurewicz, and Blakers–Massey

We emphasize that in this section all spaces have the homotopy type of path-connected CW complexes. We give the details of the proofs of the Serre and Blakers–Massey theorems which were sketched in the last section and draw some consequences from them. Our first results deal with fibrations and the Serre theorem.

We begin by proving a lemma that is needed for the proof of Serre's theorem 6.4.2. This lemma follows immediately from the Hurewicz theorem. But the proof that we will give is independent of it.

Lemma 6.4.1 *If K is a simply connected CW complex such that $H_i(K) = 0$ for all $i \geqslant 0$, then $\pi_i(K) = 0$ for $i \geqslant 0$.*

Proof. Suppose $\pi_i(K) = 0$ for $i < n$ and $\pi_n(K) \neq 0$ for some $n \geqslant 2$. Therefore there exists an abelian group G such that $\mathrm{Hom}(\pi_n(K), G) \neq 0$. Furthermore, there is a homomorphism $\eta_\pi : H^n(K;G) \to \mathrm{Hom}(\pi_n(K), G)$ defined by $\eta_\pi[f] = f_* : \pi_n(K) \to \pi_n(K(G,n))$ and, by Exercise 2.35, a homomorphism $\eta_H : H^n(K;G) \to \mathrm{Hom}(H_n(K), G)$ defined by $\eta_H[f] = f_* : H_n(K) \to H_n(K(G,n))$. If $h : \pi_n(K) \to H_n(K)$ is the Hurewicz homomorphism, then there is the following diagram

$$\begin{array}{ccc} H^n(K;G) & \xrightarrow[\cong]{\eta_\pi} & \mathrm{Hom}(\pi_n(K),G) \\ & \searrow^{\eta_H} & \uparrow h^* \\ & & \mathrm{Hom}(H_n(K),G), \end{array}$$

where h^* is induced by h. Since $G \cong \pi_n(K(G,n)) \cong H_n(K(G,n))$ (Exercise 2.34), the diagram is commutative. By Lemma 2.5.13, η_π is an isomorphism. But this contradicts the assumption that $H_n(K) = 0$. □

Using the notation of the latter part of Section 6.3, we next prove the theorem of Serre.

Theorem 6.4.2 *Let*

$$X \xrightarrow{f} Y \xrightarrow{g} Z$$

be a fiber sequence and let $C_f = Y \cup_f CX$ be the mapping cone of f. Let $\alpha : C_f \to Z$ be the excision map defined by $\alpha|CX = *$ and $\alpha|Y = g$. If X is r-connected and Z is s-connected, $r \geqslant 0$ and $s \geqslant 1$, then α is an $(r+s+2)$-equivalence.

Proof. Since g is a fiber map with fiber X, the excision map $\beta : X \to I_g = P$ defined by $\beta(x) = (f(x), *)$, for $x \in X$, is a homotopy equivalence (Proposition 3.5.10). Therefore from the exact homology sequence of

$$X \xrightarrow{\beta} P \longrightarrow C_\beta = K,$$

we conclude that $H_i(K) = 0$ for all $i \geqslant 0$. Now $\beta_* : \pi_1(X) \to \pi_1(P)$ is an isomorphism, and so it follows from Proposition B.4 that $\pi_1(K) = 0$. By Lemma 6.4.1, $\pi_i(K) = 0$ for all $i \geqslant 0$ and thus by Exercise 2.23, K is contractible. Hence the mapping cone of $\widetilde{\sigma} : K \to H = I_\alpha$ has the homotopy type of H (Proposition 1.5.18). By Corollary 6.3.6, $H \simeq X * \Omega Z$, and so H is $(r+s+1)$-connected by Proposition 6.2.11. The proposition now follows from the exact homotopy sequence of

$$I_\alpha = H \longrightarrow C_f \xrightarrow{\alpha} Z. \qquad \square$$

Remark 6.4.3 In the proof of Serre's theorem 6.4.2 it was shown that if

$$X \xrightarrow{f} Y \xrightarrow{g} Z$$

is a fiber sequence, then the map $\alpha : C_f \to Z$ which is an extension of g has homotopy fiber of the homotopy type of $X * \Omega Z$. This basic result is important in the study of Lusternik–Schnirelmann category by means of Ganea fibrations (see Section 8.3 and [21].)

We now prove a theorem of Serre which gives an exact cohomology sequence for a fibration.

Theorem 6.4.4 *Let*

$$F \xrightarrow{i} E \xrightarrow{p} B$$

be a fiber sequence with F r-connected and B s-connected, $r \geqslant 0$ and $s \geqslant 1$, and let G be an abelian group. Then there is a homomorphism $\Delta^n : H^n(F;G) \to H^{n+1}(B;G)$ for $0 \leqslant n \leqslant r+s$, called the connecting homomorphism such that the following sequence is exact

$$H^0(B;G) \xrightarrow{p^*} H^0(E;G) \xrightarrow{i^*} H^0(F;G) \xrightarrow{\Delta^0} H^1(B;G) \longrightarrow \cdots$$

$$\cdots \xrightarrow{\Delta^{N-1}} H^N(B;G) \xrightarrow{p^*} H^N(E;G) \xrightarrow{i^*} H^N(F;G),$$

where $N = r + s + 1$.

Proof. By Corollary 4.2.10, the sequence $F \xrightarrow{i} E \xrightarrow{\chi} C_i$ yields an exact cohomology sequence with coefficients in G. Since $\alpha : C_i \to B$ is an $(r + s + 2)$-equivalence by Serre's theorem, it follows from Lemma 2.5.12 that $\alpha^* : H^j(B;G) \to H^j(C_i;G)$ is an isomorphism for all $j \leqslant r + s + 1$. We then replace $H^j(C_i;G)$ by $H^j(B;G)$ in the exact cohomology sequence for all $j \leqslant r + s + 1$. The connecting homomorphism is the composition

$$H^n(F;G) \xrightarrow{\delta^n} H^{n+1}(C_i;G) \xrightarrow{\alpha^{*-1}} H^{n+1}(B;G),$$

where δ^n is the connecting homomorphism of Corollary 4.2.10. □

Remark 6.4.5 By using homology instead of cohomology in the proof of the previous theorem, we easily see that, under the hypothesis of Theorem 6.4.4, there is a homomorphism $\Delta_j : H_j(B;G) \to H_{j-1}(F;G)$ such that the following sequence is exact

$$H_N(F;G) \xrightarrow{i_*} H_N(E;G) \xrightarrow{p_*} H_N(B;G) \xrightarrow{\Delta_N} H_{N-1}(F;G) \longrightarrow \cdots$$

$$\cdots \xrightarrow{\Delta_1} H_0(F;G) \xrightarrow{i_*} H_0(E;G) \xrightarrow{p_*} H_0(B;G),$$

where $N = r + s + 1$.

Theorem 6.4.4 and Remark 6.4.5 are also consequences of deeper and more general results on the cohomology and homology of fiber spaces which are due to Leray and Serre. These results are obtained by associating a spectral sequence to a fibration that relates the cohomology of the base, fiber, and total space. For details, see [67].

We next prepare to prove the Hurewicz theorem as a consequence of Serre's theorem. Suppose

$$F \xrightarrow{i} E \xrightarrow{p} B$$

is a fiber sequence. We have seen in Proposition 4.5.18, that the map $p' : (E, F) \to (B, \{*\})$ of pairs obtained from p induces an isomorphism $p'_* : \pi_j(E, F) \to \pi_j(B)$ for all $j \geqslant 1$. We also have the excision map $\alpha : C_i = E \cup_i CF \to B$ defined by $\alpha|E = p$ and $\alpha|CF = *$ that induces a map of pairs $\alpha' : (C_i, CF) \to (B, *)$. If $k : (E, F) \to (C_i, CF)$ is the inclusion map of pairs, then we have the commutative diagram

$$\begin{array}{ccc} (E,F) & & \\ \downarrow{\scriptstyle k} & \searrow^{p'} & \\ (C_i, CF) & \xrightarrow{\alpha'} & (B,\{*\}) \end{array}.$$

This gives the following commutative diagram of groups

$$\begin{array}{ccccc} H_j(C_i) & \xrightarrow{l} & H_j(C_i, CF) & \xleftarrow{k_*} & H_j(E,F) \\ & \searrow^{\alpha_*} & \downarrow{\scriptstyle \alpha'_*} & \swarrow_{p'_*} & \\ & & H_j(B), & & \end{array}$$

where l is induced by the inclusion of pairs $(C_i, \{*\}) \to (C_i, CF)$.

Lemma 6.4.6 *If*

$$F \xrightarrow{i} E \xrightarrow{p} B$$

is a fiber sequence with F r-connected and B s-connected, $r \geqslant 0$, and $s \geqslant 1$, then

$$p'_* : H_j(E,F) \to H_j(B)$$

is an isomorphism for all $j \leqslant r+s+1$ and an epimorphism for $j = r+s+2$.

Proof. In the preceding commutative diagram l is an isomorphism because CF is contractible and k_* is an isomorphism by a standard homology excision argument. Also by Theorems 6.4.2 and 2.4.19, α_* is an isomorphism for $j \leqslant r+s+1$ and an epimorphism for $j = r+s+2$. The lemma then follows by the commutativity of the diagram. □

We now specialize to the path-space fibration of X:

$$\Omega X \xrightarrow{i} EX \xrightarrow{p} X.$$

We then have the diagram

$$\begin{array}{ccccc} \pi_j(X) & \xleftarrow{p'_*} & \pi_j(EX, \Omega X) & \xrightarrow{\partial_j} & \pi_{j-1}(\Omega X) \\ \downarrow{\scriptstyle h_j} & & \downarrow{\scriptstyle h''_j} & & \downarrow{\scriptstyle h'_{j-1}} \\ H_j(X) & \xleftarrow{p'_*} & H_j(EX, \Omega X) & \xrightarrow{\Delta_j} & H_{j-1}(\Omega X), \end{array}$$

where h_j, h'_{j-1}, and h''_j are Hurewicz homomorphisms and ∂_j and Δ_j are boundary homomorphisms. These boundary homorphisms are isomorphisms because EX is contractible. We define $\tau_{j-1} = p'_* \partial_j^{-1} : \pi_{j-1}(\Omega X) \to \pi_j(X)$ and the homology suspension $\sigma_{j-1} = p'_* \Delta_j^{-1} : H_{j-1}(\Omega X) \to H_j(X)$. We have considered τ_{j-1} and σ_{j-1} in Section 4.5 where it was shown that τ_{j-1} equals the adjoint isomorphism $\bar{\kappa}_*$ (Lemma 4.5.11) and that $h_j \tau_{j-1} = \sigma_{j-1} h'_{j-1}$,

$$\begin{array}{ccc} \pi_j(X) & \xleftarrow{\tau_{j-1}} & \pi_{j-1}(\Omega X) \\ \downarrow{h_j} & & \downarrow{h'_{j-1}} \\ H_j(X) & \xleftarrow{\sigma_{j-1}} & H_{j-1}(\Omega X), \end{array}$$

by Lemma 4.5.13.

Lemma 6.4.7 *For any n-connected space X, $n \geqslant 1$, σ_{j-1} is an isomorphism if $j \leqslant 2n$ and an epimorphism if $j = 2n + 1$.*

Proof. Because X is n-connected, ΩX is $(n-1)$-connected. Thus by Lemma 6.4.6,

$$p'_* : H_j(EX, \Omega X) \to H_j(X)$$

is an isomorphism for all $j \leqslant 2n$ and an epimorphism for $j = 2n+1$. Therefore σ_{j-1} is an isomorphism for all $j \leqslant 2n$ and an epimorphism for $j = 2n+1$. □

As a consequence we obtain the Hurewicz theorem.

Theorem 6.4.8 *If X is a 1-connected space and $H_q(X) = 0$ for all $q < n$, where $n \geqslant 2$, then $\pi_q(X) = 0$ for all $q < n$ and $h_n : \pi_n(X) \to H_n(X)$ is an isomorphism.*

Proof. We proceed by induction on n. If $n = 2$, we consider

$$\begin{array}{ccc} \pi_2(X) & \xleftarrow{\tau_1} & \pi_1(\Omega X) \\ \downarrow{h_2} & & \downarrow{h'_1} \\ H_2(X) & \xleftarrow{\sigma_1} & H_1(\Omega X). \end{array}$$

By Lemma 6.4.7, σ_1 is an isomorphism. By Proposition B.5, h'_1 is an isomorphism since $\pi_1(\Omega X)$ is abelian. Thus h_2 is an isomorphism.

For the inductive step, assume that the theorem holds for $n \geqslant 2$ and let X be a 1-connected space with $H_q(X) = 0$ for all $q < n+1$. Then $H_q(X) = 0$ for all $q < n$, and so $\pi_q(X) = 0$ for all $q < n$ and $\pi_n(X) \cong H_n(X) = 0$ by induction. Therefore $\pi_q(X) = 0$ for all $q < n+1$. Next consider the diagram

$$\begin{array}{ccc} \pi_{n+1}(X) & \xleftarrow{\tau_n} & \pi_n(\Omega X) \\ \downarrow{h_{n+1}} & & \downarrow{h'_n} \\ H_{n+1}(X) & \xleftarrow{\sigma_n} & H_n(\Omega X). \end{array}$$

By Lemma 6.4.7, σ_n is an isomorphism. Now ΩX is 1-connected and $H_q(\Omega X) = 0$ for $q < n$. We then apply the inductive assumption to ΩX and conclude that h'_n is an isomorphism. Hence h_{n+1} is an isomorphism. This completes the induction and the proof. □

Remark 6.4.9 There are several different proofs of the Hurewicz theorem. The one given here was based on the homology of a space and its loop space. Similar proofs have been given by Serre [82] and McCleary [67, p. 157]. Some of the other approaches to the Hurewicz theorem have been based on the Blakers–Massey theorem [66, p. 116], the homotopy addition theorem [14, p. 475-479], [83, p. 393-397], [91, p,178-180] and stable homotopy groups [8, p. 442], [87, p. 185].

Next we extend the notion of n-equivalence. It was defined in terms of homotopy groups and we now transfer it to homology groups.

Definition 6.4.10 A map $f : X \to Y$ is called a *homological n-equivalence*, $n \geqslant 1$, if $f_* : H_i(X) \to H_i(Y)$ is an isomorphism for $i < n$ and an epimorphism for $i = n$. For a space X, we define the *connectivity* of X, denoted $\operatorname{conn}(X)$, to be the largest nonnegative integer n (or ∞) such that $\pi_i(X) = 0$ for $i \leqslant n$. Similarly we define the *H-connectivity* of X, denoted $H_*\!-\operatorname{conn}(X)$, to be the largest nonnegative integer n (or ∞) such that $H_i(X) = 0$ for $i \leqslant n$.

The following lemma is an immediate consequence of the definition and the exact homotopy and homology sequence of a map (Corollary 4.2.17 and Remark 4.2.14).

Lemma 6.4.11 *Let $f : X \to Y$ be a map and let I_f and C_f be the homotopy fiber and mapping cone of f, respectively.*

1. *f is an n-equivalence* $\Longleftrightarrow \operatorname{conn}(I_f) \geqslant n - 1$
2. *f is a homological n-equivalence* $\Longleftrightarrow H_* - \operatorname{conn}(C_f) \geqslant n$.

Lemma 6.4.12 *If X is a space such that $\pi_1(X)$ is abelian, then* $\operatorname{conn}(X) = H_* - \operatorname{conn}(X)$.

Proof. If X is 1-connected, the result follows from the Hurewicz theorem. If X is not 1-connected, then $h_1 : \pi_1(X) \to H_1(X)$ is an isomorphism by the Hurewicz theorem (Appendix B). Thus $\operatorname{conn}(X) = 0 = H_* - \operatorname{conn}(X)$. □

Before relating the connectivity of a homotopy fiber to the connectivity of the mapping cone, we relate cohomological n-equivalences and homological n-equivalences. Recall that a map $f : X \to Y$ is called a cohomological n-equivalence if for every abelian group G, the induced homomorphism $f^* : H^i(Y; G) \to H^i(X; G)$ is an isomorphism for $i < n$ and a monomorphism for $i = n$ (Definition 4.5.22).

Lemma 6.4.13 *Let Y be a 1-connected space and let $f : X \to Y$ be a map. Then f is a homological n-equivalence if and only if f is a cohomological n-equivalence.*

Proof. Let C_f be the mapping cone of f. Then f is a homological n-equivalence if and only if $H_i(C_f) = 0$ for $i \leqslant n$, and f is a cohomological n-equivalence if and only if $H^i(C_f; G) = 0$ for $i \leqslant n$ and all groups G. Since

C_f is 1-connected by Proposition B.4, $H_i(C_f) = 0$ for $i \leqslant n$ if and only if $\pi_i(C_f) = 0$ for $i \leqslant n$, by the Hurewicz theorem 6.4.8. But the latter is equivalent to $H^i(C_f; G) = 0$ for $i \leqslant n$ and all groups G by Lemma 2.5.13. □

Now we compare the connectivity of a homotopy fiber with that of a mapping cone.

Proposition 6.4.14 *Let $f : X \to Y$ be a map of simply connected spaces, let I_f be the homotopy fiber of f, and let C_f be the mapping cone of f. Then* $\operatorname{conn}(I_f) + 1 = \operatorname{conn}(C_f)$.

Proof. Without loss of generality we assume that $I_f \xrightarrow{j} X \xrightarrow{f} Y$ is a fiber sequence (Remark 3.5.9). From the exact homotopy sequence of a fibration we have that $\pi_2(Y) \to \pi_1(I_f)$ is onto, and so $\pi_1(I_f)$ is abelian. By Lemma 6.4.12, $H_*- \operatorname{conn}(I_f) = \operatorname{conn}(I_f)$. Also C_f is simply connected, and so $H_*- \operatorname{conn}(C_f) = \operatorname{conn}(C_f)$. Thus it suffices to show that

$$H_*- \operatorname{conn}(I_f) + 1 = H_*- \operatorname{conn}(C_f).$$

Let $r = H_*- \operatorname{conn}(I_f)$ and consider the excision map $\alpha : C_j \to Y$. By Serre's theorem 6.4.2, α is an $(r + 3)$-equivalence. Therefore α is a cohomological $(r + 3)$-equivalence by Lemma 2.5.12. By Corollary 4.5.23, the excision map $\delta : \Sigma I_f \to C_f$ is a cohomological $(r+3)$-equivalence. Since C_f is 1-connected, δ is a homological $(r + 3)$-equivalence by Proposition 6.4.13. Thus

$$H_*- \operatorname{conn}(I_f) + 1 = r + 1 = H_*- \operatorname{conn}(\Sigma I_f) = H_*- \operatorname{conn}(C_f).$$ □

As a consequence we obtain Whitehead's second theorem which was alluded to in Section 2.4.

Theorem 6.4.15 [92] *Let X and Y be path-connected CW complexes, let $f : X \to Y$ be a map and let n be an integer with $n \geqslant 1$ or let $n = \infty$.*

1. *If f is an n-equivalence, then f is a homological n-equivalence.*
2. *Let X and Y be simply connected. If f is a homological n-equivalence, then f is an n-equivalence.*

Proof. We have already proved Part (1) as Proposition 2.4.19. We now prove (2) and give another proof of (1), both under the assumption that X and Y are simply connected. We must show that f is an n-equivalence if and only if f is a homological n-equivalence. By Lemma 6.4.11, it suffices to show that $\operatorname{conn}(I_f) \geqslant n - 1$ if and only if $H_*- \operatorname{conn}(C_f) \geqslant n$. But this follows at once from Proposition 6.4.14 because $H_*- \operatorname{conn}(C_f) = \operatorname{conn}(C_f)$. □

We have seen in Whitehead's first theorem (2.4.7) that a map of CW complexes which induces an isomorphism of all homotopy groups is a homotopy equivalence. Now we see from the two Whitehead theorems (2.4.7 and 6.4.15) that a map of simply connected CW complexes which induces an isomorphism of all homology groups is a homotopy equivalence. This result is often easier to apply than the former one since all the homology groups of many spaces are known, but all the homotopy groups are very rarely known.

We next discuss some of the limitations to extending the two Whitehead theorems. If X and Y are two CW complexes such that $\pi_n(X) \cong \pi_n(Y)$ for all n, it does not follow that X and Y have the same homotopy type. For example, S^2 and $S^3 \times K(\mathbb{Z}, 2)$ have isomorphic homotopy groups but different homotopy types. For an example with finite-dimensional CW complexes, consider $\mathbb{R}\mathrm{P}^m \times S^n$ and $S^m \times \mathbb{R}\mathrm{P}^n$ (Exercise 6.17). A similar situation holds for homology groups. If $X = S^2 \vee S^4$ and $Y = \mathbb{C}\mathrm{P}^2$, then $H_n(X) \cong H_n(Y)$ for all n, but X does not have the homotopy type of Y (Exercise 6.17). In these examples the homotopy groups or homology groups of the two spaces are isomorphic, but there is no map between them that induces the isomorphism as is required by Whitehead's theorems.

An immediate application of Whitehead's theorems is to Moore spaces.

Proposition 6.4.16 *If X and Y are Moore spaces of type (G, n), then $X \simeq Y$.*

Proof. Let $M(G, n)$ be the Moore space constructed in Section 2.5 and let X be any Moore space of type (G, n). By the Hurewicz theorem, $\pi_n(X) \cong G$. It follows from the universal coefficient theorem for homotopy (5.2.9) that there is a map $f : M(G, n) \to X$ such that $f_* : \pi_n(M(G, n)) \to \pi_n(X)$ is an isomorphism. By the Hurewicz theorem, f induces a homology isomorphism in all dimensions. By Whitehead's second theorem 6.4.15, f is a weak equivalence since $M(G, n)$ and X are simply connected. Then f is a homotopy equivalence according to Whitehead's first theorem 2.4.7. Thus $X \simeq M(G, n)$, and so $X \simeq Y$ if X and Y are any Moore spaces of type (G, n). □

We next restate and prove the Blakers–Massey theorem (5.6.4).

Theorem *Let*

$$X \xrightarrow{f} Y \xrightarrow{g} Z$$

be a cofiber sequence with X r-connected and Z s-connected, $r, s \geqslant 1$, let I_g be the homotopy fiber of g, and let $\beta : X \to I_g$ be the excision map defined by $\beta(x) = (f(x), *)$, for $x \in X$. Then β is an $(r + s)$-equivalence.

Proof. Because f is a cofiber map with cofiber Z, the excision map $\alpha : C_f \to Z$ is a homotopy equivalence (Proposition 3.5.4). Therefore it follows from the exact homotopy sequence of

$$I_\alpha \longrightarrow C_f \xrightarrow{\alpha} Z$$

that $\pi_n(I_\alpha) = 0$, for all $n \geqslant 1$, where I_α is the homotopy fiber of α. By Exercise 2.23, I_α is contractible. Thus if $\widetilde{\sigma} : K = C_\beta \to H = I_\alpha$ is the map defined after the proof of Lemma 6.3.4, then the mapping cone of $\widetilde{\sigma}$

is ΣC_β. By Corollary 6.3.6, $\Sigma C_\beta \simeq X * \Omega Z$. But X is r-connected and ΩZ is $(s-1)$-connected. Therefore Proposition 6.2.11 implies that ΣC_β is $(r+s+1)$-connected. By Corollary 2.4.10, $H_i(\Sigma C_\beta) = 0$ for all $i \leqslant r+s+1$, and so $H_* - \text{conn}(C_\beta) \geqslant r+s$. Thus β is a homological $(r+s)$-equivalence by Lemma 6.4.11. To apply Whitehead's second theorem to $\beta : X \to I_g$ and complete the proof, we must show that X and I_g are simply connected. This is so for X by hypothesis. For I_g, we first note that $0 = \pi_1(Z) = \pi_1(C_f)$ since $Z \simeq C_f$. But the latter group is isomorphic to $\pi_1(Y)$ (Proposition B.4). Thus Y is simply connected. To show I_g is simply connected, consider the exact homotopy sequence of the fibration

$$\Omega Z \xrightarrow{l} I_g \longrightarrow Y.$$

Since $\pi_1(Y) = 0$, it follows that $l_* : \pi_1(\Omega Z) \to \pi_1(I_g)$ is onto. However, $\pi_1(\Omega Z)$ is abelian implies that $\pi_1(I_g)$ is abelian. Furthermore, by the Hurewicz theorem B.5 and the fact that β is a homological $(r+s)$-equivalence,

$$\pi_1(I_g) \cong H_1(I_g) \cong H_1(X) = 0.$$

Hence X and I_g are simply connected, and we conclude that β is an $(r+s)$-equivalence. □

Corollary 6.4.17 *Let (X, A) be a relative CW complex, let $i : A \to X$ be the inclusion and let $q : X \to X/A$ be the projection. If A is r-connected and $\pi_j(X, A) = 0$ for $j \leqslant s$, where $r, s \geqslant 1$, then $q'_* : \pi_j(X, A) \to \pi_j(X/A)$ is an isomorphism for $j \leqslant r+s$ and an epimorphism for $j = r+s+1$.*

Proof. The sequence $A \xrightarrow{i} X \xrightarrow{q} X/A$ is a cofiber sequence. Because $\pi_j(X, A) = 0$ for $j \leqslant s$, the map $i : A \to X$ is an s-equivalence. Hence i is a homological s-equivalence (6.4.15), and so $H_j(X/A) = 0$ for $j \leqslant s$, by the exact homology sequence of a cofibration. But $\pi_1(A) \to \pi_1(X) \to \pi_1(X, A)$ is exact and $\pi_1(A) = 0 = \pi_1(X, A)$, and thus $\pi_1(X) = 0$. By Proposition B.4, $\pi_1(X \cup_i CA) = 0$. Then X/A is simply connected since $X \cup_i CA \simeq X/A$. Therefore X/A is s-connected by Lemma 6.4.12. From the Blakers–Massey theorem applied to $A \to X \to X/A$ we conclude that $\beta : A \to I_q$ is an $(r+s)$-equivalence. Therefore $q'_* : \pi_j(X, A) \to \pi_j(X/A)$ is an isomorphism for $j \leqslant r+s$ and an epimorphism for $j = r+s+1$ by Corollary 4.5.16 and Lemma 4.5.17. □

Remark 6.4.18

1. As noted in Remark 5.6.9, the Blakers–Massey theorem yields a truncated exact sequence for homotopy groups with coefficients.
2. The original proof of the Blakers–Massey theorem 5.6.4 was based on triad homotopy groups and the exact sequence of a triad (Exercise 4.22). It was proved in [10] that if $(X; A, B)$ is a triad such that (1) $X = A \cup B$ and $C = A \cap B$ and (2) $\pi_i(A, C) = 0$ for $i \leqslant m$ and $\pi_i(B, C) = 0$ for $i \leqslant n$,

then $\pi_i(X; A, B) = 0$ for $i \leqslant m + n$. The latter equality is equivalent to the assertion that the inclusion $(A, C) \to (X, B)$ induces isomorphisms of relative homotopy groups in dimensions $< m + n$ and an epimorphism in dimension $m + n$. Some restrictions on the triad are needed such as, for example, that X is a CW complex with subcomplexes A and B. More general versions of the Blakers–Massey theorem deal with the following situation [17]. Let

$$\begin{array}{ccc} A & \xrightarrow{f} & X \\ \downarrow{\scriptstyle g} & & \downarrow{\scriptstyle u} \\ Y & \xrightarrow{v} & P \end{array}$$

be a homotopy pushout square such that f is an m-equivalence and g an n-equivalence. Let Q be the homotopy pullback of u and v and let $\theta : A \to Q$ be a map determined by f and g. Then θ is an $(m + n)$-equivalence.

We next state and prove the relative Hurewicz theorem.

Theorem 6.4.19 *Let X be a space and A a 1-connected subspace such that $\pi_j(X, A) = 0$ for $j < n$, $n \geqslant 2$. Then $H_j(X, A) = 0$ for $j < n$ and $h_n : \pi_n(X, A) \to H_n(X, A)$ is an isomorphism.*

Proof. We consider the commutative diagram

$$\begin{array}{ccc} \pi_j(X, A) & \xrightarrow{q'_*} & \pi_j(X/A) \\ \downarrow{\scriptstyle h_j} & & \downarrow{\scriptstyle h'_j} \\ H_j(X, A) & \xrightarrow{q'_*} & H_j(X/A), \end{array}$$

where h_j and h'_j are Hurewicz homomorphisms and $q : X \to X/A$ is the projection. We first assume that (X, A) is a relative CW complex. By Corollary 6.4.17, $q'_* : \pi_j(X, A) \to \pi_j(X/A)$ is an isomorphism for $j \leqslant n$, and so $\pi_j(X/A) = 0$ for $j < n$ and $q'_* : \pi_n(X, A) \to \pi_n(X/A)$ is an isomorphism. By the Hurewicz theorem for X/A, we have that $H_j(X/A) = 0$ for $j < n$ and h'_n is an isomorphism (see Exercise 6.26). Furthermore, $q'_* : H_j(X, A) \to H_j(X/A)$ is an isomorphism for all j. Therefore from the commutative square above with $j = n$, we see that h_n is an isomorphism.

If (X, A) is now a pair of spaces (not necessarily a relative CW complex), then by Corollary 2.4.10(1), there exists a space K such that (K, A) is a relative CW complex and a weak equivalence $l : K \to X$ which is an extension of the inclusion $A \to X$. We apply the first part of the proof to (K, A). This gives the theorem for (K, A). But l determines a map of pairs $l' : (K, A) \to (X, A)$ that induces isomorphisms of all relative homotopy and homology groups. This completes the proof. □

The absolute and relative Hurewicz theorems assert that the nth Hurewicz homomorphism h_n is an isomorphism under certain hypotheses. We now show that under these hypotheses, h_{n+1} is an epimorphism.

Corollary 6.4.20

1. *If X is a 1-connected space and $H_q(X) = 0$ for all $q < n$, where $n \geqslant 2$, then $h_{n+1} : \pi_{n+1}(X) \to H_{n+1}(X)$ is an epimorphism.*
2. *If X is a space and A is a 1-connected subspace such that $\pi_j(X, A) = 0$ for $j < n$, $n \geqslant 2$, then $h_{n+1} : \pi_{n+1}(X, A) \to H_{n+1}(X, A)$ is an epimorphism.*

Proof. (1) Without loss of generality assume that X is a CW complex with $X^{n-1} = \{*\}$. As in Lemma 2.5.8, we attach $(n+2)$-cells to X to form a space $Y = W_{n+2}$ such that the inclusion induces isomorphisms $\pi_i(X) \cong \pi_i(Y)$, for $i \leqslant n$, and $\pi_{n+1}(Y) = 0$. Now we construct $K = K(\pi_n(X), n)$ by attaching cells of dimension $\geqslant n+3$ to Y. Thus $X \subseteq Y \subseteq K$. We consider part of the homology exact sequence of the pair (K, Y)

$$H_{n+1}(K) \longrightarrow H_{n+1}(Y) \longrightarrow H_{n+1}(K, Y).$$

Then $H_{n+1}(K, Y) = 0$ because every $(n+1)$-cell of K is in Y and $H_{n+1}(K) = 0$ by Exercise 2.32. Therefore $H_{n+1}(Y) = 0$. Now consider the commutative diagram

$$\begin{array}{ccccc}
\pi_{n+2}(Y, X) & \xrightarrow{\partial_{n+2}} & \pi_{n+1}(X) & & \\
\downarrow{\scriptstyle h_{n+2}} & & \downarrow{\scriptstyle h_{n+1}} & & \\
H_{n+2}(Y, X) & \xrightarrow{\Delta_{n+2}} & H_{n+1}(X) & \longrightarrow & H_{n+1}(Y),
\end{array}$$

where the bottom line is part of the homology exact sequence of the pair (Y, X). Since $H_{n+1}(Y) = 0$, we have that Δ_{n+2} is onto. But $H_i(Y, X) = 0$ for $i \leqslant n+1$, and so h_{n+2} is an isomorphism by the relative Hurewicz theorem 6.4.19. Hence h_{n+1} is an epimorphism.

(2) The proof is a consequence of Part (1) by the argument used in the proof of the relative Hurewicz theorem. □

The absolute Hurewicz theorem (6.4.8) for X was proved under the hypothesis that X is simply connected. In the relative Hurewicz theorem (6.4.19) for (X, A) it was assumed that A is simply connected. We now discuss how to weaken these hypotheses. In the absolute case there is an operation of $\pi_1(X)$ on $\pi_n(X)$ (Definition 5.5.3). Let H be the normal subgroup of $\pi_n(X)$ generated by all elements of the form $u \cdot \alpha - u$, for all $\alpha \in \pi_1(X)$ and $u \in \pi_n(X)$. (Of course, if $n > 1$, the word "normal" can be deleted.) We define $\pi'_n(X)$ to be $\pi_n(X)/H$. Let $\alpha = [a] \in \pi_1(X)$, let $u = [f] \in \pi_n(X)$ and let $u \cdot \alpha = [\bar{f}_a] \in \pi_n(X)$ as in Section 5.5. Then $f, \bar{f}_a : S^n \to X$ and $f \simeq_{\text{free}} \bar{f}_a : S^n \to X$ (Exercise 5.15). Therefore $f_* = \bar{f}_{a*} : H_n(S^n) \to H_n(X)$.

From this it follows that $h_n(u) = h_n(u\cdot\alpha)$, where $h_n : \pi_n(X) \to H_n(X)$ is the Hurewicz homomorphism. Hence h_n induces a homomorphism $h'_n : \pi'_n(X) \to H_n(X)$. In a similar manner, if (X, A) is a pair of spaces, then the operation of $\pi_1(A)$ on $\pi_n(X, A)$ determines a quotient group $\pi'_n(X, A)$. The relative Hurewicz homomorphism $h_n : \pi_n(X, A) \to H_n(X, A)$ yields a homomorphism $h'_n : \pi'_n(X, A) \to H_n(X, A)$. We now state without proof the generalization of the two Hurewicz theorems.

Theorem 6.4.21

1. *If X is a space and $\pi_q(X) = 0$ for all $q < n$, where $n \geqslant 1$, then $H_q(X) = 0$ for all $q < n$ and $h'_n : \pi'_n(X) \to H_n(X)$ is an isomorphism.*
2. *Let (X, A) be a pair of spaces such that $\pi_j(X, A) = 0$ for $j < n$, where $n \geqslant 2$. Then $H_j(X, A) = 0$ for $j < n$ and $h'_n : \pi'_n(X, A) \to H_n(X, A)$ is an isomorphism.*

Proofs can be found in [14, p. 478], [39, p. 371], and [83, p. 394].

We end the section by remarking on the relations between the results of this section.

Remark 6.4.22 The theorems of this section are closely interconnected and one can derive them in several different ways. We have begun with Corollary 6.3.6 and from that have obtained the Serre theorem and the Blakers–Massey theorem. The Serre theorem then gave the Hurewicz theorem and these two then yielded Whitehead's second theorem. Finally, the Hurewicz theorem and the Blakers–Massey theorem were used in the proof of the relative Hurewicz theorem. There are many other possibilities, some obvious (such as deriving Whitehead's second theorem from the relative Hurewicz theorem) and others not (such as deriving the Hurewicz theorem from the Blakers–Massey theorem [66, p. 116]). An interesting discussion of this appears in [80, pp. 67–68].

6.5 Eckmann–Hilton Duality II

In this section we present some special features of the duality, many of which have been discussed by Roitberg in [79].

(1) Most of the spaces that we study are CW complexes, often finite CW complexes. These spaces are defined by a process of successively attaching cells. More specifically, the n-skeleton of a CW complex is the mapping cone of a map from a wedge of $(n-1)$-spheres to the $(n-1)$-skeleton. Thus a finite CW complex is obtained by a finite number of these mapping cone constructions. The dual of this is a space obtained by a finite number of homotopy fibers of maps into a product of Eilenberg–Mac Lane spaces. We show in Chapter 7 that these spaces do appear (in the homotopy decomposition of a space), but they often play a secondary role. The point here is that a concept and its dual may not be of equal interest.

(2) Some definitions that we have considered are given in a form that are either not susceptible to dualization or whose dualization is obscure. One such notion is that of a fiber bundle. Since it is defined in terms of an open cover of a space, it is not clear how to dualize it. Another example is that of the Lusternik–Schnirelmann category of a space X. This was originally defined in terms of an open cover of a space and, as such, is not obviously dualizable. However, it has been possible to reformulate the definition in such a way that it can be dualized. See Section 8.3 for details.

(3) There are several pairs of results that are clearly dual to each other, but for which no dual proofs exist. For example, it can be shown that X is an H-space if and only if the canonical map $e : X \to \Omega\Sigma X$ (the adjoint of $\mathrm{id}_{\Sigma X}$) admits a left homotopy inverse. Dually, Y is a co-H-space if and only if the canonical map $\pi : \Sigma\Omega Y \to Y$ (the adjoint of $\mathrm{id}_{\Omega Y}$) admits a right homotopy inverse ([35]). The H-space result is proved either by using James's reduced product construction [50] or else by using quasi-fibrations [84, 85], neither of which have been dualized. Moreover, the co-H-space proof does not appear to be dualizable.

(4) The dual of a proven result is sometimes false. We give four examples of this. First, by Remark 6.3.8, the suspension of the loop space of a sphere is an infinite wedge of spheres. But the loop space of the suspension of an Eilenberg–Mac Lane space is not of the homotopy type of a product of Eilenberg–Mac Lane spaces. This is most easily seen by showing that the homology groups of the two spaces are different. However, as noted in Section 2.6, the duality between Moore spaces and Eilenberg–Mac Lane spaces is tenuous. Secondly, it can be shown that the set of homotopy classes of comultiplications on a space X is in one–one correspondence with the set of homotopy classes of homotopy sections of $\pi : \Sigma\Omega X \to X$. But the set of multiplications on a space Y is not in one–one correspondence with the set of homotopy classes of homotopy retractions of $e : Y \to \Omega\Sigma Y$ [35]. Thirdly, there is the Ganea conjecture that a co-H-space Y with all homology groups finitely generated has the homotopy type of a wedge $X \vee S$, where X is a 1-connected co-H-space and S is a wedge of circles [36]. The dual result for H-spaces is true. However, Iwase [48] has shown that the Ganea conjecture is false. Lastly, there is the cube theorem. We give a formal statement of it for later use.

Theorem 6.5.1 *Consider the cube diagram in which all faces are homotopy-commutative squares,*

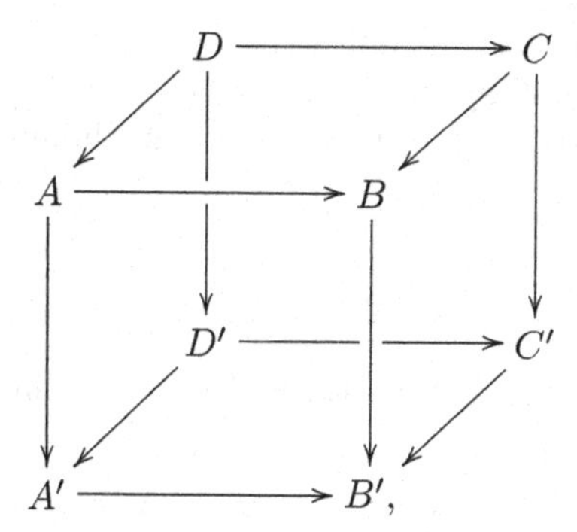

the four vertical faces are homotopy pullback squares and the bottom square is a homotopy pushout square. Then the top square is a homotopy pushout square.

This theorem has been proved in [63, Thm. 25]. It is shown in [24, p. 22] that the dual of this result is false.

(5) The dualization of a known result sometimes leads to a new proof of a well known theorem. An illustration of this is the dualization of the HELP lemma (4.5.7) given by May [66]. This coHELP lemma has led to a new proof of Whitehead's second theorem 6.4.15 which is dual to the proof of Whitehead's first theorem 2.4.7.

Exercises

Exercises marked with (∗) may be more difficult than the others. Exercises marked with (†) are used in the text.

6.1. (†) Given the homotopy-commutative square

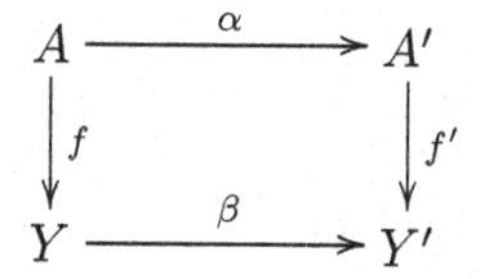

with homotopy F such that $\beta f \simeq_F f'\alpha$. By Proposition 3.2.13, there is an induced map $\Phi_F : C_f \to C_{f'}$. Now consider the diagram

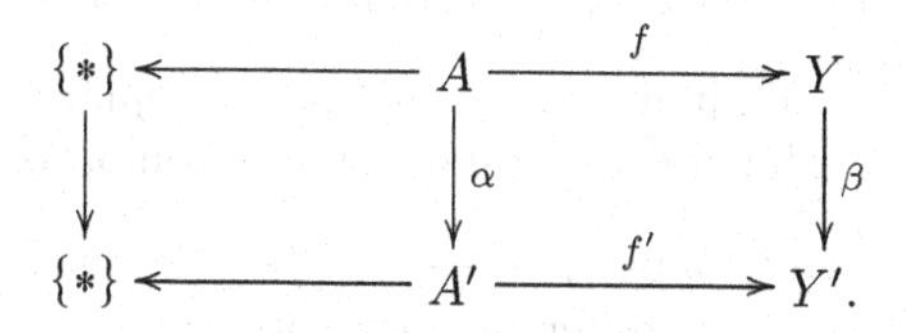

Let $S : A \times I \to \{*\}$ be the constant homotopy and let $\bar{F}$ defined by $\bar{F}(a,t) = F(a, 1-t)$ be the homotopy between $f'\alpha$ and βf. By Corollary 6.2.9, there is an induced map $\Lambda_{\bar{F},S} : C_f \to C_{f'}$ of standard homotopy pushouts. Prove that $\Phi_F \simeq \Lambda_{\bar{F},S}$.

6.2. Let

$$A \xrightarrow{i} X \xrightarrow{q} Q$$

be a cofiber sequence. We factor i through the mapping path

$$A \xrightarrow{i'} E \xrightarrow{i''} X$$

and consider the mapping cone $C_{i''}$. Show that there is a homotopy equivalence $Q \to C_{i''}$. State and prove the dual result.

6.3. Let

$$\begin{array}{ccc} R & \xrightarrow{s} & Y \\ \downarrow{\scriptstyle r} & & \downarrow{\scriptstyle g} \\ X & \xrightarrow{f} & A \end{array}$$

be a homotopy pullback square. Prove that g is a homotopy equivalence if and only if r is a homotopy equivalence. State and prove an analogous result for homotopy pushout squares.

6.4. $(*)$ Prove that

$$\begin{array}{ccc} \{*\} & \longrightarrow & Y \\ \downarrow & & \downarrow{\scriptstyle j_2} \\ X & \xrightarrow{j_1} & X \times Y \end{array}$$

is a homotopy pullback square.

6.5. Show that the homotopy pullback of

$$X \xrightarrow{f} A \xleftarrow{g} Y$$

is homeomorphic to the pullback of

$$X \times Y \xrightarrow{f \times g} A \times A \xleftarrow{\pi} A^I,$$

where $\pi(l) = (l(0), l(1))$, for $l \in A^I$. State and prove the dual result.

6.6. Give an example of a homotopy-commutative square with not all spaces trivial which is both a homotopy pushout and a homotopy pullback square.

6.7. Give an example of maps $Y \longleftarrow A \longrightarrow X$ such that the pushout and homotopy pushout have different homotopy types.

6.8. The product of two homotopy pullback squares is a homotopy pullback square. Give a precise formulation of this statement and prove it.

6.9. (Cf. Exercise 3.8) Let $q_1 : X \vee Y \to X$ and $q_2 : X \vee Y \to Y$ be the projections. Show that the standard homotopy pushout of

$$Y \xleftarrow{q_2} X \vee Y \xrightarrow{q_1} X$$

is contractible. Prove that $C_{q_1} \simeq \Sigma Y$.

6.10. Let

$$\begin{array}{ccc} A & \xrightarrow{f} & X \\ {\scriptstyle g}\downarrow & & \downarrow \\ Y & \longrightarrow & P \end{array}$$

be a homotopy pushout square. If f or g is a cofiber map, then prove that this square is equivalent to a pushout square.

6.11. Let W be a locally compact space. Prove that if

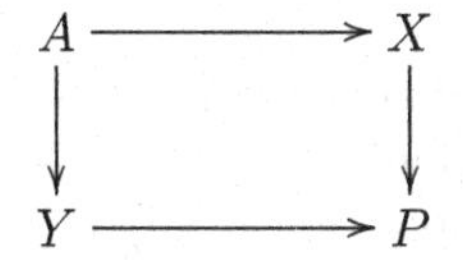

is a pushout square, then

$$\begin{array}{ccc} A \wedge W & \longrightarrow & X \wedge W \\ \downarrow & & \downarrow \\ Y \wedge W & \longrightarrow & P \wedge W \end{array}$$

is a pushout square. Prove this result for homotopy pushouts. What are the duals of these results?

6.12. Define $\pi : \Sigma\Omega X \to X$ by $\pi\langle \omega, t\rangle = \omega(t)$, for $\omega \in \Omega X$ and $t \in I$. Prove that the following diagram commutes

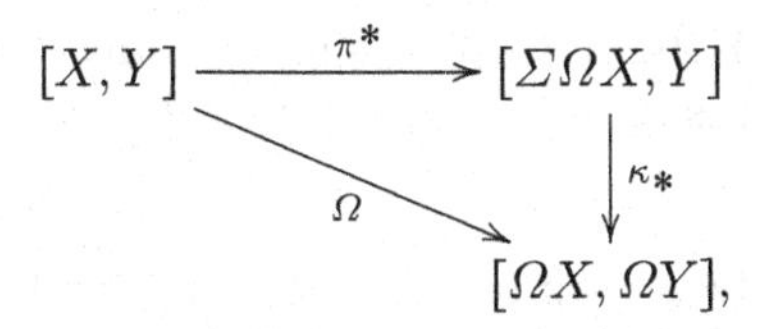

where $\Omega[\phi] = [\Omega\phi]$ and κ_* is the adjoint isomorphism.

6.13. $(*)$ (†) Consider the path-space fibration of X

$$\Omega X \xrightarrow{j} EX \xrightarrow{p} X.$$

Let C_j be the mapping cone of j with projection $q : C_j \to \Sigma\Omega X$. Let $\pi : \Sigma\Omega X \to X$ be defined by $\pi\langle\omega, t\rangle = \omega(t)$ and let $\bar{\pi} = -\pi$.

1. Prove that the following diagram is homotopy-commutative

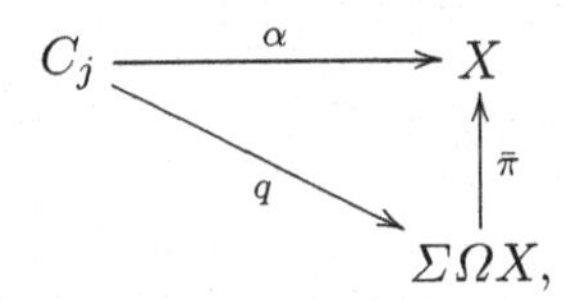

   where α is the excision map.
2. Prove that the homotopy fiber of $\pi : \Sigma\Omega X \to X$ has the homotopy type of $\Omega X * \Omega X$.
3. Let X be n-connected and let G be an abelian group. Then show that $\Omega : H^i(X;G) \to H^{i-1}(\Omega X;G)$ is an isomorphism for $i \leqslant 2n$ and a monomorphism for $i = 2n+1$.

6.14. Prove lemma E.6.

6.15. The homotopy class of the excision map $\alpha : C_f \to Z$ is dual to the homotopy class of the excision map $\beta : X \to I_g$. The homotopy class of the map $\widetilde{\sigma} : C_\beta \to I_\alpha$ (Proposition 6.3.5) is self-dual. Interpret these two statements and verify them.

6.16. Give an example of a space X such that $\text{conn}(X) \neq H_*-\text{conn}(X)$.

6.17. (*) (†) 1. Let $X = \mathbb{R}\mathrm{P}^m \times S^n$ and $Y = S^m \times \mathbb{R}\mathrm{P}^n$. Prove that there are values of m and n such that $\pi_i(X) \cong \pi_i(Y)$ for all i, but $H_i(X)$ and $H_i(Y)$ are not isomorphic for some i.

2. Let $X = S^2 \vee S^4$ and $Y = \mathbb{C}\mathrm{P}^2$. Prove that $H_n(X) \cong H_n(Y)$ for all n and that $\pi_n(X)$ and $\pi_n(Y)$ are not isomorphic for some n.

6.18. Let $f : K \to L$ be a map of simply connected, finite-dimensional CW complexes and let $N = \max(\dim K, \dim L)$. If $f_* : \pi_i(X) \to \pi_i(Y)$ is an isomorphism for all $i \leqslant N+1$, then prove that f is a homotopy equivalence.

6.19. Let $f : X \to Y$ be a map of simply connected CW complexes. If $f^* : H^n(Y) \to H^n(X)$ is an isomorphism for all $n \geqslant 0$, then prove that f is a homotopy equivalence.

6.20. Let $f : X \to Y$ be a map of CW complexes and let $\widetilde{f} : \widetilde{X} \to \widetilde{Y}$ be the induced map of universal covering spaces. Prove that if $f_* : \pi_1(X) \to \pi_1(Y)$ is an isomorphism and $\widetilde{f}_* : H_r(\widetilde{X}) \to H_r(\widetilde{Y})$ is an isomorphism for all $r \geqslant 2$, then f is a homotopy equivalence.

6.21. Let $A \xrightarrow{i} X \xrightarrow{q} Q$ be a cofiber sequence with A m-connected and Q n-connected, $m, n \geqslant 1$. If W is a CW complex with $\dim W \leqslant m+n$, then prove that the following sequence is exact

$$[W, A] \xrightarrow{i_*} [W, X] \xrightarrow{q_*} [W, Q].$$

6.22. 1. Let X and Y be 1-connected spaces and let $f : X \to Y$ be an n-equivalence. Use the Blakers–Massey theorem to prove that $\pi_i(C_f) = 0$ for $i \leqslant n$ and $\pi_{n+1}(C_f) \cong \pi_n(I_f)$.

2. Let (K, L) be a relative CW complex with K and L 1-connected and $\pi_i(K, L) = 0$ for $i \leqslant n$. Prove that $\pi_i(K/L) = 0$ for $i \leqslant n$ and $q_* : \pi_{n+1}(K, L) \to \pi_{n+1}(K/L)$ is an isomorphism, where $q : K \to K/L$ is the projection.

6.23. (*) Prove that the general triad version of the Blakers–Massey theorem in Remark 6.4.18 implies the Blakers–Massey exact sequence.

6.24. Let K be a 1-connected CW complex and $\alpha \in \pi_n(K)$, $n \geqslant 2$, such that $h(\alpha) = 0$, where $h : \pi_n(K) \to H_n(K)$ is the Hurewicz homomorphism. Prove that α can be represented by

$$S^n \xrightarrow{g} K^{n-1} \xrightarrow{i} K,$$

for some map g, where i is the inclusion.

6.25. Let $i : A \to X$ be the inclusion of a subcomplex A into a CW complex X. If X is n-dimensional and $A \simeq S^n$, $n \geqslant 1$, prove that $i_* : \pi_n(A) \to \pi_n(X)$ is a monomorphism.

6.26. (†) Prove the following variations of the Hurewicz theorems.

1. If X is a 1-connected space and $\pi_q(X) = 0$ for all $q < n$, where $n \geqslant 2$, then $H_q(X) = 0$ for all $q < n$ and $h_n : \pi_n(X) \to H_n(X)$ is an isomorphism.
2. If X is a 1-connected space and A is a 1-connected subspace such that $H_j(X, A) = 0$ for $j < n$, $n \geqslant 2$, then $\pi_j(X, A) = 0$ for $j < n$ and $h_n : \pi_n(X, A) \to H_n(X, A)$ is an isomorphism.

Chapter 7
Homotopy and Homology Decompositions

7.1 Introduction

In this chapter we discuss methods of approximating a space by simpler spaces. We assign to a space X a sequence of spaces whose nth term is called the nth section of X. As n gets larger, the nth section of X becomes a better approximation to X. There are two basic techniques for this, one in terms of the homotopy groups of X, called the homotopy decomposition, and the other in terms of the homology groups of X, called the homology decomposition. These are described in Sections 7.2 and 7.3, respectively. In the homotopy decomposition of X, the nth homotopy section $X^{(n)}$ has homotopy groups $\pi_i(X^{(n)}) \cong \pi_i(X)$ for $i \leqslant n$ and $\pi_i(X^{(n)}) = 0$ for $i > n$. Thus $X^{(n)}$ carries the homotopy groups of X in dimensions $\leqslant n$ and has trivial homotopy groups in dimensions $> n$. Furthermore, with some assumptions on X, the passage from the nth section to the $(n+1)$st section is carried out by a principal fibration $K(\pi_{n+1}(X), n+1) \to X^{(n+1)} \to X^{(n)}$. Therefore there is a map $k^{n+1} : X^{(n)} \to K(\pi_{n+1}(X), n+2)$ whose homotopy fiber is $X^{(n+1)}$. The nth homotopy sections and the homotopy classes of the k^{n+1} are the essential data of the decomposition. After proving some properties of this decomposition, we apply it to obtain some results about H-spaces and H-maps. The homology decomposition is dual to the homotopy decomposition. The nth homology section X_n has trivial homology in dimensions $> n$ and has homology isomorphic to that of X in dimensions $\leqslant n$. A principal cofibration $X_n \to X_{n+1} \to M(H_{n+1}(X), n+1)$ relates the nth and the $(n+1)$st homology sections. The principal cofibration is determined by a map $k'_{n+1} : M(H_{n+1}(X), n) \to X_{n+1}$. The homotopy class of these maps together with the homology sections are the key ingredients of the decomposition. Both of these decompositions are important techniques in homotopy theory and are often used in inductive arguments. In the last section of this chapter we generalize these decompositions from spaces to maps and obtain decompositions of maps in terms of relative homotopy groups or relative homology groups.

7.2 Homotopy Decompositions of Spaces

We begin by considering the following question. If $F \xrightarrow{i} E \xrightarrow{p} B$ is a fiber sequence, when is there a space K such that the fibration is equivalent to the principal fibration induced by some map $k : B \to K$? Of course it is necessary that $F \simeq \Omega K$. To be precise about the equivalence, we introduce the next definition.

Definition 7.2.1 A fibration $\Omega K \xrightarrow{i} E \xrightarrow{p} B$ is *equivalent to a principal fibration* if there are maps $k : B \to K$ and $\lambda : E \to I_k$ such that the diagram

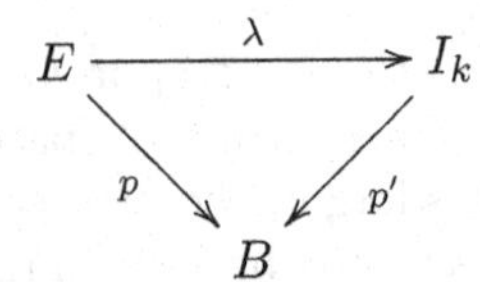

commutes and λ is a homotopy equivalence, where $p' : I_k \to B$ is the principal fiber map induced by k.

As an application of Serre's theorem 6.4.2, we prove a result that answers the question in a special case. This is used in Proposition 7.2.5.

Let G be an abelian group and let K_n be the Eilenberg–Mac Lane space $K(G, n)$. We define the *nth basic class* or *nth fundamental class* $b^n \in H^n(K_n; G) = [K_n, K_n]$ by $b^n = [\mathrm{id}_{K_n}]$ (Compare with Definition 5.3.1.)

Proposition 7.2.2 *Let* $F \xrightarrow{i} E \xrightarrow{p} B$ *be a fiber sequence with* $F = K_{m+1} = K(G, m+1)$, *let* B *be a 1-connected space, and let* $m \geqslant 2$. *Then there is a map* $k : B \to K_{m+2}$ *such that the given fiber sequence is equivalent to the principal fiber sequence*

$$F \xrightarrow{j} I_k \xrightarrow{q} B.$$

Proof. We apply Serre's theorem to the fibration $K_{m+1} \xrightarrow{i} E \xrightarrow{p} B$ and conclude that the excision map $\alpha : C_i \to B$ is an $(m+3)$-equivalence. By Corollary 4.5.23, the excision map $\delta : \Sigma K_{m+1} \to C_p$ induces an isomorphism

$$\delta^* : H^{m+2}(C_p; G) \to H^{m+2}(\Sigma K_{m+1}; G),$$

where C_p is the mapping cone of p

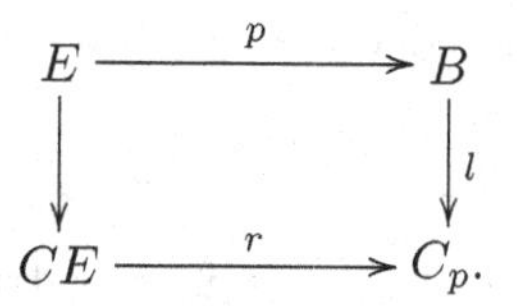

The adjoint isomorphism $H^{m+2}(\Sigma K_{m+1}; G) \xrightarrow{\kappa_*} H^{m+1}(K_{m+1}; G)$ is

$$\begin{aligned} H^{m+2}(\Sigma K_{m+1}; G) &= [\Sigma K_{m+1}, K_{m+2}] \\ &\cong [K_{m+1}, \Omega K_{m+2}] \\ &= H^{m+1}(K_{m+1}; G). \end{aligned}$$

Let $\flat^{m+1} \in H^{m+1}(K_{m+1}; G)$ be the basic class and let $\beta \in H^{m+2}(\Sigma K_{m+1}; G)$ be such that $\kappa_*(\beta) = \flat^{m+1}$. We define $[\theta] \in H^{m+2}(C_p; G)$ by $\delta^*[\theta] = \beta$ and let k be the composition

$$B \xrightarrow{l} C_p \xrightarrow{\theta} K_{m+2}.$$

We then compose the inclusion map $r : CE \to C_p = B \cup_p CE$ with $\theta : C_p \to K_{m+2}$ to obtain a map $v = \theta r : CE \to K_{m+2}$. The adjoint of v is a map $\widetilde{v} : E \to EK_{m+2}$. It follows that $\widetilde{v}$ and p determine a map λ from E into the pullback I_k,

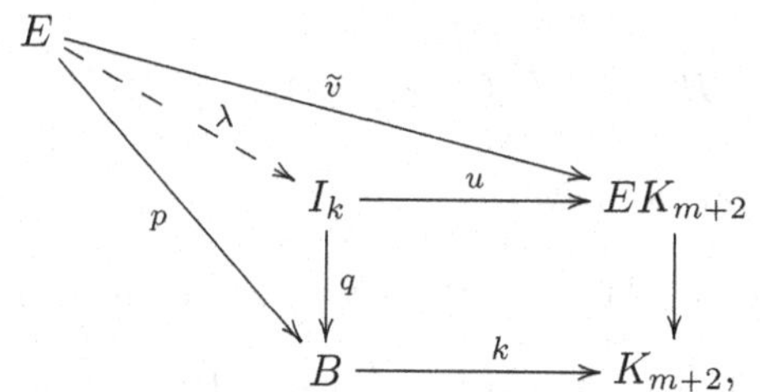

such that $u\lambda = \widetilde{v}$ and $q\lambda = p$. If $u', \lambda', \widetilde{v}' : K_{m+1} \to K_{m+1}$ are the maps of fibers induced by u, λ, and $\widetilde{v}$, respectively, then $u'\lambda' = \widetilde{v}'$. But u' is a homotopy equivalence by Proposition 3.3.12. Thus, to show λ' is a homotopy equivalence, it suffices to show that $\widetilde{v}'$ is a homotopy equivalence. But $\widetilde{v}' = \kappa(\theta\delta)$ since $\delta\langle x, t\rangle = \langle i(x), t\rangle$ for $x \in K_{m+1}$ and $t \in I$. Furthermore, $\kappa(\theta\delta) \simeq \mathrm{id}_{K_{m+1}}$. Thus $\widetilde{v}' : K_{m+1} \to K_{m+1}$ is a homotopy equivalence, and hence so is $\lambda' : K_{m+1} \to K_{m+1}$. It now follows from the exact homotopy sequence of a fibration and the five lemma that λ is a homotopy equivalence. This completes the proof. □

A generalization of Proposition 7.2.2 in which the 1-connectedness condition is weakened is given in Section 7.4. Another generalization in which B has greater connectivity and F has nontrivial homotopy groups in a range of dimensions appears in Exercise 7.1.

Next we characterize the cohomology class $[k]$ in Proposition 7.2.2. For the fibration

$$F = K(G, m+1) \xrightarrow{i} E \xrightarrow{p} B$$

that appears in Proposition 7.2.2, we have a truncated exact cohomology sequence with coefficients in G by Theorem 6.4.4. In particular, let

$$\Delta : H^{m+1}(F; G) \to H^{m+2}(B; G)$$

be the connecting homomorphism in this sequence.

Corollary 7.2.3 *Under the hypothesis of Proposition* 7.2.2, $\Delta(b^{m+1}) = -[k]$.

Proof. By Serre's theorem 6.4.4, Δ is the composition

$$H^{m+1}(F; G) \xrightarrow{\kappa_*^{-1}} H^{m+2}(\Sigma F; G) \xrightarrow{s^*} H^{m+2}(C_i; G) \xrightarrow{(\alpha^*)^{-1}} H^{m+2}(B; G),$$

where $s : C_i \to \Sigma F$ is the projection of the mapping cone and $\alpha : C_i \to B$ is the excision map. Since the following diagram is anticommutative

$$\begin{array}{ccc} H^{m+1}(F; G) & \xrightarrow{\Delta} & H^{m+2}(B; G) \\ \cong \uparrow \kappa_* & & \uparrow l^* \\ H^{m+2}(\Sigma F; G) & \xleftarrow[\cong]{\delta^*} & H^{m+2}(C_p; G), \end{array}$$

by Proposition 4.5.21, we have

$$\Delta(b^{m+1}) = -l^* \delta^{*-1}(\beta) = -l^*[\theta] = -[k].$$

□

Definition 7.2.4 Let X be a space and $n \geqslant 1$ an integer and consider the following conditions.

1. There exist spaces $X^{(n)}$ and maps $g_n : X \to X^{(n)}$ such that $g_{n*} : \pi_i(X) \to \pi_i(X^{(n)})$ is an isomorphism for $i \leqslant n$ and $\pi_i(X^{(n)}) = 0$ for $i > n$.
2. There exist maps $p_{n+1} : X^{(n+1)} \to X^{(n)}$ such that

$$K(\pi_{n+1}(X), n+1) \longrightarrow X^{(n+1)} \xrightarrow{p_{n+1}} X^{(n)}$$

is a fiber sequence.
3. The following diagram is commutative

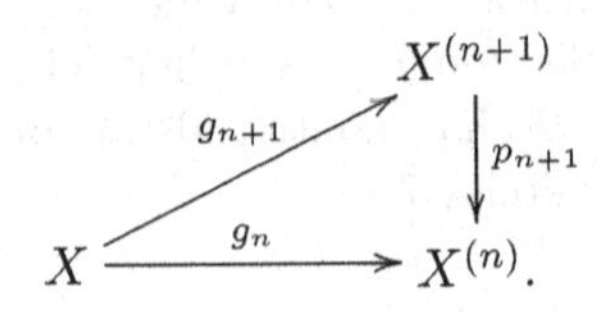

4. The fibration in (2) is equivalent to the principal fibration

$$K(\pi_{n+1}(X), n+1) \longrightarrow I_{k^{n+1}} \longrightarrow X^{(n)}$$

determined by a map k^{n+1}, where $[k^{n+1}] \in H^{n+2}(X^{(n)}; \pi_{n+1}(X))$.

A *weak homotopy decomposition* of X consists of the spaces and maps satisfying (1), (2), and (3). In addition, if (4) is satisfied, it is called a *Postnikov decomposition* or *homotopy decomposition* of X. The spaces $X^{(n)}$ are called *homotopy sections* or *Postnikov sections.* The elements $[k^{n+1}]$ or maps k^{n+1} are called *Postnikov invariants* or *k-invariants.*

The data that determine a weak homotopy decomposition of X are the homotopy sections $X^{(n)}$, the mappings $g_n : X \to X^{(n)}$, and the fibrations $p_n : X^{(n)} \to X^{(n-1)}$. We write this as $\{X^{(n)}, g_n, p_n\}$ and call it a *weak homotopy system.* The data that determines a homotopy or Postnikov decomposition of X are, in addition to the above, the Postnikov invariants k^n. We write this as $\{X^{(n)}, g_n, p_n, k^n\}$ and call it a *Postnikov system.* Finally, the sequence of fibrations

$$X^{(1)} \xleftarrow{p_2} X^{(2)} \xleftarrow{p_3} X^{(3)} \xleftarrow{p_4} \cdots$$

is called a *Postnikov tower.*

Theorem 7.2.5 *If X is a space, then X has a weak homotopy decomposition. If X is a 1-connected space, then X has a Postnikov decomposition.*

Proof. By Lemma 2.5.8, there exist spaces $W^{(n)}$ containing X such that the inclusion map $j_n : X \to W^{(n)}$ induces isomorphisms of homotopy groups in dimensions $\leqslant n$, the homotopy groups of $W^{(n)}$ are trivial in dimensions $> n$, and $W^{(n)}$ is obtained from X by attaching cells of dimensions $\geqslant n+2$. The spaces $W^{(n)}$ have property (1), and we are going to successively replace the $W^{(n)}$ by spaces $X^{(n)}$ of the same homotopy type that satisfy (2) and (3). By Proposition 2.4.13, j_n extends to a map $a_{n+1} : W^{(n+1)} \to W^{(n)}$ such that $a_{n+1}j_{n+1} = j_n$. We set $X^{(1)} = W^{(1)}$, define $X^{(2)}$ to be the mapping path E_{a_2}, and factor a_2 as

$$W^{(2)} \xrightarrow{a_2'} X^{(2)} \xrightarrow{p_2} X^{(1)} = W^{(1)},$$

where a_2' is a homotopy equivalence and p_2 is a fiber map. We obtain the homotopy-commutative diagram

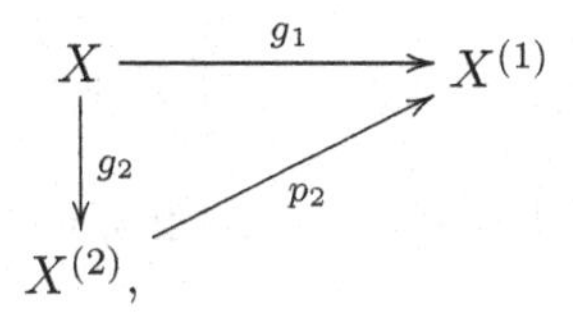

where $g_1 = j_1$ and $g_2 = a_2' j_2$. Because p_2 is a fiber map, we may assume that this diagram is commutative. We define $\bar{a}_3$ as the composition

$$W^{(3)} \xrightarrow{a_3} W^{(2)} \xrightarrow{a_2'} X^{(2)},$$

define $X^{(3)}$ to be the mapping path $E_{\bar{a}_3}$, and factor $\bar{a}_3$ as

$$W^{(3)} \xrightarrow{\bar{a}_3'} X^{(3)} \xrightarrow{p_3} X^{(2)},$$

where $\bar{a}_3'$ is a homotopy equivalence and p_3 is a fiber map. This gives the commutative diagram

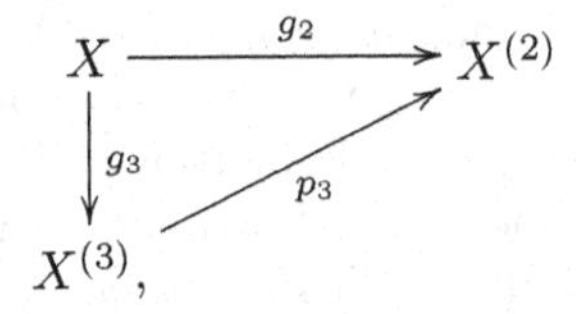

where $g_3 = \bar{a}_3' j_3$. We continue in this way. This establishes (1) and (3). The homotopy-commutative diagram of (3) and the exact homotopy sequence of a fibration imply that the fiber of p_{n+1} is a $K(\pi_{n+1}(X), n+1)$. This proves (2). Finally, (4) is an immediate consequence of Proposition 7.2.2. □

Remark 7.2.6 Let π_i denote $\pi_i(X)$.

1. By Corollary 2.4.10, $X^{(n)}$ has the homotopy type of a space obtained from X by attaching cells of dimension $\geqslant n+2$ because g_n is an $(n+1)$-equivalence.
2. We show in Remark 7.4.5 that the existence of a Postnikov decomposition holds under weaker conditions than 1-connectedness. In fact, it is not difficult to show the existence of a Postnikov decomposition under the assumption that $\pi_1(X)$ is abelian (Exercise 7.3). We also discuss in Section 7.4 the existence of a homotopy decomposition of a mapping of one space into another.
3. The Postnikov invariant $[k^{n+1}] \in H^{n+2}(X^{(n)}; \pi_{n+1})$ is sometimes denoted by $[k^{n+2}]$. However, the index $n+1$ seems more natural because $[k^{n+1}]$ determines the $n+1$ section.
4. Let $\{X^{(n)}, g_n, p_n, k^n\}$ be a Postnikov decomposition of X. If we denote by $\Delta : H^{n+1}(K(\pi_{n+1}, n+1); \pi_{n+1}) \to H^{n+2}(X^{(n)}; \pi_{n+1})$ the connecting homomorphism in the exact cohomology sequence of the fibration

$$K(\pi_{n+1}, n+1) \xrightarrow{i_{n+1}} X^{(n+1)} \xrightarrow{p_{n+1}} X^{(n)},$$

then $\Delta(b^{n+1}) = -[k^{n+1}]$ by Corollary 7.2.3, where b^{n+1} is the basic class.

5. The homotopy sections begin with $X^{(1)}$ and, if X is 1-connected, $X^{(1)} = \{*\}$. Also if $\pi_{n+1} = 0$, then $X^{(n+1)} = X^{(n)}$. Thus if X is k-connected, then $X^{(i)} = \{*\}$ for $i \leqslant k$ and $X^{(k+1)} = K(\pi_{k+1}, k+1)$. If $[k^{n+1}] = 0$, then $X^{(n+1)}$ has the homotopy type of $X^{(n)} \times K(\pi_{n+1}, n+1)$.

Before considering properties of homotopy decompositions, we give an interesting consequence of Proposition 7.2.5.

Let $\{X^{(n)}, g_n, p_n\}$ be a weak homotopy decomposition of X and let $X^{[n]}$ be the homotopy fiber of g_n. Then there is a fiber sequence

$$\Omega X^{(n)} \longrightarrow X^{[n]} \xrightarrow{v_n} X.$$

It follows from the exact homotopy sequence of this fibration that $\pi_i(X^{[n]}) = 0$ for $i \leqslant n$ and $v_{n*} : \pi_i(X^{[n]}) \to \pi_i(X)$ an isomorphism for $i > n$.

Definition 7.2.7 We call the fibrations above *n-connected fibrations* or *n-connective fibrations* of X. We say that $X^{[n]}$ is obtained by *killing the first n homotopy groups* of X.

Note that the fiber map $v_1 : X^{[1]} \to X$ is the homotopy analogue of the universal covering space map (see Exercise 7.5).

We next consider to what extent a weak homotopy decomposition of X determines the space X. We first digress to discuss the inverse limit construction. Given the following sequence (finite or infinite) of spaces and maps

$$X_l \xleftarrow{f_{l+1}} X_{l+1} \xleftarrow{f_{l+2}} X_{l+2} \longleftarrow \cdots .$$

We form the *inverse limit* $\varprojlim_{k \geqslant l} X_k$ which is the subspace of the product $\prod_{k \geqslant l} X_k$ of spaces consisting of all sequences $x = (x_l, x_{l+1}, x_{l+2}, \ldots)$ such that $x_i = f_{i+1}(x_{i+1})$ for $i = l, l+1, l+2, \ldots$. There are projection maps $q_n : \varprojlim_{k \geqslant l} X_k \to X_n$ defined by $q_n(x) = x_n$. An analogous definition can be given if the X_k are groups and the f_k are homomorphisms. Then $\varprojlim_{k \geqslant l} X_k$ is the subgroup of the product $\prod_{k \geqslant l} X_k$ of groups defined as above.

Now we return to a weak homotopy decomposition of a space X. There is a Postnikov tower

$$X^{(1)} \xleftarrow{p_2} X^{(2)} \xleftarrow{p_3} X^{(3)} \longleftarrow \cdots$$

and we form the inverse limit $\varprojlim_{k \geqslant 1} X^{(k)}$. The maps $g_n : X \to X^{(n)}$ determine a map $g : X \to \varprojlim_{k \geqslant 1} X^{(k)}$ defined by $g(x) = (g_1(x), g_2(x), g_3(x), \ldots)$ for $x \in X$. Our goal is to show that g is a weak homotopy equivalence. We begin with a lemma.

Lemma 7.2.8 *The projection maps* $q_n : \varprojlim_{k \geqslant 1} X^{(k)} \to X^{(n)}$ *induce homomorphisms* $q_{n*} : \pi_N(\varprojlim_{k \geqslant 1} X^{(k)}) \to \pi_N(X^{(n)})$ *and so there is a homomor-*

phism $\phi : \pi_N(\varprojlim_{k\geqslant 1} X^{(k)}) \to \varprojlim_{k\geqslant 1} \pi_N(X^{(k)})$, *for any* N. *Then* ϕ *is an isomorphism.*

Proof. We first show that ϕ is onto. Let $\beta = ([f_1], [f_2], [f_3], \ldots)$ be an element of $\varprojlim_{k\geqslant 1} \pi_N(X^{(k)})$, where $f_i : S^N \to X^{(i)}$ and $p_i f_i \simeq f_{i-1}$. We show inductively that if $i \geqslant 2$, there exists $f_i' : S^N \to X^{(i)}$ with $f_i' \simeq f_i$ and $p_i f_i' = f_{i-1}'$. Set $f_1' = f_1$ and assume that f_i' as above exists. Then $p_{i+1} f_{i+1} \simeq f_i \simeq f_i'$. By the covering homotopy property for the fiber map p_{i+1}, there exists $f_{i+1}' \simeq f_{i+1}$ with $p_{i+1} f_{i+1}' = f_i'$. Then the f_i' determine a map $f' : S^N \to \varprojlim_{k\geqslant 1} X^{(k)}$ such that $q_i f' = f_i'$. Therefore $\phi[f'] = ([f_1'], [f_2'], [f_3'], \ldots) = \beta$, and so ϕ is onto.

Now we show that ϕ is one–one. The inverse limit of spaces or groups is unaffected by deleting a finite number of terms at the (left) end of the sequence, thus it suffices to show that the homomorphism $\phi' : \pi_N(\varprojlim_{k\geqslant N} X^{(k)}) \to \varprojlim_{k\geqslant N} \pi_N(X^{(k)})$ obtained from ϕ is one–one. Let $[f] \in \pi_N(\varprojlim_{k\geqslant N} X^{(k)})$ be such that $\phi'[f] = 0$. We show inductively that for every $n \geqslant N$, there exists a homotopy $F_n : S^N \times I \to X^{(n)}$ such that $q_n f \simeq_{F_n} *$ and $p_n F_n = F_{n-1}$. Because $\phi'[f] = 0$, there exists $F_N : S^N \times I \to X^{(N)}$ such that $q_N f \simeq_{F_N} *$. Now assume that F_n exists with the above properties. Let $i : S^N \times \partial I \to S^N \times I$ be the inclusion map and define $a : S^N \times \partial I \to X^{(n+1)}$ by $a|S^N \times \{0\} = q_{n+1} f$ and $a|S^N \times \{1\} = *$. We then consider the diagram with commutative square

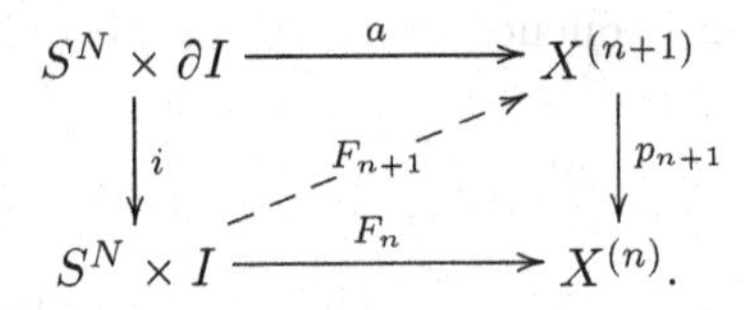

Because $\pi_N(K(G, n+1)) = 0$, it follows from Corollary 4.5.9 that there exists $F_{n+1} : S^N \times I \to X^{(n+1)}$ such that $p_{n+1} F_{n+1} = F_n$ and $q_{n+1} f \simeq_{F_{n+1}} *$. This completes the induction. Finally, the homotopies F_n determine a homotopy $F : S^N \times I \to \varprojlim_{k\geqslant N} X^{(k)}$ such that $f \simeq_F *$, and so ϕ' is one–one. □

This lemma is the main step in the following proposition.

Proposition 7.2.9 *If* $\{X^{(n)}, g_n, p_n\}$ *is a weak homotopy decomposition of* X, *then the map* $g : X \to \varprojlim_{n\geqslant 1} X^{(n)}$ *determined by the maps* $g_n : X \to X^{(n)}$ *is a weak homotopy equivalence.*

Proof. We must show that $g_* : \pi_N(X) \to \pi_N(\varprojlim_{n\geqslant 1} X^{(n)})$ is an isomorphism for all N. Choose $k > N$ and consider the diagram

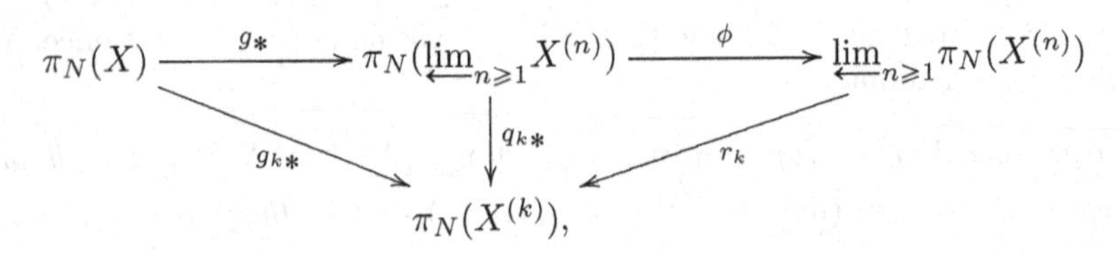

where r_k is the projection. Clearly the diagram is commutative. Now ϕ is an isomorphism by Lemma 7.2.8. Thus, since $p_{l+1*} : \pi_N(X^{(l+1)}) \to \pi_N(X^{(l)})$ is an isomorphism for all $l \geqslant k$, it follows that r_k is an isomorphism by Exercise 7.6. Therefore q_{k*} is an isomorphism. But g_{k*} is an isomorphism since $k > N$. Thus g_* is an isomorphism. □

Remark 7.2.10 From the previous proposition we see that a weak homotopy decomposition of a space X determines the weak homotopy type of X. If X is a CW complex and P is a CW approximation to $\varprojlim_{k\geqslant 2} X^{(k)}$, then by Remark 2.4.12, $X \simeq P$. Thus a weak homotopy decomposition of a CW complex X determines the homotopy type of X. In the case that X is a 1-connected CW complex, the homotopy groups $\pi_n(X)$ and the Postnikov invariants k^n determine the homotopy type of X, and these are said to be a complete set of invariants of homotopy type.

Next we want to show that a map of spaces induces a map of homotopy sections. Let $\{X^{(n)}, g_n, p_n\}$ be a weak homotopy decomposition of X and let $\{Y^{(n)}, h_n, q_n\}$ be a weak homotopy decomposition of Y. The following result and Proposition 7.2.13 were obtained by Kahn [54].

Proposition 7.2.11 *Let $f : X \to Y$ be a map. Then there exist maps $f^{(n)} : X^{(n)} \to Y^{(n)}$ such that the following diagrams are homotopy-commutative*

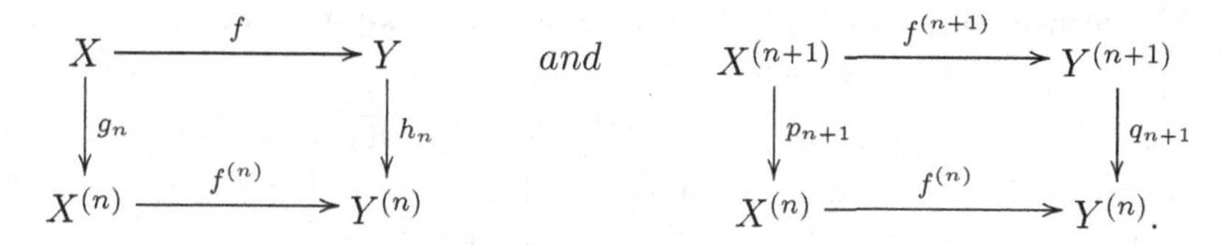

Furthermore,

1. $f \simeq f' \Longrightarrow f^{(n)} \simeq f'^{(n)}$, *for all* n.
2. *If* $l : Y \to Z$, *then* $(lf)^{(n)} \simeq l^{(n)} f^{(n)}$.
3. $(\mathrm{id}_X)^{(n)} \simeq \mathrm{id}_{X^{(n)}}$, *where the same weak homotopy decomposition is taken for the domain* X *and the codomain* X *of* id_X.
4. *If* f *is a homotopy equivalence, then* $f^{(n)}$ *is a homotopy equivalence.*

Proof. $X^{(n)}$ has the homotopy type of a space obtained by attaching cells of dimensions $\geqslant n+2$ to X, therefore there is a unique homotopy class $[f^{(n)}] \in [X^{(n)}, Y^{(n)}]$ such that $g_n^*[f^{(n)}] = [h_n f]$ by Proposition 2.4.13. This establishes the homotopy-commutativity of the first diagram and also implies (1), (2), and (3). To prove (4), assume that f is a homotopy equivalence with homotopy inverse l. Then

$$f^{(n)} l^{(n)} \simeq (fl)^{(n)} \simeq (\mathrm{id})^{(n)} \simeq \mathrm{id}.$$

A similar argument shows that $l^{(n)} f^{(n)} \simeq \mathrm{id}$.

Now we show that the second diagram in the statement of the proposition is homotopy-commutative. We have

$$q_{n+1}f^{(n+1)}g_{n+1} \simeq q_{n+1}h_{n+1}f \simeq h_n f \simeq f^{(n)}g_n \simeq f^{(n)}p_{n+1}g_{n+1},$$

and so $g_{n+1}^*[q_{n+1}f^{(n+1)}] = g_{n+1}^*[f^{(n)}p_{n+1}]$. Since $g_{n+1}^* : [X^{(n+1)}, Y^{(n)}] \to [X, Y^{(n)}]$ is a bijection, it follows that $q_{n+1}f^{(n+1)} \simeq f^{(n)}p_{n+1}$. □

We note that the second diagram in Proposition 7.2.11 can be assumed to be commutative because q_{n+1} is a fiber map.

Corollary 7.2.12 *If $A^{(n)}$ and $B^{(n)}$ are two homotopy n-sections of X (relative to two weak homotopy decompositions of X), then $A^{(n)} \simeq B^{(n)}$.*

Proof. The identity map id_X induces a homotopy equivalence $A^{(n)} \simeq B^{(n)}$ by Proposition 7.2.11. □

We next show how the Postnikov invariants behave with respect to homomorphisms induced by a map of spaces. We first introduce some notation. Let X and $\overline{X}$ be 1-connected spaces, let π_i denote $\pi_i(X)$, and let $\overline{\pi}_i$ denote $\pi_i(\overline{X})$. Let K_i denote $K(\pi_i, i)$ and $K_{n+1,n+2}$ denote $K(\pi_{n+1}, n+2)$, and similarly define $\overline{K}_i$ and $\overline{K}_{n+1,n+2}$. Let $f : X \to \overline{X}$ be a map and let $\{X^{(n)}, g_n, p_n, k^n\}$ and $\{\overline{X}^{(n)}, \overline{g}_n, \overline{p}_n, \overline{k}^n\}$ be Postnikov decompositions of X and $\overline{X}$, respectively. Because $\overline{p}_{n+1}$ is a fiber map, we may assume that the induced map $f^{(n+1)} : X^{(n+1)} \to \overline{X}^{(n+1)}$ satisfies $\overline{p}_{n+1}f^{(n+1)} = f^{(n)}p_{n+1}$. Therefore $f^{(n+1)}$ induces a map $\widetilde{f}^{(n+1)} : K_{n+1} \to \overline{K}_{n+1}$ of fibers. The homomorphism $\Omega : [K_{n+1,n+2}, \overline{K}_{n+1,n+2}] \to [K_{n+1}, \overline{K}_{n+1}]$ is an isomorphism (Exercise 6.13), thus there is a unique element $[\widehat{f}] \in [K_{n+1,n+2}, \overline{K}_{n+1,n+2}]$ such that $\Omega\widehat{f} \simeq \widetilde{f}^{(n+1)}$.

Proposition 7.2.13 *With the above notation and assumptions,*

$$\widehat{f}_*[k^{n+1}] = f^{(n)*}[\overline{k}^{n+1}],$$

where $\widehat{f}_* : H^{n+2}(X^{(n)}; \pi_{n+1}) \to H^{n+2}(X^{(n)}; \overline{\pi}_{n+1})$ is the coefficent homomorphism induced by $\widehat{f}$ and $f^{(n)*} : H^{n+2}(\overline{X}^{(n)}; \overline{\pi}_{n+1}) \to H^{n+2}(X^{(n)}; \overline{\pi}_{n+1})$ is the cohomology homomorphism induced by $f^{(n)} : X^{(n)} \to \overline{X}^{(n)}$.

Proof. Clearly $f^{(n)}$ and $f^{(n+1)}$ induce a map $\tau : C_{p_{n+1}} \to C_{\overline{p}_{n+1}}$ such that the diagram

$$\begin{array}{ccc} \Sigma K_{n+1} & \xrightarrow{\delta} & C_{p_{n+1}} \\ \downarrow{\scriptstyle \Sigma\widetilde{f}^{(n+1)}} & & \downarrow{\scriptstyle \tau} \\ \Sigma\overline{K}_{n+1} & \xrightarrow{\overline{\delta}} & C_{\overline{p}_{n+1}} \end{array}$$

is commutative, where δ and $\overline{\delta}$ are excision maps for the fibrations $K_{n+1} \to X^{(n+1)} \to X^{(n)}$ and $\overline{K}_{n+1} \to \overline{X}^{(n+1)} \to \overline{X}^{(n)}$, respectively. With the notation of Proposition 7.2.2 we have the diagram

$$\begin{array}{ccccc} X^{(n)} & \xrightarrow{l} & C_{p_{n+1}} & \xrightarrow{\theta} & K_{n+1,n+2} \\ \downarrow{\scriptstyle f^{(n)}} & & \downarrow{\scriptstyle \tau} & & \downarrow{\scriptstyle \hat{f}} \\ \overline{X}^{(n)} & \xrightarrow{\overline{l}} & C_{\overline{p}_{n+1}} & \xrightarrow{\overline{\theta}} & \overline{K}_{n+1,n+2}. \end{array}$$

The left square is commutative and we will show that the right square is homotopy-commutative. We first compose $\hat{f}\,\theta$ and $\overline{\theta}\tau$ with $\delta : \Sigma K_{n+1} \to C_{p_{n+1}}$. For $u : \Sigma K_{n+1} \to K_{n+1,n+2}$, a representative of the adjoint of the basic class $b^{n+1} \in H^{n+1}(K_{n+1}; G)$, and $\overline{u}$ similarly defined,

$$\hat{f}\,\theta\,\delta \simeq \hat{f}\,u \quad \text{and} \quad \overline{\theta}\tau\delta = \overline{\theta}\,\overline{\delta}(\Sigma \tilde{f}^{(n+1)}) \simeq \overline{u}(\Sigma \tilde{f}^{(n+1)}).$$

But

$$\kappa(\hat{f}\,u) \simeq \tilde{f}^{(n+1)}\,\mathrm{id}_{K_{n+1}} = \mathrm{id}_{\overline{K}_{n+1}}\tilde{f}^{(n+1)} \simeq \kappa(\overline{u}(\Sigma \tilde{f}^{(n+1)})).$$

Therefore

$$\delta^*[\hat{f}\,\theta] = [\hat{f}\,u] = [\overline{u}(\Sigma \tilde{f}^{(n+1)})] = \delta^*[\overline{\theta}\tau].$$

Because $\delta^* : H^{n+2}(C_{p_{n+1}}; \overline{\pi}_{n+1}) \to H^{n+2}(\Sigma K_{n+1}; \overline{\pi}_{n+1})$ is an isomorphism, $[\hat{f}\,\theta] = [\overline{\theta}\tau]$, and so the right square is homotopy-commutative. Consequently,

$$\hat{f}\,k^{n+1} = \hat{f}\theta l \simeq \overline{\theta}\,\overline{l} f^{(n)} = \overline{k}^{n+1} f^{(n)}. \qquad \square$$

Corollary 7.2.14 *If $\{X^{(n)}, g_n, p_n, k^n\}$ and $\{\overline{X}^{(n)}, \overline{g}_n, \overline{p}_n, \overline{k}^n\}$ are two Postnikov decompositions of X, then there are homotopy equivalences $h : X^{(n)} \to \overline{X}^{(n)}$ and $w : K(\pi_{n+1}(X), n+2) \to K(\pi_{n+1}(X), n+2)$ such that the following diagram homotopy-commutes*

$$\begin{array}{ccc} X^{(n)} & \xrightarrow{h} & \overline{X}^{(n)} \\ \downarrow{\scriptstyle k^{n+1}} & & \downarrow{\scriptstyle \overline{k}^{n+1}} \\ K(\pi_{n+1}(X), n+2) & \xrightarrow{w} & K(\pi_{n+1}(X), n+2). \end{array}$$

Proof. Let $\mathrm{id} : X \to X$ be the identity map, let $h = \mathrm{id}^{(n)} : X^{(n)} \to \overline{X}^{(n)}$, and let $w = \widehat{\mathrm{id}} : K(\pi_{n+1}(X), n+2) \to K(\pi_{n+1}(X), n+2)$ (using the notation of Proposition 7.2.13). Then by Proposition 7.2.13,

$$\overline{k}^{n+1} h \simeq w k^{n+1}. \qquad \square$$

Remark 7.2.15 It follows from Corollary 7.2.14 that if k^{n+1} is an $(n+1)$st Postnikov invariant of X, then any other $(n+1)$st Postnikov invariant of X has the form $wk^{n+1}l$, where $w : K(\pi_{n+1}(X), n+2) \to K(\pi_{n+1}(X), n+2)$ and $l : X^{(n)} \to X^{(n)}$ are homotopy equivalences.

Homotopy decompositions are an extremely useful tool in homotopy theory and have great theoretical value. Our main application is in Chapter 9 where we develop obstruction theory based on homotopy decompositions. In addition, the homotopy groups of some spaces have been computed using homotopy decompositions (or n-connected fibrations). The papers of Serre ([81], [82]) give results on the homotopy groups of spheres using n-connected fibrations and McCleary [67] has described how to calculate $\pi_4(S^2)$ using Postnikov systems. But for these computations it is necessary to know something about the spectral sequence of a fibration and the cohomology of Eilenberg–Mac Lane spaces. However, we next calculate the first Postnikov invariant of the 2-sphere without using these results.

Example 7.2.16 If $X = S^2$, then $\pi_2(X) = \mathbb{Z}$ and $\pi_3(X) = \mathbb{Z}$. Thus a Postnikov decomposition of X starts with the fibration

$$K(\mathbb{Z}, 3) \longrightarrow X^{(3)} \xrightarrow{p_3} X^{(2)} = K(\mathbb{Z}, 2)$$

and k-invariant $k^3 : K(\mathbb{Z}, 2) \to K(\mathbb{Z}, 4)$. Now $K(\mathbb{Z}, 2) = \mathbb{C}\mathrm{P}^\infty$, and so $[k^3] \in H^4(\mathbb{C}\mathrm{P}^\infty)$. If $b \in H^2(\mathbb{C}\mathrm{P}^\infty)$ is the basic class, then it is known [39, Thm. 3.12] that the cup product $b^2 \in H^4(\mathbb{C}\mathrm{P}^\infty) \cong \mathbb{Z}$ is a generator of $\mathbb{Z}$. Therefore $[k^3] = mb^2$ for some integer m. We will show that $m = \pm 1$. The Serre exact cohomology sequence of the above fibration yields the exact sequence

$$H^3(K(\mathbb{Z}, 3)) \xrightarrow{\Delta} H^4(\mathbb{C}\mathrm{P}^\infty) \longrightarrow H^4(X^{(3)}) \longrightarrow H^4(K(\mathbb{Z}, 3)).$$

By Exercise 7.11, $H^4(K(\mathbb{Z}, 3)) = 0$. Also, the basic class $\beta \in H^3(K(\mathbb{Z}, 3)) \cong \mathbb{Z}$ is a generator of $\mathbb{Z}$. Furthermore, $\Delta(\beta) = -[k^3] = -mb^2$ by Remark 7.2.6. Therefore $H^4(X^{(3)}) = \mathbb{Z}_m$. On the other hand, $g_3 : X \to X^{(3)}$ is a 4-equivalence, and so g_3 is a cohomological 4-equivalence by Whitehead's second theorem 6.4.15 and Lemma 6.4.13. Therefore $g_3^* : H^4(X^{(3)}) \to H^4(X)$ is a monomorphism. Since $H^4(X) = 0$, it follows that $H^4(X^{(3)}) = 0$, and we have $\mathbb{Z}_m = 0$. Thus $m = \pm 1$, and so $k^3 = \pm b^2$.

We next record a useful result about weak homotopy decompositions.

Proposition 7.2.17 *Let X be a space with weak homotopy decomposition $\{X^{(n)}, g_n, p_n\}$, let A be a CW complex, and let $g_{n*} : [A, X] \to [A, X^{(n)}]$. If $\dim A \leqslant n$, then g_{n*} is a bijection. If $\dim A = n+1$, then g_{n*} is a surjection.*

Proof. This follows at once from Proposition 2.4.6. □

If X is an H-complex and $\{X^{(n)}, g_n, p_n\}$ is a weak homotopy decomposition of X, then by Exercise 7.8, $X^{(n)}$ is an H-space and $g_n : X \to X^{(n)}$ and $p_{n+1} : X^{(n+1)} \to X^{(n)}$ are H-maps. We next consider additional results that relate homotopy decompositions and H-spaces. We first discuss some generalities regarding H-maps.

Let (X, m) be an H-complex, let (Y, n) be a grouplike space, and let $f : X \to Y$ be a map. Consider the diagram

$$\begin{array}{ccc} X \times X & \xrightarrow{f \times f} & Y \times Y \\ \downarrow m & & \downarrow n \\ X & \xrightarrow{f} & Y \end{array}$$

which is not necessarily homotopy-commutative. Clearly

$$n(f \times f)|X \vee X \simeq fm|X \vee X : X \vee X \to Y.$$

Using the grouplike structure of Y, we form the difference

$$d_f = n(f \times f) - fm : X \times X \to Y$$

and have that $d_f|X \vee X \simeq *$. Now consider the cofiber sequence

$$X \vee X \xrightarrow{j} X \times X \xrightarrow{q} X \wedge X$$

and the resulting exact sequence of groups

$$[\Sigma(X \vee X), Y] \xrightarrow{\partial^*} [X \wedge X, Y] \xrightarrow{q^*} [X \times X, Y] \xrightarrow{j^*} [X \vee X, Y].$$

Since $\Sigma j : \Sigma(X \vee X) \to \Sigma(X \times X)$ has a homotopy retraction by Lemma 5.4.11, $(\Sigma j)^* : [\Sigma(X \times X), Y] \to [\Sigma(XvX), Y]$ is onto, and so q^* is one–one. But $j^*[d_f] = 0$, and so there is a unique homotopy class

$$[D_f] \in [X \wedge X, Y]$$

such that $q^*[D_f] = [d_f]$.

Definition 7.2.18 The homotopy class $[D_f] \in [X \wedge X, Y]$ is called the *H-map deviation* of f.

Remark 7.2.19 For the definition of the H-map deviation it is not necessary that Y be a grouplike space. In fact, it was only the multiplication and the inverse that were used in the definition. We show in Chapter 8 that an H-complex always admits inverses.

The proof of the following result is straightforward, and hence omitted.

Proposition 7.2.20 *Let (X, m) be an H-complex, let (Y, n) be a grouplike space, and let $f : X \to Y$ be a map.*
1. $f : (X, m) \to (Y, n)$ is an H-map if and only if $[D_f] = 0$.
2. If (X', m') is an H-complex and $g : (X', m') \to (X, m)$ is an H-map, then

$$[D_{fg}] = (g \wedge g)^*[D_f],$$

where $g \wedge g : X' \wedge X' \to X \wedge X$ is the map induced by $g \times g : X' \times X' \to X \times X$.

As a consequence we obtain the following proposition.

Proposition 7.2.21 *If X is an H-complex and $\{X^{(n)}, g_n, p_n, k^n\}$ is a Postnikov decomposition of X, then $X^{(n)}$ is an H-space and $g_n : X \to X^{(n)}$, $p_{n+1} : X^{(n+1)} \to X^{(n)}$, and $k^{n+1} : X^n \to K(\pi_{n+1}(X), n+2)$ are all H-maps.*

Proof. By our earlier remarks, it is only necessary to show that k^{n+1} is an H-map. We assume without loss of generality that the 0-skeleton of X is $\{*\}$. Let $k = k^{n+1}$, let $p = p_{n+1}$, and let $G = \pi_{n+1}(X)$. We show that the homotopy deviation $[D_k] = 0$. Since $kp \simeq *$,

$$(p \wedge p)^*[D_k] = [D_{kp}] = 0$$

by Proposition 7.2.20. Therefore it suffices to show that $(p \wedge p)^* : H^{n+2}(X^{(n)} \wedge X^{(n)}; G) \to H^{n+2}(X^{(n+1)} \wedge X^{(n+1)}; G)$ is a monomorphism. To do this we apply Corollary 2.4.10 to the $(n+1)$-equivalence $p : X^{(n+1)} \to X^{(n)}$ and so regard $X^{(n)}$ as obtained from $X^{(n+1)}$ by attaching positive-dimensional cells of dimension $\geqslant n+2$ to $X^{(n+1)}$. Thus, as a CW complex, $X^{(n)} \wedge X^{(n)}$ is $X^{(n+1)} \wedge X^{(n+1)}$ with positive-dimensional cells of dimension $\geqslant n+3$ attached. Hence the cofiber of $X^{(n+1)} \wedge X^{(n+1)} \to X^{(n)} \wedge X^{(n)}$ has positive-dimensional cells in dimensions $\geqslant n+3$. Consequently, $(p \wedge p)^* : H^{n+2}(X^{(n)} \wedge X^{(n)}; G) \to H^{n+2}(X^{(n+1)} \wedge X^{(n+1)}; G)$ is a monomorphism. □

It is also possible to show that H-structures on the homotopy sections of X induce an H-structure on X.

Proposition 7.2.22 *Let $\{X^{(n)}, g_n, p_n\}$ be a weak homotopy decomposition of an n-dimensional CW complex X. If $X^{(2n-1)}$ is an H-space, then X is an H-space and $g_{2n-1} : X \to X^{(2n-1)}$ is an H-map.*

Proof. Let $g = g_{2n-1} : X \to X^{(2n-1)}$ and let m' be the multiplication on $X^{(2n-1)}$. The map g is a $2n$-equivalence and $\dim(X \times X) = 2n$. Therefore by Proposition 2.4.6, $g_* : [X \times X, X] \to [X \times X, X^{(2n-1)}]$ is a surjection. Hence there exists an $[m] \in [X \times X, X]$ such that $g_*[m] = [m'(g \times g)]$. We show that m is a multiplication. Let $j_1 : X \to X \times X$ and $j_1' : X^{(2n-1)} \to X^{(2n-1)} \times X^{(2n-1)}$ be inclusions into the first factor. Then

$$gmj_1 \simeq m'(g \times g)j_1 = m'j_1'g \simeq g.$$

Thus $g_*[mj_1] = g_*[\mathrm{id}_X]$, where $g_* : [X, X] \to [X, X^{(2n-1)}]$. Because g_* is a bijection by Proposition 2.4.6, $mj_1 \simeq \mathrm{id}_X$. Similarly $mj_2 \simeq \mathrm{id}_X$. Therefore m is a multiplication. □

See [55] for this proposition in the case when X is not finite-dimensional.

7.3 Homology Decompositions of Spaces

In this section we discuss the approximation of a space X by means of homology sections X_n. The step from X_n to X_{n+1} is carried out by attaching a cone on a Moore space $M(H_{n+1}(X), n)$ to X_n. This construction is similar to the construction of the $(n+1)$st skeleton from the nth skeleton in a CW complex, although here we adjoin the cone on a Moore space instead of $(n+1)$-cells. The homology decomposition has some, but not all, of the analogues of the properties of homotopy decompositions. We give an example to show that the homotopy type of the nth homology section is not determined by the homotopy type of X.

Definition 7.3.1 Let X be a 1-connected CW complex and write $H_r = H_r(X)$. Suppose there exists a sequence of 1-connected CW complexes $X_2, X_3, \ldots$ and maps $j_n : X_n \to X$ and $i_n : X_n \to X_{n+1}$ that satisfy the following conditions.

1. $j_{n*} : H_r(X_n) \to H_r(X)$ is an isomorphism for $r \leqslant n$ and $H_r(X_n) = 0$ for $r > n$.
2. $X_n \xrightarrow{i_n} X_{n+1} \longrightarrow M(H_{n+1}, n+1)$ is a principal cofibration induced by a map $k'_{n+1} : M(H_{n+1}, n) \to X_n$.
3. The following diagram commutes

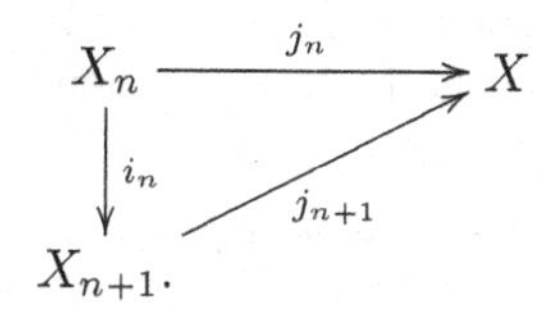

Then the collection $\{X_n, j_n, i_n, k'_n\}$ is called a *homology decomposition* of X. The spaces X_n are called *nth homology sections* of X and the maps $k'_{n+1} : M(H_{n+1}, n) \to X_n$ or homotopy classes $[k'_{n+1}] \in \pi_n(X_n; H_{n+1})$ are called *(n+1)st k′-invariants.*

Theorem 7.3.2 *If X is a 1-connected CW complex, then X has a homology decomposition.*

Proof. There are some difficulties in dualizing the proof of Theorem 7.2.5, and we give a proof based on pairs of spaces (but see Exercise 7.3). We show the existence of X_n, j_n, i_n, and k'_n by induction on n. When $n = 2$ we need only construct X_2 and j_2. We write M_r for $M(H_r, r)$, set $X_2 = M_2$, and consider the epimorphism

$$[M_2, X] = \pi_2(X; H_2) \xrightarrow{\eta} \mathrm{Hom}(H_2, \pi_2(X))$$

of the universal coefficient theorem for homotopy 5.2.9. We choose $j_2 : X_2 \to X$ such that $\eta[j_2] = h^{-1} : H_2 \to \pi_2(X)$, the inverse of the Hurewicz homomorphism. Since $j_{2*} : \pi_2(X_2) \to \pi_2(X)$ is an isomorphism, $j_{2*} : H_2(X_2) \to H_2(X)$ is an isomorphism. Now assume the result for n. We factor $j_n : X_n \to X$ as

$$X_n \xrightarrow{j'_n} X' \xrightarrow{j''_n} X,$$

where X' is the mapping cylinder of j_n, j'_n is an inclusion, and j''_n is a homotopy equivalence. For the pair (X', X_n), we have

$$H_r(X', X_n) \cong \begin{cases} 0 & \text{if } r \leqslant n \\ H_r & \text{if } r \geqslant n+1, \end{cases}$$

and so $\pi_{n+1}(X', X_n) \cong H_{n+1}$ by the relative Hurewicz theorem. We consider the relative homotopy group with coefficients $\pi_{n+1}(X', X_n; H_{n+1})$, which is the set of homotopy classes of maps of pairs $(CM, M) \to (X', X_n)$, where $M = M(H_{n+1}, n)$ (see the discussion after Definition 4.5.3). There is an epimorphism

$$\eta : \pi_{n+1}(X', X_n; H_{n+1}) \to \mathrm{Hom}(H_{n+1}, \pi_{n+1}(X', X_n)) \cong \mathrm{Hom}(H_{n+1}, H_{n+1}),$$

by the universal coefficient theorem for homotopy, and hence there is a map of pairs $u : (CM, M) \to (X', X_n)$ such that $u_* : H_{n+1}(CM, M) \to H_{n+1}(X', X_n)$ is an isomorphism. Let $k' = u|M : M \to X_n$, let $X_{n+1} = C_{k'}$, and let $i_n : X_n \to X_{n+1}$ be the inclusion. Then in the diagram

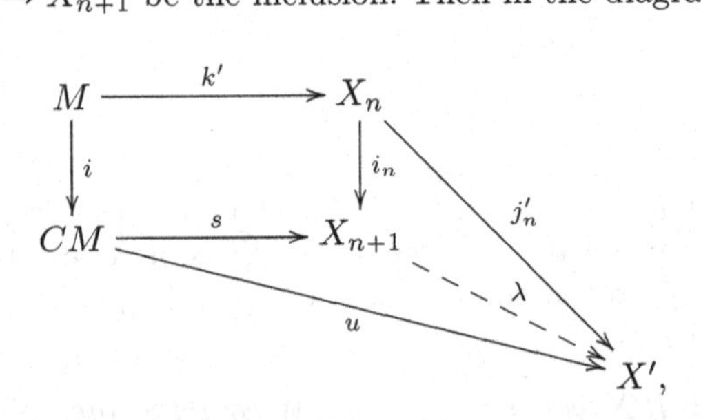

the square is a pushout square and $ui = j'_n k'$. Therefore there exists $\lambda : X_{n+1} \to X'$ such that $\lambda i_n = j'_n$ and $\lambda s = u$. Without loss of generality we assume that λ is an inclusion map. We denote $\lambda_* : H_r(X_{n+1}) \to H_r(X')$

by λ_{*r} and show that $H_r(X_{n+1}) = 0$ for $r > n+1$ and that λ_{*r} is an isomorphism for $r \leqslant n+1$. Because $H_r(X_{n+1}, X_n) \cong H_r(M_{n+1})$, we have

$$H_r(X_{n+1}, X_n) \cong \begin{cases} 0 & \text{if } r \neq n+1 \\ H_{n+1} & \text{if } r = n+1. \end{cases}$$

By considering the exact homology sequence of the triple (X', X_{n+1}, X_n) (Appendix C) and the results stated above for $H_r(X', X_n)$ and $H_r(X_{n+1}, X_n)$, we obtain

$$H_r(X', X_{n+1}) \cong \begin{cases} 0 & \text{if } r \leqslant n+1 \\ H_r & \text{if } r \geqslant n+2. \end{cases}$$

It now follows from the exact homology sequence of the pair (X', X_{n+1}) that $H_r(X_{n+1}) = 0$ for $r > n+1$ and λ_{*r} is an isomorphism for $r \leqslant n+1$. To complete the induction we set $k'_{n+1} = k'$ and compose $\lambda : X_{n+1} \to X'$ with the homotopy equivalence $j''_n : X' \to X$ to obtain $j_{n+1} : X_{n+1} \to X$ with the desired properties. □

We next make a few elementary observations on homology decompositions.

Remark 7.3.3 We use the notation $H_i = H_i(X)$ and $M_i = M(H_i, i)$.

1. If X is a space with no torsion in its homology, then all the Moore spaces that appear in a homology decomposition of X are wedges of spheres, that is, $M(H_{n+1}, n) = S^n \vee \cdots \vee S^n$, where the number of spheres equals the rank of H_{n+1}. In this case, X_n is the n-skeleton of a CW decomposition of X, and the homology decomposition and the CW decomposition of X are the same.
2. The homology decomposition could start with $X_1 = \{*\}$ instead of with X_2. Furthermore, if $H_{n+1} = 0$, then $X_n \simeq X_{n+1}$. Thus if $H_i = 0$ for $i < N$, then $X_i = \{*\}$ for $i < N$ and $X_N \simeq M_N$. In addition, if some $[k'_{n+1}] = 0$, then $X_{n+1} \simeq X_n \vee M_{n+1}$.
3. The k′-invariant is trivial on homology groups, that is, the homomorphism $k'_{n+1*} = 0 : H_n(M(H_{n+1}, n)) \to H_n(X_n)$, for all n. This follows from the exact homology sequence of

$$M(H_{n+1}, n) \xrightarrow{k'_{n+1}} X_n \xrightarrow{i_n} X_{n+1}$$

since $i_{n*} : H_n(X_n) \to H_n(X_{n+1})$ is an isomorphism. Let H_N be the first nonzero homology group of X, so $X_N = M_N$. If $[k'_{N+1}] \in \pi_N(M_N; H_{N+1})$ is the first k′-invariant, then $k'_{N+1*} = 0 : \pi_N(M(H_{N+1}, N)) = H_{N+1} \to \pi_N(M_N)$. Thus if $\eta : \pi_N(M_N; H_{N+1}) \to \mathrm{Hom}(H_{N+1}, \pi_N(M_N))$ is the homomorphism in the universal coefficient theorem for homotopy, then $[k'_{N+1}] \in \operatorname{Ker} \eta \cong \mathrm{Ext}(H_{N+1}, \pi_{N+1}(M_N))$.

The analogues of some of the properties of homotopy decompositions do not hold for homology decompositions. We next illustrate this with examples.

Example 7.3.4

1. We give a simple example to show that induced maps of homology sections do not necessarily exist. Let $M_n = M(\mathbb{Z}_2, n)$ and consider the space $X = M_n \vee S^{n+1}$. The nth homology section of X is M_n and the cofiber map of the homology decomposition is the inclusion

$$X_n = M_n \xrightarrow{i_1} M_n \vee S^{n+1} = X = X_{n+1}.$$

Since $[M_n, S^{n+1}] = \pi_n(S^{n+1}; \mathbb{Z}_2) \cong \operatorname{Ext}(\mathbb{Z}_2, \mathbb{Z}) \cong \mathbb{Z}_2$, by the universal coefficient theorem for homotopy, we can choose a map $q : M_n \to S^{n+1}$ such that $q \not\simeq *$. (In Exercise 7.14, q is given explicitly.) Define $f : X \to X$ by $f = \{i_2 q, *\} : M_n \vee S^{n+1} \to M_n \vee S^{n+1}$, where $i_2 : S^{n+1} \to M_n \vee S^{n+1}$ is the inclusion. With the above homology decomposition for X, suppose there exists an induced map $f_n : M_n \to M_n$. Thus $i_1 f_n \simeq f i_1 : M_n \to M_n \vee S^{n+1}$. Then if $q_2 : M_n \vee S^{n+1} \to S^{n+1}$ is the projection,

$$q = q_2 i_2 q = q_2 f i_1 \simeq q_2 i_1 f_n = *.$$

This contradicts $q \not\simeq *$.
In [4] conditions are given for induced maps to exist.

2. Here we show, following [15], that two homology n-sections of the same space do not necessarily have the same homotopy type. Let $[h] \in \pi_n(S^m)$ be an element that is a generator of $\mathbb{Z}_p \subseteq \pi_n(S^m)$, where p is an odd prime and $n > m + 1$, such that the suspension homomorphism $\Sigma : \pi_n(S^m) \to \pi_{n+1}(S^{m+1})$ is an isomorphism. For example, let m be odd, set $n = m + 2p - 3$ (see the end of Section 5.6) and assume that $2p - 2 < m$ so that Σ is an isomorphism by the Freudenthal theorem 5.6.7. We write $M_r = M(\mathbb{Z}_p, r)$. Then, because $p[h] = 0$, h extends to a map $h' : M_n \to S^m$. Thus there is a commutative diagram

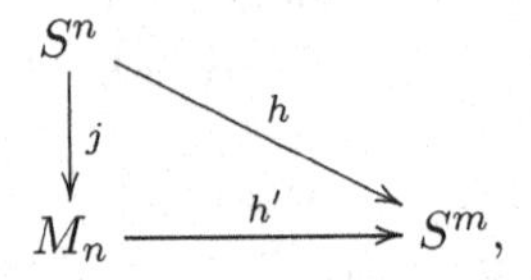

where j is the inclusion. Let $q : M_{n-1} \to M_{n-1}/S^{n-1} = S^n$ be the collapsing map and define

$$A_n = S^m \vee \Sigma M_{n-1}, \quad A'_n = S^m \cup_{hq} CM_{n-1},$$

and

$$X = (S^m \cup_{h'} CM_n) \vee \Sigma M_{n-1}.$$

Then the nontrivial homology groups of X are $\mathbb{Z}$ in degree m and $\mathbb{Z}_p$ in degrees n and $n+1$. The nontrivial homology groups of A_n and A'_n are

$\mathbb{Z}$ in degree m and $\mathbb{Z}_p$ in degree n. If $j_n : A_n \to X$ is the inclusion, then $j_{n*} : H_i(A_n) \to H_i(X)$ is an isomorphism for $i \leqslant n$. Therefore A_n is a homology n-section of X. Now define $j'_n : A'_n \to X$ by $j'_n\langle x\rangle = (\langle x\rangle, *)$, for $x \in S^m$ and

$$j'_n\langle y,t\rangle = \begin{cases} (\langle jq(y), 2t\rangle, *) \text{ if } 0 \leqslant t \leqslant \frac{1}{2} \\ (*, \langle y, 2t-1\rangle) \text{ if } \frac{1}{2} \leqslant t \leqslant 1, \end{cases}$$

for $\langle y,t\rangle \in CM_{n-1}$. We show that j'_n induces a homology isomorphism in degree n. Let

$$p : A_n \to \Sigma M_{n-1}, \ \ p' : A'_n \to \Sigma M_{n-1}, \ \text{ and } \bar{p} : X \to \Sigma M_{n-1}$$

be the projections obtained by respectively shrinking S^m, S^m, and $S^m \cup_{h'} CM_n$ to a point. Then, because $\bar{p}j'_n \simeq (* + \mathrm{id})p' : A'_n \to \Sigma M_{n-1}$, the following diagram is commutative

$$\begin{array}{ccc} H_n(A'_n) & \xrightarrow{j'_{n*}} & H_n(X) \\ \big\downarrow{\scriptstyle p'_*} & & \big\downarrow{\scriptstyle \bar{p}_*} \\ H_n(\Sigma M_{n-1}) & \xrightarrow{\mathrm{id}} & H_n(\Sigma M_{n-1}). \end{array}$$

p'_* and $\bar{p}_*$ are isomorphisms, therefore j'_{n*} is an isomorphism. Hence A'_n is a homology n-section of X.

We now show that A_n and A'_n do not have the same homotopy type. Suppose that there is a homotopy equivalence $f : A_n \to A'_n$. By the cellular approximation theorem, f is homotopic to a cellular map g and g induces $g' : S^m \to S^m$ which is clearly a homotopy equivalence. Thus g and g' induce a homotopy equivalence $\tilde{g} : \Sigma M_{n-1} \to \Sigma M_{n-1}$. There is then a diagram

$$\begin{array}{ccc} A_n & \xrightarrow{g} & A'_n \\ \big\downarrow{\scriptstyle p} & & \big\downarrow{\scriptstyle p'} \\ \Sigma M_{n-1} & \xrightarrow{\tilde{g}} & \Sigma M_{n-1} \\ \big\downarrow{\scriptstyle *} & & \big\downarrow{\scriptstyle \Sigma(hq)} \\ \Sigma S^m & \xrightarrow{\Sigma g'} & \Sigma S^m \end{array}$$

with top square commutative and bottom square homotopy-commutative, where the vertical maps are the continuation of the coexact sequences

$$M_{n-1} \xrightarrow{*} S^m \longrightarrow A_n \ \text{ and } \ M_{n-1} \xrightarrow{hq} S^m \longrightarrow A'_n.$$

We have that $\Sigma(hq)\,\tilde{g} \simeq *$, and so $(\Sigma h)\,q' \simeq *$, where $q' = \Sigma q : M_n \to S^{n+1}$ is the projection. Finally, we consider the coexact sequence of the defining cofibration for M_n,

$$S^n \xrightarrow{a} S^n \xrightarrow{j} M_n \xrightarrow{q'} S^{n+1} \xrightarrow{a} S^{n+1},$$

where a is a map of degree p. There is then the exact sequence

$$\pi_{n+1}(S^{m+1}) \xrightarrow{a^*} \pi_{n+1}(S^{m+1}) \xrightarrow{q'^*} [M_n, S^{m+1}],$$

where a^* is multiplication by p. Since $q'^*[\Sigma h] = 0$, we have that $[\Sigma h] = p\alpha$, for some $\alpha \in \pi_{n+1}(S^{m+1})$, contradicting the choice of $[h]$ as a generator of $\mathbb{Z}_p$.

We next present a result that relates the k′-invariant of a homology decomposition of a space with the Hurewicz homomorphism of the space.

Proposition 7.3.5 *Let $\{X_n, j_n, i_n, k'_n\}$ be a homology decomposition of X. Then $k'_{n+1*} = 0 : \pi_n(M(H_{n+1}(X), n)) \to \pi_n(X_n)$ if and only if the Hurewicz homomorphism $h_{n+1} : \pi_{n+1}(X) \to H_{n+1}(X)$ is an epimorphism.*

Proof. For notational convenience we write $H_i = H_i(X)$, $M_i = M(H_i, i)$, $M = M(H_{n+1}, n)$, and $k' = k'_{n+1}$. We consider

$$M \xrightarrow{k'} X_n \xrightarrow{i_n} X_{n+1}$$

which can be regarded as a cofiber sequence. Since M is $(n-1)$-connected and X_{n+1} is 1-connected, we obtain an exact sequence

$$\pi_n(M) \xrightarrow{k'_*} \pi_n(X_n) \xrightarrow{i_{n*}} \pi_n(X_{n+1}),$$

by the Blakers–Massey exact homotopy sequence. Therefore $k'_* = 0 : \pi_n(M) \to \pi_n(X_n) \iff i_{n*} : \pi_n(X_n) \to \pi_n(X_{n+1})$ is a monomorphism. Consider the cofiber sequence

$$X_n \xrightarrow{i_n} X_{n+1} \xrightarrow{q} M_{n+1}$$

and let $q_{**} : H_{n+1}(X_{n+1}) \to H_{n+1}(M_{n+1})$ and $i_{n**} : H_n(X_n) \to H_n(X_{n+1})$ be the induced homology homomorphisms. Then there is a commutative diagram

$$\begin{array}{ccccccccc}
 & & \pi_{n+1}(X_{n+1}) & \xrightarrow{q_*} & \pi_{n+1}(M_{n+1}) & \longrightarrow & \pi_n(X_n) & \xrightarrow{i_{n*}} & \pi_n(X_{n+1}) \\
 & & \downarrow{\scriptstyle h'_{n+1}} & & \downarrow{\scriptstyle h''_{n+1}} & & \downarrow{\scriptstyle h'''_n} & & \\
0 & \longrightarrow & H_{n+1}(X_{n+1}) & \xrightarrow{q_{**}} & H_{n+1}(M_{n+1}) & \longrightarrow & H_n(X_n) & \xrightarrow{i_{n**}} & H_n(X_{n+1}),
\end{array}$$

where the top line is the Blakers–Massey exact homotopy sequence, the bottom line is the exact homology sequence of a cofibration and h'_{n+1}, h''_{n+1} and

h_n''' are Hurewicz homomorphisms. Since i_{n**} is an isomorphism, q_{**} is an isomorphism. But h_{n+1}'' is an isomorphism, and so

$$h_{n+1}' \text{ is onto} \iff q_* \text{ is an onto} \iff i_{n*} \text{ is one} - \text{one} \iff k_*' = 0.$$

Therefore it suffices to show that h_{n+1}' is onto $\iff h_{n+1} : \pi_{n+1}(X) \to H_{n+1}(X)$ is onto. But there is the commutative diagram

$$\begin{array}{ccc} \pi_{n+1}(X_{n+1}) & \xrightarrow{j_{n+1*}} & \pi_{n+1}(X) \\ \downarrow h_{n+1}' & & \downarrow h_{n+1} \\ H_{n+1}(X_{n+1}) & \xrightarrow{j_{n+1**}} & H_{n+1}(X), \end{array}$$

with j_{n+1**} an isomorphism. Since j_{n+1} is an $(n+1)$-equivalence, j_{n+1*} is an epimorphism, and the proposition is proved. □

We conclude this section with a simple calculation of a homology decomposition. Let m be an odd integer $\geqslant 3$ and let M_i be the Moore space $M(\mathbb{Z}_m, i)$. We form the smash product $X = M_r \wedge M_s$. It then easily follows from the Künneth theorem [83, p. 235] that $H_i(X) = \mathbb{Z}_m$ for $i = r+s$ and $r+s+1$ and $H_i(X) = 0$ for other values of i. Thus there is one k′-invariant $[k'] \in \pi_{r+s}(M_{r+s}; \mathbb{Z}_m)$ for X.

Lemma 7.3.6 *The k′-invariant $[k']$ of $X = M_r \wedge M_s$ is zero and hence $X \simeq M_{r+s} \vee M_{r+s+1}$.*

Proof. By Remark 7.3.3, $[k'] \in \operatorname{Ext}(\mathbb{Z}_m, \pi_{r+s+1}(M_{r+s}))$. By Proposition 5.6.11, $\pi_{r+s+1}(M_{r+s})) = 0$, and so $[k'] = 0$. □

As a consequence of this lemma a binary operation on homotopy groups with coefficients in $\mathbb{Z}_m$ can be defined, for m an odd positive integer $\geqslant 3$. We sketch this construction. Let (Y, m, i) be a grouplike space and define a commutator map $c : Y \times Y \to Y$ by $c(x, y) = (x+y) + ((-x) + (-y))$, where $m(x, y) = x + y$ and $i(x) = -x$, for $x, y \in Y$ (see Exercise 2.3). Then $c|Y \vee Y \simeq *$ and so c induces a map $\tilde{c} : Y \wedge Y \to Y$. If $\alpha = [f] \in \pi_r(Y; \mathbb{Z}_m)$ and $\beta = [g] \in \pi_s(Y; \mathbb{Z}_m)$, then define $\langle \alpha, \beta \rangle \in \pi_{r+s+1}(Y; \mathbb{Z}_m)$ to be the homotopy class of the composition

$$M_{r+s+1} \xrightarrow{i_2} M_{r+s} \vee M_{r+s+1} \simeq M_r \wedge M_s \xrightarrow{f \wedge g} Y \wedge Y \xrightarrow{\tilde{c}} Y.$$

The element $\langle \alpha, \beta \rangle \in \pi_{r+s+1}(Y; \mathbb{Z}_m)$ is called the *Samelson product* of α and β. The existence of a nontrivial Samelson product implies that the H-space Y is not homotopy-commutative (for if it were, we would have $c \simeq *$ and hence $\tilde{c} \simeq *$). For an arbitrary space X, we can form a product of $\alpha \in \pi_p(X; \mathbb{Z}_m)$ with $\beta \in \pi_q(X; \mathbb{Z}_m)$ by taking Samelson products of the adjoints. More precisely, if $\kappa_* : \pi_i(X; \mathbb{Z}_m) \to \pi_{i-1}(\Omega X; \mathbb{Z}_m)$ is the adjoint isomorphism, then define

$$[\alpha, \beta] = \kappa_*^{-1}\langle \kappa_*(\alpha), \kappa_*(\beta) \rangle \in \pi_{p+q}(X; \mathbb{Z}_m),$$

for $p, q \geqslant 3$. These products and others like them have been studied in [38], [40], and [75].

7.4 Homotopy and Homology Decompositions of Maps

We begin this section with a lemma that gives a condition for a fibration to be principal. This leads to Proposition 7.4.2 which is a generalization of Proposition 7.2.2. Proposition 7.4.2 is then applied to obtain k-invariants for the weak homotopy decomposition of a map. Then we discuss the homology decomposition of a map. The homotopy and homology decomposition of a map have appeared in [30].

Lemma 7.4.1 [39, Lemma 4.70] *Let (X, A) be a CW pair such that the homotopy fiber of the inclusion map $u : A \to X$ is a $K(G, n)$, $n \geqslant 1$, and the action of $\pi_1(A)$ on $\pi_{n+1}(X, A)$ is trivial. Then there is a map $k : X \to K(G, n+1)$ with the following property. If*

$$I_k \xrightarrow{j} E_k \xrightarrow{p} K(G, n+1)$$

is the mapping path fibration of k, there exist weak homotopy equivalences $\mu : X \to E_k$ and $\theta : A \to I_k$ such that the following diagram commutes

$$\begin{array}{ccc} A & \xrightarrow{u} & X \\ \downarrow{\scriptstyle \theta} & & \downarrow{\scriptstyle \mu} \\ I_k & \xrightarrow{j} & E_k. \end{array}$$

Proof. If I_u is the homotopy fiber of u, then

$$\pi_r(X, A) = \pi_{r-1}(I_u) = \begin{cases} G \text{ if } r = n+1 \\ 0 \text{ if } r \neq n+1. \end{cases}$$

The operation of $\pi_1(A)$ on $\pi_{n+1}(X, A)$ is trivial, thus the Hurewicz homomorphism $h_{n+1} : \pi_{n+1}(X, A) \to H_{n+1}(X, A)$ is an isomorphism (Theorem 6.4.21). Therefore

$$H_r(X/A) = H_r(X, A) = \begin{cases} G \text{ if } r = n+1 \\ 0 \text{ if } r < n+1. \end{cases}$$

Moreover, the homomorphism $\eta_\pi : H^{n+1}(X/A; G) \to \operatorname{Hom}(\pi_{n+1}(X/A), G) = \operatorname{Hom}(G, G)$ which assigns to a homotopy class its induced homotopy homomorphism is an isomorphism by Lemma 2.5.13. Thus there is a map

$\bar{k} : X/A \to K(G, n+1)$ such that $\bar{k}_* : \pi_{n+1}(X/A) \to \pi_{n+1}(K(G, n+1))$ is an isomorphism. Now let $k : X \to K(G, n+1)$ be the composition

$$X \xrightarrow{\quad q \quad} X/A \xrightarrow{\quad \bar{k} \quad} K(G, n+1),$$

where q is the projection. We then have the commutative diagram

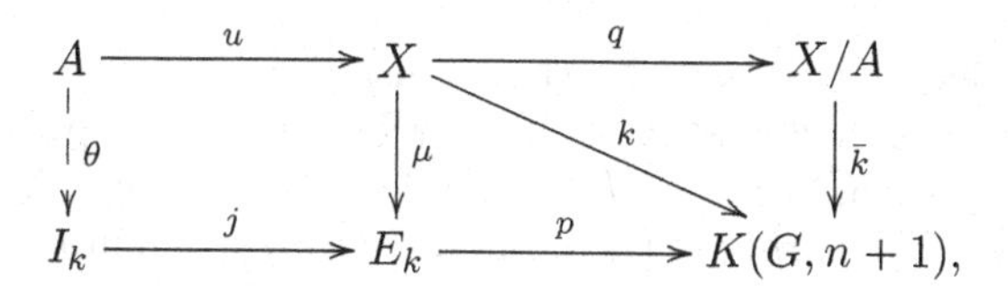

where k is factored as $p\mu$ for p a fiber map and μ a homotopy equivalence. Therefore there is a map $\theta = \mu | A : A \to I_k$ which makes the left square commute. We show that the map of pairs $\mu' : (X, A) \to (E_k, I_k)$ determined by μ induces isomorphisms of homotopy groups. Consider the commutative diagram

$$\begin{array}{ccc} (X, A) & \xrightarrow{\quad q' \quad} & (X/A, \{*\}) \\ \downarrow{\scriptstyle \mu'} & & \downarrow{\scriptstyle \bar{k}} \\ (E_k, I_k) & \xrightarrow{\quad p' \quad} & (K(G, n+1), \{*\}), \end{array}$$

where q' and p' are maps of pairs obtained from q and p. We show that the maps q', $\bar{k}$, and p' all induce isomorphisms of homotopy groups in dimension $n+1$. Now $p'_* : \pi_i(E_k, I_k) \to \pi_i(K(G, n+1))$ is an isomorphism for all i by Proposition 4.5.18 and $\bar{k}_* : \pi_{n+1}(X/A) \to \pi_{n+1}(K(G, n+1)$ is an isomorphism by the choice of $\bar{k}$. Since the inclusion $u : A \to X$ is an n-equivalence, $n \geqslant 1$, it follows that $C_u \simeq X/A$ is 1-connected, and so X/A is n-connected by the Hurewicz theorem. Hence by the commutativity of the diagram

$$\begin{array}{ccc} \pi_{n+1}(X, A) & \xrightarrow{\quad q'_* \quad} & \pi_{n+1}(X/A) \\ \cong \downarrow{\scriptstyle h_{n+1}} & & \cong \downarrow{\scriptstyle h'_{n+1}} \\ H_{n+1}(X, A) & \xrightarrow[\cong]{\quad q'_* \quad} & H_{n+1}(X/A), \end{array}$$

$q'_* : \pi_{n+1}(X, A) \to \pi_{n+1}(X/A)$ is an isomorphism. Thus $\mu'_* : \pi_{n+1}(X, A) \to \pi_{n+1}(E_k, I_k)$ is an isomorphism. Because $\pi_i(X, A) = 0 = \pi_i(E_k, I_k)$, for all $i \neq n+1$, we have that $\mu'_* : \pi_i(X, A) \to \pi_i(E_k, I_k)$ is an isomorphism for all i. It is now follows from the exact homotopy sequence of the pairs (X, A) and (E_k, I_k) that $\theta : A \to I_k$ is a weak homotopy equivalence. This completes the proof. □

For a map $p : E \to B$, the statement that $\pi_1(E)$ operates on $\pi_r(B, E)$ means that $\pi_1(E)$ operates on $\pi_r(M_p, E)$, where M_p is the mapping cylinder of p.

Proposition 7.4.2 *Let* $F \xrightarrow{i} E \xrightarrow{p} B$ *be a fibration with* $F = K(G,n)$ *and let* $\pi_1(E)$ *operate trivially on* $\pi_{n+1}(B,E)$. *Then there exists a map* $l : B \to K(G,n+1)$ *such that the given fibration is equivalent to the principal fibration*

$$F \longrightarrow I_l \xrightarrow{v} B$$

induced by l.

Proof. We replace $p : E \to B$ by the inclusion $u : A \to X$ of the mapping cylinder, that is, $A = E$ and $X = M_p$. By Lemma 7.4.1 applied to the pair (X, A), there is a map $k : X \to K(G, n+1)$ and the square in Lemma 7.4.1 is commutative. Furthermore, there are homotopy-commutative squares

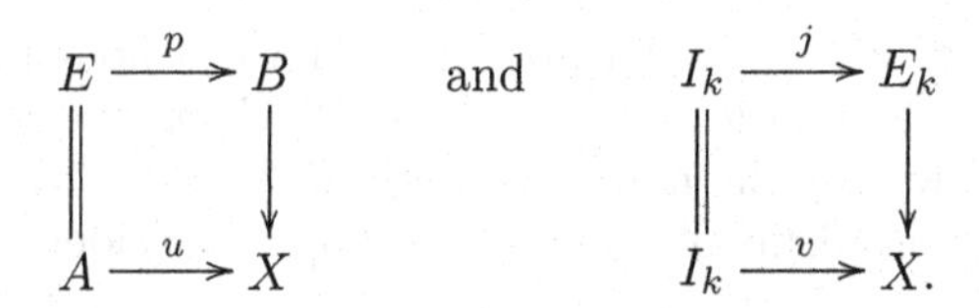

Each of these three squares induces a map of homotopy fibers by Proposition 3.3.15 and so we have the homotopy-commutative diagram

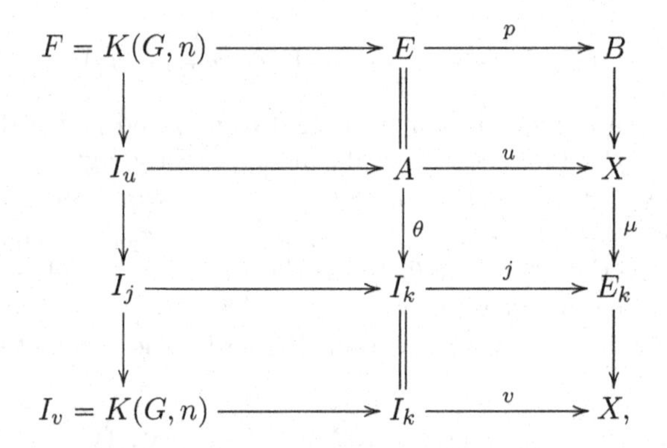

where j is the inclusion and v is the fiber map. All of the vertical maps in the second and third columns are weak homotopy equivalences and thus so are all of the maps in the first column. Therefore we obtain a homotopy-commutative diagram

$$\begin{array}{ccccc}
K(G,n) & \longrightarrow & E & \xrightarrow{p} & B \\
\downarrow{\scriptstyle \alpha} & & \downarrow{\scriptstyle \lambda} & & \downarrow{\scriptstyle \gamma} \\
K(G,n) & \longrightarrow & I_k & \xrightarrow{v} & X,
\end{array}$$

where α, λ, and γ are weak homotopy equivalences. Then, with $l = k\gamma : B \to K(G, n+1)$, it follows from Exercises 3.15 and 7.16 that the fibrations

$$F \xrightarrow{i} E \xrightarrow{p} B \quad \text{and} \quad F \longrightarrow I_l \longrightarrow B$$

are equivalent. □

Next we consider the homotopy decomposition of a map.

Definition 7.4.3 Let $f : X \to Y$ be a map. Then a *weak homotopy decomposition of* f consists of spaces $Z^{(n)}$ for every $n \geqslant 1$ and maps $a_n : X \to Z^{(n)}$ and $b_n : Z^{(n)} \to Y$ such that

1. $b_n a_n = f$.
2. a_n is an n-equivalence.
3. $b_{n*} : \pi_i(Z^{(n)}) \to \pi_i(Y)$ is an isomorphism for $i > n$ and a monomorphism for $i = n$.
4. There are maps $p_{n+1} : Z^{(n+1)} \to Z^{(n)}$ such that $p_{n+1}a_{n+1} = a_n$ and $b_n p_{n+1} = b_{n+1}$

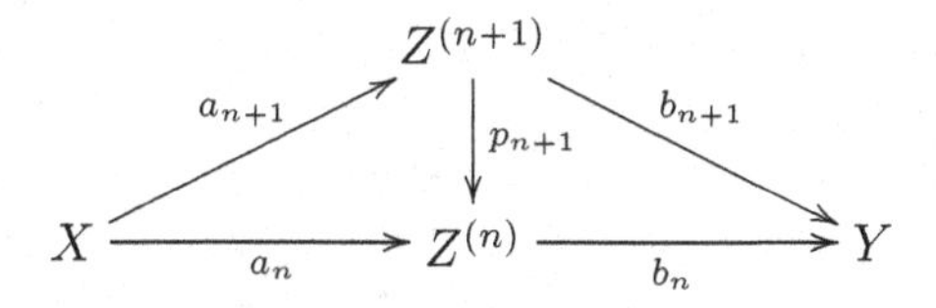

5. The homotopy fiber of $p_{n+1} : Z^{(n+1)} \to Z^{(n)}$ is $K(\pi_n(I_f), n)$, where I_f is the homotopy fiber of f.

If, in addition, p_{n+1} is a principal fibration induced by a map $k^{n+1} : Z^{(n)} \to K(\pi_n(I_f), n+1)$, then we have a *homotopy decomposition of* f or a *Moore–Postnikov decomposition of* f. The weak homotopy decomposition of f is denoted $\{Z^{(n)}, a_n, b_n, p^n\}$ and the homotopy decomposition of f is denoted $\{Z^{(n)}, a_n, b_n, p^n\, k^n\}$. The spaces $Z^{(n)}$ are called the *Postnikov sections of the map* f or the *homotopy sections of the map* f. The maps k^n or homotopy classes $[k^n]$ are called the *Postnikov invariants of the map* f or *k-invariants of the map* f.

Intuitively, the spaces $Z^{(n)}$ are like X in lower dimensions and like Y in higher dimensions. As n increases, the spaces $Z^{(n)}$ "are more like" X and "less like" Y. We see next that weak homotopy decompositions of maps always exist.

Theorem 7.4.4 *For any map* $f : X \to Y$, *a weak homotopy decomposition exists. If* $\pi_1(X)$ *operates trivially on* $\pi_n(Y, X)$, *for all* n, *then* f *has a Moore–Postnikov decomposition.*

Proof. [39, p. 414] By Theorem 2.4.9, for every $n \geqslant 1$, there exists a space $Z'^{(n)}$ obtained from X by attaching cells of dimension $\geqslant n+1$ and maps $a'_n : X \to Z'^{(n)}$ and $b'_n : Z'^{(n)} \to Y$ such that a'_n is the inclusion map, $b'_n a'_n = f$ and $b'_{n*} : \pi_i(Z'^{(n)}) \to \pi_i(Y)$ is an isomorphism for $i > n$ and a monomorphism for $i = n$. For each $n \geqslant 1$, there is a mapping path factorization of b'_n (Definition 3.5.7)

$$Z'^{(n)} \xrightarrow{w} \overline{Z}^{(n)} \xrightarrow{\overline{b}_n} Y,$$

where w is a homotopy equivalence and $\overline{b}_n$ is a fiber map. We consider the n-equivalence $\overline{a}_n = wa'_n$ and the commutative diagram

$$\begin{array}{ccc} X & \xrightarrow{\overline{a}_n} & \overline{Z}^{(n)} \\ \Big\downarrow{\scriptstyle \overline{a}_{n+1}} & & \Big\downarrow{\scriptstyle \overline{b}_n} \\ \overline{Z}^{(n+1)} & \xrightarrow{\overline{b}_{n+1}} & Y. \end{array}$$

By Corollary 4.5.9, there exists a map $q_{n+1} : \overline{Z}^{(n+1)} \to \overline{Z}^{(n)}$ such that $q_{n+1}\overline{a}_{n+1} \simeq \overline{a}_n$ and $\overline{b}_n q_{n+1} \simeq \overline{b}_{n+1}$. Therefore we have the homotopy-commutative diagram

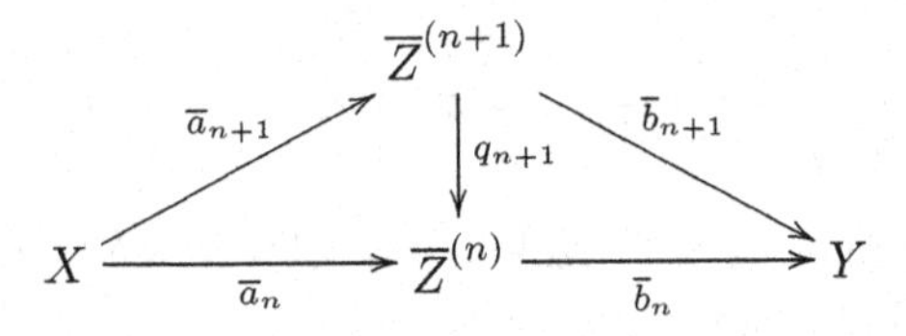

which we will replace with a similar one that is strictly commutative. For this we inductively replace each q_{n+1} by a fiber map p_{n+1} as follows. For $n = 1$, set $Z^{(1)} = \overline{Z}^{(1)}$ and factor $q_2 : \overline{Z}^{(2)} \to \overline{Z}^{(1)} = Z^{(1)}$ through the mapping path as

$$\overline{Z}^{(2)} \xrightarrow{v_2} Z^{(2)} \xrightarrow{p_2} Z^{(1)},$$

where v_2 is a homotopy equivalence and p_2 is a fiber map. Next assume that such a factorization holds for n,

$$\overline{Z}^{(n)} \xrightarrow{v_n} Z^{(n)} \xrightarrow{p_n} Z^{(n-1)}$$

and consider

$$\overline{Z}^{(n+1)} \xrightarrow{q_{n+1}} \overline{Z}^{(n)} \xrightarrow{v_n} Z^{(n)}$$

which is factored as

$$\overline{Z}^{(n+1)} \xrightarrow{v_{n+1}} Z^{(n+1)} \xrightarrow{p_{n+1}} Z^{(n)},$$

where v_{n+1} is a homotopy equivalence and p_{n+1} is a fiber map. This completes the induction. Now define $a_n = v_n\overline{a}_n$ and $\widehat{b}_n = \overline{b}_n\widetilde{v}_n$, for all $n \geqslant 2$, where $\widetilde{v}_n$

is the homotopy inverse of v_n. We have the diagram

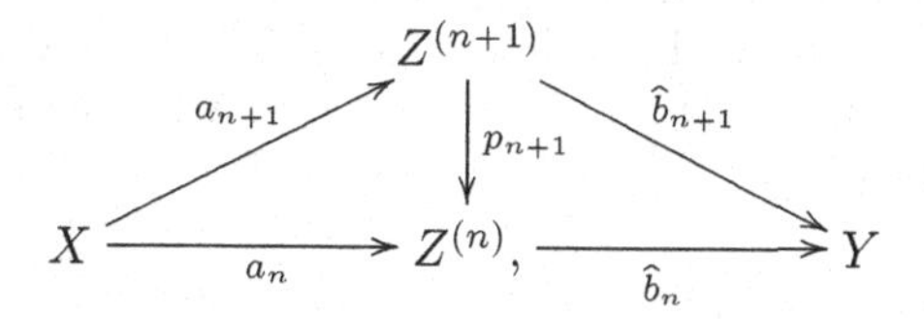

with $p_{n+1}a_{n+1} \simeq a_n$ and $\widehat{b}_n p_{n+1} \simeq \widehat{b}_{n+1}$. Using the CHP, we successively replace each a_n, $n \geqslant 2$, by a homotopic map (also called a_n) such that the left triangles are commutative. We then successively replace each $\widehat{b}_{n+1}$ by $b_{n+1} = b_n p_{n+1}$. Then the right triangles commute. It follows that (2), (3), and (4) of Definition 7.4.3 hold. Furthermore, (1) holds because $b_n a_n = b_1 a_1 = \bar{b}_1 \bar{a}_1 = f$. Therefore to show that we have a weak homotopy decomposition of f, it only remains to show that the homotopy fiber of p_{n+1} is $K(\pi_n(I_f), n)$. We consider the diagram

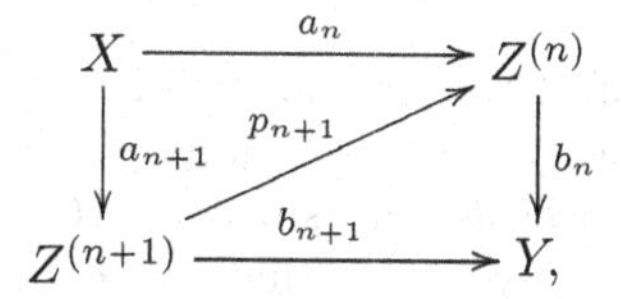

and replace maps by inclusions via mapping cylinders so as to be able to take relative groups. Specifically, we replace a_{n+1} with an inclusion $X \subseteq Z'^{(n+1)}$, where $Z'^{(n+1)} \simeq Z^{(n+1)}$ and $\pi_i(Z'^{(n+1)}, X) = 0$ for $i \leqslant n+1$. Next replace the map $Z'^{(n+1)} \simeq Z^{(n+1)} \to Z^{(n)}$ with an inclusion $Z'^{(n+1)} \subseteq Z'^{(n)}$, where $Z^{(n)} \simeq Z'^{(n)}$. Finally replace $Z'^{(n)} \simeq Z^{(n)} \to Y$ with an inclusion $Z'^{(n)} \subseteq Y'$, where $Y \simeq Y'$ and $\pi_i(Y', Z'^{(n)}) = 0$ for $i \geqslant n+1$. Then we drop the primes for notational convenience and so assume that $X \subseteq Z^{(n+1)} \subseteq Z^{(n)} \subseteq Y$ and that the relative homotopy groups have the properties above (without the primes). By definition, $\pi_r(I_{p_{n+1}}) \cong \pi_{r+1}(Z^{(n)}, Z^{(n+1)})$ and $\pi_r(I_f) \cong \pi_{r+1}(Y, X)$, for all r (Definition 4.5.3). Therefore it suffices to show that

$$\pi_r(Z^{(n)}, Z^{(n+1)}) = \begin{cases} 0 & \text{if } r \neq n+1 \\ \pi_{n+1}(Y, X) & \text{if } r = n+1. \end{cases}$$

We next consider the homotopy exact sequence of the triple $Z^{(n+1)} \subseteq Z^{(n)} \subseteq Y$ (Exercise 4.22)

$$\pi_{i+1}(Y, Z^{(n)}) \longrightarrow \pi_i(Z^{(n)}, Z^{(n+1)}) \longrightarrow \pi_i(Y, Z^{(n+1)}) \longrightarrow \pi_i(Y, Z^{(n)}).$$

If $i \geqslant n+1$, then the first and fourth groups are zero, and so

$$\pi_i(Z^{(n)}, Z^{(n+1)}) \cong \pi_i(Y, Z^{(n+1)}).$$

Therefore $\pi_i(Z^{(n)}, Z^{(n+1)}) = 0$ for $i > n+1$ and $\pi_{n+1}(Z^{(n)}, Z^{(n+1)}) \cong \pi_{n+1}(Y, Z^{(n+1)})$. Similarly, from the homotopy exact sequence of the triple $X \subseteq Z^{(n+1)} \subseteq Z^{(n)}$ we conclude that $\pi_i(Z^{(n)}, Z^{(n+1)}) = 0$ for $i \leqslant n$. Thus it suffices to prove that $\pi_{n+1}(Y, Z^{(n+1)}) \cong \pi_{n+1}(Y, X)$. For this we consider the commutative diagram

$$\begin{array}{ccccccccc}
\pi_{n+1}(X) & \longrightarrow & \pi_{n+1}(Y) & \longrightarrow & \pi_{n+1}(Y, X) & \longrightarrow & \pi_n(X) & \longrightarrow & \pi_n(Y) \\
\downarrow & & \| & & \downarrow & & \downarrow & & \| \\
\pi_{n+1}(Z^{(n+1)}) & \longrightarrow & \pi_{n+1}(Y) & \longrightarrow & \pi_{n+1}(Y, Z^{(n+1)}) & \longrightarrow & \pi_n(Z^{(n+1)}) & \longrightarrow & \pi_n(Y),
\end{array}$$

where the rows are the exact homotopy sequence of a pair. The first vertical arrow (on the left) is an epimorphism and the fourth vertical arrow is an isomorphism. Therefore, by the five lemma (Proposition C.1), the middle arrow is an isomorphism. Hence

$$\pi_{n+1}(Z^{(n)}, Z^{(n+1)}) \cong \pi_{n+1}(Y, Z^{(n+1)}) \cong \pi_{n+1}(Y, X).$$

This completes the proof of the first assertion of the theorem. The remainder of the theorem is easily proved as follows. Since $\pi_1(X)$ operates trivially on $\pi_n(Y, X)$ by hypothesis, and since $\pi_{n+1}(Z^{(n)}, Z^{(n+1)}) \cong \pi_{n+1}(Y, X)$ and $\pi_1(Z^{(n+1)}) \cong \pi_1(X)$, we have that $\pi_1(Z^{(n+1)})$ operates trivially on $\pi_{n+1}(Z^{(n)}, Z^{(n+1)})$ (see Exercise 4.17). By Proposition 7.4.2, p_{n+1} is a principal fibration induced by some map $k^{n+1} : Z^{(n)} \to K(\pi_n(I_f), n+1)$. □

It is possible to prove Theorem 7.4.4 without replacing maps by inclusion maps as was done in the previous proof [40]. However, this would require a lengthy digression on the homotopy groups and cohomology groups of a map, and so we have not done it.

Remark 7.4.5 Many of the topics considered in Section 7.2 for the homotopy decomposition of a space could be investigated for the homotopy decomposition of a map (see [83, 440-444] and [91, 443-449]). Instead we briefly comment on some special cases of Theorem 7.4.4. If $f : X \to Y$ is a fiber map with fiber F, then $\pi_n(I_f) \cong \pi_n(F)$, and so the fibrations $p_{n+1} : Z^{(n+1)} \to Z^{(n)}$ have fiber $K(\pi_n(F), n)$. Also, if we take for f the constant map $X \to \{*\}$, then $\pi_i(Z^{(n)}) = 0$ for $i \geqslant n$ and $K(\pi_i(I_f), i) = K(\pi_i(X), i)$. Therefore, if we set $X^{(n-1)} = Z^{(n)}$, for all n, we obtain a Postnikov decomposition for X under the assumption that $\pi_1(X)$ operates trivially on $\pi_n(X)$, for all n. Finally, consider the map $\{*\} \to Y$. Then the Postnikov decomposition of this map essentially gives the tower of n-connective fibrations (Definition 7.2.7 and Exercise 7.4).

Next we turn to the homology decomposition of a map. We begin with the following basic lemma.

Lemma 7.4.6 *Let $f : X \to Y$ be a map, where X and Y are 1-connected spaces. We regard f as an inclusion map and suppose that $H_i(Y, X) = 0$ for $i \leqslant n$, where $n \geqslant 1$. If $G = H_{n+1}(Y, X)$, then there exists a space X' and a factorization of f as*

$$X \xrightarrow{j} X' \xrightarrow{f'} Y$$

where j and f' are inclusions such that

1. *j is a principal cofiber map induced by some $k : M(G, n) \to X$, that is, $X' = X \cup_k CM(G, n)$.*
2. *$H_i(Y, X') = 0$ for $i \leqslant n + 1$.*
3. *$H_i(Y, X') \cong H_i(Y, X)$ for $i > n + 1$.*

Proof. To avoid considering Moore spaces of type $(G, 1)$, we assume in the proof that $n \geqslant 2$. In Remark 7.4.7 we discuss the necessary modifications for $n = 1$. Recall that $\pi_{n+1}(Y, X; G)$ consists of homotopy classes of maps of pairs $(CM(G, n), M(G, n)) \to (Y, X)$. By the relative Hurewicz theorem (see Exercise 6.26), $G \cong \pi_{n+1}(Y, X)$. Furthermore, the universal coefficient theorem for homotopy applied to the space of paths $E(Y; X, \{*\})$ yields an epimorphism

$$\eta : \pi_{n+1}(Y, X; G) \to \mathrm{Hom}(G, \pi_{n+1}(Y, X)) \cong \mathrm{Hom}(G, G).$$

Therefore there exists $\theta : (CM(G, n), M(G, n)) \to (Y, X)$ such that $\theta_* : H_{n+1}(CM(G, n), M(G, n)) \to H_{n+1}(Y, X)$ is an isomorphism. Now let $k = \theta|M(G, n) : M(G, n) \to X$ and consider the commutative diagram

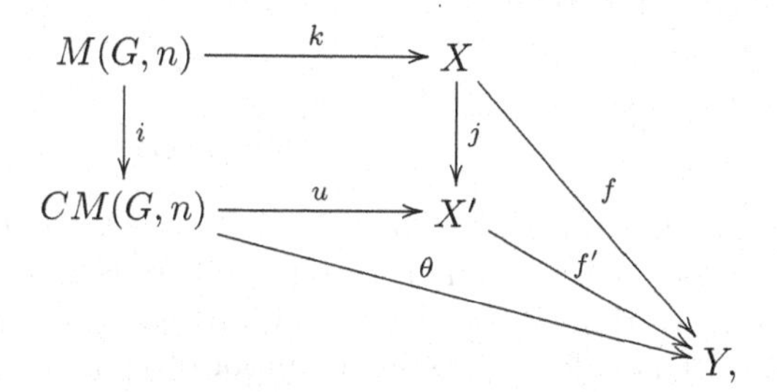

where X' is the pushout of i and k and f' is the map determined by θ and f. Now j is an inclusion and we regard f' as an inclusion (by replacing Y by the mapping cylinder $M_{f'}$). Then $u : (CM(G, n), M(G, n)) \to (X', X)$ induces a homology isomorphism since the pairs have homeomorphic cofibers. Furthermore, $\theta_* : H_{n+1}(CM(G, n), M(G, n)) \to H_{n+1}(Y, X)$ is an isomorphism by the choice of θ. Therefore the inclusion $(X', X) \to (Y, X)$ induces an isomorphism of $(n + 1)$-dimensional relative homology groups. Next consider the exact homology sequence of the triple $X \subseteq X' \subseteq Y$ (Appendix C),

$$H_i(X', X) \longrightarrow H_i(Y, X) \longrightarrow H_i(Y, X') \longrightarrow H_{i-1}(X', X).$$

Since $H_i(X', X) = 0$ for $i \neq n+1$, we have

1. $H_i(Y, X') \cong H_i(Y, X)$ for $i \neq n+1, n+2$.
2. $H_{n+2}(Y, X) \to H_{n+2}(Y, X')$ is a monomorphism.
3. $H_{n+1}(Y, X) \to H_{n+1}(Y, X')$ is an epimorphism.

Then by (2) and (3) and the fact that $H_{n+1}(X', X) \to H_{n+1}(Y, X)$ is an isomorphism, we conclude from the exact homology sequence of the triple $X \subseteq X' \subseteq Y$ that

$$H_{n+1}(Y, X') = 0 \quad \text{and} \quad H_{n+2}(Y, X') \cong H_{n+2}(Y, X).$$

(This is Exercise 7.19.) This completes the proof. □

Remark 7.4.7 We indicate the proof of Lemma 7.4.6 in the case $n = 1$. We assume $H_1(Y, X) = 0$ and set $G = H_2(Y, X)$. We write $G = F/R$ where F is free-abelian with basis $\{x_\alpha \mid \alpha \in A\}$ and R is free-abelian with basis $\{y_\beta \mid \beta \in B\}$. Let $M^1 = \bigvee_{\alpha \in A} S^1_\alpha$ and so $R \subseteq F = H_1(M^1)$. For each $\beta \in B$ we choose $[g_\beta] \in \pi_1(M^1)$ such that $h[g_\beta] = y_\beta$, where h is the Hurewicz homomorphism. We then form M by attaching 2-cells to M^1 by the maps g_β. We set $\Gamma = \pi_1(M)$, so $\Gamma/[\Gamma, \Gamma] = G$ by Proposition B.5. If $E = E(Y; X, \{*\})$, we identify $\pi_2(Y, X; G) = [M, E]$ with $[(CM, M), (Y, X)]$, the homotopy classes of maps of the pair (CM, M) into the pair (Y, X). For every $\phi : \Gamma \to G = \pi_1(E)$, there exists $f : M \to E$ such that $f_* = \phi : \pi_1(M) \to \pi_1(E)$ by Lemma 2.5.1. From this it follows that the function $\eta_\pi : [M, E] \to \mathrm{Hom}(\Gamma, G)$ is onto, where $\eta_\pi[f] = f_* : \pi_1(M) \to \pi_1(E)$. Next consider the commutative diagram

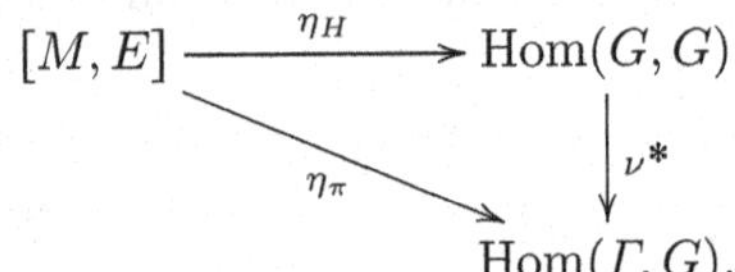

where $\eta_H[k] = k_* : H_1(M) \to H_1(E)$ and $\nu : \Gamma \to G$ is the projection onto the quotient. Because η_π is onto and ν^* is one–one, η_H is onto. Therefore $\eta_H : [(CM, M), (Y, X)] \to \mathrm{Hom}(G, G)$ defined by $\eta_H[g] = g_* : H_2(CM, M) \to H_2(Y, X)$ is onto. It now follows as in the proof of Lemma 7.4.6 that there exists $\theta : (CM, M) \to (Y, X)$ such that $\theta_* : H_2(CM, M) \to H_2(Y, X)$ is an isomorphism. If we write $M = M(G, 1)$, the rest of the proof proceeds as in Lemma 7.4.6.

The following theorem gives the homology decomposition of a map.

Theorem 7.4.8 *Let X and Y be 1-connected spaces and let $f : X \to Y$ be a map. Then for every $k \geqslant 1$, there exist spaces and maps*

$$X \xrightarrow{j_k} W_k \xrightarrow{f_k} Y$$

with j_k an inclusion map such that

1. $f = f_k j_k$.
2. $j_{k*} : H_i(X) \to H_i(W_k)$ *is an isomorphism for* $i > k$ *and a monomorphism for* $i = k$.
3. $f_{k*} : H_i(W_k) \to H_i(Y)$ *is an isomorphism for* $i < k$ *and an epimorphism for* $i = k$.
4. *If we regard* f *and* f_k *as inclusions, then* $H_i(Y, W_k) \cong H_i(Y, X)$ *for* $i > k$.
5. *There exists a map* $l_{k+1} : M(H_{k+1}(Y, X), k) \to W_k$ *such that* $W_{k+1} = C_{l_{k+1}} = W_k \cup_{l_{k+1}} CM(H_{k+1}(Y, X), k)$
6. *If* $i_k : W_k \to W_{k+1}$ *is the inclusion, then the following diagram commutes*

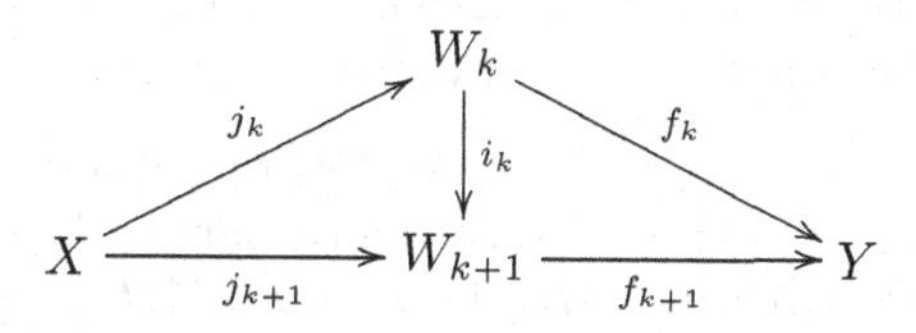

Proof. We define the spaces W_k and the maps j_k and f_k inductively on k. For $k = 1$, we set $W_1 = X$, $j_1 = \mathrm{id}$, and $f_1 = f$. Now assume the theorem for k, so $H_i(Y, W_k) = 0$ for $i \leqslant k$. We then apply Lemma 7.4.6 to $f_k : W_k \to Y$ and obtain a factorization of f_k as

$$W_k \xrightarrow{i_k} W_{k+1} \xrightarrow{f_{k+1}} Y.$$

Therefore from Lemma 7.4.6 we conclude that $f_{k+1*} : H_i(W_{k+1}) \to H_i(Y)$ is an isomorphism for $i < k+1$ and an epimorphism for $i = k+1$ and $H_i(Y, W_{k+1}) \cong H_i(Y, X)$ for $i > k+1$. We then define $j_{k+1} = i_k j_k$ and note that it only remains to prove (2) for $k+1$, that is, $H_i(W_{k+1}, X) = 0$ for $i \geqslant k+2$. From the exact homology sequence of the triple $X \subseteq W_k \subseteq W_{k+1}$ and the fact that $H_i(W_{k+1}, W_k) = 0$ for $i \neq k+1$, we see that $H_i(W_{k+1}, X) \cong H_i(W_k, X)$ for $i \neq k, k+1$. Hence for $i \geqslant k+2$, it follows that $H_i(W_{k+1}, X) \cong H_i(W_k, X) = 0$. This establishes (2) for $k+1$ and completes the proof. □

The spaces W_k together with the maps j_k, f_k, p_k, and l_k that satisfy (1)–(6) of Theorem 7.4.8 constitute a *homology decomposition of the map* f. The spaces W_k are *homology sections of the map* f and the maps l_k or homotopy classes $[l_k]$ are the *k′-invariants of the map* f.

It is interesting to compare the factorization of f in Theorem 7.4.8 as $f = f_k j_k$ with that of f in Theorem 2.4.9 which we write as $f = \overline{f}_k i_k$

$$X \xrightarrow{i_k} T_k \xrightarrow{\overline{f}_k} Y.$$

In the first case, as k gets larger, $f_{k*} : \pi_r(W_k) \to \pi_r(Y)$ is an isomorphism for more values of r and W_k "looks more like" Y. In the second case, as k gets larger, $i_{k*} : \pi_r(X) \to \pi_r(T_k)$ is an isomorphism for more values of r and T_k "looks more like" X.

We observe that if we take the map $\{*\} \to Y$ for f, then the homology decomposition of the map f reduces to the homology decomposition of the space Y.

Corollary 7.4.9 *If X and Y are 1-connected spaces such that $H_i(X) = 0$ for $i \geqslant N$ and $H_i(Y) = 0$ for $i \geqslant N+1$ and $f : X \to Y$ is a map, then $f_N : W_N \to Y$ is a homotopy equivalence.*

Proof. The hypothesis implies that $H_{N+1}(Y, X) = 0$ and so $W_N = W_{N+1}$ and $f_N = f_{N+1}$. But $j_N : X \to W_N$ induces homology isomorphisms in dimensions $> N$. Therefore $H_i(W_N) = 0$ for $i > N$. Also $f_N = f_{N+1} : W_{N+1} \to Y$ induces homology isomorphisms in dimensions $< N+1$. Because W_N and Y are 1-connected, $f_N = f_{N+1}$ is a homotopy equivalence. □

This corollary can be interpreted as asserting that if X and Y are 1-connected finite-dimensional CW complexes, then any map $f : X \to Y$ can be factored as an inclusion $X \to X \cup CM(H_2, 1) \cup \cdots \cup CM(H_N, N-1)$ followed by a homotopy equivalence, for some N, where $H_i = H_i(Y, X)$.

Exercises

Exercises marked with (∗) may be more difficult than the others. Exercises marked with (†) are used in the text.

7.1. (∗) Let $F \xrightarrow{i} E \xrightarrow{p} B$ be a fibration such that (1) B is $(n-1)$-connected, $n \geqslant 2$, (2) if $i \leqslant m-1$ or $i \geqslant m+n-1$, then $\pi_i(F) = 0$, and (3) $F = \Omega X$, for some space X, then prove that there is a map $k : B \to X$ such that the principal fibration $\Omega X \longrightarrow I_k \longrightarrow B$ is equivalent to the given fibration.

7.2. (∗) 1. Let X be a space and let $n : X \times X \to X$ be a map such that $nj_1, nj_2 : X \to X$ have right homotopy inverses. Prove that there is a map $m : X \times X \to X$ such that (X, m) is an H-space.

2. Consider the fibration $\Omega X \longrightarrow X^I \xrightarrow{p} X \times X$, where X is a CW complex and $p(l) = (l(0), l(1))$. Prove that if this fibration is equivalent to a principal fibration, then X is an H-space.

7.3. Show the existence of a Postnikov decomposition for a space with abelian fundamental group by dualizing the proof of Theorem 7.3.2.

7.4. (†) Let $v_n : X^{[n]} \to X$ be the n-connected fibration of a CW complex X. Prove that there is a map $\theta_{n+1} : X^{[n+1]} \to X^{[n]}$ such that $v_n \theta_{n+1} = v_{n+1}$. What is the homotopy fiber of θ_{n+1}?

7.5. Let $v_1 : X^{[1]} \to X$ be the 1-connected fibration of X. Prove that $X^{[1]}$ has the homotopy type of the universal cover of X.

7.6. (†) Let

$$G_l \xleftarrow{f_{l+1}} G_{l+1} \xleftarrow{f_{l+2}} G_{l+2} \longleftarrow \cdots$$

be a sequence of groups and homomorphisms and let $r_i : \varprojlim_{s \geqslant l} G_s \to G_i$ be the projection of the inverse limit. Prove: If there is a $k \geqslant l$ such that f_i is an isomorphism for $i = k+1, k+2, \ldots$, then r_i is an isomorphism for $i \geqslant k$.

7.7. Give a definition of the product of weak homotopy systems $\{X^{(n)}, g_n, p_n\}$ and $\{Y^{(n)}, h_n, q_n\}$. If the first system is a weak homotopy decomposition of X and the second is a weak homotopy decomposition of Y, then prove that this product is a weak homotopy decomposition of $X \times Y$. Given a Postnikov system for X and one for Y, what is the relation between the Postnikov invariants of X, Y and $X \times Y$?

7.8. (†) If $\{X^{(n)}, g_n, p_n\}$ is a weak homotopy decomposition of X and (X, μ) is an H-space, show that there is an H-space multiplication μ_n on $X^{(n)}$ such that $g_n : (X, \mu) \to (X^{(n)}, \mu_n)$ and $p_{n+1} : (X^{(n+1)}, \mu_{n+1}) \to (X^{(n)}, \mu_n)$ are H-maps.

7.9. If X is a space and $m \leqslant n$, prove that $(X^{[m]})^{(n)} \simeq (X^{[n]})^{(m)}$.

7.10. (*) If $\{X^{(n)}, g_n, p_n\}$ is a weak homotopy decomposition of a 1-connected space X and $\dim X \leqslant n$, then prove that $H_{n+1}(X^{(n)}) = 0$ and $H_{n+2}(X^{(n)}) \cong \pi_{n+1}(X)$. Use this result to determine $\pi_3(S^2)$.

7.11. (†) Prove that $H^{n+1}(K(\mathbb{Z}, n)) = 0$.

7.12. (*) Let (X, m) be an H-complex and let (Y, n) be a grouplike space. Let $f : X \to Y$ be a map and let $[D_f] \in [X \wedge X, Y]$ be the H-map deviation of f. Prove that there is a multiplication $\tilde{n}$ on Y such that $f : (X, m) \to (X', \tilde{n})$ is an H-map if and only if $[D_f]$ is in the image of $(f \wedge f)^* : [Y \wedge Y, Y] \to [X \wedge X, Y]$.

7.13. If X is a 1-connected space, show that there exist spaces $X_{[n]}$ and maps $q_n : X \to X_{[n]}$ such that $H_i(X_{[n]}) = 0$ for $i \leqslant n$ and $q_{n*} : H_i(X) \to H_i(X_{[n]})$ is an isomorphism for $i > n$. Show that there exist maps $X_{[n]} \to X_{[n+1]}$ whose mapping cone is an $M(H_{n+1}(X), n+2)$.

7.14. (*) (†) Let $M = M(\mathbb{Z}_m, n) = S^n \cup e^{n+1}$, $m \geqslant 2$, and let $q : M \to M/S^n = S^{n+1}$ be the projection. Prove that $q \not\simeq *$.

7.15. Prove the following dual of Proposition 7.3.5. Let $\{X^{(n)}, g_n, p_n, k^n\}$ be a Postnikov decomposition for a 1-connected space X. Then $k^{n+1}_* = 0 : H_{n+2}(X^{(n)}) \to H_{n+2}(K(\pi_{n+1}(X), n+2))$ if and only if the Hurewicz homomorphism $h_{n+1} : \pi_{n+1}(X) \to H_{n+1}(X)$ is a monomorphism.

7.16. (†) Consider a commutative diagram

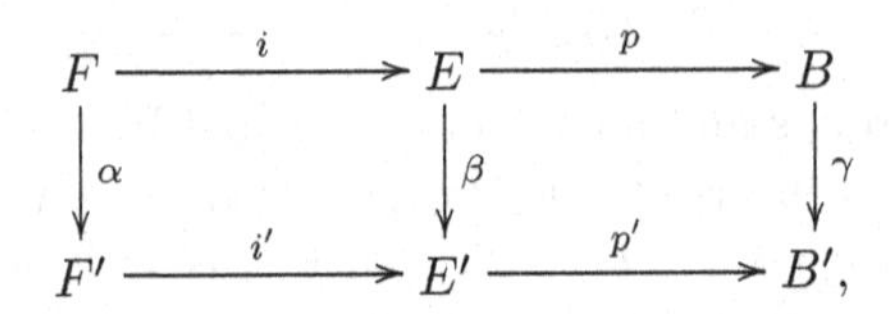

where the top line and bottom line are fiber sequences and α, β, and γ are weak homotopy equivalences. Let P be the pullback of p' and γ and let $F' \to P \to B$ be the resulting fiber sequence. Show that this fiber sequence is equivalent to $F \to E \to B$.

7.17. Prove that Proposition 7.4.2 implies Proposition 7.2.2 by showing that $\pi_1(E)$ operates trivially on $\pi_{n+1}(B, E)$, under the hypothesis of Proposition 7.2.2.

7.18. Let $f : X \to Y$ be a map with homotopy fiber $F = I_f$. Assume that a weak homotopy decomposition for f exists as in Definition 7.4.3 and let F_n be the homotopy fiber of b_n. Show that there exists a map $\phi_n : F \to F_n$ which is an n-equivalence and that $\pi_i(F_n) = 0$ for $i \geqslant n$. (Thus F_n could be taken to be the $(n-1)$-homotopy section of F.)

7.19. Prove the assertion made at the end of the proof of Lemma 7.4.6.

7.20. Let X be 1-connected and consider a homology decomposition of the map $X \to \{*\}$. What is the relation of the space X to the spaces W_k of Theorem 7.4.8?

Chapter 8
Homotopy Sets

8.1 Introduction

In this chapter we consider $[X, Y]$, the collection of morphisms from X to Y in the homotopy category $HoTop_*$. For arbitrary spaces X and Y, $[X, Y]$ is just a set and there is not too much to say. We could ask if it is finite or infinite. If finite, how many elements does it have? If infinite, is it countable or not? We answer some of these in the next section. However, by giving $[X, Y]$ natural group structure we can then study properties of the group $[X, Y]$. But before doing this we discuss the category of a space in Section 8.3. This is a nonnegative integer invariant (or ∞) associated to a space X and denoted $\operatorname{cat} X$. We showed in Chapter 2 that $[X, Y]$ has a natural binary operation with two-sided identity if Y is an H-space or if X is a co-H-space. In Section 8.4 we prove that this set with binary operation is an algebraic loop if Y is an H-complex or if X is a 1-connected co-H-complex. From this we deduce that H-complexes and 1-connected co-H-complexes always have left and right homotopy inverses. We then study the nilpotency of the group $[X, Y]$ when Y is a grouplike space. We prove that the group is nilpotent whenever X has finite category and that the nilpotency class of $[X, Y]$ is bounded above by $\operatorname{cat} X$.

8.2 The Set $[X, Y]$

In the preceding chapters we have proved several results that give conditions for an induced map of homotopy sets to be a bijection. In order to summarize, we state the most general of these results next.

Proposition 8.2.1 (Proposition 2.4.6) *Let X be a CW complex, let B and Y be spaces (not necessarily of the homotopy type of CW complexes), and let $e : B \to Y$ be an n-equivalence, $n < \infty$. Then $e_* : [X, B] \to [X, Y]$ is* *an injection if* $\dim X < n$ *and a surjection if* $\dim X \leqslant n$. *If* $n = \infty$, *then* $e_* : [X, B] \to [X, Y]$ *is a bijection.*

Special cases of this proposition are Proposition 1.5.24 (where e is the inclusion $(Z, C)^n \to Z$ of the n-skeleton of a relative CW complex (Z, C)) and Proposition 7.2.17 (where e is the map $g_{n-1} : Y \to Y^{(n-1)}$ of a weak homotopy decomposition of Y).

Proposition 8.2.2 *Let* $f : X \to Y$ *be an* n*-equivalence, let* Z *be a space, and let* $f^* : [Y, Z] \to [X, Z]$ *be the induced map. If* $\pi_i(Z) = 0$ *for* $i \geqslant n$, *then* f^* *is a surjection. If* $\pi_i(Z) = 0$ *for* $i \geqslant n + 1$, *then* f^* *is an injection.*

This proposition has not been proved explicitly, but it follows immediately from Corollary 2.4.10 and Proposition 2.4.13.

We next give conditions for $[X, Y]$ to be finite. We first introduce some terminology and notation. If G is a finitely generated abelian group, then $r(G)$ denotes the *rank of* G. If we write $G = \mathbb{Z} \oplus \cdots \oplus \mathbb{Z} \oplus T$, where there are r copies of $\mathbb{Z}$ and T is a finite group, then $r(G) = r$. For a space X with $H^n(X)$ finitely generated, we define the *nth Betti number* $\beta_n(X)$ *of* X to be $r(H^n(X))$. If $\pi_n(X)$ is a finitely generated abelian group, we define $\gamma_n(X)$ to be $r(\pi_n(X))$.

Lemma 8.2.3 *Let* X *be a space and let* G *be an abelian group such that* $H^n(X)$, $H^{n+1}(X)$, *and* G *are finitely generated. If* $\beta_n(X)\, r(G) = 0$, *then* $H^n(X; G)$ *is a finite group.*

Proof. Consider the exact sequence of the universal coefficient theorem for cohomology (Theorem 5.2.4)

$$0 \longrightarrow H^n(X) \otimes G \longrightarrow H^n(X; G) \longrightarrow H^{n+1}(X) * G \longrightarrow 0.$$

Because $\beta_n(X) r(G) = 0$, $H^n(X) \otimes G$ is finite. Because $H^{n+1}(X)$ and G are finitely generated, $H^{n+1}(X) * G$ is finite. Therefore $H^n(X; G)$ is finite. □

We next give an easily verifiable criterion for the set $[X, Y]$ to be finite.

Proposition 8.2.4 *Let* X *be a finite CW complex of dimension* N *and let* Y *be a simply connected space with finitely generated homotopy groups* $\pi_i(Y)$ *for* $i \leqslant N$. *If* $\beta_n(X)\gamma_n(Y) = 0$ *for all* $n \leqslant N$, *then* $[X, Y]$ *is finite.*

Proof. Let $\{Y^{(n)}, g_n, p_n, k^n\}$ be a Postnikov system for Y. By Proposition 8.2.1, $g_{N*} : [X, Y] \to [X, Y^{(N)}]$ is a bijection, and so it suffices to prove that $[X, Y^{(N)}]$ is finite. We do this by showing by induction on n that $[X, Y^{(n)}]$ is finite for all $n \leqslant N$. Let $\pi_i = \pi_i(Y)$ and consider the case when $n = 2$. Then $Y^{(2)} = K(\pi_2, 2)$ and $[X, Y^{(2)}] = H^2(X; \pi_2)$. But $\beta_2(X)\gamma_2(Y) = 0$, and so $[X, Y^{(2)}]$ is finite by Lemma 8.2.3. Now assume that $[X, Y^{(n-1)}]$ is finite for $n \leqslant N$. We have the principal fiber sequence

$$K(\pi_n, n) \longrightarrow Y^{(n)} \xrightarrow{p_n} Y^{(n-1)}$$

and the resulting exact sequence

$$[X, K(\pi_n, n)] \longrightarrow [X, Y^{(n)}] \xrightarrow{p_{n*}} [X, Y^{(n-1)}].$$

Because $[X, Y^{(n-1)}]$ is finite by induction, the image set $p_{n*}[X, Y^{(n)}]$ is finite. Next consider the operation of $H^n(X; \pi_n) = [X, K(\pi_n, n)]$ on $[X, Y^{(n)}]$. If $x, y \in [X, Y^{(n)}]$, then $p_{n*}(x) = p_{n*}(y)$ if and only if $y = x^\alpha$, for some $\alpha \in [X, K(\pi_n, n)]$ by Theorem 4.4.4. But $\beta_n(X)\gamma_n(Y) = 0$, and so the group $[X, K(\pi_n, n)]$ is finite. Therefore for $w \in p_{n*}[X, Y^{(n)}]$, each preimage set $p_{n*}^{-1}(w)$ is finite. Thus $[X, Y^{(n)}]$ is finite. This completes the induction and shows that $[X, Y]$ is finite. □

Proposition 8.2.4 holds if the condition $\beta_n(X)\gamma_n(Y) = 0$ for all $n \leqslant N$ is replaced by the condition that $H^n(X)$ is a finite group for all $n \leqslant N$ or by the condition that $\pi_n(X)$ is a finite group for all $n \leqslant N$. Furthermore, the converse of Proposition 8.2.4 does not hold as we show by example. There are many such examples, but the one we give uses an interesting fact about the Hopf map. For any sphere S^n, we denote a map of degree k by $\mathrm{k}\colon S^n \to S^n$.

Lemma 8.2.5 *The following diagram is homotopy-commutative*

$$\begin{array}{ccc} S^3 & \xrightarrow{\mathrm{k}^2} & S^3 \\ \downarrow \phi & & \downarrow \phi \\ S^2 & \xrightarrow{\mathrm{k}} & S^2, \end{array}$$

for any integer k, where ϕ is the Hopf map.

Proof. The map $\mathrm{k}\phi$ represents an element in $\pi_3(S^2)$ and so $\mathrm{k}\phi \simeq n\phi$ for some integer n. Therefore there is a map $\mathrm{n}\colon S^3 \to S^3$ of degree n such that $\mathrm{k}\phi = \phi\mathrm{n}$. Then n and k induce a map $\theta : \mathbb{C}\mathrm{P}^2 \to \mathbb{C}\mathrm{P}^2$ because $\mathbb{C}\mathrm{P}^2$ is the mapping cone C_ϕ. If $u \in H^2(\mathbb{C}\mathrm{P}^2) \cong \mathbb{Z}$ is a generator, then $\theta^*(u) = ku$. Furthermore, $u^2 \in H^4(\mathbb{C}\mathrm{P}^2) \cong \mathbb{Z}$ is a generator and

$$nu^2 = \theta^*(u^2) = \theta(u)\theta(u) = k^2u^2.$$

Hence $n = k^2$. □

The following example provides a counterexample to the converse of Proposition 8.2.4.

Example 8.2.6 (Cf. Exercise 8.4) The set $[\mathbb{C}\mathrm{P}^2, S^2]$ is a finite set.

Proof. Consider the mapping cone sequence

$$S^3 \xrightarrow{\phi} S^2 \xrightarrow{j} \mathbb{C}P^2 \xrightarrow{q} S^4,$$

where j is the inclusion and q is the projection, and the exact sequence of sets

$$[S^4, S^2] \xrightarrow{q^*} [\mathbb{C}P^2, S^2] \xrightarrow{j^*} [S^2, S^2] \xrightarrow{\phi^*} [S^3, S^2].$$

Then $\phi^*(\mathrm{k}) = \mathrm{k}\phi = k^2\phi$ by Lemma 8.2.5, and so $\phi^*(\mathrm{k}) = 0 \iff k = 0$. Therefore $\mathrm{Im} j^* = 0$ and thus q^* is onto. Because $[S^4, S^2] = \pi_4(S^2) \cong \pi_4(S^3) \cong \mathbb{Z}_2$ by Theorem 5.6.10, $[\mathbb{C}P^2, S^2]$ is finite. □

Next we consider when $[X, Y]$ is a countable set. For this we use some basic facts about simplicial complexes that can be found in [73]. In addition, a countable CW complex is one in which the number of cells is finite or countably infinite.

Proposition 8.2.7 *If X is a finite CW complex and Y is a countable CW complex, then $[X, Y]$ is a countable set.*

Proof. By [39, p. 182], a finite CW complex has the homotopy type of a finite simplicial complex and a countable CW complex has the homotopy type of a countable simplicial complex. Therefore it suffices to show that $[K, L]$ is countable where K is a finite simplicial complex and L is a countable simplicial complex. We begin by assuming that both K and L are finite simplicial complexes. Any simplicial map $\phi : K \to L$ carries the vertices of a simplex of K into the vertices of some simplex in L and is completely determined by its values on the vertices. Thus there are finitely many simplicial maps $K \to L$. Next recall that the simplicial approximation theorem states that any map $f : K \to L$ is homotopic to a simplicial map $\phi : K^{(n)} \to L$, where $K^{(n)}$ is the nth barycentric subdivision of K. Since $K^{(n)}$ is a finite complex, there are finitely many simplicial maps $K^{(n)} \to L$, and we denote the set of these maps by $\mathcal{S}_n$. Then

$$\mathcal{S} = \bigcup_{n \geqslant 0} \mathcal{S}_n$$

is a countable set. Since every map $f : K \to L$ is homotopic to some element of $\mathcal{S}$, the set $[K, L]$ is countable. If L is countably infinite, the Simplicial Approximation still holds and there are countably many simplicial maps $K \to L$. The rest of the preceding argument then goes through. □

8.3 Category

In preparation for studying the group structure on the homotopy set $[X, Y]$ when Y is a grouplike space, we discuss the notion of category in this section and establish a few of its basic properties. Category is a numerical invariant of the homotopy type of a space that is interesting in its own right.

Recall that a co-H-space consists of a space X and a map $c : X \to X \vee X$, called the comultiplication, such that $q_1 c \simeq \text{id} \simeq q_2 c$, where $q_1, q_2 : X \vee X \to X$ are the two projections. This can be restated as follows (Exercise 2.6). If X is a space, then there is a comultiplication on X if and only if the diagonal map $\Delta : X \to X \times X$ is homotopic to a map that factors through $X \vee X$. This latter characterization can be easily generalized.

Definition 8.3.1 For a space X, the *nth fat wedge* is defined by

$$T^n(X) = \{(x_1, x_2, \dots, x_n) \mid x_i \in X, \text{ some } x_i = *\} \subseteq X^n.$$

If $j_{n+1} : T^{n+1}(X) \to X^{n+1}$ is the inclusion, then X has *category* $\leqslant n$, written $\operatorname{cat} X \leqslant n$, if there is a map $\phi : X \to T^{n+1}(X)$ such that $j_{n+1}\phi \simeq \Delta : X \to X^{n+1}$, where $\Delta = \Delta_X^{n+1}$ is the diagonal map of X. We say $\operatorname{cat} X = n$ if $\operatorname{cat} X \leqslant n$ and $\operatorname{cat} X \nleqslant n - 1$.

Remark 8.3.2 For a space X, clearly $\operatorname{cat} X = 0$ if and only if X is contractible, and $\operatorname{cat} X \leqslant 1$ if and only if there is a map $c : X \to X \vee X$ such that (X, c) is a co-H-space. The category of a space has also been called the *Lusternik–Schnirelmann category* or the *LS category.* The original definition given in [58] which is equivalent to Definition 8.3.1 is as follows. A space X has LS category $\leqslant n$ if there is an open cover $\{U_0, U_1, \dots, U_n\}$ of X such that each U_i is freely contractible in X to a point. The motivation for studying category came from a theorem of Lusternik–Schnirelmann on smooth compact manifolds M [58]. This asserts that a lower bound for the number of critical points of any smooth function on M is $\operatorname{cat} X + 1$. With some restrictions, this result holds for infinite-dimensional manifolds, and so has applications to the calculus of variations. For these and many other results on category, see [21].

In general, it is difficult to calculate the category of a space. We next prove two basic results that will enable us to make this calculation for some spaces.

Proposition 8.3.3 *If* $f : X \to Y$ *is a map and* $C_f = Y \cup_f CX$ *is the mapping cone of* f*, then*

$$\operatorname{cat} C_f \leqslant \operatorname{cat} Y + 1.$$

Proof. Using the mapping cylinder, we can regard $f : X \to Y$ as a cofiber map. The inclusion map $i : CX \to C_f$ is then a cofiber map (Proposition 3.2.10). We denote C_f by Z and consider the diagram

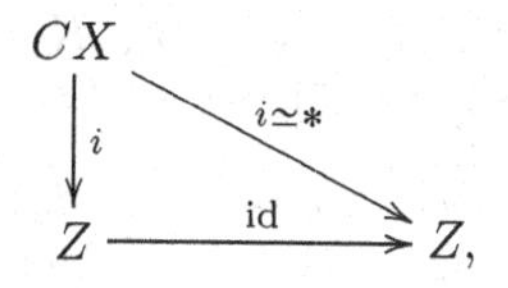

where $i \simeq *$ since CX is contractible. Because i is a cofiber map, there is homotopy $r_t : Z \to Z$ such that $r_0 = \mathrm{id}$ and $r_1|CX = *$. Now suppose $\operatorname{cat} Y = n - 1$ with map $\phi : Y \to T^n(Y)$ such that $\Delta_Y \simeq j\phi : Y \to Y^n$, where $j = j_n : T^n(X) \to X^n$. Thus there is a homotopy $h_t : Y \to Y^n$ with $h_0 = \Delta_Y$ and $h_1 = j\phi$. Consider the diagram

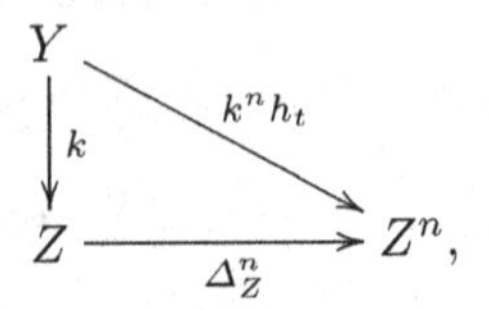

where $k : Y \to Z$ is the inclusion. Since k is a cofiber map, there exists a homotopy $l_t : Z \to Z^n$ such that $l_0 = \Delta_Z^n$ and $l_1 k = k^n j\phi$. Now define $m_t : Z \to Z^{n+1}$ by $m_t(z) = (r_t(z), l_t(z))$ for $z \in Z$. Then $m_0 = \Delta_Z^{n+1}$. Furthermore $m_1\langle x, t\rangle \in \{*\} \times Z^n$ and $m_1\langle y\rangle \in Z \times T^n(Z)$, for $x \in X$, $t \in I$, and $y \in Y$. Therefore $m_1(Z) \subseteq T^{n+1}(Z)$, and so $\operatorname{cat} Z \leqslant n$. □

Corollary 8.3.4 *If X is a finite-dimensional CW complex, then* $\operatorname{cat} X \leqslant \dim X$. *If X is 1-connected and N is the number of nontrivial positive-dimensional homology groups of X, then* $\operatorname{cat} X \leqslant N$.

Proof. For the first assertion, apply Proposition 8.3.3 to the skeletal decomposition of X. For the second assertion, apply Proposition 8.3.3 to a homology decomposition of X. □

In order to prove the next result, we digress to discuss cup products in cohomology, where cohomology is regarded as homotopy classes of maps into an Eilenberg–Mac Lane space. Let G_i be abelian groups, let $n_i \geqslant 1$ be integers and let K_i denote the Eilenberg–Mac Lane space $K(G_i, n_i)$, $i = 1, \ldots, k$. We define $T(K_1, \ldots, K_k) \subseteq K_1 \times \cdots \times K_k$ to consist of all k-tuples $(x_1, \ldots, x_k)$ such that some $x_i = *$. Thus $T(X, \ldots, X)$ with k copies of X is just $T^k(X)$. We set $K_1 \wedge \cdots \wedge K_k = K_1 \times \cdots \times K_k / T(K_1, \ldots, K_k)$.

Lemma 8.3.5 *With the above notation,*

$$\pi_N(K_1 \wedge \cdots \wedge K_k) \cong G_1 \otimes \cdots \otimes G_k,$$

where $N = n_1 + \cdots + n_k$.

Proof. Because the $(n_i - 1)$-skeleton $K_i^{n_i - 1} = \{*\}$, the $(N-1)$-skeleton $T(K_1, \ldots, K_k)^{N-1} = (K_1 \times \cdots \times K_k)^{N-1}$. Therefore $(K_1 \wedge \cdots \wedge K_k)^{N-1} = \{*\}$, and so $K_1 \wedge \cdots \wedge K_k$ is $(N-1)$-connected. Thus $\pi_N(K_1 \wedge \cdots \wedge K_k) \cong H_N(K_1 \wedge \cdots \wedge K_k)$ by the Hurewicz theorem. But $H_N(K_1 \wedge \cdots \wedge K_k) \cong G_1 \otimes \cdots \otimes G_k$ by the Künneth theorem [83, p. 235]. □

Lemma 8.3.6 *If G is an abelian group and $m : G_1 \otimes \cdots \otimes G_k \to G$ is any homomorphism, then there exists a unique element $[\theta_m] \in H^N(\bigwedge_{i=1}^n K_i; G)$ such that $\eta_\pi[\theta_m] = m$, where*

$$\eta_\pi : H^N(\bigwedge_{i=1}^n K_i; G) \to \mathrm{Hom}(\pi_N(\bigwedge_{i=1}^n K_i), G) \cong \mathrm{Hom}(G_1 \otimes \cdots \otimes G_k, G)$$

is the homomorphism that assigns to a homotopy class the induced homotopy homomorphism in dimension N and $\bigwedge_{i=1}^n K_i = K_1 \wedge \cdots \wedge K_k$.

Proof. By Proposition 2.5.13, η_π is an isomorphism. □

Now we define the cup product of k cohomology elements. Given $\alpha_i = [f_i] \in H^{n_i}(X; G_i)$ and a homomorphism $m : G_1 \otimes \cdots \otimes G_k \to G$, we consider the maps

$$X \xrightarrow{\Delta} X^k \xrightarrow{f_1 \times \cdots \times f_k} \prod_{i=1}^n K_i \xrightarrow{q} \bigwedge_{i=1}^n K_i \xrightarrow{\theta_m} K(G, N),$$

where q is the projection, θ_m is defined in Lemma 8.3.6, and $N = n_1 + \cdots + n_k$. The *k-fold cup product* $\alpha_1 \cup \cdots \cup \alpha_k \in H^N(X; G)$ is the homotopy class $[\theta_m q(f_1 \times \cdots \times f_k)\Delta]$. If R is a commutative ring and $G_i = R$ for all i, we choose m to be the multiplication homomorphism $m : R \otimes \cdots \otimes R \to R$ defined by $m(x_1, \ldots, x_k) = x_1 \cdots x_k$ for $x_1, \ldots, x_k \in R$. In this case we obtain the k-fold cup product as a function $H^{n_1}(X; R) \times \cdots \times H^{n_k}(X; R) \to H^N(X; R)$.

Definition 8.3.7 Let X be a space and let R be a commutative ring. The *R-cup length of X*, denoted $\mathrm{cup}_R X$, is the least integer k such that all $(k+1)$-fold cup products of positive-dimensional elements of $H^*(X; R)$ are zero.

Proposition 8.3.8 *For any commutative ring R and space X,*

$$\mathrm{cup}_R X \leqslant \mathrm{cat}\, X.$$

Proof. We suppose that $\mathrm{cat}\, X = k - 1$ and so there is a map $\phi : X \to T^k(X)$ such that $j_k \phi \simeq \Delta : X \to X^k$, where $j_k : T^k(X) \to X^k$ is the inclusion. If $\alpha_i = [f_i] \in H^{n_i}(X; R)$, with $i = 1, \ldots, k$ and $n_i > 0$, then we must show $\alpha_1 \cup \cdots \cup \alpha_k = 0$. There is a homotopy-commutative diagram

$$\begin{array}{ccccccccc}
 & & T^k(X) & \xrightarrow{f} & T(K_i, \ldots, K_k) & & & & \\
 & {}^{\phi}\nearrow & \downarrow{\scriptstyle j_k} & & \downarrow{\scriptstyle j} & & & & \\
X & \xrightarrow{\Delta} & X^k & \xrightarrow{f_1 \times \cdots \times f_k} & \prod_{i=1}^n K_i & \xrightarrow{q} & \bigwedge_{i=1}^n K_i & \xrightarrow{\theta_m} & K(G, N),
\end{array}$$

where j is the inclusion and f is induced by $f_1 \times \cdots \times f_k$. Then

$$\begin{aligned}
\theta_m q(f_1 \times \cdots \times f_k)\Delta &\simeq \theta_m q(f_1 \times \cdots \times f_k) j_k \phi \\
&= \theta_m q j f \phi \\
&= *
\end{aligned}$$

because $qj = *$. Therefore $\alpha_1 \cup \cdots \cup \alpha_k = 0$. □

Corollary 8.3.4 and Proposition 8.3.8 enable us to make some computations.

Corollary 8.3.9 *For the projective spaces,*

$$\operatorname{cat} \mathbb{R}\mathrm{P}^n = \operatorname{cat} \mathbb{C}\mathrm{P}^n = \operatorname{cat} \mathbb{H}\mathrm{P}^n = n.$$

Proof. We have $\operatorname{cup}_{\mathbb{Z}_2} \mathbb{R}\mathrm{P}^n = n$, $\operatorname{cup}_{\mathbb{Z}} \mathbb{C}\mathrm{P}^n = n$, and $\operatorname{cup}_{\mathbb{Z}} \mathbb{H}\mathrm{P}^n = n$. The result for $\mathbb{R}\mathrm{P}^n$ now follows from the first assertion of Corollary 8.3.4 and the results for $\mathbb{C}\mathrm{P}^n$ and $\mathbb{H}\mathrm{P}^n$ follow from the second assertion of Corollary 8.3.4. □

There is another definition of category that we now describe and which is equivalent to Definition 8.3.1. Let

$$F \xrightarrow{i} E \xrightarrow{p} B$$

be a fibration that we denote by $\mathcal{F}$. The excision map $\alpha : C_i \to B$ is defined by $\alpha | CF = *$ and $\alpha | E = p$ (Definition 4.5.20). Furthermore, by Corollary 6.3.6 and the proof of the Serre theorem 6.4.2, the homotopy fiber $I_\alpha \simeq F * \Omega B$. By replacing α by a fiber map, we obtain a fibration

$$F' \longrightarrow E' \longrightarrow B$$

with fiber $F' \simeq F * \Omega B$. We denote this fiber sequence by $\mathcal{F}'$ and call this construction, which assigns the fibration $\mathcal{F}'$ to the fibration $\mathcal{F}$, the *cofiber–fiber construction.*

Definition 8.3.10 For a space X, the *nth Ganea fibration*

$$F_n(X) \xrightarrow{i_n} G_n(X) \xrightarrow{p_n} X ,$$

$n \geqslant 0$, denoted $\mathcal{F}_n$, is inductively defined as follows: $\mathcal{F}_0$ is the path space fibration $\Omega X \to EX \to X$. Assuming that $\mathcal{F}_{n-1}$ has been defined, we define $\mathcal{F}_n$ to be $\mathcal{F}'_{n-1}$, that is, we apply the cofiber–fiber construction to $\mathcal{F}_{n-1}$.

As we have seen, the fiber $F_n(X)$ has the homotopy type of $\Omega X * \cdots * \Omega X$, the join of $n+1$ copies of ΩX. In addition, the nth Ganea fibratiion is easily seen to be natural, that is, a map $f : X \to X'$ yields a map $G_n(f) : G_n(X) \to G_n(X')$ which is compatible with f and the Ganea fiber maps.

Next we consider the space $\widetilde{G}_n(X)$ defined by the homotopy pullback

$$\begin{array}{ccc} \widetilde{G}_n(X) & \longrightarrow & T^{n+1}(X) \\ \Big\downarrow {\scriptstyle \widetilde{p}_n} & & \Big\downarrow {\scriptstyle j_{n+1}} \\ X & \xrightarrow{\Delta} & X^{n+1}. \end{array}$$

Theorem 8.3.11 *There is a homotopy equivalence $\theta : G_n(X) \to \widetilde{G}_n(X)$ such that the following diagram is homotopy-commutative*

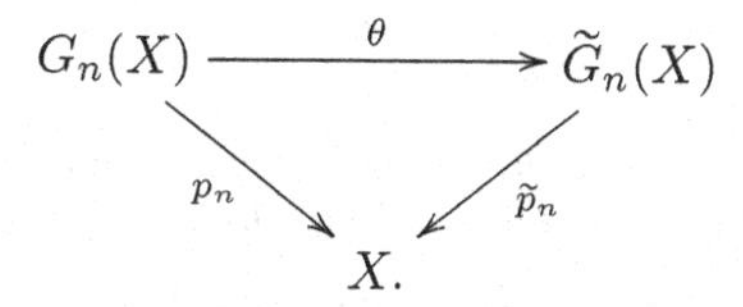

The proof of this theorem is long and we omit it. It can be found in [5] and [21, pp. 29-31].

It now follows that $p_n : G_n(X) \to X$ has a homotopy section if and only if $\widetilde{p}_n : \widetilde{G}_n(X) \to X$ has a homotopy section. But from the definition of $\widetilde{G}_n(X)$ as a homotopy pullback, we see that $\widetilde{p}_n$ has a homotopy section if and only if the diagonal map $\Delta : X \to X^{n+1}$ factors up to homotopy through $T^{n+1}(X)$. Thus we have the following result.

Proposition 8.3.12 *For a space X, $\operatorname{cat} X \leqslant n$ if and only if the nth Ganea fiber map $p_n : G_n(X) \to X$ has a homotopy section.*

The condition that p_n has a homotopy section is sometimes taken as the definition of $\operatorname{cat} X \leqslant n$.

8.4 Loop and Group Structure in $[X, Y]$

We begin with the definition of an algebraic loop.

Definition 8.4.1 Let S be a set with a binary operation that is written additively. Then S is called an *algebraic loop* if for every $a, b \in S$, the equations $a + x = b$ and $y + a = b$ have unique solutions $x, y \in S$.

Lemma 8.4.2 *Let S be a set with a binary operation and define $\sigma, \mu : S \times S \to S \times S$ by $\sigma(a, b) = (a, a + b)$ and $\mu(a, b) = (b, a + b)$ for $a, b \in S$. Then S is an algebraic loop if and only if σ and μ are bijections.*

Proof. The assertion that σ is a bijection is equivalent to the assertion that $a + x = b$ has a unique solution. Similarly for μ and $y + a = b$. □

Next let X, Y, A, and B be spaces and consider the projections $p_1 : A \times B \to A$ and $p_2 : A \times B \to B$ and the inclusions $i_1 : A \to A \vee B$ and $i_2 : B \to A \vee B$. We proved the following results in Corollary 1.3.7. There is a bijection $\lambda : [X, A \times B] \to [X, A] \times [X, B]$ defined by

$$\lambda(\alpha) = (p_{1*}(\alpha), p_{2*}(\alpha))$$

and a bijection $\tau : [A \vee B, Y] \to [A, Y] \times [B, Y]$ defined by

$$\tau(\alpha) = (i_1^*(\alpha), i_2^*(\alpha)).$$

The first part of the following proposition is due to James [52].

Proposition 8.4.3

1. *If (Y, m) is an H-complex and X is a space, then $[X, Y]$, with the binary operation induced by m, is an algebraic loop.*
2. *If (X, c) is a 1-connected co-H-complex and Y is a space, then $[X, Y]$, with the binary operation induced by c, is an algebraic loop.*

Proof. (1) By Lemma 8.4.2, it suffices to show that $\sigma, \mu : [X, Y] \times [X, Y] \to [X, Y] \times [X, Y]$ are bijections. We do this only for σ because the argument for μ is similar. Consider the square diagram (actually rectangular)

$$\begin{array}{ccc} [X, Y \times Y] & \xrightarrow{\sigma'_*} & [X, Y \times Y] \\ \downarrow{\scriptstyle\lambda} & & \downarrow{\scriptstyle\lambda} \\ [X, Y] \times [X, Y] & \xrightarrow{\sigma} & [X, Y] \times [X, Y], \end{array}$$

where λ is the bijection above and σ' is defined by $\sigma' = (p_1, m) : Y \times Y \to Y \times Y$. If $[f] \in [X, Y \times Y]$, then

$$\lambda\sigma'_*[f] = ([p_1 f], [mf])$$

and

$$\sigma\lambda[f] = ([p_1 f], [p_1 f] + [p_2 f]).$$

Because $mf \simeq (p_1 + p_2)f \simeq p_1 f + p_2 f$, the square diagram is commutative. We now show that σ'_* is a bijection by showing that σ' is a homotopy equivalence. For this, it suffices to prove that σ' induces an isomorphism of all homotopy groups by Whitehead's first theorem. Consider the square diagram with $X = S^n$. Because $\pi_n(Y)$ is a group, $\sigma : \pi_n(Y) \times \pi_n(Y) \to \pi_n(Y) \times \pi_n(Y)$ is a bijection by Lemma 8.4.2. Therefore $\sigma'_* : \pi_n(Y \times Y) \to \pi_n(Y \times Y)$ is an isomorphism. This completes the proof of (1).

For (2) we show that the map $\chi : X \vee X \to X \vee X$ defined by $\chi = \{c, i_2\}$ is a homotopy equivalence. For this we prove that χ induces an isomorphism of all homology groups (Exercise 8.7). Since X is 1-connected, so is $X \vee X$, and thus χ is a homotopy equivalence by Whitehead's second theorem 6.4.15. □

Hence if (Y, m) is an H-complex, then $[Y, Y]$ is an algebraic loop. Thus there are maps $l, r : Y \to Y$ such that $[l] + [\mathrm{id}] = 0$ and $[\mathrm{id}] + [r] = 0$. This yields $m(l, \mathrm{id}) \simeq *$ and $m(\mathrm{id}, r) \simeq *$, and we call l a *left homotopy inverse* and r a *right homotopy inverse* of m (see Definition 2.2.1). If m is homotopy associative, then $[Y, Y]$ is associative. Therefore

$$l \simeq l + (\mathrm{id} + r) \simeq (l + \mathrm{id}) + r \simeq r.$$

Similar considerations hold for co-H-spaces. Thus we have the following.

Proposition 8.4.4 *If (Y, m) is a homotopy-associative H-complex, then (Y, m) is a grouplike space. If (X, c) is a 1-connected homotopy-associative co-H-complex, then (X, c) is a cogroup.*

We next consider the group $[X, Y]$ when X is a cogroup. We first introduce some simple facts about (not necessarily abelian) groups. A group G is said to satisfy the *maximal condition* if G and all its subgroups are finitely generated. The following facts are not difficult to prove (see Exercise 8.13).

- If G satisfies the maximal condition, then so does every subgroup of G and every quotient group of G.
- If

$$A \longrightarrow B \longrightarrow C$$

is an exact sequence of groups and A and C satisfy the maximal condition, then B satisfies the maximal condition.

Proposition 8.4.5 *Let X and Y be spaces such that X is a finite complex and a cogroup and $\pi_i(Y)$ is a finitely generated abelian group for all $i \geqslant 1$. Then $[X, Y]$ satisfies the maximal condition.*

Proof. Let $\{Y^{(n)}, g_n, p_n\}$ be a weak homotopy decomposition of Y and suppose that $\dim X \leqslant N$. Because $g_{N*} : [X, Y] \to [X, Y^{(N)}]$ is an isomorphism by Proposition 8.2.1, it suffices to prove that $[X, Y^{(n)}]$ satisfies the maximal condition for all n, and we do this by induction on n. If $\pi_i(Y) = 0$ for $i < k$, where $k \geqslant 1$, then the weak homotopy decomposition of Y starts with $Y^{(k)} = K(\pi_k(Y), k)$. Because $[X, Y^{(k)}] = H^k(X; \pi_k(Y))$, it follows from the universal coefficient theorem for cohomology (Theorem 5.2.4) that $[X, Y^{(k)}]$ satisfies the maximal condition. Now assume that $[X, Y^{(n)}]$ satisfies the maximal condition for some $n \geqslant 1$ and consider the principal fibration

$$K(\pi_{n+1}(Y), n+1) \longrightarrow Y^{(n+1)} \longrightarrow Y^{(n)}.$$

We obtain the exact sequence of groups

$$H^{n+1}(X; \pi_{n+1}(Y)) \longrightarrow [X, Y^{(n+1)}] \longrightarrow [X, Y^{(n)}].$$

Because $H^{n+1}(X; \pi_{n+1}(Y))$ and $[X, Y^{(n)}]$ satisfy the maximal condition, so does $[X, Y^{(n+1)}]$. This completes the induction. □

A similar result holds when Y is a grouplike space.

Proposition 8.4.6 *If X is a 1-connected finite complex and Y is a grouplike space with $\pi_i(Y)$ a finitely generated group for all $i \geqslant 1$, then $[X, Y]$ satisfies the maximal condition.*

The proof is left as an exercise (Exercise 8.14).

We now consider the group $[X, Y]$ in more detail when Y is a grouplike space (or by Proposition 8.4.4, a homotopy-associative H-complex). We first recall some group theory. If G is a group, we denote the commutator $xyx^{-1}y^{-1}$ of $x, y \in G$ by $[x, y]$. A k-fold commutator $[x_1, x_2, \ldots, x_k]$ of elements $x_1, x_2, \ldots, x_k \in G$ is inductively defined as follows. A 1-fold commutator is a group element and a 2-fold commutator is just a commutator. If $(k-1)$-fold commutators have been defined, set $[x_1, x_2, \ldots, x_k] = [[x_1, x_2, \ldots, x_{k-1}], x_k]$. For subgroups H and K of G, we let $[H, K]$ equal the subgroup of G generated by all commutators $[h, k]$ for $h \in H$ and $k \in K$. We then inductively define subgroups $\Gamma_n(G)$ as follows. Let $\Gamma_1(G) = G$. We assume $\Gamma_{n-1}(G)$ defined and set $\Gamma_n(G) = [\Gamma_{n-1}(G), G]$. Then there is a chain of subgroups

$$\cdots \subseteq \Gamma_n(G) \subseteq \Gamma_{n-1}(G) \subseteq \cdots \subseteq \Gamma_2(G) \subseteq \Gamma_1(G) = G$$

called the *lower central series of* G. If there is an n such that $\Gamma_n(G) = 1$, the trivial group, then G is called *nilpotent*. The smallest $k \geqslant 0$ such that $\Gamma_{k+1}(G) = 1$ is called the *nilpotency* of G and is denoted $\operatorname{nil} G$. Clearly $\operatorname{nil} G \leqslant k$ if and only if all $(k+1)$-fold commutators $[x_1, x_2, \ldots, x_{k+1}] = 1$. In particular, $\operatorname{nil} G \leqslant 1$ if and only if G is abelian. An exact sequence of groups

$$H \xrightarrow{j} G \xrightarrow{q} K$$

is called a *central sequence* if $j(H)$ is contained in the center of G.

We next state and prove two simple lemmas that lead to a short proof of a result of G. W. Whitehead that relates the category of X to the nilpotency of $[X, Y]$. This proof has appeared in [6].

Lemma 8.4.7 *If*

$$H \xrightarrow{j} G \xrightarrow{q} K$$

is a central sequence of groups and $\operatorname{nil} K \leqslant k-1$, *then* $\operatorname{nil} G \leqslant k$.

Proof. If $[[x_1, x_2, \ldots, x_k], x_{k+1}]$ is a $(k+1)$-fold commutator in G, then $q[x_1, x_2, \ldots, x_k] = [q(x_1), q(x_2), \ldots, q(x_k)] = 1$. Hence $[x_1, x_2, \ldots, x_k] = j(x)$, for some $x \in H$. Therefore $[[x_1, x_2, \ldots, x_k], x_{k+1}] = [j(x), x_{k+1}] = 1$ because $j(x)$ is in the center of G. Therefore $\operatorname{nil} G \leqslant k$. □

The next lemma deals with category. We take Definition 8.3.10 using the Ganea fibrations

$$F_n(X) \xrightarrow{i_n} G_n(X) \xrightarrow{p_n} X$$

as the definition of category.

Lemma 8.4.8 *If* X *is a space and* Y *is a grouplike space, then the nilpotency* $\operatorname{nil}[G_k(X), Y] \leqslant k$.

Proof. This is proved by induction on k. Clearly $\operatorname{nil}[G_0(X), Y] = 0$ because $G_0(X)$ is contractible. Suppose the result true for $k-1$. Recall that $G_k(X) \simeq C_{i_{k-1}}$, the mapping cone of the inclusion $i_{k-1} : F_{k-1}(X) \to G_{k-1}(X)$. Therefore it suffices to show $\operatorname{nil}[C_{i_{k-1}}, Y] \leqslant k$. Consider the mapping cone sequence

$$F_{k-1}(X) \xrightarrow{i_{k-1}} G_{k-1}(X) \xrightarrow{j_{k-1}} C_{i_{k-1}} \xrightarrow{q_k} \Sigma F_{k-1}(X),$$

where j_{k-1} is the inclusion and q_k is the projection. This gives an exact sequence of groups

$$[\Sigma F_{k-1}(X), Y] \xrightarrow{q_k^*} [C_{i_{k-1}}, Y] \xrightarrow{j_{k-1}^*} [G_{k-1}(X), Y].$$

By Proposition 5.4.6, this is a central sequence. But $\operatorname{nil}[G_{k-1}(X), Y] \leqslant k-1$ by induction. Therefore, $\operatorname{nil}[G_k(X), Y] \leqslant k$, by Lemma 8.4.7. □

Theorem 8.4.9 [91, Chapter X, §3] *If X is a space of finite category and Y is a grouplike space, then the group $[X, Y]$ is nilpotent and*

$$\operatorname{nil}[X, Y] \leqslant \operatorname{cat} X.$$

Proof. Let $\operatorname{cat} X = n$. Then the Ganea fiber map $p_n : G_n(X) \to X$ has a homotopy section $s : X \to G_n(X)$. Therefore $s^* : [G_n(X), Y] \to [X, Y]$ is onto. Since $\operatorname{nil}[G_n(X), Y] \leqslant n$ by Lemma 8.4.8, $\operatorname{nil}[X, Y] \leqslant n$. □

Remark 8.4.10

1. For a product of k spheres $S^{n_1} \times \cdots \times S^{n_k}$ it can be shown that $\operatorname{cat}(S^{n_1} \times \cdots \times S^{n_k}) \leqslant k$ [21, p. 18]. Thus $\operatorname{nil}[S^{n_1} \times \cdots \times S^{n_k}, Y] \leqslant k$ if Y is a grouplike space. In [91, p. 467] an explicit lower central series is given for $[S^{n_1} \times \cdots \times S^{n_k}, Y]$ whose successive quotients are homotopy groups of Y.
2. Several papers have been written that dualize the notion of category. One such is [74, pp. 323–347] which contains references to many others. The definition of category in terms of open covers or of the fat wedge seems less amenable to dualization than the definition in terms of the Ganea fibration. Although many of the results on category have been dualized, the notion of cocategory is of lesser interest than that of category. This may be partially due to the fact that, aside from contractible spaces, H-spaces, and spaces with a finite number of nontrivial homotopy groups, it appears that the cocategory of no other space has been calculated.

Exercises

Exercises marked with (*) may be more difficult than the others. Exercises marked with (†) are used in the text.

8.1. Let X and Y be CW complexes with weak homotopy decompositions $\{X^{(n)}, g_n, p_n\}$ and $\{Y^{(n)}, h_n, q_n\}$, respectively. Let $f : X \to Y$ be a map and let $f^{(n)} : X^{(n)} \to Y^{(n)}$ be the induced map (Proposition 7.2.11). Prove that if $\dim X \leqslant n$, then the function $\rho : [X,Y] \to [X^{(n)}, Y^{(n)}]$ defined by $\rho[f] = [f^{(n)}]$ is a bijection.

8.2. Let X be an $(n-1)$-connected H-space, $n \geqslant 2$, and let Y be a grouplike space such that $\pi_i(Y) = 0$, for $i \geqslant 2n$. Prove that every map $f : X \to Y$ is an H-map (Cf. Exercise 2.7).

8.3. Determine for which integers $k, l \geqslant 1$ the set $[\mathbb{R}\mathrm{P}^k, S^l]$ is finite and for which integers it is infinite. (For this problem it is necessary to know the Betti numbers of $\mathbb{R}\mathrm{P}^k$.)

8.4. Prove that $[\mathbb{C}\mathrm{P}^2, S^2] = 0$.

8.5. How many elements are there in the sets $[\mathbb{R}\mathrm{P}^{2k}, \mathbb{C}\mathrm{P}^k]$, $[\mathbb{R}\mathrm{P}^{2k+1}, \mathbb{C}\mathrm{P}^{k+1}]$, and $[\mathbb{R}\mathrm{P}^\infty, \mathbb{C}\mathrm{P}^\infty]$?

8.6. Let X, Y, A, and B be spaces and consider the projections $q_1 : A \vee B \to A$ and $q_2 : A \vee B \to B$ and the inclusions $j_1 : A \to A \times B$ and $j_2 : B \to A \times B$. By Corollary 1.3.7 the function $\theta : [X,A] \times [X,B] \to [X, A \times B]$ defined by $\theta([f],[g]) = [(f,g)]$ is a bijection. Prove that if X is a co-H-space, then $\theta : [X,A] \times [X,B] \to [X, A \times B]$ is given by $\theta(\beta,\gamma) = j_{1*}(\beta) + j_{2*}(\gamma)$. Dually, show that the function $\rho : [A,Y] \times [B,Y] \to [A \vee B, Y]$ defined by $\rho([f],[g]) = [\{f,g\}]$ satisfies $\rho(\beta,\gamma) = q_1^*(\beta) + q_2^*(\gamma)$ if Y is an H-space.

8.7. (†) In the proof of Proposition 8.4.3, show that $\chi : X \vee X \to X \vee X$ induces the following homomorphism of homology groups: $\chi_*(\alpha,\beta) = (\alpha, \alpha + \beta)$, where we identify $H_*(X \vee X)$ with $H_*(X) \oplus H_*(X)$.

8.8. Let Y be an H-complex and $l, r : Y \to Y$ left and right homotopy inverses respectively.

1. Prove that $lr \simeq \mathrm{id}$ and $rl \simeq \mathrm{id}$.
2. Prove that l and r are homotopy equivalences without using part (1).
3. Dualize (1) and (2).

8.9. If X is a CW complex of dimension $< 2n-1$, then show that $[X, S^n]$ has abelian group structure with the following property. If X' is a CW complex of dimension $< 2n-1$ and $f : X \to X'$ is a map, then $f^* : [X', S^n] \to [X, S^n]$ is a homomorphism. The group $[X, S^n]$ is called the *cohomotopy group* of X.

8.10. Let X be a $(k-1)$-connected CW complex of dimension $\leqslant nk-1$, where $k \geqslant 2$. Prove that $\operatorname{cat} X \leqslant n-1$. If $\dim X < nk-1$ and $\phi, \psi : X \to T^n(X)$ are maps such that $j\phi \simeq \Delta \simeq j\psi : X \to X^n$, then prove that $\phi \simeq \psi$.

8.11. By using the covering definition of category in Remark 8.3.2, prove Proposition 8.3.3.

8.12. Let $p_1 : G_1(X) \to X$ be the first Ganea fiber map and let $\pi : \Sigma\Omega X \to X$ be the map defined by $\pi(\omega, t) = \omega(t)$, for $\omega \in \Omega X$ and $t \in I$. Use Exercise 6.13 to show that there is a homotopy equivalence $\theta : G_1(X) \to \Sigma\Omega X$ such that the diagram

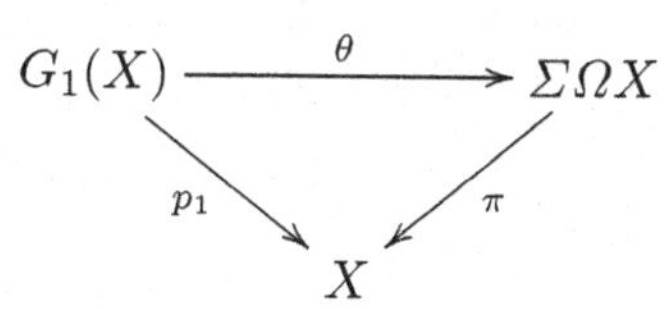

is homotopy-commutative. Conclude that X is a co-H-space if and only if π has a homotopy section.

8.13. (†) Prove the following assertions which appear in Section 8.4.

1. If G satisfies the maximal condition, then so does every subgroup of G and every quotient group of G.
2. If $A \longrightarrow B \longrightarrow C$ is an exact sequence of groups and A and C satisfy the maximal condition, then B satisfies the maximal condition.

8.14. (†) Prove Proposition 8.4.6.

8.15. By dualizing the Ganea fibrations, give a definition of $\operatorname{cocat} X \leqslant n$, where cocat is the dual of cat.

Chapter 9
Obstruction Theory

9.1 Introduction

There are two basic topological problems regarding mappings that are called the *extension problem* and the *lifting problem.* In algebraic topology, these problems are often treated by means of obstruction theory. In this chapter we give an introduction to the main ideas of this theory.

In the extension problem we are given a relative CW complex (X, A) with inclusion $i : A \to X$, a space Y, and a map $g : A \to Y$ and we ask if there is a map $h : X \to Y$, called an extension of g, such that the diagram

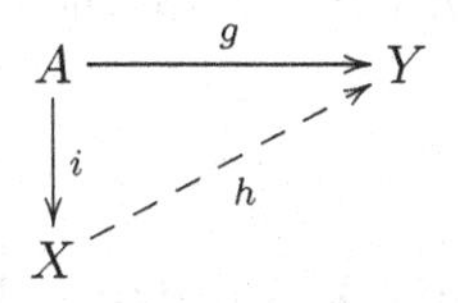

commutes. In the obstruction theory for the extension problem discussed in Section 9.3, we consider the homotopy sections $Y^{(n)}$ of Y and assume that there is a map $X \to Y^{(n)}$ compatible with g. This determines an element in $H^{n+1}(X, A; \pi_n(Y))$ called the obstruction element. The vanishing of the obstruction element is a necessary and sufficient condition for the existence of a compatible map $X \to Y^{(n+1)}$. If $\dim(X, A)$ is finite, the existence of compatible maps $X \to Y^{(n)}$ for large n implies the existence of an extension h. Thus there is a systematic procedure to determine if g has an extension.

In the lifting problem there is a fiber map $p : E \to B$ and a map $f : X \to B$ and we ask if there is an $h : X \to E$, called a lift, such that the following diagram commutes

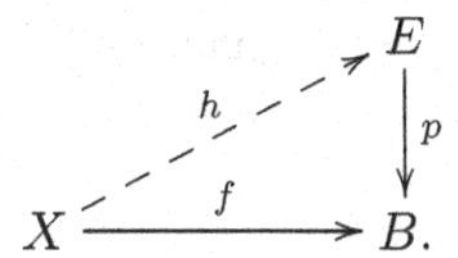

Obstruction elements in cohomology are also defined for this problem (in 9.3) so that the nth stage lift yields an $(n+1)$st stage lift if and only if the nth obstruction vanishes.

We combine both of these problems into a single problem by considering the following diagram

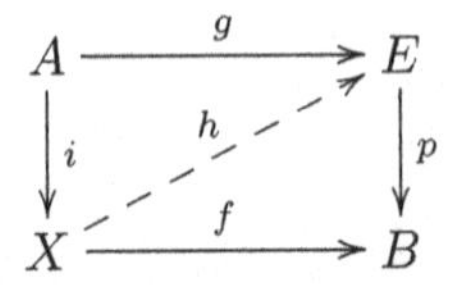

with commutative square called the *extension-lifting square.* We assume that (X, A) is a relative CW complex and that p is a fiber map. The problem is to determine if there exists a map $h : X \to E$ such that $hi = g$ and $ph = f$. This is called the *extension-lifting problem* and the solution h is called a *relative lift.* By taking $B = \{*\}$, we obtain the extension problem, and by taking $A = \{*\}$, we obtain the lifting problem.

The approach in this chapter uses a homotopy decomposition of the map p ([14], [39]) or a homology decomposition of the map i. In the former case, we develop an obstruction theory for the extension-lifting problem with obstruction elements in $H^{n+1}(C_i; \pi_n(F))$, where C_i is the mapping cone of i and F is the fiber of p. We specialize these results in Section 9.3 to obtain obstructions for the extension and lifting problems. In the last section we briefly discuss some additional topics in obstruction theory and indicate how to obtain obstructions to the extension-lifting problem by using a homology decomposition of the map i. In this case the obstructions lie in homotopy groups with coefficients.

9.2 Obstructions Using Homotopy Decompositions

We begin with a key lemma that is needed to define the obstruction.

Suppose that (X, A) is a relative CW complex and that there is a commutative diagram

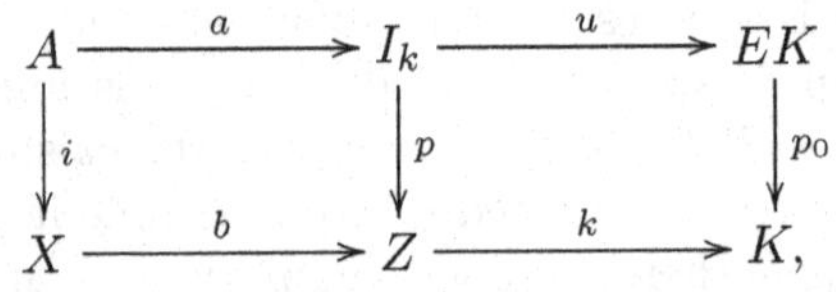

where the square on the right is the pullback diagram that defines I_k, p and p_0 are fiber maps and i is the inclusion map. Let $\widetilde{ua} : CA \to K$ be the adjoint of $ua : A \to EK$. We then have the diagram

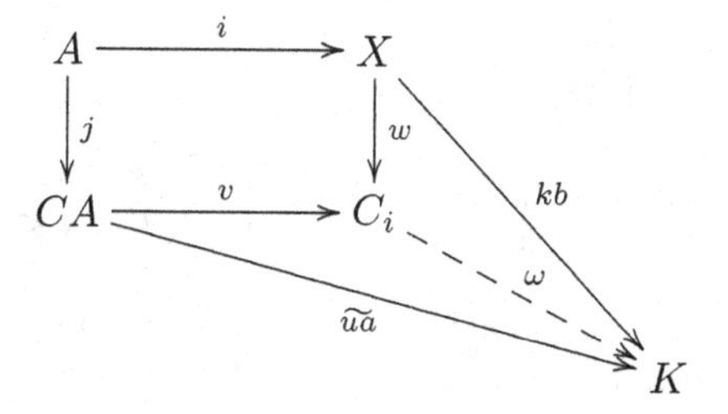

in which the square is the pushout diagram defining the mapping cone $C_i = X \cup_i CA$. Because $kbi = \widetilde{ua}j$, there is a map $\omega : C_i \to K$ such that $\omega v = \widetilde{ua}$ and $\omega w = kb$.

The following lemma gives necessary and sufficient conditions for the existence of a relative lift in the case when p is a principal fiber map.

Lemma 9.2.1 [14, p. 500] *With the above notation, there exists a map $b' : X \to I_k$ such that $pb' = b$ and $b'i = a$,*

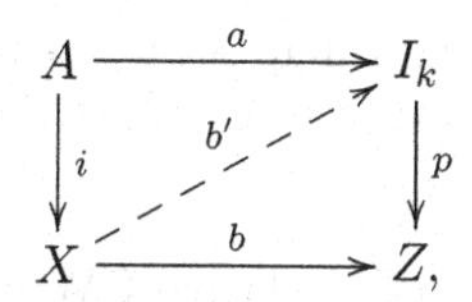

if and only if $* \simeq \omega : C_i \to K$.

Proof. Suppose that $* \simeq \omega$ with homotopy $\omega_t : C_i \to K$, and so $* \simeq_{\omega_t w} kb : X \to K$ and $* \simeq_{\omega_t v} \widetilde{ua} : CA \to K$. Thus the adjoint $\widetilde{\omega_t v} : A \to EK$ is a homotopy between $*$ and ua, and we have the diagram

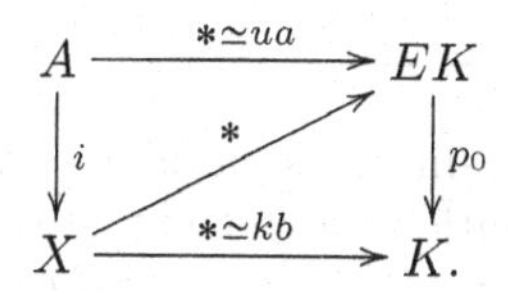

Because $p_0\widetilde{\omega_t v} = \omega_t wi$, by the CHEP (Corollary 3.3.6), there is a homotopy $h_t : X \to EK$ such that $h_0 = *$, $p_0 h_t = \omega_t w$, and $h_t i = \widetilde{\omega_t v}$. Then $h_1 : X \to EK$ and $b : X \to Z$ determine a map $b' : X \to I_k$ such that $ub' = h_1$ and $pb' = b$,

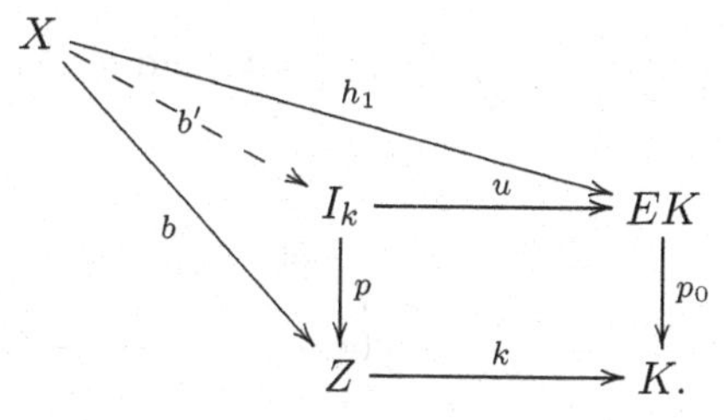

It follows that $b'i = a$.

Conversely, suppose there exists $b' : X \to I_k$ such that $pb' = b$ and $b'i = a$. We define $l : CA \to EK$ by

$$l\langle x, t\rangle(s) = \begin{cases} ua(x)(s+t) & \text{if } 0 \leqslant s \leqslant 1-t \\ * & \text{if } 1-t \leqslant s \leqslant 1, \end{cases}$$

for $x \in A$ and $t, s \in I$. Note that $l\langle x, t\rangle$ is the path along $ua(x)$ from $ua(x)(t)$ to $ua(x)(1) = *$ followed by the constant path and thus $lj = ua$. We consider the diagram

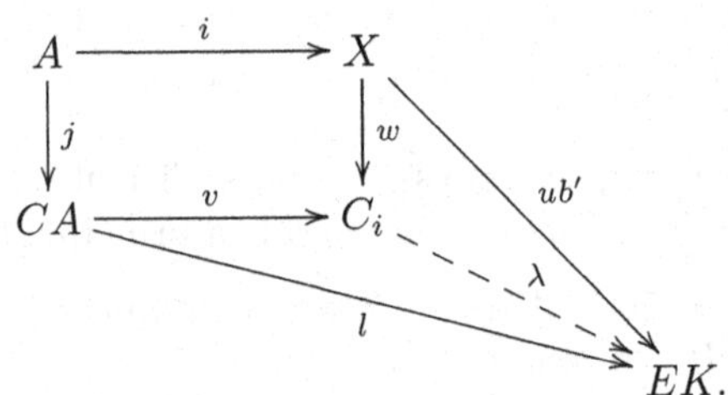

Because $lj = ub'i$, there exists a map $\lambda : C_i \to EK$ such that $\lambda w = ub'$ and $\lambda v = l$. It then easily follows that $p_0\lambda = \omega : C_i \to K$. Therefore $* \simeq \omega$ since EK is contractible. □

Now suppose that (X, A) is a relative CW complex with inclusion map $i : A \to X$ and that $p : E \to B$ is a fiber map with fiber F. We assume that there are maps $g : A \to E$ and $f : X \to B$ such that the following diagram commutes

$$\begin{array}{ccc} A & \xrightarrow{g} & E \\ {\scriptstyle i}\downarrow & & \downarrow{\scriptstyle p} \\ X & \xrightarrow{f} & B, \end{array}$$

so that this is an extension-lifting square. We also assume that there is a Moore–Postnikov decomposition for the map p. By Theorem 7.4.4 this occurs if $\pi_1(E)$ acts trivially on $\pi_n(B, E)$ (see Exercise5.17), for all n. Thus by Definition 7.4.3, there are spaces $Z^{(n)}$ and maps $a_n : E \to Z^{(n)}$ and $b_n : Z^{(n)} \to B$ such that $p = b_n a_n$, for all $n \geqslant 1$. Furthermore, there are principal fibrations

$$K_n \longrightarrow Z^{(n+1)} \xrightarrow{p_{n+1}} Z^{(n)},$$

where $K_n = K(\pi_n(F), n)$ and $p_{n+1}a_{n+1} = a_n$ and $b_n p_{n+1} = b_{n+1}$,

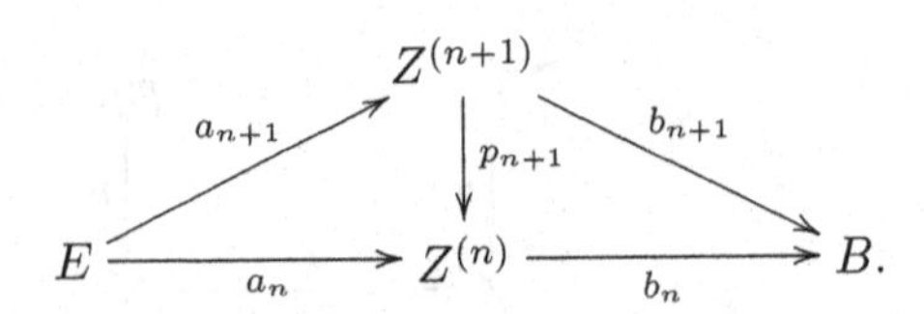

Now suppose that there is a map $h_n : X \to Z^{(n)}$ such that the following diagram commutes

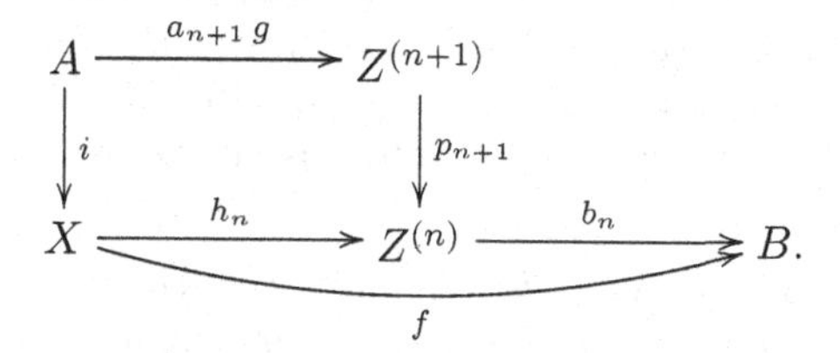

We can then define an obstruction to a relative lift $X \to Z^{(n+1)}$.

Definition 9.2.2 The *obstruction* to the existence of a map $h_{n+1} : X \to Z^{(n+1)}$ such that $h_{n+1}i = a_{n+1}g$ and $p_{n+1}h_{n+1} = h_n$ is the element $\omega = \omega_{h_n} \in H^{n+1}(C_i; \pi_n(F))$ defined just before Lemma 9.2.1, where C_i is the mapping cone of i.

By Lemma 9.2.1, h_{n+1} exists if and only if $\omega_{h_n} = 0$. In this case, $h_{n+1}i = p_{n+2}a_{n+2}g$ and $b_{n+1}h_{n+1} = f$, and so the obstruction $\omega_{h_{n+1}}$ can be defined.

Now suppose that $\dim(X, A) \leqslant N$ and h_N exists. Then there is the commutative diagram

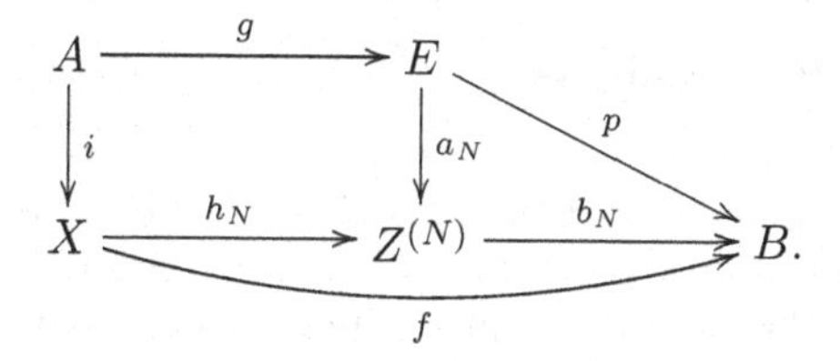

Since a_N is an N-equivalence, the HELP lemma (4.5.7) implies that there exists a map $h' : X \to E$ such that $h'i = g$ and $a_N h' \simeq_F h_N$, where $F : X \times I \to Z^{(N)}$ is a homotopy which is stationary on $A \times I$. Thus $ph' = b_N a_N h' \simeq b_N h_N = f$, and this homotopy is stationary on $A \times I$. By the CHEP applied to the fiber map p, there is a map $h : X \to E$ with $h \simeq h'$, $hi = g$, and $ph = f$. Therefore h is a relative lift.

The map h_N can be obtained by requiring that a sequence of obstructions is zero. We do this next and in the process we add a few restrictive assumptions. We begin by defining a map $h_1 : X \to Z^{(1)}$. If we assume $\pi_1(B) = 0$, then we can take $Z^{(1)} = B$ because b_1 is a homotopy equivalence. Then $b_1 = \mathrm{id}$, $a_1 = p$, and $h_1 = f$. An alternative assumption would be to take $Z^{(1)}$ to be the covering space of B which corresponds to the subgroup $p_*\pi_1(E) \subseteq \pi_1(B)$. Hence $b_1 : Z^{(1)} \to B$ is the covering map and we require that $f_*\pi_1(X) \subseteq p_*\pi_1(E)$ (see [39, p. 418]). Then f lifts to $h_1 : X \to Z^{(1)}$, and by the uniqueness of lifts, $h_1|A = a_1 g$. Thus we obtain $\omega_{h_1} \in H^2(C_i; \pi_1(F))$, where $\pi_1(F)$ is assumed to be abelian. We suppose that this element is zero and get $h_2 : X \to Z^{(2)}$. We continue in this way until we reach $\omega_{h_{N-1}}$ which is assumed to be zero. Thus the map h_N exists, and so we have the desired relative lift $h : X \to E$.

In general, there are problems in carrying out this procedure. This is because if $\omega_{h_{n-1}} = 0$, then the resulting map h_n is neither unique nor described explicitly. Hence it may be difficult then to determine ω_{h_n}. However, with the above assumptions, there is one obvious situation in which all the obstructions vanish.

Proposition 9.2.3 *Let* (X, A) *be a relative CW complex with* $\dim(X, A) \leqslant N$*, let* $p : E \to B$ *be a fiber map with fiber* F*, and suppose that the cohomology group* $H^{j+1}(C_i; \pi_j(F)) = 0$ *for all* $j < N$*. Then for any maps* $g : A \to E$ *and* $f : X \to B$ *such that* $pg = fi$*, there is a relative lift* $h : X \to E$*.*

This proposition also holds when $N = \infty$. We give the proof of this in the next section for the case when $B = \{*\}$ (the extension problem).

If H^*_{sing} denotes singular cohomology, then

$$H^{j+1}(C_i; \pi_j(F)) \cong H^{j+1}_{\text{sing}}(C_i; \pi_j(F)) \cong H^{j+1}_{\text{sing}}(X, A; \pi_j(F)),$$

and so the obstructions can be regarded as elements of $H^{j+1}_{\text{sing}}(X, A; \pi_j(F))$.

9.3 Lifts and Extensions

In this section we consider special cases of the results of Section 9.2 in order to develop an obstruction theory for lifts and extensions.

We begin with obstructions of lifts which we discuss very briefly because it easily follows from Section 9.2. The extension-lifting square with $A = \{*\}$ yields a diagram

$$\begin{array}{ccc} & & E \\ & \overset{h}{\nearrow} & \downarrow p \\ X & \xrightarrow{f} & B, \end{array}$$

where X is a CW complex, f is a map, and p is a fiber map with fiber F. We seek a map $h : X \to E$ such that $ph = f$. It is assumed that there is a Moore–Postnikov decomposition of p as in Section 9.2 and that there is a map $h_n : X \to Z^{(n)}$ such that the following diagram commutes

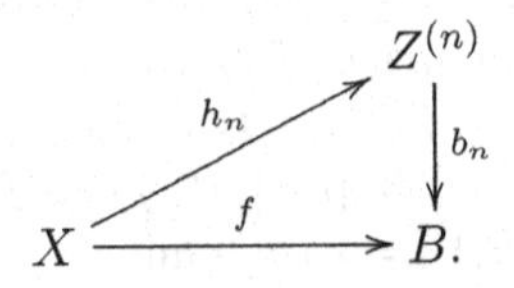

Assuming the hypotheses of Section 9.2, we have $\omega_{h_n} \in H^{n+1}(X; \pi_n(F))$ by Definition 9.2.2. Then a map h_{n+1} exists such that $b_{n+1}h_{n+1} = f$ if and only

if $\omega_{h_n} = 0$. In this case, the obstruction $\omega_{h_{n+1}}$ can be defined. The following proposition is a special case of Proposition 9.2.3.

Proposition 9.3.1 *Let X be a CW complex with* $\dim X \leqslant N$, *let* $p : E \to B$ *be a fiber map with fiber F, and suppose that the cohomology group* $H^{j+1}(X; \pi_j(F)) = 0$ *for all* $j \leqslant N$. *Then for any map* $f : X \to B$, *there is a lift* $h : X \to E$, *that is,* $ph = f$.

Next we turn to the obstruction theory of extensions. We consider the extension-lifting square of Section 9.1 in the case $B = \{*\}$ and $E = Y$. Thus there is the diagram

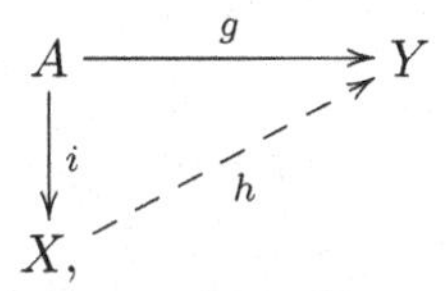

where (X, A) is a relative CW complex with inclusion i, and we seek a map h that makes the triangle commutative. The results of Section 9.2 are then applied to the fibration $Y \to Y \to \{*\}$. The Moore–Postnikov decomposition of $Y \to \{*\}$ is just the Postnikov decomposition of Y which we write as $\{Y^{(n)}, g_n, p_n, k^n\}$. We suppose that there is a map $h_n : X \to Y^{(n)}$ such that the following square is commutative

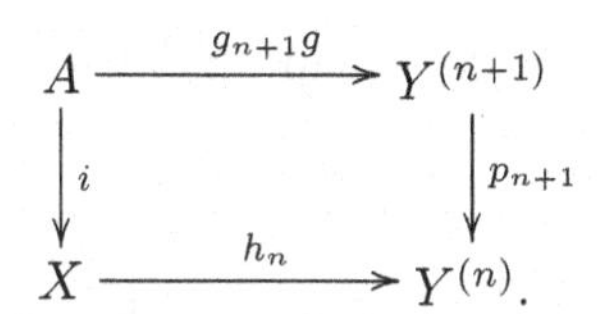

There is then an obstruction element.

Definition 9.3.2 The *obstruction* to the existence of a relative lift $h_{n+1} : X \to Y^{(n+1)}$ in the above square is the element $\omega_{h_n} \in H^{n+1}(C_i; \pi_n(Y))$ given by Definition 9.2.2, where C_i is the mapping cone of i.

As in the previous section, if there is a map $h_n : X \to Y^{(n)}$ such that $h_n i = g_n g$, and $\omega_{h_n} = 0$, then a relative lift $h_{n+1} : X \to Y^{(n+1)}$ exists. Therefore the obstruction $\omega_{h_{n+1}}$ exists, and this process can be repeated. We then obtain the following result.

Proposition 9.3.3 *Let (X, A) be a relative CW complex, let Y be a simple space, and suppose that the cohomology group* $H^{j+1}(C_i; \pi_j(Y)) = 0$ *for all* $j \geqslant 0$. *Then any map* $f : A \to Y$ *has an extension* $h : X \to Y$.

Proof. Most of the proof is a straightforward consequence of Proposition 9.2.3. The assumptions of Proposition 9.2.3 hold because the simplicity of Y implies that a Postnikov decomposition of Y exists (Remark 7.4.5) and that $\pi_1(Y)$ is abelian. It only remains to prove the proposition when $\dim(X, A) = \infty$. Suppose for every $n \geqslant 1$, there exist $h_n : X \to Y^{(n)}$ such that $h_n i = g_n g$

and $p_n h_n = h_{n-1}$. The h_n determine a map $\rho : X \to \varprojlim Y^{(n)}$ and the $g_n : Y \to Y^{(n)}$ determine a weak homotopy equivalence $\gamma : Y \to \varprojlim Y^{(n)}$ (Proposition 7.2.9). Then there is a commutative diagram

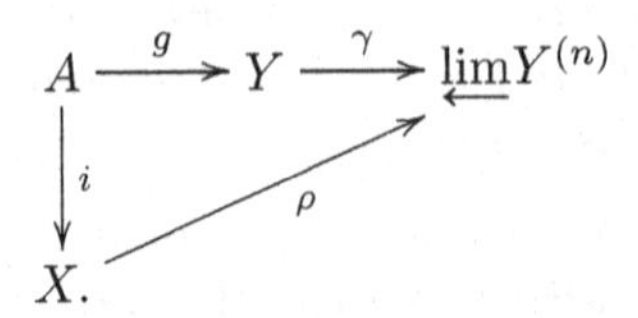

If $M = M_\gamma$ is the mapping cylinder of γ, then γ factors as

$$Y \xrightarrow{\gamma'} M \xrightarrow{\gamma''} \varprojlim Y^{(n)},$$

where γ' is a weak homotopy equivalence and γ'' is a homotopy equivalence with homotopy inverse $l : \varprojlim Y^{(n)} \to M$. Since $l\rho i \simeq \gamma' g$,

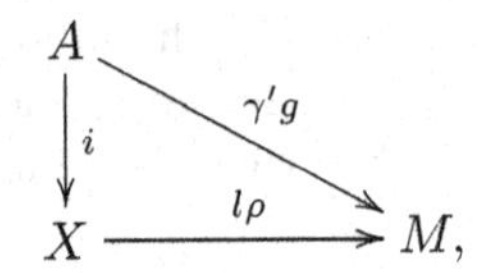

and i is a cofiber map, there is a map $s : X \to M$ such that the following square is commutative

$$\begin{array}{ccc} A & \xrightarrow{g} & Y \\ \downarrow{\scriptstyle i} & & \downarrow{\scriptstyle \gamma'} \\ X & \xrightarrow{s} & M. \end{array}$$

By the HELP lemma, there exists $h : X \to Y$ such that $hi = g$. □

The following corollary is a generalization of the surjective part of Proposition 2.4.13 and follows immediately from Proposition 9.3.3. (The injective part is generalized in Proposition 9.4.2.)

Corollary 9.3.4 *Under the hypotheses of Proposition* 9.3.3, *the function* $i^* : [X, Y] \to [A, Y]$ *is surjective, where* $i : A \to X$ *is the inclusion map.*

We complete the discussion of obstructions to an extension by characterizing the first nontrivial obstruction. We assume that $\pi_j(Y) = 0$ for $j < n$, where $n \geqslant 2$, and let $\pi_i = \pi_i(Y)$. Then $h_{n-1} : X \to Y^{(n-1)} = \{*\}$ and the obstruction $\omega_{h_{n-1}} \in H^n(C_i; \pi_{n-1})$ is zero. Thus there exists $h_n : X \to Y^{(n)}$ and an obstruction $\omega_{h_n} \in H^{n+1}(C_i; \pi_n)$. This element is denoted by ω and is called the *primary obstruction.* Let $\eta_\pi : H^n(Y; \pi_n) \to \mathrm{Hom}(\pi_n, \pi_n)$ be the isomorphism that assigns to a homotopy class its induced homomorphism of n-dimensional homotopy groups (Lemma 2.5.13). Let $b_n \in H^n(Y; \pi_n)$ be defined by $\eta_\pi(b_n) = \mathrm{id}$.

Lemma 9.3.5 *With the notation and assumptions above,*

$$\omega = q^* \bar{\kappa}_* g^*(b^n),$$

where $q : C_i \to \Sigma A$ is the projection and $\bar{\kappa}_*$ is the adjoint isomorphism,

$$H^n(Y;\pi_n) \xrightarrow{g^*} H^n(A;\pi_n) \xrightarrow{\bar{\kappa}_*} H^{n+1}(\Sigma A;\pi_n) \xrightarrow{q^*} H^{n+1}(C_i;\pi_n).$$

The proof is left as an exercise (Exercise 9.3). Note that by Corollary 4.2.10, $q^*\bar{\kappa}_* = \delta^n : H^n(A;\pi_n) \to H^{n+1}(C_i;\pi_n)$, the connecting homomorphism in the exact cohomology sequence of the map $i : A \to X$ (Cf. [14, p. 508]).

9.4 Obstruction Miscellany

We begin this section by showing how the obstruction theory of the previous sections can be applied to some common problems in homotopy theory.

Sections and Retracts

Suppose that

$$F \longrightarrow E \xrightarrow{p} B$$

is a fiber sequence. A section of p is a map $s : B \to E$ such that $ps = \text{id}$. Thus a section is just a lift of the identity map $\text{id} : B \to B$. There is an obstruction theory for this which is a special case of the obstruction theory for lifts, and so the obstructions lie in the groups $H^{n+1}(B;\pi_n(F))$. Dually, suppose (X, A) is a relative CW complex with inclusion map $i : A \to X$. A retraction $r : X \to A$ is a map such that $ri = \text{id}$. The retraction r is just an extension of the identity map $\text{id} : A \to A$ to X. Therefore the obstruction theory for extensions yields obstructions to a retraction that lie in the groups $H^{n+1}(X, A;\pi_n(A))$.

Obstructions to Homotopy

We next show how to define obstructions for homotopy of maps. A homotopy between two given maps $f_0, f_1 : X \to Y$ is an extension of the map $X \times \partial I \cup \{*\} \times I \to Y$ determined by f_0 and f_1 to the space $X \times I$, therefore we can use the previously defined obstruction theory to obtain obstructions for homotopy. However, we proceed more generally by considering homotopy

between two relative lifts. Let

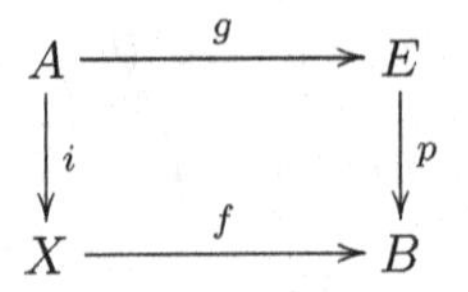

be an extension-lifting square, where (X, A) is a pair of CW complexes and $p : E \to B$ is a fiber map with fiber F, and suppose that $h_0, h_1 : X \to E$ are two relative lifts for this diagram. Then h_0, h_1, g, and f determine continuous functions $\Psi : X \times \partial I \cup A \times I \to E$ defined by

$$\Psi(x,t) = \begin{cases} h_t(x) & \text{if } t = 0, 1 \text{ and } x \in X \\ g(x) & \text{if } x \in A \end{cases}$$

and $\Phi : X \times I \to B$ defined by

$$\Phi(x,t) = f(x)$$

for $x \in X$ and $t \in I$. Then there is a commutative square

$$\begin{array}{ccc} X \times \partial I \cup A \times I & \xrightarrow{\Psi} & E \\ \downarrow j & & \downarrow p \\ X \times I & \xrightarrow{\Phi} & B, \end{array}$$

where j is inclusion. A relative lift $\Lambda : X \times I \to E$ for this square is a homotopy between h_0 and h_1 such that $\Lambda(a,t) = g(a)$ and $p\Lambda(x,t) = f(x)$, for $a \in A$, $x \in X$, and $t \in I$. But the obstructions to this relative lift lie in

$$\begin{aligned} H^{n+1}(C_j; \pi_n(F)) &\cong H^{n+1}(X \times I/(X \times \partial I \cup A \times I); \pi_n(F)) \\ &\cong H^{n+1}(\Sigma X/\Sigma A; \pi_n(F)) \\ &\cong H^{n+1}(\Sigma(X/A); \pi_n(F)) \\ &\cong H^n(C_i; \pi_n(F)). \end{aligned}$$

Thus we have an obstruction theory for homotopy between relative lifts.

This can be generalized as follows. Let

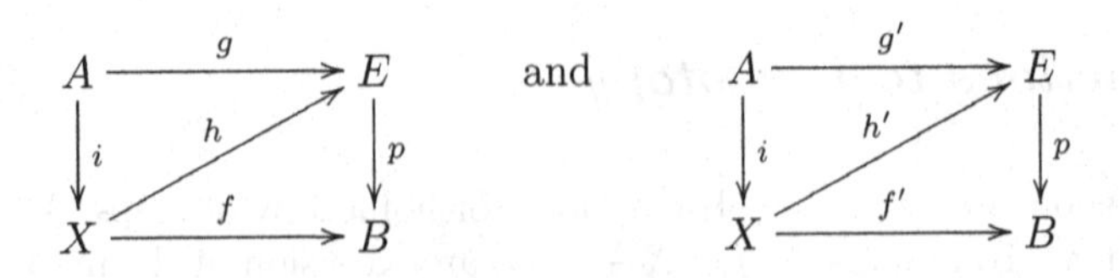

be two extension-lifting squares with relative lifts h and h'. Suppose that $f \simeq_F f'$ and $g \simeq_G g'$ such that $pG = F|A \times I$. Using h, h', F, and G, we obtain as before an extension-lifting square

$$\begin{array}{ccc} X \times \partial I \cup A \times I & \xrightarrow{\Psi} & E \\ \downarrow j & & \downarrow p \\ X \times I & \xrightarrow{\Phi = F} & B. \end{array}$$

Thus the obstructions to a homotopy H between h and h' such that $H|A \times I = G$ and $pH = F$ lie in the groups $H^n(C_i; \pi_n(F))$.

Remark 9.4.1 There are other situations that come up frequently to which obstruction theory can be applied. We list two of them.

- Suppose $A = \{*\}$ and $h, h' : X \to E$ are both lifts of f. Then we ask if $h \simeq_H h'$ such that $pH(x,t) = f(x)$, for $x \in X$ and $t \in I$. If f is a fiber map, then h and h' are fiber-preserving maps and we are asking if they are fiber homotopic (Definition 3.3.18).
- Suppose that $h, h' : X \to Y$ are maps such that $h|A = h'|A$. Are h and h' homotopic rel A? This question deals with a special case of homotopy of relative lifts with $E = Y$ and $B = \{*\}$. More generally, suppose that $h, h' : X \to Y$ are maps such that $h|A \simeq_F h'|A$. Then is there a homotopy H between h and h' such that $H|A \times I = F$?

A consequence of the last remark is the following generalization of Proposition 2.4.13 whose proof is omitted.

Proposition 9.4.2 *Let (X, A) be a relative CW complex with* $\dim(X, A) \leqslant N$, *let Y be a simple space, and let the cohomology group $H^j(C_i; \pi_j(Y)) = 0$ for all $j \leqslant N$. Then the function $i^* : [X, Y] \to [A, Y]$ is an injection, where $i : A \to X$ is the inclusion map.*

We end this section by sketching an obstruction theory for the extension-lifting problem by using a homology decomposition of the map $i : A \to X$.

Obstructions Using Homology Decompositions

We begin with the dual of Lemma 9.2.1. Consider the commutative diagram

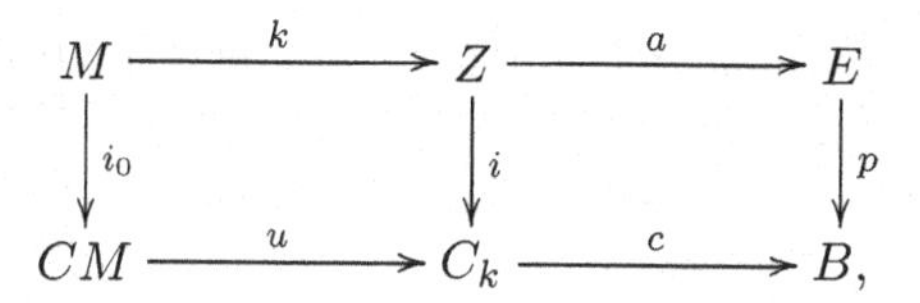

where the left square is the pushout diagram defining the mapping cone C_k, i and i_0 are inclusion maps, and p is a fiber map. Let $\widetilde{cu} : M \to EB$ be the adjoint of $cu : CM \to B$. Then there is the diagram

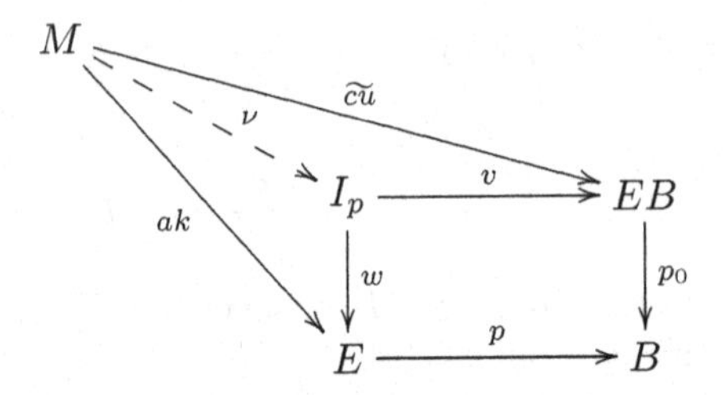

in which the square is the pullback diagram that defines the homotopy fiber I_p. Because $pak = p_0\widetilde{cu}$, there is a map $\nu : M \to I_p$ such that $v\nu = \widetilde{cu}$ and $w\nu = ak$.

We then have the following lemma.

Lemma 9.4.3 *With the above notation, there exists a map $c' : C_k \to E$ such that $pc' = c$ and $c'i = a$,*

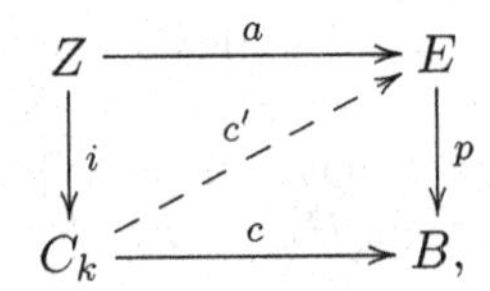

if and only if $* \simeq \nu : M \to I_p$.

The proof is a straightforward dualization of the proof of Lemma 9.2.1, and is left as an exercise (Exercise 9.5). Next assume that there is an extension-lifting square

$$\begin{array}{ccc} A & \xrightarrow{g} & E \\ {\scriptstyle i}\downarrow & & \downarrow{\scriptstyle p} \\ X & \xrightarrow{f} & B, \end{array}$$

where (X, A) is a relative CW complex and $p : E \to B$ is a fiber map with fiber F. We further assume that A and X are 1-connected and so the inclusion map $i : A \to X$ has a homology decomposition. By Theorem 7.4.8, there exist spaces and maps

$$A \xrightarrow{j_n} W_n \xrightarrow{k_n} X$$

such that $i = k_n j_n$, for all $n \geqslant 1$. Furthermore, there are principal cofibrations

$$W_n \xrightarrow{i_n} W_{n+1} \longrightarrow \Sigma M_n,$$

where $M_n = M(H_{n+1}(X, A), n)$ and $W_{n+1} = C_{l_{n+1}}$ for some map $l_{n+1} : M_n \to W_n$. Finally, the following diagram commutes

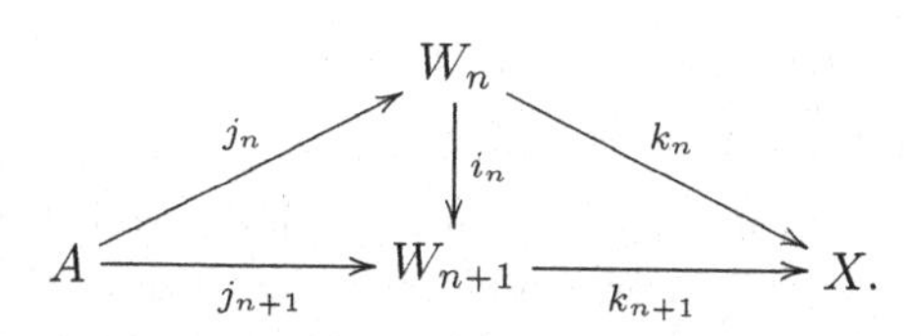

Now suppose that there is a map $h_n : W_n \to E$ such that the diagram

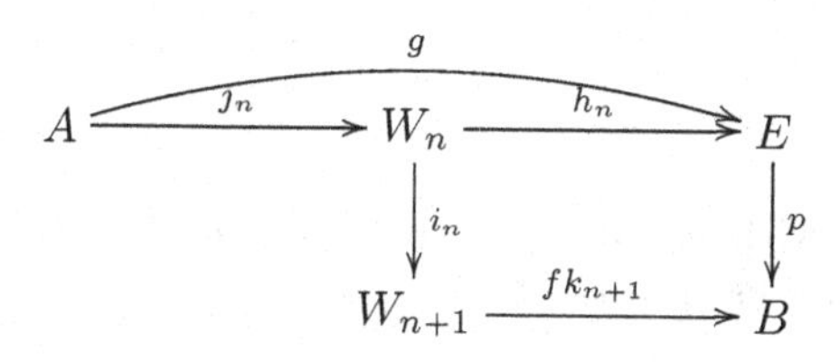

commutes. We then define an obstruction to a relative lift $W_{n+1} \to E$.

Definition 9.4.4 The *obstruction* to the existence of a map $h_{n+1} : W_{n+1} \to E$ such that $h_{n+1}i_n = h_n$ and $ph_{n+1} = fk_{n+1}$ is the element $\nu = \nu_{h_n}$ of the homotopy group with coefficients $\pi_n(I_p; H_{n+1}(X, A))$ defined just before Lemma 9.4.3.

Because $I_p \simeq F$, the obstruction can be regarded as an element of $\pi_n(F; H_{n+1}(X, A))$. By Lemma 9.4.3, h_{n+1} exists if and only if $\nu_{h_n} = 0$. In this case, $ph_{n+1} = fk_{n+2}i_{n+1}$ and $h_{n+1}j_{n+1} = g$, and so the obstruction $\nu_{h_{n+1}}$ can be defined.

Now suppose that a relative lift $h_N : W_N \to E$ exists by requiring that a sequence of obstructions is successively zero. If (X, A) is a relative CW complex and $H_i(A) = 0$ for $i \geqslant N$ and $H_i(X) = 0$ for $i \geqslant N + 1$, then by Corollary 7.4.9, $f_N : W_N \to X$ is a homotopy equivalence. Thus we obtain a relative lift $h' : X \to E$ up to homotopy, that is, $h'i \simeq g$ and $ph' \simeq f$. It can be shown that there is a relative lift $h : X \to E$ by May's coHELP lemma [65] and the CHEP. However there are two important special cases where the existence of a relative lift can be shown directly.

Remark 9.4.5

1. We consider the extension-lifting square with $B = \{*\}$, and so the relative lift is just an extension of $g : A \to E$ to X. The obstructions lie in $\pi_n(E; H_{n+1}(X, A))$ and are assumed to be trivial for $n < N$. If $H_i(A) = 0$ for $i \geqslant n$ and $H_i(X) = 0$ for $i \geqslant N + 1$, we obtain $h' : X \to E$ such that $h'i \simeq g$ as above. Then by the homotopy extension property for the pair (X, A), we conclude that there is an extension $h : X \to E$.
2. We consider the extension-lifting square with $A = \{*\}$, and so the relative lift is a lift of $f : X \to B$ to E. The obstructions lie in $\pi_n(F; H_{n+1}(X))$ and are assumed to be trivial for $n < N$. If $H_i(X) = 0$ for $i \geqslant N + 1$, we obtain $h' : X \to E$ such that $ph' \simeq f$ as above. Then, either directly or by

the covering homotopy property for the fiber map $p : E \to B$, we conclude that there is a lift $h : X \to E$.

In general, the obstructions that are obtained from a homology decomposition of i are less well known and often less useful than those that are obtained from a homotopy decomposition of p.

Exercises

Exercises marked with (*) may be more difficult than the others. Exercises marked with (†) are used in the text.

9.1. Use obtruction theory to prove Corollary 4.5.9.

9.2. Given an extension-lifting square

$$\begin{array}{ccc} A & \xrightarrow{g} & E \\ {\scriptstyle i}\downarrow & & \downarrow{\scriptstyle p} \\ X & \xrightarrow{f} & B, \end{array}$$

with the fiber F of p an $(n-1)$-connected space, $n \geqslant 2$. Assume that there is a Moore–Postnikov decomposition of p for which we take $Z^{(n)} = B$, $a_n = p$, $b_n = \mathrm{id}$, and $h_n = f$. Let $\omega = \omega_{h_n} \in H^{n+1}(C_i; \pi_n(F))$ be the obstruction to the existence of a relative lift $h_{n+1} : X \to Z^{(n+1)}$ and let $w : X \to C_i$ be the inclusion. Prove

$$w^*(\omega) = f^*(k^{n+1}).$$

9.3. Prove Lemma 9.3.5.

9.4. (*) If $A_1, A_2, \ldots, A_r$ are spaces, define the r-fold join inductively by $A_1 * \cdots * A_{r-1} * A_r = (A_1 * \cdots * A_{r-1}) * A_r$.

1. Prove that if the space A_i is $(p_i - 1)$-connected, $i = 1, \ldots, r$, then $A_1 * \cdots * A_{r-1} * A_r$ is $(p_1 + \cdots + p_r + r - 2)$-connected.
2. Let X be a $(p-1)$-connected space, $p \geqslant 1$, of dimension $< (k+1)p$. Give two independent proofs that $\operatorname{cat} X \leqslant k$ using the two characterizations of category (Definition 8.3.1 and Proposition 8.3.12).

9.5. (†) Prove Lemma 9.4.3.

9.6. Let (X, A) be a relative CW complex with $i : A \to X$ the inclusion map and let $h_0, h_1 : X \to Y$ be maps such that Y is simple and $h_0|A = h_1|A$. The obstructions to h_0 and h_1 being homotopic rel A lie in the groups $H^k(C_i; \pi_k(Y))$ (see Remark 9.4.1). Suppose that Y is $(n-1)$-connected and that $\omega \in H^n(C_i; \pi_n(Y))$ is the first nontrivial obstruction. Prove that

$$l^*(\omega) = h_0^*(b^n) - h_1^*(b^n),$$

where $l : X \to C_i$ is the inclusion and $b^n \in H^n(Y; \pi_n(Y))$ is the basic class.

9.7. Let Y be a simple space. Prove that $\pi_r(Y) = 0$ for $r \leqslant n-1$, where $n \geqslant 1$, if and only if for every CW complex K of dim $\leqslant n-1$, any map $f : K \to Y$ is nullhomotopic.

9.8. [39, p. 418] ($*$) Let X and Y be simple CW complexes and let $f : X \to Y$ be a map that induces isomorphisms on all homology groups. Show that f is a homotopy equivalence.

9.9. Let (X, A) be a relative CW complex with $\dim(X, A) \leqslant n$ and let Y be an $(n-1)$-connected space. Then it follows from Proposition 9.2.3 that every map $g : A \to Y$ has an extension. Prove this result using the obstructions obtained from a homology decomposition (Section 9.4).

Appendix A
Point-Set Topology

In this appendix we present some definitions and results on point-set topology that are used throughout the book. The topological spaces that we consider here are unbased and Hausdorff. The terms "map" and "continuous function" are used synonymously in this appendix.

Weak Topology

Definition A.1 Let X be a set and let $W_\alpha \subseteq X$ be subsets for $\alpha \in A$ such that each W_α is a topological space. The *weak topology on* X with respect to $\{W_\alpha\}_{\alpha \in A}$ is the largest topology on X such that the inclusion functions $W_\alpha \to X$ are continuous. This is defined explicitly as follows. A set $U \subseteq X$ is open if and only if $U \cap W_\alpha$ is open in W_α for every $\alpha \in A$. It is easily verified that this is a topology on X and that any topology on X such that the inclusion functions $W_\alpha \to X$ are continuous is contained in the weak topology.

The following simple example of a weak topology is referred to several times in the text.

Example A.2 Given topological spaces W_α for $\alpha \in A$. We form their disjoint union X written $\bigsqcup_{\alpha \in A} W_\alpha$. (If the W_α are not disjoint, we replace them with homeomorphic spaces that are.) Then X is given the weak topology with respect to $\{W_\alpha\}_{\alpha \in A}$.

Quotient Spaces

Definition A.3 Let X be a topological space, let Q be a set, and let $q : X \to Q$ be a surjection. The *quotient topology*, also called the *identification* *topology*, on Q is defined as follows. $U \subseteq Q$ is open $\iff q^{-1}(U)$ is open in X. This is equivalent to the condition that $F \subseteq Q$ is closed $\iff q^{-1}(F)$ is closed in X. This topology is denoted $\mathcal{T}(q)$. If X and Y are topological spaces and $p : X \to Y$ is a surjection, then p is called a *quotient map* or *identification map* if the topology on Y is $\mathcal{T}(p)$.

The following example of a quotient space occurs throughout the book.

Example A.4 Let X be a topological space, let $\sim$ be an equivalence relation on X, and let $X/\sim$ be the set of equivalence classes. Let $q : X \to X/\sim$ be the projection defined by $q(x) = \langle x \rangle$, where $x \in X$ and $\langle x \rangle$ is the equivalence class containing x. Then $X/\sim$ is given the quotient topology $\mathcal{T}(q)$.

There is a special case that arises frequently. If X is a topological space and $A \subseteq X$ is a subspace, then we define an equivalence relation on X as follows. If $x, x' \in A$, then $x \sim x'$, and if $x \notin A$, then x is equivalent only to itself. The resulting quotient space $X/\sim$ is denoted X/A and called the *space obtained from X by identifying (or shrinking) A to a point.*

The following are some basic properties of the quotient topology.

- Every continuous open surjection or continuous closed surjection is a quotient map.
- Let $q : X \to Q$ be a quotient map, let Y be a topological space and let $f : X \to Y$ be a function that is constant on each set of the form $q^{-1}(z)$, for $z \in Q$. Then there is a function $\widetilde{f} : Q \to Y$ such that $\widetilde{f}q = f$. If f is continuous, then $\widetilde{f}$ is continuous.
- If $f : X \to Y$ and $g : Y \to Z$ are quotient maps, then $gf : X \to Z$ is a quotient map, that is, $\mathcal{T}(g) = \mathcal{T}(gf)$.
- If $f : X \to X'$ is a quotient map, $\mathrm{id} : Y \to Y$ is the identity map, and Y is locally compact, then $f \times \mathrm{id} : X \times Y \to X' \times Y$ is a quotient map. More generally, if $f : X \to X'$ and $g : Y \to Y'$ are quotient maps and X' and Y' are locally compact, then $f \times g : X \times Y \to X' \times Y'$ is a quotient map.

The first three of these properties are easily proved. The fourth is more difficult and we refer to [72, p. 186].

Compact–Open Topology

Definition A.5 Let X and Y be topological spaces and let $M(X,Y)$ denote the set of continuous functions $X \to Y$. We give $M(X,Y)$ a topology by specifying a subbasis. Let C be a compact subset of X and let U be an open subset of Y. Then $W(C,U)$ denotes the subset of $M(X,Y)$ consisting of all continuous functions f such that $f(C) \subseteq U$. The sets $W(C,U)$ as C runs over all compact subsets of X and U runs over all open subsets of Y

is a subbasis for the *compact–open topology* on $M(X,Y)$. Unless otherwise stated, $M(X,Y)$ has the compact–open topology and is often denoted Y^X. For details on the compact–open topology, see [72, pp. 285-289].

Now let X,, Y, and Z be topological spaces and define a function $T : M(X,Y) \times M(Y,Z) \to M(X,Z)$ by $T(f,g) = gf$.

Proposition A.6 *If Y is locally compact, then the function T is continuous.*

A special case occurs when X is a one-point space. Then the function T becomes the *evaluation map* $e : Y \times M(Y,Z) \to Z$ given by $e(y,g) = g(y)$, for $y \in Y$.

Corollary A.7 *If Y is locally compact, then the evaluation map e is continuous.*

Next let $f : X \times Y \to Z$ and $g : X \to M(Y,Z)$ be functions such that

$$f(x,y) = g(x)(y),$$

for $x \in X$ and $y \in Y$.

Proposition A.8 [72, p. 287]

1. *If f is continuous, then g is continuous.*
2. *If g is continuous and Y is locally compact, then f is continuous.*

Thus there is a function $M(X \times Y, Z) \to M(X, M(Y,Z))$ when Y is locally compact.

Proposition A.9 [14, p. 438] *If X and Y are locally compact, then*

$$M(X \times Y, Z) \cong M(X, M(Y,Z)).$$

Two Useful Results

The following simple result, called the pasting lemma, is surprisingly useful.

Proposition A.10 [72, p. 108] *Consider the space $X = F_1 \cup \cdots \cup F_n$, where F_i is a closed subset of X, $i = 1, \ldots, n$. Suppose that there are continuous functions $f_i : F_i \to Y$ such that $f_i|F_i \cap F_j = f_j|F_i \cap F_j$, for all $i, j = 1, \ldots, n$. Then the function $f : X \to Y$ defined by $f(x) = f_i(x)$, for $x \in F_i$ is continous.*

The next result is also referred to in the text. Let X be a metric space with metric d. For a subset $A \subseteq X$, the diameter of A is $\sup\{d(x,y) \mid x, y \in A\} \leqslant \infty$.

Proposition A.11 (Lebesgue covering lemma)[72, p. 175] *If (X,d) is a compact metric space and $\mathcal{U}$ is an open cover of X, then there is a number ϵ, $0 < \epsilon < \infty$, such that any subset of X which has diameter $< \epsilon$ is completely contained in some member of $\mathcal{U}$.*

The number ϵ (or any positive number less than it) is sometimes called the *Lebesgue number* of the covering.

Compactly Generated Spaces

We can delete the hypothesis of local compactness in the previous results by working with compactly generated spaces. We give a brief discussion of this without details. A space X is called *compactly generated* if X has the weak topology with respect to all compact subsets. It can be shown that locally compact spaces, metric spaces, and CW complexes are all compactly generated. Furthermore, for any space X, we can introduce a compactly generated topology on X by giving it the weak topology with respect to compact subspaces. This space is denoted by $k(X)$ and is closely related to X (for example, their homotopy and homology groups are isomorphic). However, if X and Y are compactly generated, their cartesian product $X \times Y$ need not be compactly generated. To get around this, we consider compactly generated spaces $k(X \times Y)$, denoted $X \times_k Y$. This is called the k-product and it is the categorical product in the category of compactly generated spaces. In the preceding results on quotient spaces and the compact–open topology, we can assume that all the spaces are compactly generated. If we replace "$\times$" with "$\times_k$", then these results hold without the hypothesis of local compactness. We do not do this in the text because all of our spaces have the homotopy type of CW complexes and the CW product agrees with the k-product. For more information about compactly generated spaces, see [23] and [37].

Appendix B
The Fundamental Group

We begin with some basic group theory.

Free Products of Groups

Let $\{G_\alpha\}_{\alpha \in A}$ be an indexed collection of groups with identity element $e_\alpha \in G_\alpha$. The groups are not necessarily abelian and the index set A is not necessarily finite.

A *word* is an m-tuple of elements from the disjoint union $\bigsqcup_{\alpha \in A} G_\alpha$, for some $m \geqslant 0$. The m-tuple $(g_1, \ldots, g_m)$ is called a word of length m, and for $m = 0$, we take the empty word. The set of all words is denoted $\mathcal{W}$ and a multiplication in $\mathcal{W}$ is defined by juxtaposition

$$(g_1, \ldots, g_m)(h_1, \ldots, h_n) = (g_1, \ldots, g_m, h_1, \ldots, h_n).$$

There are elementary operations in $\mathcal{W}$ that send one word to another. These are one of the following.

$$(g_1, \ldots, g_i, g_{i+1}, \ldots, g_m) \mapsto (g_1, \ldots, g_i g_{i+1}, \ldots, g_m) \text{ if } g_i, g_{i+1} \in G_\alpha$$
$$(g_1, \ldots, g_{i-1}, g_i, g_{i+1}, \ldots, g_m) \mapsto (g_1, \ldots, g_{i-1}, g_{i+1}, \ldots, g_m) \text{ if } g_i = e_\alpha \in G_\alpha$$

or its inverse operation. This gives an equivalence relation on $\mathcal{W}$ as follows. If words $w, w' \in \mathcal{W}$, then $w \sim w'$ if w can be transformed into w' by a sequence of elementary operations. The set $\mathcal{W}/\!\sim$ is called the *free product* of the groups G_α and is denoted $\ast_{\alpha \in A} G_\alpha$ or just $\ast_\alpha G_\alpha$. If A is a finite set, say $A = \{1, 2, \ldots, n\}$, then the free product is written $G_1 * \cdots * G_n$. The multiplication in $\mathcal{W}$ induces a multiplication in $\ast_\alpha G_\alpha$.

Proposition B.1 [39, p. 41] *With the multiplication defined above, $\ast_\alpha G_\alpha$ is a group whose identity is the class of the empty word.*

A word $(g_1, \dots, g_m)$ is called *reduced* if its length cannot be shortened by any sequence of elementary operations. Then it can be shown that the equivalence class of any word in $\mathcal{W}$ contains a unique *reduced word* [56, p. 196]. We write the equivalence class of $(g_1, \dots, g_m)$ as $g_1 \cdots g_m$ and also call it a word, reduced if $(g_1, \dots, g_m)$ is reduced. Note that there is a monomorphism $i_\alpha : G_\alpha \to *_\alpha G_\alpha$ defined by setting $i_\alpha(g)$ equal to the equivalence class of the word g, for $g \in G_\alpha$. The following property of the free product is easily obtained. If $\varphi_\alpha : G_\alpha \to H$ are homomorphisms, then there exists a unique homomorphism $\Phi : *_\alpha G_\alpha \to H$ such that $\Phi i_\alpha = \varphi_\alpha$. Thus $(*_\alpha G_\alpha, i_\alpha)$ is the categorical sum of the groups G_α in the category of groups (Appendix F). The homomorphism Φ is denoted $\{\varphi_\alpha\}$.

The Seifert-van Kampen Theorem

The free product of fundamental groups appears in the Seifert–van Kampen theorem which we discuss next.

Suppose that X is a space containing open subsets U and V such that (1) $X = U \cup V$, (2) U, V, and $U \cap V$ are path-connected, and (3) the spaces X, U, V, and $U \cap V$ have a common basepoint which is the base point for their fundamental groups. We consider the inclusion maps

$$i : U \cap V \to U, \quad j : U \cap V \to V, \quad k : U \to X, \quad \text{and} \quad l : V \to X.$$

These induce homomorphisms of fundamental groups

$$i_* : \pi_1(U \cap V) \to \pi_1(U), \quad j_* : \pi_1(U \cap V) \to \pi_1(V)$$

$$k_* : \pi_1(U) \to \pi_1(X) \quad \text{and} \quad l_* : \pi_1(V) \to \pi_1(X).$$

We define a function $F : \pi_1(U \cap V) \to \pi_1(U) * \pi_1(V)$ by $F(w) = i_*(w)^{-1} j_*(w)$, for $w \in \pi_1(U \cap V)$, and let $\overline{F(\pi_1(U \cap V))}$ be the normal closure of $F(\pi_1(U \cap V))$ in $\pi_1(U) * \pi_1(V)$ (that is, the intersection of all normal subgroups of $\pi_1(U) * \pi_1(V)$ that contain $F(\pi_1(U \cap V))$.) Furthermore, let $\Phi : \pi_1(U) * \pi_1(V) \to \pi_1(X)$ be the unique homomorphism determined by k_* and l_*.

Proposition B.2 (Seifert–van Kampen) *The homomorphism $\Phi : \pi_1(U) * \pi_1(V) \to \pi_1(X)$ is surjective and its kernel is $\overline{F(\pi_1(U \cap V))}$. Thus*

$$\pi_1(X) \cong (\pi_1(U) * \pi_1(V))/\overline{F(\pi_1(U \cap V))}.$$

For a proof, see [56, pp. 221-226].

We give two applications of Theorem B.2 that are used in the text. First consider the wedge of circles $\bigvee_\alpha S_\alpha$ for $\alpha \in A$, where A is any index set and $S_\alpha = S^1_\alpha$, and let $i_\alpha : S_\alpha \to \bigvee_\alpha S_\alpha$ be the inclusion.

Proposition B.3 $\{i_{\alpha *}\} : \ast_{\alpha \in A}\ \pi_1(S_\alpha) \to \pi_1(\bigvee_{\alpha \in A} S_\alpha)$ *is an isomorphism.*

Proof. We assume that each circle is a CW complex with a 0-cell and a 1-cell and that the 0-cell is the basepoint. We first prove the proposition for the wedge of two circles, $X = S_1 \vee S_2$. Let $*_i$ denote the basepoint of S_i, let $*_i' = -*_i$ be the antipode of $*_i$, and consider $W_i = S_i - \{*_i'\}$. Then W_i is an open subset of S_i and $*_i$ is a strong deformation retract of W_i. If $U = S_1 \cup W_2$ and $V = S_2 \cup W_1$, then $X = U \cup V$ and $U \cap V = W_1 \cap W_2$. Furthermore S_1 is a strong deformation retract of U, S_2 is a strong deformation retract of V and $U \cap V$ is contractible. We then apply Theorem B.2 and obtain

$$\pi_1(S_1 \vee S_2) \cong \pi_1(S_1) * \pi_1(S_2).$$

The case of $n > 2$ circles follows by induction from Theorem B.2 because we can easily find open subsets U and V of $S_1 \vee \cdots \vee S_n$ that contain $S_1 \vee \cdots \vee S_{n-1}$ and S_n, respectively, as strong deformation retracts and such that $U \cap V$ is contractible. Finally, if A is an infinite set, we show that $\theta_A = \{i_{\alpha *}\} : \ast_{\alpha \in A}\ \pi_1(S_\alpha) \to \pi_1(\bigvee_{\alpha \in A} S_\alpha)$ is a bijection. If $f : S^1 \to \bigvee_{\alpha \in A} S_\alpha$, then $f(S^1)$ is compact and so is contained in $\bigvee_{\alpha \in E} S_\alpha$, for some finite set $E \subseteq A$, by Lemma 1.5.6. Therefore $[f] \in \pi_1(\bigvee_{\alpha \in E} S_\alpha)$ and so is in the image of $\theta_E : \ast_{\alpha \in E}\ \pi_1(S_\alpha) \to \pi_1(\bigvee_{\alpha \in E} S_\alpha)$. Because $\ast_{\alpha \in E}\ \pi_1(S_\alpha) \subseteq \ast_{\alpha \in A}\ \pi_1(S_\alpha)$, it follows that θ_A is onto. Now we show that the kernel of θ_A is trivial. Suppose $w \in \ast_{\alpha \in A}\ \pi_1(S_\alpha)$ with $\theta_A(w) = [f]$ and $f \simeq_H * : S^1 \to \bigvee_{\alpha \in A} S_\alpha$. Then there exists a finite set $E \subseteq A$ such that $w \in \ast_{\alpha \in E}\ \pi_1(S_\alpha)$. Furthermore, $H(S^1 \times I)$ is compact, thus there exists a finite set $F \subseteq A$ such that $f \simeq * : S^1 \to \bigvee_{\alpha \in F} S_\alpha$. If $\theta_{E \cup F} : \ast_{\alpha \in E \cup F}\ \pi_1(S_\alpha) \to \pi_1(\bigvee_{\alpha \in E \cup F} S_\alpha)$, then $\theta_{E \cup F}(w) = [f] = 0$. Because $E \cup F$ is finite, $\theta_{E \cup F}$ is an isomorphism, and so $w = 0$. □

Proposition B.4 *Let X and Y be spaces, let $f : X \to Y$ be a map, and let C_f be the mapping cone of f. Then*

$$\pi_1(C_f) \cong \pi_1(Y)/\overline{f_* \pi_1(X)}.$$

Proof. Let $*_X$ be the basepoint of X and $*_Y$ the basepoint of Y. Instead of C_f, we consider the unreduced mapping cone $K_f = (X \times I \sqcup Y)/\sim$ with equivalence relation defined by $(x, 1) \sim (x', 1)$ and $(x, 0) \sim f(x)$, for $x, x' \in X$, basepoint $\bar{*} = \langle *_X, 1/2 \rangle$ and quotient map $q : X \times I \sqcup Y \to K_f$. Let $U = q(X \times [0, 2/3) \sqcup Y)$ and $V = q(X \times (1/3, 1])$. Then V is contractible and it follows that $\pi_1(V, \bar{*}) = 0$ (see Exercise 5.11). We obtain from Theorem B.2,

$$\pi_1(K_f, \bar{*}) \cong \pi_1(U, \bar{*})/\overline{i_* \pi_1(U \cap V, \bar{*})},$$

where i is the inclusion of $U \cap V$ into U. Because $U \simeq Y$ and $U \cap V \simeq X$, we have

$$\pi_1(K_f, *) \cong \pi_1(Y, *)/\overline{f_* \pi_1(X, *)},$$

where $* = \langle *_X, 0\rangle = \langle *_Y\rangle$. The result for $\pi_1(C_f)$ follows from Proposition 1.5.18. □

The Hurewicz Homomorphism

We conclude this appendix with a result on the first Hurewicz homomorphism. Let X be a space with basepoint $*$ and let Δ_n be the standard ordered n-simplex in $\mathbb{R}^{n+1}$. We consider the chain complex $C'_n(X)$ which is the free-abelian group generated by all singular n-simplexes $T : \Delta_n \to X$ such that $T(v) = *$, for every vertex v of Δ_n. The nth homology group of this chain complex is denoted by $H'_n(X)$. If $i : C'_n(X) \to C_n(X)$ is the inclusion of this chain complex into the standard singular chain complex, then it can be shown that $i_* : H'_n(X) \to H_n(X)$ is an isomorphism([31, p. 440], [80, p. 68]). (We need this result only for $n = 1$.) From this we see that the definition of the first Hurewicz homomorphism $h_1 : \pi_1(X) \to H_1(X)$ in Section 2.4 is equivalent to the following definition of $h'_1 : \pi_1(X) \to H'_1(X)$. Let $\alpha = [f] \in \pi_1(X)$, where $f : (I, \partial I) \to (X, *)$, then $h'_1(\alpha) = \langle f\rangle \in H'_1(X)$, the homology class of the 1-cycle $f : I = \Delta_1 \to X$.

Proposition B.5 *The Hurewicz homomorphism $h_1 : \pi_1(X) \to H_1(X)$ is an epimorphism whose kernel is the commutator subgroup of $\pi_1(X)$.*

Proof. We prove the proposition for h'_1. If $f : (I, \partial I) \to (X, *)$, we can regard f as a representative of a homotopy class $[f] \in \pi_1(X)$ or as a 1-chain in $C'_1(X) = Z'_1(X)$ whose homology class is denoted $\langle f\rangle \in H'_1(X)$. Note that $h'_1[f] = \langle f\rangle$, so that h'_1 is onto. Furthermore, since $H'_1(X)$ is abelian, the commutator subgroup $[\pi_1(X), \pi_1(X)]$ is contained in $\operatorname{Ker} h'_1$. Thus it suffices to prove the reverse inclusion.

Let $\pi_1^*(X) = \pi_1(X)/[\pi_1(X), \pi_1(X)]$ be the fundamental group abelianized and let $\nu : \pi_1(X) \to \pi_1^*(X)$ be the projection. We define $\chi : C'_1(X) \to \pi_1^*(X)$ by $\chi(f) = \nu[f]$. We show $\chi(B'_1(X)) = 0$. Let $T : \Delta_2 \to X$, be a singular 2-simplex in $C'_1(X)$ and let $\partial' : C'_2(X) \to C'_1(X)$ be the boundary operator. Then $\partial'(T) = T^{(0)} - T^{(1)} + T^{(2)}$, where the vertices of Δ are written 0, 1, and 2 and $T^{(i)}$ denotes T restricted to the edge opposite the vertex i. Then

$$\begin{aligned}\chi\partial'(T) &= \chi(T^{(0)} - T^{(1)} + T^{(2)})\\ &= \chi(T^{(2)}) + \chi(T^{(0)}) - \chi(T^{(1)})\\ &= \nu[T^{(2)}]\nu[T^{(0)}]\nu[T^{(1)}]^{-1}\\ &= 0,\end{aligned}$$

because $[T^{(1)}] = [T^{(2)}][T^{(0)}]$. Since $(B'_1(X))$ is generated by the $\partial'(T)$, we have $\chi(B'_1(X)) = 0$, and so χ induces a homomorphism $\chi^* : H'_1(X) \to \pi_1^*(X)$. Clearly $\chi^* h'_1 = \nu$, and so $\operatorname{Ker} h'_1 \subseteq [\pi_1(X), \pi_1(X)]$. □

Appendix C
Homology and Cohomology

We assume that the reader is familiar with the rudiments of homology and cohomology theory. There are many excellent books that present this material such as [83], [39], [61], [43], [90], and [14]. In this section we briefly discuss a number of topics in homology and cohomology theory that are used in the text. We consider singular homology and cohomology and CW homology and cohomology. All of our spaces have the homotopy type of CW complexes (unless otherwise stated), thus the singular theory and the CW theory are isomorphic. Therefore we write $H_n(X, A; G)$ (respectively, $H^n(X, A; G)$) for the nth singular or CW homology group (respectively, cohomology group).

Definition of CW homology

We now recall how CW homology is derived from singular homology theory. Let K be a CW complex and consider the relative singular homology group $H_n(K^n, K^{n-1})$. This group is known to be the free-abelian group with a basis in one–one correspondance with the n-cells of K and we write

$$C_n(K) = H_n(K^n, K^{n-1})$$

for the nth CW chain group of K. The boundary homomorphism $\partial_n : C_n(K) \to C_{n-1}(K)$ is defined to be the composition

$$H_n(K^n, K^{n-1}) \xrightarrow{\Delta} H_{n-1}(K^{n-1}) \xrightarrow{j_*} H_{n-1}(K^{n-1}, K^{n-2}),$$

where Δ is the connecting homomorphism in the exact homology sequence of the pair (K^n, K^{n-1}) and j is the inclusion of K^{n-1} into (K^{n-1}, K^{n-2}). This defines the CW chain complex $\{C_n(K), \partial_n\}$ whose homology is the *CW homology* of K. These homology groups are isomorphic to the singular homology groups of the space K.

Exact Homology Sequence of a Triple

If X is a space with subspace A and B is a subspace of A, that is, $B \subseteq A \subseteq X$, then there is a long exact sequence

$$\cdots \longrightarrow H_n(A, B; G) \xrightarrow{i_*} H_n(X, B; G) \xrightarrow{j_*} H_n(X, A; G) \longrightarrow \cdots,$$

where $i : (A, B) \to (X, B)$ and $j : (X, B) \to (X, A)$ are inclusion functions. This sequence is called the *exact homology sequence of the triple* (X, A, B). There is a similar result for cohomology. Note that in Exercise 4.22 there is an exact sequence for homotopy groups that is analogous to the exact homology sequence of a triple.

The Five Lemma

This is a very useful algebraic result dealing with a homomorphism from one exact sequence to another. Consider the diagram of abelian groups and homomorphisms

$$\begin{array}{ccccccccc} A & \longrightarrow & B & \longrightarrow & C & \longrightarrow & D & \longrightarrow & E \\ \downarrow{\scriptstyle a} & & \downarrow{\scriptstyle b} & & \downarrow{\scriptstyle c} & & \downarrow{\scriptstyle d} & & \downarrow{\scriptstyle e} \\ A' & \longrightarrow & B' & \longrightarrow & C' & \longrightarrow & D' & \longrightarrow & E', \end{array}$$

where both rows are exact and each square is commutative.

Proposition C.1 *1. If b and d are surjective and e is injective, then c is surjective.*

2. If b and d are injective and a is surjective, then c is injective.
Consequently, if a, b, d and e are isomorphisms, then c is an isomorphism.

This proposition can be easily proved.

Ext and Tor

We first define a function of a pair of abelian groups A and B called $\mathrm{Ext}(A, B)$. We omit details and do not state the definition in full generality. However, this is sufficient for our purposes. We write $A = F/R$, where F is a free-abelian group and $R \subseteq F$ is a subgroup. Then there is a short exact sequence

$$0 \longrightarrow R \xrightarrow{i} F \xrightarrow{p} A \longrightarrow 0$$

which is called a *free resolution* of A. We denote the group of homomorphisms $A \to B$ by $\mathrm{Hom}(A,B)$ and note that $\mathrm{Hom}(-,B)$ is left exact [42, p. 17]. Thus

$$0 \longrightarrow \mathrm{Hom}(A,B) \xrightarrow{p^{\#}} \mathrm{Hom}(F,B) \xrightarrow{i^{\#}} \mathrm{Hom}(R,B)$$

is an exact sequence, where $p^{\#}$ and $i^{\#}$ are induced by p and i. Then

$$\mathrm{Ext}(A,B) = \mathrm{Hom}(R,B)/i^{\#}\mathrm{Hom}(F,B).$$

For this definition to be well-defined, it would be necessary to show that $\mathrm{Ext}(A,B)$ is independent of the choice of resolution of A. We next list some well-known properties of Ext (see [42, pp. 89-98] or [59, Chap. III]).

Proposition C.2 *1. For a fixed abelian group B, a homomorphism $f : A \to A'$ induces a homomorphism $f^* : \mathrm{Ext}(A',B) \to \mathrm{Ext}(A,B)$. Furthermore, $\mathrm{Ext}(-,B)$ is a contravariant functor (Appendix F).*

2. For a fixed abelian group A, a homomorphism $g : B \to B'$ induces a homomorphism $g_* : \mathrm{Ext}(A,B) \to \mathrm{Ext}(A,B')$. In fact, $\mathrm{Ext}(A,-)$ is a covariant functor (Appendix F).

3. If A is free-abelian, then $\mathrm{Ext}(A,B) = 0$ for all B.

4. $\mathrm{Ext}(\mathbb{Z}_m, B) \cong B/mB$ and so $\mathrm{Ext}(\mathbb{Z}_m, \mathbb{Z}_n) \cong \mathbb{Z}_{(m,n)}$, where (m,n) is the greatest common divisor of m and n.

5. If A_i is a collection of abelian groups for $i \in I$, then

$$\mathrm{Ext}\Big(\sum_i A_i, B\Big) \cong \prod_i \mathrm{Ext}(A_i, B),$$

where $\sum$ denote direct sum of abelian groups and $\prod$ denotes direct product.

6. If B_j is a collection of abelian groups for $j \in J$, then

$$\mathrm{Ext}\Big(A, \prod_j B_j\Big) \cong \prod_j \mathrm{Ext}(A, B_j).$$

It follows that if A and B are finitely generated abelian groups which are written as direct sums of cyclic groups, then $\mathrm{Ext}(A,B)$ can be determined as a direct sum of cyclic groups.

We next consider the functor Tor. For abelian groups A and B, we define an abelian group written $A * B$ or $\mathrm{Tor}(A,B)$ called the *torsion product* of A and B. There are some analogies with Ext, and we begin with a free resolution of A,

$$0 \longrightarrow R \xrightarrow{i} F \xrightarrow{p} A \longrightarrow 0.$$

We apply the tensor product $-\otimes B$, which is right exact [59, p. 148], to this exact sequence and obtain the exact sequence

$$R \otimes B \xrightarrow{i_\#} F \otimes B \xrightarrow{p_\#} A \otimes B \longrightarrow 0,$$

where $p_\#$ and $i_\#$ are induced by p and i. Then

$$A * B = \text{Kernel}\, i_\#.$$

As before, it is necessary to show that $A * B$ is independent of the choice of resolution of A. We list some well-known properties of Tor [42, pp. 112–115] and [59, pp. 150–153].

Proposition C.3 *1. For a fixed abelian group B, a homomorphism $f : A \to A'$ induces a homomorphism $f_* : A * B \to A' * B$. Furthermore, $- * B$ is a covariant functor.*

2. For a fixed abelian group A, a homomorphism $g : B \to B'$ induces a homomorphism $g_* : A * B \to A * B'$. Moreover, $A * -$ is a covariant functor.

3. $A * B \cong B * A$.

4. If A or B is torsion-free, then $A * B = 0$.

5. $\mathbb{Z}_m * B \cong \{x \in B \mid mx = 0\}$ and so $\mathbb{Z}_m * \mathbb{Z}_n \cong \mathbb{Z}_{(m,n)}$.

6. If A_i is a collection of abelian groups for $i \in I$, then

$$\Big(\sum_i A_i\Big) * B \cong \sum_i A_i * B,$$

where $\sum$ denotes direct sum.

7. If B_j is a collection of abelian groups for $j \in J$, then

$$A * \Big(\sum_j B_j\Big) = \sum_j A * B_j.$$

It follows that if A and B are finitely generated abelian groups which are written as direct sums of cyclic groups, then $A * B$ can be determined as a direct sum of cyclic groups.

Universal Coefficient Theorems

We begin with the universal coefficient theorem for cohomology. This theorem expresses cohomology with coefficients in an abelian group in terms of integral homology.

Theorem C.4 *Let X is be a space and let G be an abelian group. Then there is a short exact sequence*

$$0 \longrightarrow \operatorname{Ext}(H_{n-1}(X), G) \xrightarrow{\alpha} H^n(X; G) \xrightarrow{\beta} \operatorname{Hom}(H_n(X), G) \longrightarrow 0.$$

The sequence splits and α and β are compatible with homomorphisms induced by continuous functions from one pair to another. Furthermore, the result holds if the space X is replaced by a pair of spaces (X, A).

This theorem could be compared to the universal coefficient theorem 5.2.4. The latter result gives a short exact sequence that expresses cohomology with coefficients in an abelian group in terms of integral cohomology.

Next we state the universal coefficient theorem for homology.

Theorem C.5 *Let X be a space and let G be an abelian group. Then there is a short exact sequence*

$$0 \longrightarrow H_n(X) \otimes G \xrightarrow{\alpha} H_n(X; G) \xrightarrow{\beta} H_{n-1}(X) * G \longrightarrow 0.$$

The sequence splits and α and β are compatible with homomorphisms induced by continuous functions from one space to another. Furthermore, the result holds if the space X is replaced by a pair of spaces (X, A).

Appendix D
Homotopy Groups of the n-Sphere

In this appendix we determine $\pi_n(S^n)$ and $\pi_{n+1}(S^n)$.

$\pi_n(S^n)$

Proof of Proposition 2.4.16 We show that the degree function $\deg : \pi_n(S^n) \to \mathbb{Z}$ is an isomorphism, for all $n \geqslant 1$. We have seen that deg is an epimorphism, thus it remains to prove that deg is one–one. For this it suffices to show that $\pi_n(S^n) \cong \mathbb{Z}$. We do this by induction on n. For $n = 1$, the Hurewicz homomorphism $h_1 : \pi_1(S^1) \to H_1(S^1)$ is an epimorphism whose kernel is the commutator subgroup by Theorem B.5. Because S^1 is an H-space, $\pi_1(S^1)$ is abelian, and so $\pi_1(S^1) \cong H_1(S^1) \cong \mathbb{Z}$. For $n = 2$, we consider the exact homotopy group sequence of the Hopf fibration $S^1 \to S^3 \to S^2$. We obtain the exact sequence

$$0 \longrightarrow \pi_2(S^2) \longrightarrow \pi_1(S^1) \longrightarrow 0\,,$$

and hence $\pi_2(S^2) \cong \mathbb{Z}$. Next, by the Freudenthal theorem (5.6.7), there is a sequence of isomorphisms

$$\pi_2(S^2) \xrightarrow{\cong} \pi_3(S^3) \xrightarrow{\cong} \pi_4(S^4) \xrightarrow{\cong} \cdots\,,$$

and so $\pi_n(S^n) \cong \mathbb{Z}$, for all $n \geqslant 1$. □

Remark D.1 There are other proofs that $\deg : \pi_n(S^n) \to \mathbb{Z}$ is a monomorphism, for all $n \geqslant 1$. A direct and elementary proof that is much longer than the proof above appears in [91, pp. 13–17] and [14, pp. 301–304]. In addition, there is a short, elegant proof that $\deg : \pi_1(S^1) \to \mathbb{Z}$ is an isomorphism using covering spaces (see [14, p. 142]).

$\pi_{n+1}(S^n)$

We next prove $\pi_{n+1}(S^n) \cong \mathbb{Z}_2$ for $n \geqslant 3$ (Theorem 5.6.10(1)) and identify the generator. For this we use the Wang sequence which is a long exact cohomology sequence of a fibration in which the base is a sphere. We state this result without proof. It follows easily from the spectral sequence of a fibration, and proofs can be found in [45, p. 282] and [91, p. 319].

Theorem D.2 (Wang) *Let $F \to X \to S^n$ be a fibration with $n \geqslant 2$. Then there is an exact sequence of unreduced integral cohomology groups*

$$\cdots \longrightarrow H^{q-1}(X) \longrightarrow H^{q-1}(F) \xrightarrow{\theta_{q-1}} H^{q-n}(F) \longrightarrow H^q(X) \longrightarrow \cdots.$$

Furthermore, $\theta_r : H^r(F) \to H^{r+1-n}(F)$ is a homomorphism and a derivation, that is,

$$\theta_{p+q}(xy) = \theta_p(x)\, y + (-1)^{p(n-1)} x\, \theta_q(y),$$

for $x \in H^p(F)$ and $y \in H^q(F)$.

Proof of Theorem 5.6.10(1) It suffices to show $\pi_4(S^3) \cong \mathbb{Z}_2$. For then, since $\Sigma : \pi_3(S^2) \to \pi_4(S^3)$ is onto and $\Sigma : \pi_n(S^{n-1}) \to \pi_{n+1}(S^n)$ is an isomorphism for $n \geqslant 4$, by the Freudenthal theorem, it follows that $\pi_{n+1}(S^n) \cong \mathbb{Z}_2$ with generator $[\Sigma^{n-2}\phi]$. We consider the 3-connective fibration of S^3 (see Definition 7.2.7). Thus we have a fiber sequence $F \to X \to S^3$, where $\pi_i(X) = 0$ for $i \leqslant 3$ and $\pi_i(X) \cong \pi_i(S^3)$ for $i \geqslant 4$. Also $\pi_i(F) = 0$ for $i \geqslant 3$ and $\pi_i(F) \cong \pi_{i+1}(S^3)$ for $i \leqslant 2$. Hence F is an Eilenberg–Mac Lane space $K(\mathbb{Z}, 2) \simeq \mathbb{C}\mathrm{P}^\infty$. Because X is 3-connected, $H_4(X) \cong \pi_4(X) \cong \pi_4(S^3)$ by the Hurewicz theorem, and so it suffices to prove that $H_4(X) \cong \mathbb{Z}_2$. For this we apply the Wang sequence to the fibration above and obtain the exact sequence

$$\cdots \longrightarrow H^{q-1}(X) \longrightarrow H^{q-1}(F) \xrightarrow{\theta_{q-1}} H^{q-3}(F) \longrightarrow H^q(X) \longrightarrow \cdots$$

with $F = \mathbb{C}\mathrm{P}^\infty$. If $q = 3$ in the sequence above, then $H^2(X) = 0 = H^3(X)$, and so $\theta_2 : H^2(F) \to H^0(F)$ is an isomorphism. We choose a generator $x \in H^2(F)$ such that $\theta_2(x) = 1$. Then x^n is a generator of $H^{2n}(F) \cong \mathbb{Z}$. Because θ is a derivation, $\theta_4(x^2) = 2x$. Now consider the previous sequence with $q = 5$. We obtain the exact sequence

$$H^1(F) \longrightarrow H^4(X) \longrightarrow H^4(F) \xrightarrow{\theta_4} H^2(F) \longrightarrow H^5(X) \longrightarrow H^5(F)$$

with $H^1(F) = 0 = H^5(F)$ and $H^4(F) \cong H^2(F) \cong \mathbb{Z}$. Hence the homomorphism $\theta_4 : \mathbb{Z} \to \mathbb{Z}$ is multiplication by 2. Thus $H^4(X) = 0$ and $H^5(X) \cong \mathbb{Z}_2$, and so $H_4(X) \cong \mathbb{Z}_2$ by Theorem C.3. □

Appendix E
Homotopy Pushouts and Pullbacks

In this appendix we prove two theorems on homotopy pushouts and homotopy pullbacks. We give detailed proofs of these important results. Because the proofs are long and technical they have been relegated to an appendix. In addition, we show how the cube theorem 6.5.1 can be used to give another proof of Lemma 6.3.4. We begin by introducing some useful notation.

NOTATION Let W and Z be spaces and let $F_i : W \times I \to Z$ be unbased continuous functions, $i = 1, \dots, n$. Suppose $F_i|(W \times \{1\}) = F_{i+1}|(W \times \{0\})$, for $i = 1, \dots, n-1$. Then we define an unbased continuous function $F_1 + F_2 + \cdots + F_n : W \times I \to Z$ as follows

$$(F_1 + F_2 + \cdots + F_n)(w,t) = F_i(w, nt - i + 1) \text{ for } \tfrac{i-1}{n} \leqslant t \leqslant \tfrac{i}{n},$$

where $w \in W$ and $t \in I$. Furthermore, for $F : W \times I \to Z$, we define $-F : W \times I \to Z$ by

$$(-F)(w,t) = F(w, 1-t),$$

where $w \in W$ and $t \in I$.

By reparametrizing the interval I, it is easily seen that there is general associativity up to relative homotopy, that is, $F_1 + F_2 + \cdots + F_n$ is homotopic rel $(W \times \partial I)$ to any bracketing of $F_1, F_2, \dots, F_n$. For example,

$$(F_1 + F_2) + F_3 \simeq F_1 + F_2 + F_3 \text{ rel } (W \times \partial I) \quad \text{and}$$

$$F_1 + (F_2 + F_3) \simeq F_1 + F_2 + F_3 \text{ rel } (W \times \partial I).$$

The Induced Map

The first result gives conditions for the induced map of homotopy pushouts to be a homotopy equivalence. We recall Definition 6.2.7.

For spaces and maps

$$Y \xleftarrow{g} A \xrightarrow{f} X,$$

we denote the standard homotopy pushout square by

$$\begin{array}{ccc} A & \xrightarrow{f} & X \\ \downarrow{\scriptstyle g} & & \downarrow{\scriptstyle u} \\ Y & \xrightarrow{v} & O. \end{array}$$

If we are given

$$Y' \xleftarrow{g'} A' \xrightarrow{f'} X',$$

then we obtain a standard homotopy pushout square as above with all spaces and maps primed. Assume that there are maps $\alpha : A \to A'$, $\beta : X \to X'$, and $\gamma : Y \to Y'$ such that the following diagram is homotopy-commutative

$$\begin{array}{ccccc} Y & \xleftarrow{g} & A & \xrightarrow{f} & X \\ \downarrow{\scriptstyle \gamma} & & \downarrow{\scriptstyle \alpha} & & \downarrow{\scriptstyle \beta} \\ Y' & \xleftarrow{g'} & A' & \xrightarrow{f'} & X' \end{array}$$

with $f'\alpha \simeq_F \beta f$ and $g'\alpha \simeq_G \gamma g$. Then $\Lambda = \Lambda_{F,G} : O \to O'$ is defined by

$$\begin{aligned} \Lambda u &= u'\beta, \\ \Lambda v &= v'\gamma, \quad \text{and} \\ \Lambda w &= -(u'F) + w'(\alpha \times \mathrm{id}) + v'G, \end{aligned}$$

where $w : A \times I \to O$ is the inclusion defined by $w(a,t) = \langle *, (a,t), * \rangle$, for $a \in A$ and $t \in I$, and $w' : A' \times I \to O'$ is similarly defined.

Theorem 6.2.8 *If α, β, and γ are homotopy equivalences, then the induced map Λ is a homotopy equivalence.*

This theorem is proved by three lemmas that are of independent interest. Suppose that we are given the previous homotopy-commutative diagram and the following similar one

$$\begin{array}{ccccc} Y' & \xleftarrow{g'} & A' & \xrightarrow{f'} & X' \\ \downarrow{\scriptstyle \gamma'} & & \downarrow{\scriptstyle \alpha'} & & \downarrow{\scriptstyle \beta'} \\ Y'' & \xleftarrow{g''} & A'' & \xrightarrow{f''} & X'', \end{array}$$

with $f''\alpha' \simeq_{F'} \beta' f'$ and $g''\alpha' \simeq_{G'} \gamma' g'$. Define $F \circ F' : A \times I \to X''$ by

$$F \circ F' = F'(\alpha \times \mathrm{id}) + \beta' F$$

and define $G \circ G' : A \times I \to Y''$ similarly. Then $f''\alpha'\alpha \simeq_{F \circ F'} \beta'\beta f$ and $g''\alpha'\alpha \simeq_{G \circ G'} \gamma'\gamma g$. Therefore the function $\Lambda_{F \circ F', G \circ G'} : O \to O''$ exists, where O'' is the standard homotopy pushout of f'' and g''. For notational simplicity we write $\Lambda = \Lambda_{F,G}$, $\Lambda' = \Lambda_{F',G'}$, and $\Lambda'' = \Lambda_{F \circ F', G \circ G'}$.

Lemma E.1 $\Lambda_{F',G'} \Lambda_{F,G} \simeq \Lambda_{F \circ F', G \circ G'} : O \to O''$.

Proof. Since O is a quotient of $X \vee (A \times I) \vee Y$, we first define the homotopy on each of the three summands of the wedge. Now

$$(\Lambda' \Lambda)u = \Lambda'' u'' \quad \text{and} \quad (\Lambda' \Lambda)v = \Lambda'' v''.$$

Therefore we take the stationary homotopy between $\Lambda' \Lambda$ and Λ'' on X and on Y. Furthermore,

$$\Lambda' \Lambda w = -u''\beta' F + \Big(-u'' F'(\alpha \times \mathrm{id}) + w''(\alpha'\alpha \times \mathrm{id}) + v'' G'(\alpha \times \mathrm{id})\Big) + v''\gamma' G$$

and

$$\Lambda'' w = \Big(-u''\beta' F - u'' F'(\alpha \times \mathrm{id})\Big) + w''(\alpha'\alpha \times \mathrm{id}) + \Big(v'' G'(\alpha \times \mathrm{id})) + v''\gamma' G\Big).$$

Thus these two maps of $A \times I$ into O'' are homotopic rel $A \times \partial I$. We put these homotopies together on $X \vee (A \times I) \vee Y$. This induces a homotopy $O \times I \to O''$ which is the desired homotopy. □

For the next lemma we assume that there is a homotopy-commutative diagram as before with $f'\alpha \simeq_F \beta f$ and $g'\alpha \simeq_G \gamma g$. We further assume that there are maps $\bar{\alpha} : A \to A'$, $\bar{\beta} : X \to X'$, and $\bar{\gamma} : Y \to Y'$ and homotopies E, B, and C such that $\alpha \simeq_E \bar{\alpha}$, $\beta \simeq_B \bar{\beta}$, and $\gamma \simeq_C \bar{\gamma}$. We define $\bar{F} : A \times I \to X'$ by

$$\bar{F} = -f'E + F + B(f \times \mathrm{id})$$

and we define $\bar{G} : A \times I \to Y'$ similarly. Then $f'\bar{\alpha} \simeq_{\bar{F}} \bar{\beta} f$ and $g'\bar{\alpha} \simeq_{\bar{G}} \bar{\gamma} g$. Thus there is a map $\Lambda_{\bar{F},\bar{G}} : O \to O'$.

Lemma E.2 $\Lambda_{F,G} \simeq \Lambda_{\bar{F},\bar{G}} : O \to O'$.

Proof. In the proof we will sometimes write the homotopies E, B, and C as α_t, β_t, and γ_t, respectively. For fixed $s \in I$, define $\Phi^s : A \times I \to X'$ by

$$\Phi^s(a,t) = \begin{cases} f'E(a, s - 3t) & \text{for } 0 \leqslant t \leqslant \frac{s}{3} \\ F\Big(a, \dfrac{3t - s}{3 - 2s}\Big) & \text{for } \frac{s}{3} \leqslant t \leqslant 1 - \frac{s}{3} \\ B(f(a), 3t + s - 3) & \text{for } 1 - \frac{s}{3} \leqslant t \leqslant 1, \end{cases}$$

for $a \in A$ and $t \in I$. Then $\Phi^s(a,0) = f'\alpha_s(a)$ and $\Phi^s(a,1) = \beta_s f(a)$. Similarly we define $\Psi^s : A \times I \to Y'$ and have that $\Psi^s(a,0) = g'\alpha_s(a)$ and $\Psi^s(a,1) =$

$\gamma_s g(a)$. Thus in the diagram

$$\begin{array}{ccccc} Y & \xleftarrow{g} & A & \xrightarrow{f} & X \\ \downarrow{\scriptstyle \gamma_s} & & \downarrow{\scriptstyle \alpha_s} & & \downarrow{\scriptstyle \beta_s} \\ Y' & \xleftarrow{g'} & A' & \xrightarrow{f'} & X', \end{array}$$

the squares are homotopy-commutative with homotopies Φ^s and Ψ^s. For each $s \in I$, we form $\Lambda_{\Phi^s,\Psi^s} : O \to O'$. Then $\Lambda_{\Phi^0,\Psi^0} = \Lambda_{F,G}$ and $\Lambda_{\Phi^1,\Psi^1} = \Lambda_{\bar{F},\bar{G}}$. Therefore Λ_{Φ^s,Ψ^s} is the desired homotopy. □

Next consider the diagram

$$\begin{array}{ccccc} Y & \xleftarrow{g} & A & \xrightarrow{f} & X \\ \downarrow{\scriptstyle \mathrm{id}} & & \downarrow{\scriptstyle \mathrm{id}} & & \downarrow{\scriptstyle \mathrm{id}} \\ Y & \xleftarrow{g} & A & \xrightarrow{f} & X, \end{array}$$

with homotopies $H : A \times I \to X$ and $K : A \times I \to Y$ such that $f \simeq_H f$ and $g \simeq_K g$.

Lemma E.3 *$\Lambda_{H,K}$ is a homotopy equivalence.*

Proof. We have $f \simeq_{(-H)} f$ and $g \simeq_{(-K)} g$, and so by Lemma E.1,

$$\Lambda_{H,K}\,\Lambda_{(-H),(-K)} \simeq \Lambda_{H\circ(-H),K\circ(-K)} = \Lambda_{(-H)+H,(-K)+K}.$$

We fix $s \in I$ and define $\Phi^s : A \times I \to X$ by

$$\Phi^s(a,t) = \begin{cases} H(a, 1-2st) & \text{for } 0 \leqslant t \leqslant \frac{1}{2} \\ H(a, 1-2s(1-t))) & \text{for } \frac{1}{2} \leqslant t \leqslant 1, \end{cases}$$

where $a \in A$ and $t \in I$. For fixed s, this function starts at $H(a,1)$ and goes backward along H to $H(a,1-s)$ for $0 \leqslant t \leqslant \frac{1}{2}$ and then goes from $H(a,1-s)$ forward along H to $H(a,1)$ for $\frac{1}{2} \leqslant t \leqslant 1$. If $p_1 : A \times I \to A$ is the projection, then $\Phi^0 = fp_1$ and $\Phi^1 = (-H)+H$. Furthermore, $\Phi^s(a,0) = f(a) = \Phi^s(a,1)$, and so for every $s \in I$, we have $f \simeq_{\Phi^s} f$. Similarly, there is a $\Psi^s : A \times I \to Y$ with $\Psi^0 = gp_1$, $\Psi^1 = (-K)+K$ and $g \simeq_{\Psi^s} g$. Therefore Λ_{Φ^s,Ψ^s} is a homotopy between Λ_{fp_1,gp_1} and $\Lambda_{H\circ(-H),K\circ(-K)}$. But it is clear that $\Lambda_{fp_1,gp_1} \simeq \mathrm{id}$. Thus $\Lambda_{H,K}\Lambda_{(-H),(-K)} \simeq \mathrm{id}$ and similarly $\Lambda_{(-H),(-K)}\Lambda_{H,K} \simeq \mathrm{id}$. Therefore $\Lambda_{H,K}$ is a homotopy equivalence. □

Proof of Theorem 6.2.8 We apply Lemma E.1 to the case where α, β, and γ are homotopy equivalences and their homotopy inverses are α', β', and γ'. Then by Lemma E.2,

$$\Lambda_{F',G'}\,\Lambda_{F,G} \simeq \Lambda_{F\circ F',G\circ G'} \simeq \Lambda_{H,K},$$

where H and K are homotopies such that $f \simeq_H f$ and $g \simeq_K g$. By Lemma E.3, $\Lambda_{H,K}$ is a homotopy equivalence, and so $\Lambda_{F,G}$ has a left homotopy inverse. Similarly, $\Lambda_{F,G}$ has a right homotopy inverse. By Exercise 1.11, $\Lambda_{F,G}$ is a homotopy equivalence. This completes the proof. □

Remark E.4 1. This theorem appears in the literature in many places. Some of these are [13, Chap. 12, Sect. 4.2], [60, Thm. 4.9], and [63, Cor. 9]. Our proof follows Nomura [76, Thm. 2.6].

2. In Definition 6.2.15, we have defined induced maps for homotopy pullbacks and stated the dual of Theorem 6.2.8 as Theorem 6.2.16. We remark that the proof of the latter theorem is obtained by dualizing the proof of Theorem 6.2.8 that we have just given.

3. We briefly discuss a shorter, but less explicit and general proof of Theorem 6.2.8. By Remark 6.2.2, the standard homotopy pushout squares of Theorem 6.2.8 can be replaced by pushout squares

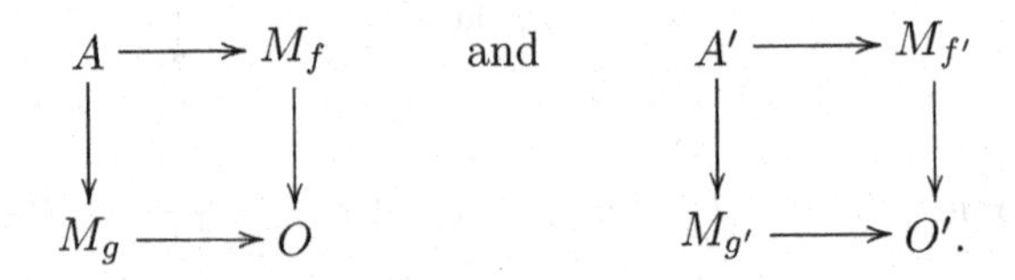

Thus O is a space with subspaces M_f, M_g, and A such that $O = M_f \cup M_g$ and $A = M_f \cap M_g$, and similarly for O'. Then, with mild restrictions on the spaces, we apply standard Mayer–Vietoris methods [61, p. 207] to the triads $(O; M_f, M_g)$ and $(O'; M_{f'}, M_{g'})$. The map of the first triad into the second gives a map of the Mayer–Vietoris sequence of the first triad into the second. This together with the five lemma shows that $\Lambda_* : H_q(O) \to H_q(O')$ is an isomorphism for all q. If X, X', Y, and Y' are all 1-connected CW complexes, then O and O' are 1-connected. We then apply Whitehead's theorem 2.4.7 to Λ to conclude that it is a homotopy equivalence.

The Prism Theorem

We state and prove the theorem. 6.3.3

Theorem E.5 *Given a homotopy-commutative diagram*

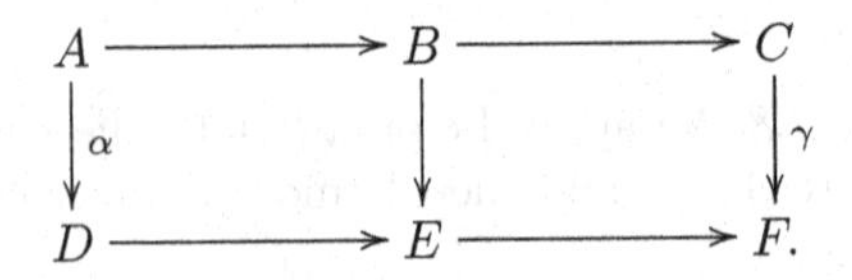

1. *If the right square is a homotopy pullback square, then the left square is a homotopy pullback square if and only if A-C-F-D is a homotopy pullback square.*
2. *If the left square is a homotopy pushout square, then the right square is a homotopy pushout square if and only if A-C-F-D is a homotopy pushout square.*

Proof. We only prove (1) since the proofs of (1) and (2) are dual. We first assume that the left and right squares are homotopy pullback squares and prove that the rectangle is. By Proposition 3.5.8, we factor γ as

$$C \xrightarrow{\gamma'} C' \xrightarrow{\gamma''} F,$$

where γ' is a homotopy equivalence and γ'' is a fibration. We then form the *pullback*

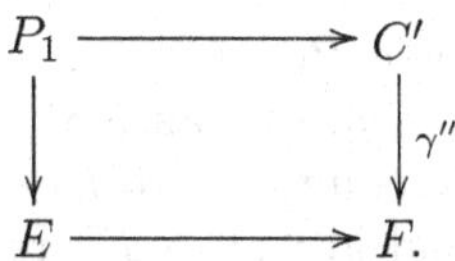

Because $\gamma'' : C' \to F$ is a fiber map, this pullback square is a homotopy pullback square by Proposition 6.2.14. Then by Definition 6.2.15, the maps $\gamma' : C \to C'$, $\mathrm{id} : F \to F$, and $\mathrm{id} : E \to E$ induce a map $B \to P_1$ which is a homotopy equivalence by Theorem 6.2.16. We then have the following diagram with homotopy-commutative faces

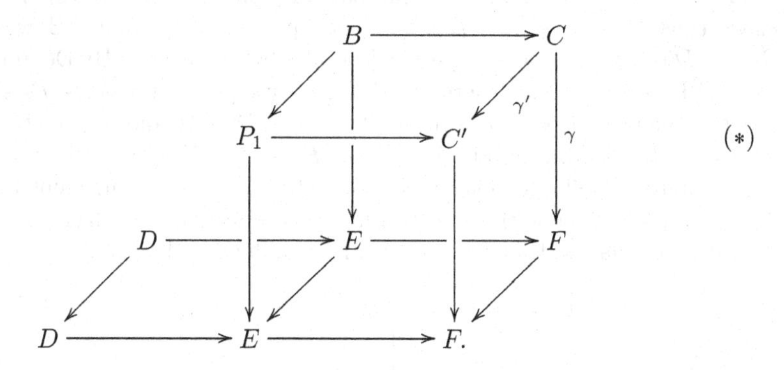

We now form the pullback

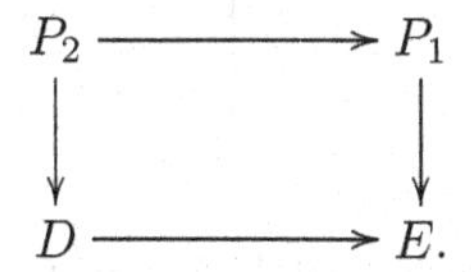

Since γ'' is a fiber map, by Proposition 3.3.12, $P_1 \to E$ is a fiber map, and so the previous pullback square is a homotopy pullback square. Then by

Definition 6.2.15, the maps $B \to P_1$, id : $E \to E$ and id : $D \to D$ induce a map $A \to P_2$ which is a homotopy equivalence by Theorem 6.2.16. We then have the following diagram with homotopy-commutative faces

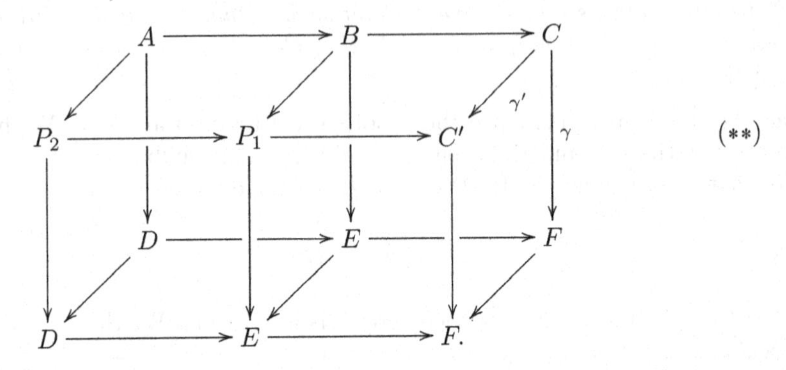

By Exercise 3.14, the rectangle $P_2 - C' - F - D$ is a pullback square and hence a homotopy pullback square because $\gamma'' : C' \to F$ is a fibration. But the rectangle $A-C-F-D$ is equivalent to $P_2-C'-F-D$ because $A \to P_2$, $\gamma' : C \to C'$, id : $F \to F$, and id : $D \to D$ are homotopy equivalences. By Proposition 6.3.2, $A - C - F - D$ is a homotopy pullback square.

The proof of the other direction is similar. We first prove it under the assumption that the left square is commutative. The right square $B - C - F - E$ is a homotopy pullback square, thus we have diagram $(*)$ with $B \to P_1$ a homotopy equivalence. We take the pullback P_2 of $D \longrightarrow E \longleftarrow P_1$ and note as before that it is a homotopy pullback. By the first part of this proof, it follows that $P_2 - C' - F - D$ is a homotopy pullback square. However, $A-C-F-D$ is also a homotopy pullback square by hypothesis. By Definition 6.2.15 and Theorem 6.2.16, there is a homotopy equivalence $A \to P_2$ such that in the diagram $(**)$, the square $A - P_2 - D - D$ and the rectangle $A - C - C' - P_2$ are commutative. Because $P_2 - P_1 - E - D$ is a homotopy pullback square, it suffices to show that $A - B - E - D$ is equivalent to it. For this it suffices to show that the top face of the left cube in diagram $(**)$ is commutative, that is, that the square in the following diagram

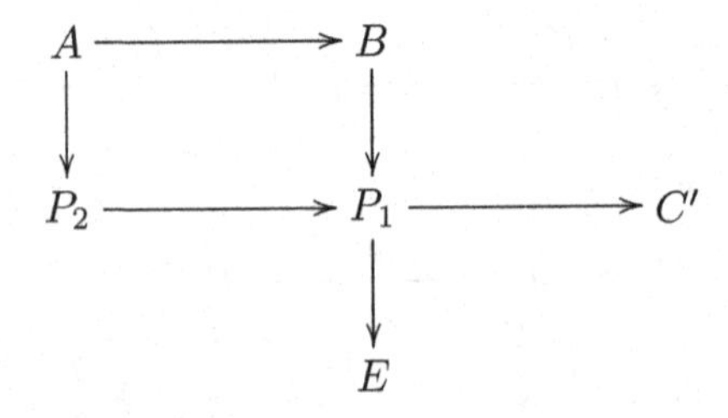

is commutative. But it is easily verified that the two maps from A to P_1 agree when composed with $P_1 \to C'$. Furthermore, because the square $A-B-E-D$

is commutative, the two maps agree when composed with $P_1 \to E$. Because P_1 is a pullback, the result follows.

In the general case we replace the left square $A - B - E - D$ by an equivalent one which is commutative. To do this, we first consider the homotopy-commutative square

$$\begin{array}{ccc} A & \longrightarrow & B \\ {\scriptstyle \alpha'}\downarrow & & \downarrow \\ M_\alpha & \xrightarrow{\alpha''} D \longrightarrow & E, \end{array}$$

where M_α is the mapping cylinder of α and α is factored as $\alpha''\alpha'$. Because α' is a cofiber map, we are able to replace the lower horizontal map $M_\alpha \to D \to E$ by one that is homotopic to it with the property that the following square is commutative

$$\begin{array}{ccc} A & \longrightarrow & B \\ {\scriptstyle \alpha'}\downarrow & & \downarrow \\ M_\alpha & \longrightarrow & E \end{array}$$

and equivalent to the left square $A - B - E - D$. This completes the proof of the prism theorem. $\square$

Finally, we give a proof of Lemma 6.3.4 based on Theorem 6.5.1, the cube theorem. We first present a lemma that is will be needed. This lemma may be of interest in its own right.

Lemma E.6 *Let*

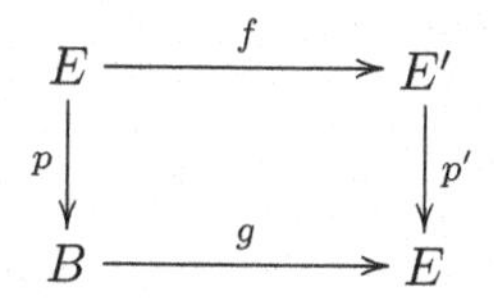

be a commutative square and let p and p' be fiber maps with fibers F and F', respectively. Let $\tilde{f} : F \to F'$ be the map induced by f. If $\tilde{f}$ is a homotopy equivalence, then the given square is a homotopy pullback square.

The proof is left as an exercise (Exercise 6.14).

Now we prove Lemma 6.3.4, and we freely use the notation at the end of Section 6.3. Consider the following cube diagram, where each of the four side squares is commutative and each vertical map in the cube is a fiber map,

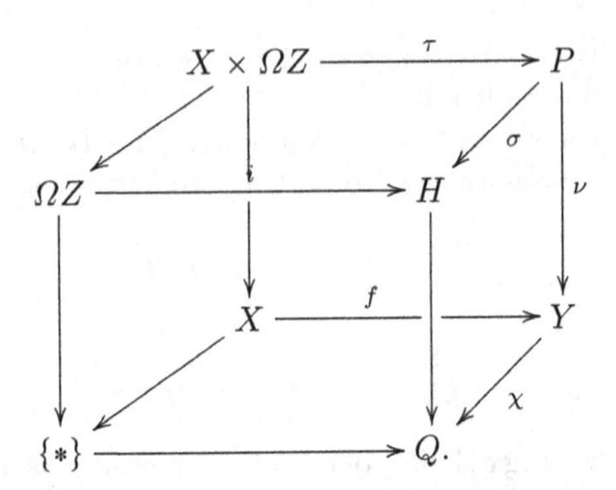

Since $Q = C_f$, the bottom square is a homotopy pushout square. The top square is homotopy-commutative as was shown in the beginning of the proof of Lemma 6.3.4. Furthermore, it is easily seen that each of horizontal maps in the top square induces a homotopy equivalence on the fibers. Therefore by Lemma E.6, each side square is a homotopy pullback square. By the cube theorem, the top square is a homotopy pushout square, and thus this proves Lemma 6.3.4.

Appendix F
Categories and Functors

In this appendix we give some basic definitions and examples regarding categories and functors. Although the reader has most likely seen much of this material, we present it as a reference for parts of the text, especially for our discussion of the Eckmann–Hilton duality.

Definitions

Definition F.1 A *category* $\mathcal{C}$ consists of

1. A class $\mathrm{obj}(\mathcal{C})$ of *objects* and for each ordered pair $A, B \in \mathrm{obj}(\mathcal{C})$, a set $\mathrm{Morph}(A, B) = \mathrm{Morph}_{\mathcal{C}}(A, B)$ of *morphisms* from A to B.
2. A function $\mathrm{Morph}(A, B) \times \mathrm{Morph}(B, C) \to \mathrm{Morph}(A, C)$ which assigns to $f \in \mathrm{Morph}(A, B)$ and $g \in \mathrm{Morph}(B, C)$, a morphism $gf \in \mathrm{Morph}(A, C)$.

We require that

(i) The morphism sets $\mathrm{Morph}(A, B)$ be pairwise disjoint.

(ii) For every $A \in \mathrm{obj}(\mathcal{C})$, there exists a morphism $\mathrm{id}_A \in \mathrm{Morph}(A, A)$ such that $f\,\mathrm{id}_A = f$ and $\mathrm{id}_A\, g = g$, for every $f \in \mathrm{Morph}(A, B)$ and every $g \in \mathrm{Morph}(C, A)$.

(iii) If $f \in \mathrm{Morph}(A, B)$, $g \in \mathrm{Morph}(B, C)$, and $h \in \mathrm{Morph}(C, D)$, then $h(gf) = (hg)f$.

If $f \in \mathrm{Morph}(A, B)$, we sometimes write $f : A \to B$. This notation is used because in many examples the morphisms are functions. The operation in (iii) is called composition.

Definition F.2 Let $\mathcal{C}$ and $\mathcal{D}$ be categories. A *(covariant) functor* F from $\mathcal{C}$ to $\mathcal{D}$, written $F : \mathcal{C} \to \mathcal{D}$, is a function that assigns to each object A in $\mathcal{C}$, an object $F(A)$ in $\mathcal{D}$ and assigns to each morphism $f : A \to B$ in $\mathcal{C}$, a morphism $F(f) : F(A) \to F(B)$ in $\mathcal{D}$ such that $F(\mathrm{id}_A) = \mathrm{id}_{F(A)}$ and

$F(gf) = F(g)F(f)$. A *contravariant functor* $F : \mathcal{C} \to \mathcal{D}$ is defined in the same way except we require that for $f : A \to B$ in $\mathcal{C}$, there is a morphism $F(f) : F(B) \to F(A)$ in $\mathcal{D}$ such that $F(gf) = F(f)F(g)$.

Definition F.3 If $\mathcal{C}$ and $\mathcal{D}$ are categories and every object of $\mathcal{C}$ is an object of $\mathcal{D}$ and every morphism of $\mathcal{C}$ is a morphism of $\mathcal{D}$, then we say that $\mathcal{C}$ is a *subcategory* of $\mathcal{D}$ and write $\mathcal{C} \subseteq \mathcal{D}$. If, in addition, for all objects A and B in $\mathcal{C}$, $\mathrm{Morph}_{\mathcal{C}}(A, B) = \mathrm{Morph}_{\mathcal{D}}(A, B)$, then $\mathcal{C}$ is a *full subcategory* of D.

Examples

We next consider many familiar examples of categories and functors with particular emphasis on those appearing in this book.

- The category *Sets* of sets and functions. Thus obj(*Sets*) consists of all sets and Morph(A, B) consists of all functions from the set A to the set B. There is also $Sets_*$, the category of based sets and functions.
- The category *Gr* of groups and homomorphisms
- The category *Ab* of abelian groups and homomorphisms
- The category *Top* of topological spaces and continuous functions and the category Top_* of based topological spaces and maps (continuous, based functions)
- The category $\mathcal{P}$ of pairs of spaces (X, A), where $A \subseteq X$ and continuous functions (of pairs)
  In all of the preceding examples, the morphisms are functions. This need not be the case as the following important example shows.
- The category $HoTop_*$ of based topological spaces and based homotopy classes of maps called the (based) homotopy category
- The category $\mathcal{H}$ of H-spaces and homotopy classes of H-maps and the category $\mathcal{HG}$ of grouplike spaces and homotopy classes of H-maps
- The category $\mathcal{CH}$ of co-H-spaces and homotopy classes of co-H-maps and the category $\mathcal{CG}$ of cogroups and homotopy classes of co-H-maps
- The category of CW complexes and continuous functions and the category of CW complexes and cellular functions
- The *forgetful functor* $F : Gr \to Sets$ that assigns to a group the underlying set of the group and to a homomorphism the function which is the homomorphism. There are many examples of such forgetful functors: $Top \to Sets$, $Ab \to Gr$, and so on.
- The nth cohomology group functor $H^n : Top \to Ab$ which is a contravariant functor.
- The fundamental group functor $\pi_1 : HoTop_* \to Gr$.
- The loop space functor $\Omega : HoTop_* \to \mathcal{HG}$ and the suspension functor $\Sigma : HoTop_* \to \mathcal{CG}$.

Adjoint Functors

We define adjoint functors after introducing some notation. Let $\mathcal{C}$ be a category, let B be an object of $\mathcal{C}$, and let $f : A \to A'$ be a morphism in $\mathcal{C}$. Define $f^* : \text{Morph}(A', B) \to \text{Morph}(A, B)$ by $f^*(h) = hf$, for $h \in \text{Morph}(A', B)$. Now let A be an object in $\mathcal{C}$ and $g : B \to B'$ a morphism in $\mathcal{C}$. Define $g_* : \text{Morph}(A, B) \to \text{Morph}(A, B')$ by $g_*(k) = gk$, for $k \in \text{Morph}(A, B)$.

Definition F.4 Let $F : \mathcal{C} \to \mathcal{D}$ and $G : \mathcal{D} \to \mathcal{C}$ be functors. Suppose that for every $A \in \text{obj}(\mathcal{C})$ and every $B \in \text{obj}(\mathcal{D})$, there is a bijection $\eta_{A,B,}$: $\text{Morph}_{\mathcal{C}}(A, G(B)) \to \text{Morph}_{\mathcal{D}}(F(A), B)$ such that

1. If $f : A \to A'$ is a morphism in $\mathcal{C}$, then for every object B in $\mathcal{D}$, the following diagram commutes

$$\begin{array}{ccc} \text{Morph}_{\mathcal{C}}(A', G(B)) & \xrightarrow{\eta_{A',B}} & \text{Morph}_{\mathcal{D}}(F(A'), B) \\ \Big\downarrow{\scriptstyle f^*} & & \Big\downarrow{\scriptstyle F(f)^*} \\ \text{Morph}_{\mathcal{C}}(A, G(B)) & \xrightarrow{\eta_{A,B}} & \text{Morph}_{\mathcal{D}}(F(A), B); \end{array}$$

2. If $g : B \to B'$ is a morphism in $\mathcal{D}$, then for every object A in $\mathcal{C}$, the following diagram commutes

$$\begin{array}{ccc} \text{Morph}_{\mathcal{C}}(A, G(B)) & \xrightarrow{\eta_{A,B}} & \text{Morph}_{\mathcal{D}}(F(A), B) \\ \Big\downarrow{\scriptstyle G(g)_*} & & \Big\downarrow{\scriptstyle g_*} \\ \text{Morph}_{\mathcal{C}}(A, G(B')) & \xrightarrow{\eta_{A,B'}} & \text{Morph}_{\mathcal{D}}(F(A), B'). \end{array}$$

Then F and G are called *adjoint functors.*

We sometimes say that F is the adjoint of G and G is the adjoint of F. We occasionally stretch the terminology by calling $f : A \to G(B)$ and $\eta_{A,B}(f) : F(A) \to B$ *adjoint morphisms.*

We next give some examples of adjoint functors from the text. The main example, which is the source of the others, is the following.

- In the category *Top* of topological (unbased) spaces, the set of morphisms from X to Y consists of all continuous functions $X \to Y$ which we write as $C(X, Y)$. Consider the functors $F, G : Top \to Top$ defined by $F(X) = X \times I$ and $G(X) = Y^I$, where the latter space is the space of continuous functions $Y \to I$ with the compact–open topology (Appendix A). Then these are adjoint functors, that is, there is a bijection $C(X \times I, Y) \to C(X, Y^I)$ such that the two commutative diagrams above hold (Proposition 1.3.4). This has enabled us to have two equivalent definitions of a homotopy: one as a function defined on a cylinder $X \times I$ and the other as a function into a path space Y^I. In addtion, many examples of adjoint functors earlier in the book are derived from the previous example. We next mention two of the most common ones.

- In the category Top_* of based spaces and maps the cone functor CX and the path-space functor EY are adjoint (Section 1.4).
- The most useful pair of adjoint functors that we consider are the suspension and loop space functors. The adjointness of these two functors is a key result in many places. In more detail, let $HoTop_*$ be the based homotopy category and let $\Sigma, \Omega : HoTop_* \to HoTop_*$ be the suspension and the loop space functors, respectively. We show in Proposition 2.3.5 that there is a bijection $\kappa_* : [\Sigma X, Y] \to [X, \Omega Y]$, for spaces X and Y, and it follows that Σ and Ω are adjoint functors. Moreover, these two homotopy sets have group structure and we prove that κ_* is an isomorphism.

Products and Coproducts

We have discussed products and coproducts in $HoTop_*$ previously. Now we give the general definitions in a category.

Definition F.5 Let $\mathcal{C}$ be a category and let A_1 and A_2 be two objects in $\mathcal{C}$.

1. A *product* of these objects consists of an object P in $\mathcal{C}$ together with two morphisms $p_1 : P \to A_1$ and $p_2 : P \to A_2$ satisfying the following condition. If $f : D \to A_1$ and $g : D \to A_2$ are any morphisms, then there exists a unique morphism $h : D \to P$ such that $p_1 h = f$ and $p_2 h = g$. If for every two objects A_1 and A_2, the product P exists, we say that $\mathcal{C}$ is a *category with (finite) products.*
2. A *coproduct* of the objects A_1 and A_2 consists of an object C in $\mathcal{C}$ together with two morphisms $i_1 : A_1 \to C$ and $i_1 : A_2 \to C$ satisfying the following condition. If $f : A_1 \to D$ and $g : A_2 \to D$ are any morphisms, then there exists a unique morphism $k : C \to D$ such that $ki_1 = f$ and $ki_2 = g$. If for every two objects A_1 and A_2, the coproduct C exists, we say that $\mathcal{C}$ is a *category with (finite) coproducts.*

We make a few remarks about these definitions. It can be shown that the objects P and D are essentially unique. We write $P = A_1 \times A_2$. and $C = A_1 \sqcup A_2$. Clearly, the definitions can be extended to a product or coproduct of any number of objects A_i in $\mathcal{C}$. The two definitions in F.5 are more than just similar. They are dual in a formal categorical sense (see Section 2.6).

We conclude this section with a few relevant examples.

- In the group categories *Gr* and *Ab*, the product of two groups A_1 and A_2 is just the cartesian product $A_1 \times A_2$ of the two groups (also called the external direct sum $A_1 \oplus A_2$). However, the coproduct is different in these two categories. In *Ab* the coproduct is the cartesian product. But in *Gr* the coproduct is the free product $A_1 * A_2$ (see Appendix B).

- In the topological category *Top*, the product of two spaces A_1 and A_2 is just the cartesian product $A_1 \times A_2$ (with the product topology) and projection functions p_1 and p_2. The coproduct is the disjoint union $A_1 \sqcup A_2$ with the weak topology (Example A.2) and inclusion functions i_1 and i_2.
- In the based topological category Top_*, the product of two based spaces A_1 and A_2 is just the cartesian product $A_1 \times A_2$ with projection maps p_1 and p_2. The coproduct is the wedge $A_1 \vee A_2$ with inclusion maps i_1 and i_2. Of particular interest for homotopy theory are the product and coproduct in the homotopy category $HoTop_*$. The product of two based spaces A_1 and A_2 is just the cartesian product $A_1 \times A_2$ with the homotopy classes of the projection maps $[p_1]$ and $[p_2]$. The coproduct is the wedge $A_1 \vee A_2$ with the homotopy classes of the inclusion maps $[i_1]$ and $[i_2]$. These two statements follow from Lemma 1.3.6.

Hints to Some of the Exercises

Chapter One

1.10 Let x be a limit point of A with $r(x) \neq x$ and separate x and $r(x)$ by disjoint neighborhoods U and V. Consider $r^{-1}(V)$.

1.17 First embed G in a symmetric group.

1.19 See Proposition 3.2.15.

1.23 Let $X = \mathbb{R}^{n+1} - \{0\}$ and $\mathbb{R}\mathrm{P}^n = X/\!\sim$. Define $f : X \times X \to \mathbb{R}$ by

$$f(x,y) = \sum_{i \neq j} (x_i y_j - x_j y_i)^2,$$

where $x = (x_1, \ldots, x_{n+1})$ and $y = (y_1, \ldots, y_{n+1})$. Then $f^{-1}(0) = \{(x,y) \mid x \sim y\}$ is closed in $X \times X$.

1.24 It suffices to show that a finite union of open cells L is contained in a finite subcomplex. Argue by induction on $\dim L$, the dimension of the largest dimensional cell in L.

1.26 Consider the union of S^2 and an arc from $\{*\}$ to the South Pole which is "outside" S^2 and does not meet S^2 elsewhere.

1.27 $X \simeq L/(K \cap L) \simeq L$.

Chapter Two

2.10 Let $\rho', \tau' : I^2 \to I^2$ be defined by $\rho'(x,y) = (1-x, y)$ and $\tau'(x,y) = (y,x)$. These are reflections across lines that become reflections across diameters when $I^2 \cong E^2$. They are then homotopic by a family of rotations of E^2 that carry S^1 to itself.

2.15

$$H(l,s)(t) = \begin{cases} * & \text{if } 0 \leqslant t \leqslant \frac{s}{2} \\ l(2t-s) & \text{if } \frac{s}{2} \leqslant t \leqslant \frac{1}{2} \\ l(2-2t-s) & \text{if } \frac{1}{2} \leqslant t \leqslant 1-\frac{s}{2} \\ * & \text{if } 1-\frac{s}{2} \leqslant t \leqslant 1. \end{cases}$$

2.30 Consider the Hurewicz homomorphism h_1.

2.18 Map $I \times I \to I \times I$ such that (i) $I \times \{0\}$ goes to $[0, \frac{1}{2}] \times \{0\}$, (ii) $\{1\} \times I$ goes to the point $(\frac{1}{2}, 0)$, (iii) $I \times \{1\}$ goes to $[\frac{1}{2}, 1] \times \{0\}$ (in reverse direction) and (iv) $\{0\} \times I$ goes to $\{0\} \times I \cup I \times \{1\} \cup \{1\} \times I$. This is best done by taking E^2 instead of $I \times I$.

2.32 If K is a CW complex, the inclusion $K^i \to K$ induces an epimorphism $H_i(K^i) \to H_i(K)$. Let $K = K(G,n)$ with $K(G,n)^{n+1} = M(G,n)$.

2.35 Let $\Delta : X \to X \times X$ be the diagonal map and let j_i and j_2 be the two inclusions of X into $X \times X$. Show that for $x \in H_i(X)$, $\Delta_*(x) = j_{1*}(x) + j_{2*}(x)$.

Chapter Three

3.14 (2): Denote the maps $a : A \to X$, $b : Y \to P$, $u : X \to Z$, $h : A \to Y$ and $g : X \to P$. Let $\alpha : W \to X$ and $\beta : W \to Y$ be such that $g\alpha = b\beta$. There exists $\theta : W \to A$ with $ua\theta = u\alpha$ and $h\theta = \beta$. Then $g\alpha = ga\theta$, and so $\alpha = a\theta$.

3.20 (1) If U is open in G, then $p^{-1}p(U)$ is the union of open sets Uh, for every $h \in H$. (2) Suppose $aH \neq bH$ and define $F : G \times G \to G$ by $F(x,y) = x^{-1}y$. Then $F(a,b) \in H'$, the complement of H. There exist open neighborhoods U and V of a and b such that $F(U \times V) \subseteq H'$. (3) Suppose $G = U \cup V$, where U and V are open nonempty subsets of G. Then $G/H = p(U) \cup p(V)$, and so there is some $aH \in p(U) \cap p(V)$. Therefore $aH \cap U$ is nonempty and $aH \cap V$ is nonempty. Now $aH = (aH \cap U) \cup (aH \cap V)$. Therefore $(aH \cap U) \cap (aH \cap V)$ is nonempty, and hence $U \cap V$ is nonempty.

3.24 (2) Prove this inductively using Exercise 3.23 with $k = 1$.

3.25 For $w \in \mathbb{F}^n$ and $X \in G_k(\mathbb{F}^n)$, let $\rho_w(X)$ be the square of the distance from w to X. Then $\rho_w : G_k(\mathbb{F}^n) \to \mathbb{R}$ is continuous. If X and Y are distinct k-planes, choose $w \in X$ and $w \notin Y$, and so $\rho_w(X) \neq \rho_w(Y)$. By separating $\rho_w(X)$ and $\rho_w(Y)$ with disjoint neighborhoods, we obtain disjoint neighborhoods of X and Y.

Chapter Four

4.1 Consider $F : \Omega B \times I \to I_p$ defined by $F(\omega, t) = (\lambda(*, \omega)(t), \omega_{t,1})$.

4.2 It suffices to show that the inclusion $i : \Omega B \to I_p$ is homotopic to $\beta(-\mathrm{id}) : \Omega B \to I_p$, where $i(\omega) = (*, \omega)$ and $\beta(\omega) = (\omega, *)$. Then $F : \Omega B \times I \to I_p$ is given by $F(\omega, t)(s) = (\omega((1-s)t), \omega((1-s)t + s))$.

4.3 (1) If $j : K \to K \cup CL$ is the inclusion, then C_j is the pushout of $CL \longleftarrow L \longrightarrow CK$. Hence $\Sigma L \simeq C_j/CK \simeq C_j \simeq C_j/CL \simeq CK/L$. (2) Consider the commutative diagram

$$\begin{array}{ccc} K \cup CL & \xrightarrow{k} & C_j \\ \big\downarrow p & & \big\downarrow q \\ K/L = K \cup CL/CL & \xrightarrow{\partial} & C_j/CK = \Sigma L, \end{array}$$

where p and q are homotopy equivalences and k is inclusion. Thus $\partial \simeq * \iff k \simeq *$. Next consider

$$\begin{array}{ccc} K \cup CL & \xrightarrow{k} & C_j \\ \big\downarrow p & & \big\downarrow s \\ K/L = K \cup CL/CL & \xrightarrow{i} & C_j/CL = CK/L. \end{array}$$

(3) If $l : L \to K$ is the inclusion, consider the exact sequence

$$[\Sigma K, \Sigma L] \xrightarrow{(\Sigma l)^*} [\Sigma L, \Sigma L] \xrightarrow{\partial^*} [K/L, \Sigma L] .$$

4.4 It suffices to show $\beta m \simeq \phi_0(\beta \times \mathrm{id})$. Consider the homotopy $F(\omega, \nu, t) = (\omega + \nu_{0,t}, \nu_{t,1})$.

4.5 The function $j_* : [W, F] \to [W, E]$ is a bijection by Corollary 4.2.19(1) and Theorem 4.4.5 because B is contractible.

4.7 The operation $+_i$ induces group structure on $\pi_n(X)$, for every space X, such that h_* is a homomorphism for every $h : X \to X'$. By Proposition 2.2.11 there is a comultiplication c_i on S^n obtained from $+_i$. But $c \simeq c_i$, where c is the standard comultiplication on S^n by Exercise 2.7.

4.14 Let ϕ be the standard comultiplication on S^n and let $E = E(X; A, \{*\})$. If $f, g : S^n \to E$, then $f + g = \nabla(f \vee g)\phi$, where ∇ is the folding map of E. Take adjoints of f and g and regard them as maps $\hat{f}, \hat{g} : (CS^n, S^n) \to (X, A)$. The adjoint of $f + g$ is $\nabla'(\hat{f} \vee \hat{g})C\phi$, where $\nabla' : (X \vee X, A \vee A) \to (X, A)$ is the folding map of pairs. Then show $(\hat{f} + \hat{g})_* = \hat{f}_* + \hat{g}_*$.

4.19 Let $h_t : A \to X$ be such that $h_0 = i$ and $h_1 = *$. If $f : (I^{n-1}, \partial I^{n-1}) \to (A, *)$, define $f' : (I^n, I^{n-1}, J^{n-1}) \to (X, A, \{*\})$ by setting $f'(t_1, \ldots, t_n) = h_{t_n}(f(t_1, \ldots, t_{n-1}))$. This gives a homomorphism $\mu : \pi_{n-1}(A) \to \pi_n(X, A)$ such that $\partial_n \mu = \text{id}$.

4.24 By Proposition 4.5.15 and Lemma 4.5.17, the left square is commutative and the right square is anti-commutative

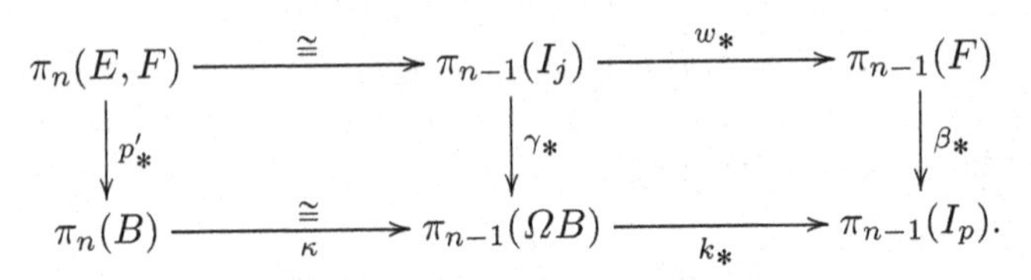

Chapter Five

5.6 (1) Consider $S^1 \xrightarrow{j_1} S^1 \times S^1 \longrightarrow (S^1 \times S^1)/(S^1 \times \{*\})$. If m is the multiplication on S^1, then $mi_1 = \text{id}$. But $S^1 \vee (S^1 \times S^1)/(S^1 \times \{*\})$ does not have the same homotopy type as $S^1 \times S^1$. (2) If X is a co-H-space and $i_1 : A \to A \vee Q$ and $i_2 : Q \to A \vee Q$ are inclusions, then $i_1 r + i_2 q : X \to A \vee Q$ induces homology isomorphisms. (3) Use Corollary 5.4.7.

5.11 Modify the proofs in Section 5.5.

5.12 Let g be the homotopy inverse of f, let $y_0 = f(x_0)$ and let $x_1 = g(y_0)$. Then id $\simeq_F gf : X \to X$, and if $[h] \in \pi_n(X, x_0)$, then $h \simeq_{F(h \times \text{id})} gfh$. For $x \in \partial I^n, F(h \times \text{id})(x,t) = F(x_0, t)$ is a path $a(t)$ from x_0 to x_1. Then $a^{\#}[h] = g_* f_* [h]$.

5.14 Let X be the comb space and let Y be a point.

5.18 Consider $p_Z : \Sigma\Omega Z \to Z$.

5.19 $\pi_i(\mathbb{R}\text{P}^\infty) \cong \pi_i(\mathbb{R}\text{P}^{i+1})$ for $i \geqslant 2$.

Chapter Six

6.4 See Exercise 3.8.

6.13 (1) Define $F : C_j \times I \to X$ by $F(\langle \omega, s \rangle, t) = \omega(t(1-s))$ and $F(\langle \lambda \rangle, t) = \lambda(t)$. (2) The map q is a homotopy equivalence and $I_\alpha \simeq \Omega X * \Omega X$ by Remark 6.4.3. (3) Use Exercise 6.12.

6.17 (1) Let $m > n > 1$. Then X and Y have the same fundamental group and the same universal covering space. By Theorem C.6, they have different homology groups. (2) S^4 is a retract of X so $\pi_4(X) \neq 0$. But $\pi_4(Y) = 0$.

6.23 Let $X \xrightarrow{j} Y \xrightarrow{q} Z$ be a cofiber sequence with j an inclusion, let X be $(n-1)$-connected and let Z be m-connected and conclude that $\pi_i(Y, X) \to \pi_i(C_j, CX)$ is an isomorphism for $i < m + n$. Then consider the exact homotopy sequence of the pair (Y, X).

Chapter Seven

7.1 See Proposition 7.2.2.

7.2 (2) Consider the commutative diagram

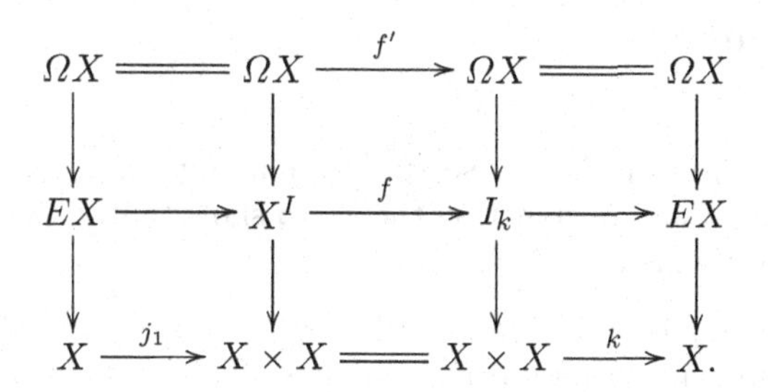

Because f' is a weak equivalence, kj_1 is also, and hence it is a homotopy equivalence. Similarly, kj_2 is a homotopy equivalence. Now apply Part (1).

7.10 Because g_n is an $(n+1)$-equivalence, $g_{n*} : H_{n+1}(X) \to H_{n+1}(X^{(n)})$ is onto. For $H_{n+2}(X^{(n)})$, consider $K(\pi_{n+1}, n+1) \longrightarrow X^{(n+1)} \longrightarrow X^{(n)}$ and its exact homology sequence.

7.12 For the implication $\Longleftarrow$, write $[D_f] = (f \wedge f)^*[w]$ and define $\tilde{n} = -wq + n$, where $q : Y \times Y \to Y \wedge Y$ is the projection.

7.14 The defining cofibration of M yields the exact sequence

$$[S^{n+1}, S^{n+1}] \xrightarrow{f^*} [S^{n+1}, S^{n+1}] \xrightarrow{q^*} [M, S^{n+1}],$$

where f^* is multiplication by m.

7.16 We obtain a commutative diagram

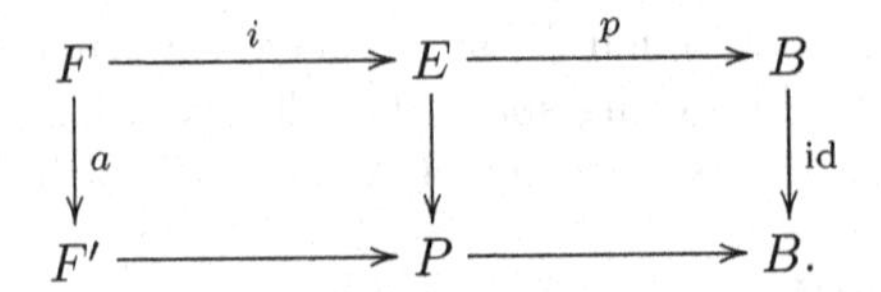

Show that $a \simeq \alpha$.

Chapter Eight

8.2 Consider $j : X \vee X \to X \times X$.

8.9 The generalized Freudenthal theorem.

Chapter Nine

9.4 (2) For Definition 8.3.1, use the cellular approximation theorem. For Proposition 8.3.12, use obstruction theory to show that a section of the fiber map $G_k(X) \to X$ exists.

9.8 Assume that f is an inclusion and show that $\pi_i(Y, X) = 0$ for all i using the relative Hurewicz theorem. For this it suffices to show that $\pi_1(X)$ acts trivially on $\pi_i(Y, X)$. Use obstruction theory to obtain a retraction $Y \to X$ and conclude that $\pi_i(Y) \to \pi_i(Y, X)$ is onto.

References

1. Adams, J.F.: On the non-existence of elements of Hopf invariant one. Ann. of Math. (2) **72**, 20–104 (1960)
2. Adams, J.F.: Vector fields on spheres. Ann. of Math. (2) **75**, 603–632 (1962)
3. Arkowitz, M.: The group of self-homotopy equivalences—a survey. In: Groups of Self-Equivalences and Related Topics (Montreal, PQ, 1988), *Lecture Notes in Math.*, vol. 1425, pp. 170–203. Springer, Berlin (1990)
4. Arkowitz, M.: Induced mappings of homology decompositions. In: Homotopy and Geometry (Warsaw, 1997), *Banach Center Publ.*, vol. 45, pp. 225–233. Polish Acad. Sci., Warsaw (1998)
5. Arkowitz, M.: Equivalent definitions of the Ganea fibrations and cofibrations. Manu. Math. **100**(2), 221–229 (1999)
6. Arkowitz, M.: On Whitehead's inequality, nil$[X, G] \leqslant \operatorname{cat} X$. Int. J. Math. and Math. Sci. **25**(5), 311–313 (2001)
7. Arkowitz, M., Gutierrez, M.: Comultiplications on free groups and wedges of circles. Trans. Amer. Math. Soc. **350**(4), 1663–1680 (1998)
8. Baues, H.J.: Algebraic Homotopy, *Cambridge Studies in Advanced Mathematics*, vol. 15. Cambridge University Press, Cambridge (1989)
9. Berstein, I.: A note on spaces with non-associative co-multiplication. Proc. Cambridge Philos. Soc. **60**, 353–354 (1964)
10. Blakers, A.L., Massey, W.S.: The homotopy groups of a triad. II. Ann. of Math. (2) **55**, 192–201 (1952)
11. Bott, R.: The stable homotopy of the classical groups. Ann. of Math. (2) **70**, 313–337 (1959)
12. Bott, R., Tu, L.W.: Differential Forms in Algebraic Topology, *Graduate Texts in Mathematics*, vol. 82. Springer-Verlag, New York (1982)
13. Bousfield, A.K., Kan, D.M.: Homotopy Limits, Completions and Localizations. Springer-Verlag, Berlin (1972). Lecture Notes in Mathematics, Vol. 304
14. Bredon, G.E.: Topology and Geometry, *Graduate Texts in Mathematics*, vol. 139. Springer-Verlag, New York (1993)
15. Brown Jr., E.H., Copeland Jr., A.H.: An homology analogue of Postnikov systems. Michigan Math. J **6**, 313–330 (1959)
16. Brown, R.: Two examples in homotopy theory. Proc. Cambridge Philos. Soc. **62**, 575–576 (1966)
17. Chachólski, W.: A generalization of the triad theorem of Blakers-Massey. Topology **36**(6), 1381–1400 (1997)
18. Chevalley, C.: Theory of Lie Groups. I. Princeton University Press, Princeton, NJ (1946, 1957)

19. Cohen, F.R., Moore, J.C., Neisendorfer, J.A.: The double suspension and exponents of the homotopy groups of spheres. Ann. of Math. (2) **110**(3), 549–565 (1979)
20. Cooke, G.E., Finney, R.L.: Homology of Cell Complexes. Based on lectures by Norman E. Steenrod. Princeton University Press, Princeton, NJ (1967)
21. Cornea, O., Lupton, G., Oprea, J., Tanré, D.: Lusternik-Schnirelmann Category, *Mathematical Surveys and Monographs*, vol. 103. American Mathematical Society, Providence, RI (2003)
22. Coxeter, H.S.M.: Projective Geometry. Springer-Verlag, New York (1994). Revised reprint of the second (1974) edition
23. Davis, J.F., Kirk, P.: Lecture Notes in Algebraic Topology, *Graduate Studies in Mathematics*, vol. 35. American Mathematical Society, Providence, RI (2001)
24. Doeraene, J.P.: Homotopy pull backs, homotopy push outs and joins. Bull. Belg. Math. Soc. Simon Stevin **5**(1), 15–37 (1998)
25. Dowker, C.H.: Topology of metric complexes. Amer. J. Math. **74**, 555–577 (1952)
26. Dugundji, J.: Topology. Allyn and Bacon, Boston, MA (1966)
27. Eckmann, B., Hilton, P.J.: Groupes d'homotopie et dualité. C. R. Acad. Sci. Paris **246**, 2444–2447, 2555–2558, 2991–2993 (1958)
28. Eckmann, B., Hilton, P.J.: Transgression homotopique et cohomologique. C. R. Acad. Sci. Paris **247**, 620–623 (1958)
29. Eckmann, B., Hilton, P.J.: Décomposition homologique d'un polyèdre simplement connexe. C. R. Acad. Sci. Paris **248**, 2054–2056 (1959)
30. Eckmann, B., Hilton, P.J.: On the homology and homotopy decomposition of continuous maps. Proc. Nat. Acad. Sci. U.S.A. **45**, 372–375 (1959)
31. Eilenberg, S.: Singular homology theory. Ann. of Math. (2) **45**, 407–447 (1944)
32. Eilenberg, S., Steenrod, N.: Foundations of Algebraic Topology. Princeton University Press, Princeton, NJ (1952)
33. Fritsch, R., Piccinini, R.A.: Cellular Structures in Topology, *Cambridge Studies in Advanced Mathematics*, vol. 19. Cambridge University Press, Cambridge (1990)
34. Ganea, T.: On the loop spaces of projective spaces. J. Math. Mech. **16**, 853–858 (1967)
35. Ganea, T.: Cogroups and suspensions. Invent. Math. **9**, 185–197 (1969/1970)
36. Ganea, T.: Some problems on numerical homotopy invariants. In: Symposium on Algebraic Topology (Battelle Seattle Res. Center, Seattle Wash., 1971), pp. 23–30. Lecture Notes in Math., Vol. 249. Springer, Berlin (1971)
37. Gray, B.: Homotopy Theory. Academic Press [Harcourt Brace Jovanovich Publishers], New York (1975). An introduction to algebraic topology, Pure and Applied Mathematics, Vol. 64
38. Green, P.: Products in homotopy groups with certain prime coefficients. Proc. Cambridge Philos. Soc. **64**, 11–14 (1968)
39. Hatcher, A.: Algebraic Topology. Cambridge University Press, Cambridge (2002)
40. Hilton, P.: Homotopy Theory and Duality. Gordon and Breach Science, New York (1965)
41. Hilton, P.J.: An Introduction to Homotopy Theory. Cambridge Tracts in Mathematics and Mathematical Physics, no. 43. Cambridge, at the University Press (1953)
42. Hilton, P.J., Stammbach, U.: A Course in Homological Algebra, *Graduate Texts in Mathematics*, vol. 4, second edn. Springer-Verlag, New York (1997)
43. Hilton, P.J., Wylie, S.: Homology Theory: An Introduction to Algebraic Topology. Cambridge University Press, New York (1960)
44. Hopf, H.: Ueber die Abbildungen der dreidimensionalen Sphaere auf die Kugelflaeche. Math. Ann. **104**, 637–665 (1931)
45. Hu, S.t.: Homotopy Theory. Pure and Applied Mathematics, Vol. VIII. Academic Press, New York (1959)
46. Huber, P.J.: Homotopical cohomology and Cech cohomology. Math. Ann. **144**, 73–76 (1961)
47. Husemoller, D.: Fibre Bundles, second edn. Springer-Verlag, New York (1975). Graduate Texts in Mathematics, No. 20

48. Iwase, N.: Co-H-spaces and the Ganea conjecture. Topology **40**(2), 223–234 (2001)
49. Jacobson, N.: Basic Algebra. I, second edn. W. H. Freeman, New York (1985)
50. James, I.M.: Reduced product spaces. Ann. of Math. (2) **62**, 170–197 (1955)
51. James, I.M.: Multiplication on spheres. II. Trans. Amer. Math. Soc. **84**, 545–558 (1957)
52. James, I.M.: On H-spaces and their homotopy groups. Quart. J. Math. Oxford Ser. (2) **11**, 161–179 (1960)
53. James, I.M.: The Topology of Stiefel Manifolds. Cambridge University Press, Cambridge (1976). London Mathematical Society Lecture Note Series, No. 24
54. Kahn, D.W.: Induced maps for Postnikov systems. Trans. Amer. Math. Soc. **107**, 432–450 (1963)
55. Klaus, S.: H-space structures and Postnikov approximations. Math. Proc. Cambridge Philos. Soc. **135**(1), 133–135 (2003)
56. Lee, J.M.: Introduction to Topological Manifolds, *Graduate Texts in Mathematics*, vol. 202. Springer-Verlag, New York (2000)
57. Lundell, A., Weingram, S.: The Topology of CW Complexes, *Mathematical Surveys and Monographs*, vol. 63. Van Nostrand Reinhold, New York (1969)
58. Lusternik, L., Schnirelmann, L.: Méthodes Topologique dans les Próblemes Variationnels. Herman, Paris (1934)
59. Mac Lane, S.: Homology. Die Grundlehren der mathematischen Wissenschaften, Bd. 114. Academic Press Inc., Publishers, New York (1963)
60. Marcum, H.J.: Parameter constructions in homotopy theory. An. Acad. Brasil. Ci. **48**(3), 387–402 (1976)
61. Massey, W.S.: A Basic Course in Algebraic Topology, *Graduate Texts in Mathematics*, vol. 127. Springer-Verlag, New York (1991)
62. Mather, M.: A generalisation of Ganea's theorem on the mapping cone of the inclusion of a fibre. J. London Math. Soc. (2) **11**(1), 121–122 (1975)
63. Mather, M.: Pull-backs in homotopy theory. Canad. J. Math. **28**(2), 225–263 (1976)
64. Maunder, C.R.F.: Algebraic Topology. Dover, Mineola, NY (1996). Reprint of the 1980 edition
65. May, J.P.: The dual Whitehead theorems. In: Topological Topics, *London Math. Soc. Lecture Note Ser.*, vol. 86, pp. 46–54. Cambridge Univ. Press, Cambridge (1983)
66. May, J.P.: A Concise Course in Algebraic Topology. Chicago Lectures in Mathematics. University of Chicago Press, Chicago, IL (1999)
67. McCleary, J.: A user's Guide to Spectral Sequences, *Cambridge Studies in Advanced Mathematics*, vol. 58, second edn. Cambridge University Press, Cambridge (2001)
68. McGibbon, C.A., Neisendorfer, J.A.: On the homotopy groups of a finite-dimensional space. Comment. Math. Helv. **59**(2), 253–257 (1984)
69. Milnor, J.: The geometric realization of a semi-simplicial complex. Ann. of Math. (2) **65**, 357–362 (1957)
70. Milnor, J.: On spaces having the homotopy type of a CWcomplex. Trans. Amer. Math. Soc. **90**, 272–280 (1959)
71. Milnor, J.: Morse Theory. Based on lecture notes by M. Spivak and R. Wells. Annals of Mathematics Studies, No. 51. Princeton University Press, Princeton, N.J. (1963)
72. Munkres, J.R.: Topology: A First Course. Prentice-Hall Inc., Englewood Cliffs, NJ (1975)
73. Munkres, J.R.: Elements of Algebraic Topology. Addison-Wesley, Menlo Park, CA (1984)
74. Murillo, A., Viruel, A.: Lusternik-Schnirelmann cocategory: a Whitehead dual approach. In: Cohomological methods in homotopy theory (Bellaterra, 1998), *Progr. Math.*, vol. 196, pp. 323–347. Birkhäuser, Basel (2001)
75. Neisendorfer, J.: Primary Homotopy Theory. Mem. Amer. Math. Soc. **25**(232), iv+67 (1980)
76. Nomura, Y.: An application of the path-space technique to the theory of triads. Nagoya Math. J. **22**, 169–188 (1963)

77. Radó, T.: Ueber den Begriff der Riemannschen Flaeche. Acta Litt. Sci. Szeged **2**, 101–121 (1925)
78. Ravenel, D.C.: Complex Cobordism and Stable Homotopy Groups of Spheres, *Pure and Applied Mathematics*, vol. 121. Academic Press, Orlando, FL (1986)
79. Roitberg, J.: Eckmann-Hilton duality. In: Encyclopaedia of Mathematics, Supplement II, pp. 178–180. Kluwer Academic (2000)
80. Selick, P.: Introduction to Homotopy Theory, *Fields Institute Monographs*, vol. 9. American Mathematical Society, Providence, RI (1997)
81. Serre, J.P.: Homologie singulière des espaces fibrés. Applications. Ann. of Math. (2) **54**, 425–505 (1951)
82. Serre, J.P.: Groupes d'homotopie et classes de groupes abéliens. Ann. of Math. (2) **58**, 258–294 (1953)
83. Spanier, E.H.: Algebraic Topology. Springer-Verlag, New York (1981). Corrected reprint
84. Stasheff, J.: On homotopy Abelian H-spaces. Proc. Cambridge Philos. Soc. **57**, 734–745 (1961)
85. Stasheff, J.D.: Homotopy associativity of H-spaces. I, II. Trans. Amer. Math. Soc. 108 (1963), 275-292; ibid. **108**, 293–312 (1963)
86. Strom, J.A.: Two special cases of Ganea's conjecture. Trans. Amer. Math. Soc. **352**(2), 679–688 (2000)
87. Switzer, R.M.: Algebraic Topology—Homotopy and Homology. Classics in Mathematics. Springer-Verlag, Berlin (2002)
88. Toda, H.: Composition Methods in Homotopy Groups of Spheres. Annals of Mathematics Studies, No. 49. Princeton University Press, Princeton, NJ (1962)
89. Varadarajan, K.: Groups for which Moore spaces $M(\pi, 1)$ exist. Ann. of Math. (2) **84**, 368–371 (1966)
90. Vick, J.W.: Homology Theory, *Graduate Texts in Mathematics*, vol. 145, second edn. Springer-Verlag, New York (1994). An introduction to algebraic topology
91. Whitehead, G.W.: Elements of Homotopy Theory, *Graduate Texts in Mathematics*, vol. 61. Springer-Verlag, New York (1978)
92. Whitehead, J.H.C.: Combinatorial homotopy. I. Bull. Amer. Math. Soc. **55**, 213–245 (1949)

Index